Modern Physics for Scientists and Engineers

Modern Physics for Scientists and Engineers

R.R. Yadav

Professor, Department of Physics
University of Allahabad
Allahabad

Devraj Singh

Assistant Professor and Head
Department of Applied Physics
Amity School of Engineering and Technology
(Guru Gobind Singh Indraprastha University)
New Delhi

Sunil P. Singh

Reader, Department of Physics
Kamla Nehru Institute of Physical and Social Sciences
(Dr. Ram Manohar Lohiya Awadh University)
Sultanpur

Dharmendra K. Pandey

Assistant Professor, Department of Physics
P.P.N. College
(Chhatrapati Shahu Jee Maharaj University)
Kanpur

PHI Learning Private Limited

Delhi-110092
2014

₹ 395.00

MODERN PHYSICS FOR SCIENTISTS AND ENGINEERS

R.R. Yadav, Devraj Singh, Sunil P. Singh, and Dharmendra K. Pandey

ISBN-978-81-203-4858-5

The export rights of this book are vested solely with the publisher.

Published by Asoke K. Ghosh, PHI Learning Private Limited, Rimjhim House, 111, Patparganj Industrial Estate, Delhi-110092 and Printed by Raj Press, New Delhi-110012.

Contents

v

2.   Quantum Mechanics I   38–71

3.   Quantum Mechanics II   72–106

4. Atomic Physics

5.   Molecular Physics   201–266

6. Nuclear Physics 267–323

7.   Solid State Physics

8.   Superconductivity   378–396

11. Optical Fibers 449–472

12. Motion of Charged Particles in EM Fields 473–517

Preface

We feel extremely happy in presenting this standard treatise to students and professors of various universities.

We have developed this book from our lecture notes prepared for teaching the undergraduate level courses over past several years. The book is suitable as a text for various topics on Modern Physics in B.Sc., B.E., B. Tech., A.M.I.E., M.Sc., M. Phil. and M. Tech. degree programs. It will also serve as a reference to those preparing for competition examinations.

The book is divided into 12 chapters. Chapter 1 introduces the concepts of special theory of relativity. Chapter 2 is devoted to origin of quantum mechanics. Chapter 3 is focused on Schrödinger's matter wave equation. Chapter 4 is concerned on Atomic Physics. Chapter 5 discusses various aspects of Molecular Physics. Chapter 6 concentrates on Nuclear Physics. Chapter 7 explains several phenomena of Solid State Physics. Chapter 8 includes superconductivity with its various experimental and theoretical facts. Chapter 9 describes X-rays with its applications for human beings. Chapter 10 covers fundamentals of lasers along with Lasers types and applications. Chapter 11 contains Fiber optics in lucid manner. Chapter 12 elaborates motion of charged particles in electromagnetic fields.

Salient Features of the Book

- Simplified development of each topic to improve the subject understanding, which helps in answering conceptual and other questions
- Presentation of fundamentals concepts in simplified language
- Methodical and systematic treatment in all the chapters
- Concise and simplified derivations of important formulae
- Well illustrated diagrams
- Chapter wise multiple choice questions
- Topic wise numerical problems to reinforce the understanding of the subject matter

- Numerous solved numerical problems
- Exercises to recognize particular concepts so as to strengthen their problem solving ability

We feel highly obliged to numerous authors and publishers of various books which we have often referred for preparing the manuscript.

We express our gratitude to our family members, friends and well-wishers for their love, cooperation and constant encouragement without which this work would have never been accomplished.

We would like to thank whole team of PHI Learning for their skilled and enthusiastic help throughout the project.

We are sure that both students and professors will find this book useful. Constructive suggestions, criticism and intimation of errors and misprints for the improvement of the book would be most welcome and highly appreciated.

R.R. Yadav
Devraj Singh
Sunil P. Singh
Dharmendra K. Pandey

CHAPTER 1

The Special Theory of Relativity

CHAPTER OUTLINES

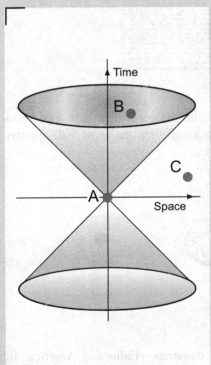

1.1 INTRODUCTION

The concept of relativity can be explained with the help of following examples which show that position, size, time, motion and length all are relative.

(i) Let us consider a river (Figure 1.1), and two brothers (Shobhit and Partha) are standing on the banks of the river facing towards each other. As per Shobhit, house is on the right side, while Partha will say that house is on the left side. But in reality the position of house is fixed. This indicates that position is relative.

Figure 1.1 Position is relative.

(ii) Refer to Figure 1.2. If we compare the ball with the earth then ball appears to be small, but if we compare the ball with an atom, then the same ball appears to be large. This indicates that size is relative.

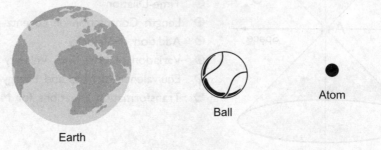

Earth

Ball

Atom

Figure 1.2 Size is relative.

(iii) Figure 1.3 shows current local time in two countries—India and America. In India, if it is 3.00 p.m. then in America it would be 7.00 p.m. Therefore, both the countries have different time at the same instant. Without referring the place, we cannot tell the exact time. Hence time is relative.

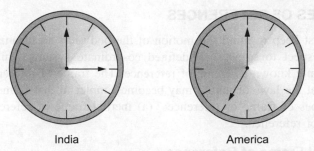

India   America

Figure 1.3 Time is relative.

(iv) Figure 1.4 depicts three sticks A, B and C. For an observer, the length of stick B is smaller than A. But in comparison to C, stick B is longer. This indicates length is relative.

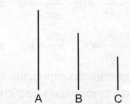

A   B   C

Figure 1.4 Length is relative.

(v) Suppose a person A is moving with speed 60 km/hr (A is observer) and car B is moving with speed 20 km/hr in the same direction, while car C is moving with speed 30 km/hr in the opposite direction.

A
60 km/hr

B
20 km/hr

C
30 km/hr

Figure 1.5 Motion is relative.

Now, for an observer A (it is our reference), the speed of A is 60 km/hr, the speed of car B is (60 – 20 = 40 km/hr), i.e., 40 km/hr, and the speed of car C is (60 + 30 = 90 km/hr), i.e., 90 km/hr. So we can say that motion is relative.

Thus it is obvious that the motion of a body has no meaning unless it is described with respect to which the velocity of the body is measured.

The theory which deals with the relativity of motion and rest is called the theory of relativity and is divided into two parts special theory of relativity and general theory of relativity. The former one deals with objects and systems, which are moving at constant speed with respect to each other or at rest. The later one deals with objects and systems, which are accelerating or decelerating with respect to one another.

▌1.2   FRAMES OF REFERENCES

We must keep in mind the motion of the body has no meaning unless it is described with respect to some well defined co-ordinate system. This co-ordinate system is commonly known as frame of reference. The frame of reference is selected in such a way that the laws of nature may become simpler in that frame of reference. There are two types of frame of reference: (a) inertial frame of reference and (b) non-inertial frame of reference.

1.2.1   Inertial Frame of Reference

Let us consider any co-ordinate system relative to which a body in motion has co-ordinates (x, y, z). These co-ordinates are functions of time. Since the body is not being under the action of force, therefore, we have

$$\frac{d^2x}{dt^2} = 0; \quad \frac{d^2y}{dt^2} = 0 \quad \text{and} \quad \frac{d^2z}{dt^2} = 0$$

This gives

$$\frac{dx}{dt} = u_x, \quad \frac{dy}{dt} = u_y \quad \text{and} \quad \frac{dz}{dt} = u_z$$

where u_x, u_y and u_z are the three components of velocity along x, y and z directions respectively. These velocity components are constant, i.e., without the application of force, a body in motion continues in motion with uniform velocity in a straight line. This is Newton's first law. Thus, we may always choose a frame of reference with respect to which the body is at rest or in uniform motion. When a body is not subjected to any external force, there exists a frame of reference with respect to which the body is at rest or moving with same constant linear velocity, i.e., with respect to which the body is unaccelerated. Such a frame is known as an inertial frame of reference.

In brief, we may say that an inertial frame is the one in which bodies obey Newton's law of inertia and other laws of Newtonian Mechanics.

1.2.2   Non-inertial Frame of Reference

A frame of reference is said to be non-inertial frame when a body, not acted upon by an external force, is accelerated and Newton's laws are not applicable.

Let us consider two frames of reference S and S', where S' is moving relative to a frame S with an acceleration a' (Figure 1.6).

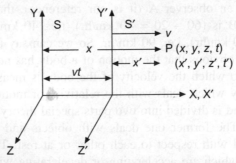

Figure 1.6   Relations of observations of position and time.

Personal Documents can be Emaile to

Kindle	cappearson_39@kindle.com
iPad	cappearson@kindle.com
iPhon	cappearson_48@Kindle.com

-6-

SPOKANE COUNTY
ASSESSOR
1116 W BROADWAY
SPOKANE, WA   99260

Comcast 1-800-266-2278

BARRY'S CABIN   107 LORENZ ROAD KPN,
LAKEBAY.

Pre APP Meeting PL.

477326 HIGHWAY 95 NORTH
PONDERAY, IDAHO

A particle, moving with constant velocity v with respect to an observer in frame of reference S will appear to be moving with an acceleration $-a'$ to an observer in S'. Therefore, the observer in S' will measure a force $-ma'$ on particle, m being the mass of the particle. This force is known as the fictitious force as it is observed due to motion of S' with respect to S, i.e., relative acceleration. Thus, the frame of reference S' is non-inertial as the Newton's law does not hold good in it. Suppose the particle is moving with an acceleration a in frame of reference S, then the observed acceleration in S' will be $(a - a')$. Due to this acceleration, the force on the particle in S' will be $m(a - a')$, i.e.,

$$\vec{F} = ma - ma' = ma + (-ma')$$
$$= \text{Real force in S} + \text{Fictitious force in S'}$$

The force $\vec{F}$ is known as apparent force in S'. Because of this fictitious component a man inside the lift (when moving up with a uniform acceleration) feels more weight than his real weight and feels less weight when lift is moving down with uniform acceleration.

1.3 GALILEAN TRANSFORMATIONS

If the co-ordinates of a body or an event are known in one frame of reference, then its co-ordinates in the second frame of reference moving uniformly relative to first may be calculated. The equations showing the correlation between the co-ordinates in two reference frames are known as Galilean transformations.

Thus by Galilean transformation equation co-ordinates, velocity and acceleration can be transformed from one inertial frame of reference to another. If the laws of motion of a body are known in one inertial frame, then the laws of motion of the same body in another inertial frame of reference can be derived.

When second frame moves relative to first in positive direction of X-axis

Let the origins of two frames S and S' coincide initially. S is at rest and S' is moving with a constant velocity v relative to S along the positive direction of X-axis (Figure 1.6).

If the event happening at P be denoted by (x, y, z, t) in frame S and by (x', y', z', t') in S', then the Galilean transformation may be defined as:

$$\left.\begin{array}{l} x' = x - vt \\ y' = y \\ z' = z \\ t' = t \end{array}\right\} \tag{1.1}$$

When the second frame is moving along a straight line relative to first in any direction

If the second frame is moving with a velocity $v = iv_x + jv_y + kv_z$ as shown in Figure 1.7, and if the two frames coincide initially, we have

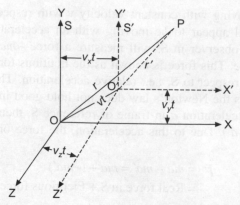

Figure 1.7 Relations of observations of position and time.

$$\left.\begin{array}{l} x' = x - v_x t \\ y' = y - v_y t \\ z' = z - v_z t \\ r' = r - vt \\ t' = t \end{array}\right\} \qquad (1.2)$$

Now, differentiating Eq. (1.2) i.e., $r' = r - vt$ with respect to time t and t', keeping v as constant, we get

$$\frac{d\vec{r}'}{dt} = \frac{d\vec{r}}{dt} - v \qquad [dt' = dt]$$

or $$u' = u - v \qquad (1.3)$$

where u' and u are the velocities of the particle relative to S′ and S respectively.

If a and a' be the accelerations of the particle in S and S′ frames respectively, then

$$\vec{a}' = \vec{a} \qquad (1.4)$$

Thus, the particles possess the same acceleration in inertial frames moving with constant velocity v relative to each other or in other words, we can say that Newton's second law is valid in every inertial frame of reference.

▍1.4 CONCEPT OF ETHER

It is well known that sound waves require material medium for their propagation. Therefore, it was thought, after the development of wave theory of light, that the light waves also require some medium for its propagation. This medium (known as ether) supposed to be massless, invisible, perfectly transparent, and perfectly non-resistive and must be prevailing over whole of the universe. It was also assumed at that time that all the bodies in the universe including our planet earth move freely through this medium. Now, two cases arise:

When material body is moving in space carried along with ether, then there will be no relative motion between ether and moving body. As a result, the velocity of light in ether frame and body frame will be same.

On the other hand, it ether is at rest and material body is moving through it, there will be relative motion between the two and the velocity of light in the body frame will be different from ether frame, consequently the absolute motion of the moving body can be measured. Thus, a large number of experiments were performed in order to prove the existence of such hypothetical medium. For this an experiment was performed by Michelson and Morley to measure the velocity of earth relative to ether frame.

1.5 MICHELSON–MORLEY EXPERIMENT

Albert Michelson and Edward Morley performed an experiment in 1887 to measure the motion of the earth through the ether. Ether was assumed to be a hypothetical medium as discussed in Section 1.4, pervading the universe and supposed to be invisible, massless, perfectly transparent, perfectly non-resistive and at absolute rest. It was without producing any disturbance. So if the ether hypothesis is correct, then it is possible to determine the absolute velocity of earth with respect to ether frame.

1.5.1 Experimental Arrangement

Light from a monochromatic extended source S after being rendered parallel by a collimating lens L falls on a semi-silvered glass plate G inclined at an angle 45° to the beam as shown in Figure 1.8. It is divided into two parts. The two rays, i.e., ray 1 and ray 2 fall normally on mirrors M_1 and M_2 and are reflected back along their original paths. The reflected rays again meet at the semi-silvered surface of glass plate G and enter the telescope where interference pattern is obtained. The optical distances of mirrors M_1 and M_2 from G are made equal with the help of compensating plate G′ as shown in Figure 1.8.

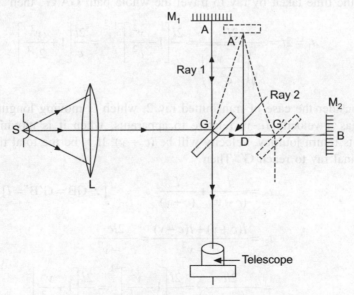

Figure 1.8 Michelson–Morley experiment.

If the apparatus is at rest in ether, the two reflected rays would take equal time to return to the glass plate G. But actually the whole apparatus is moving along with earth. Let the direction of motion of earth is in the direction of initial beam. Due to motion of earth, the optical paths traversed by both the rays are not the same.

1.5.2 Theoretical Analysis

Let the two mirrors M_1 and M_2 be at a distance l from the glass plate G. Further let c and v be the velocities of light and apparatus or earth respectively. Reflected ray 1 from glass plate G strikes the mirror M_1 and A' and not at A due to the motion of earth. The total path of ray from G to A' and back to G' will be GA'G'. But by the law of reflection,

$$GG' = 2GD = 2AA$$

$$\therefore \quad GA'G' = GA' + A'G' = 2G'A'$$

But

$$(GA')^2 = (GD)^2 + (A'D)^2$$

$$= (AA')^2 + (A'D)^2 \tag{1.5}$$

If t be the time taken by the ray to move from G to A', then from Eq. (1.5), we have

$$(ct)^2 = (vt)^2 + l^2$$

$$(c^2 - v^2)t^2 = l^2 \qquad t^2 = \frac{l^2}{(c^2 - v^2)}$$

$$\Rightarrow \qquad t = \frac{l}{\sqrt{(c^2 - v^2)}} = \frac{l}{c\sqrt{\left(1 - \dfrac{v^2}{c^2}\right)}}$$

$$c^2\left(1 - \frac{v^2}{c^2}\right)$$

If t_1 be the time taken by ray to travel the whole path GA'G', then

$$t_1 = 2t = \frac{2l}{c\sqrt{\left(\dfrac{1 - v^2}{c^2}\right)}} = \frac{2l}{c}\left[1 - \frac{v^2}{c^2}\right]^{-1/2} = \frac{2l}{c}\left[1 + \frac{v^2}{2c^2}\right] \tag{1.6}$$

Now, consider the case of transmitted ray 2, which is moving longitudinally towards M_2. It has a velocity $(c-v)$ relative to apparatus, when it is moving from G to B. During its return journey, velocity will be $(c + v)$. If t_2 be the total time taken by the longitudinal ray to reach G'. Then

$$t_2 = \frac{l}{(c - v)} + \frac{l}{(c + v)} \qquad [\because \ GB = G'B' = l]$$

$$\therefore \qquad t_2 = \frac{l(c + v) + l(c - v)}{c^2 - v^2} = \frac{2lc}{c^2 - v^2}$$

$$= \frac{2lc}{c^2\left(1 - \dfrac{v^2}{c^2}\right)} = \frac{2l}{c}\left[1 - \frac{v^2}{c^2}\right]^{-1} = \frac{2l}{c}\left[1 + \frac{v^2}{c^2}\right] \tag{1.7}$$

Thus, the time difference of travel of longitudinal and transverse journeys is given as:

$$\Delta t = t_2 - t_1$$

$$= \frac{2l}{c}\left[1 + \frac{v^2}{c^2}\right] - \frac{2l}{c}\left[1 + \frac{v^2}{2c^2}\right] = \frac{2l}{c}\frac{v^2}{2c^2} = \frac{lv^2}{c^3} \qquad (1.8)$$

Hence the path difference between two rays is given by

$$\delta = c\Delta t = \frac{lv^2}{c^2} \qquad (1.9)$$

Michelson and Morley performed the experiment in two steps, (i) by setting apparatus as shown in Figure 1.8, and (ii) by turning the apparatus through 90°. Now, the path difference is in opposite direction, i.e., it is $-\dfrac{lv^2}{c^2}$. The resultant path difference becomes

$$\delta' = \frac{lv^2}{c^2} - \left(-\frac{lv^2}{c^2}\right) = \frac{2lv^2}{c^2} \qquad (1.10)$$

We know that a change in optical path by λ in an interferometer corresponds to a shift of one fringe, and hence the path difference $2\dfrac{lv^2}{c^2}$ will correspond to a shift of $\dfrac{lv^2}{c^2\lambda}$ fringes.

In Michelson–Morley experiment, the effective path l was taken to be 11 metres, and if velocity of earth through ether is considered to be the orbital velocity of earth, i.e., $v = 3 \times 10^4$ m/s, the fringe shift for a visible light, say, $\lambda = 5500$ Å is given by

$$\frac{2lv^2}{c^2} = \frac{2 \times 11 \times (3 \times 10^4)^2}{(3 \times 10^8)^2} \times \frac{1}{5.5 \times 10^{-7}} = 0.4$$

This displacement could be very easily detected by Michelson–Morley apparatus as it could detect fringe shift up to 0.01 of a fringe. But in reality no fringe was observed. The experiment was repeated at different times and at different places, but no change was observed. This concludes that the relative velocity between ether and earth is zero.

1.5.3 Explanation of Negative Results

Following separate interpretations were given to explain the negative results of Michelson–Morley experiment:

Ether Drag Hypothesis

The moving earth drags the ether along with it and hence there is no relative motion between the earth and ether. This explanation of Michelson himself, however, is not reasonable, because it goes against the observed phenomenon of aberration of starlight.

Lorentz–Fitzerald Contraction Hypothesis

Fitzerald and Lorentz independently offered a more radical suggestion that a material body moving through is contracted in the direction of motion by a factor $\sqrt{1-(v^2/c^2)}$. It would equalize the time t_1 and t_2 in Michelson–Morley experiment and hence no fringe shift can be expected. But this contraction phenomenon is purely mathematical without any experimental confirmation.

Constancy of Speed of Light

The appropriate explanation of negative results was given by Einstein. He proposed that the speed of light is an invariant quantity, i.e., the speed of light is constant and does not depend upon the motion of the source, observer, or medium. This was one of the two postulates of the special theory of relativity.

1.6   POSTULATES OF SPECIAL THEORY OF RELATIVITY

The special theory of relativity is called special because it is concerned only with inertial frame of reference out of more general frames, inertial as well as non-inertial. The one dealing with non-inertial frames is called the General theory of relativity (Section 1.1).

Albert Einstein published an article entitled 'On the Electrodynamics of moving Bodies' in the Journal *Annals of Physik* in 1905, where in he gave two postulates of the special theory of relativity.

1. All the basic laws of physics are the same in all the inertial frames of reference.
2. The measured speed of light in empty space has always the same value c, irrespective the relative motion of the source and the observer.

Explanation

The first postulate is the special principle of relativity. It is only an extension of the Newtonian principle of relativity from the laws of mechanics to all the laws of physics. The laws of physics, other than that of mechanics, are the laws of electrodynamics, optical wave propagation, etc. (we find that the e.m. wave equation, which does not retain in its form under the Galilean transformations, does retain it under Lorentz transformation given by special theory of relativity.)

The second postulate is observable evidence, the speed of light emitted by a stationary source is c, but from a moving source it will not become $c + v$. It will again found to be c. Also, no material particle can attain the speed of light. In an ultimate speed experiment performed by William Bertozzi in 1962 in Massachusetts, electrons were accelerated by giving them up to 0.5 MeV energy. By noting their travel time along the length of the accelerator, their speed was measured. According to the classical formula for kinetic energy KE = 1/2 mv^2, the velocity of the 0.5 MeV electrons should become nearly $2.7c$, but it was actually found to be not exceeding c.

It is remarkable to note that a whole new dynamics can be based on two postulates. The Galilean transformations will not be applicable under the second postulate of special theory or relativity (as the Galilean law of addition of velocities fails here), and there appears a need of some new sets of transformation. This need was fulfilled by Lorentz transformations.

▌1.7 LORENTZ TRANSFORMATION EQUATIONS

H.A. Lorentz algebraically searched for these transformation equations, which relates the observation of position and time made by two observers sitting in two different inertial frames, are known as Lorentz transformation equations.

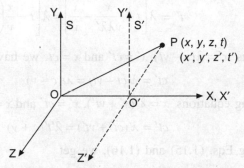

Figure 1.9 Two inertial frames of reference.

Consider there are two inertial frames of reference S and S' as shown in Figure 1.9. S' having uniform velocity v relative to S in the positive direction of X-axis of two-co-ordinate systems coincide at $t = t' = 0$. Let event P be a light signal generated at time $t = 0$ at the origin, which grows in the space as sphere. Event P is determined by co-ordinates (x, y, z, t) and (x', y', z', t') by O and O' respectively in frames S and S'. When the light is observed from S, then we have

Velocity of light signal, $c = \dfrac{\text{Distance}}{\text{Time}} = \dfrac{\sqrt{x^2 + y^2 + z^2}}{t}$

or
$$x^2 + y^2 + z^2 = c^2 t^2 \qquad (1.11)$$

Similarly, for observer O' in S' system, we have

$$x'^2 + y'^2 + z'^2 = c^2 t'^2 \qquad (1.12)$$

According to second postulate of special theory of relativity, c is constant. Therefore,

$$y = y^2 \text{ and } z = z^2$$

From above equation, we have

$$x = ct \text{ and } x' = ct'$$

Now, the transformation equations between x and x' can be represented by letting $x' = \lambda(x - vt)$ where λ being independent of x and t.

Since motion is always relative, we can assume that S is moving with respect to S' with velocity $-v$ along positive X-axis, thus

$$x = \lambda'(x' - vt')$$

$$x = \lambda'[\lambda(x - vt) + vt'] \qquad \text{[Using } x' = \lambda(x - vt)] \tag{1.13}$$

or
$$\frac{x}{\lambda'} = \lambda x - \lambda vt + vt'$$

or
$$vt' = \frac{x}{\lambda'} + \lambda vt - \lambda x$$

$$= \lambda \left[\frac{x}{\lambda\lambda'} + vt - x \right]$$

or
$$t' = \lambda \left[t + \frac{x}{v\lambda\lambda'} - \frac{x}{v} \right] = \lambda \left[t - \frac{x}{v}\left(1 - \frac{1}{\lambda\lambda'}\right) \right] \tag{1.14}$$

From equations $x' = \lambda(x - vt)$, $x' = ct'$ and $x = ct$, we have

$$ct' = \lambda(ct - vt) = \lambda t(c - v) \tag{1.15}$$

Similarly, using equations $x = \lambda'(x' + vt')$, $x' = ct'$ and $x = ct$, we have

$$ct' = \lambda'(ct' + vt') = \lambda' t'(c + v) \tag{1.16}$$

On multiplying Eqs. (1.15) and (1.16), we get

$$c^2 tt' = \lambda\lambda' tt'(c^2 - v^2)$$

or
$$c^2 = \lambda\lambda'(c^2 - v^2) \tag{1.17}$$

Suppose $\lambda = \lambda'$, we have

$$c^2 = \lambda^2(c^2 - v^2)$$

$$\lambda^2 = \frac{c^2}{c^2 - v^2} \quad \text{or} \quad \lambda = \frac{1}{\sqrt{\left(1 - \dfrac{v^2}{c^2}\right)}} \tag{1.18}$$

Further from t' and x', we have

$$t' = \lambda \left[t - \frac{x}{v}\left(1 - \frac{1}{\lambda\lambda'}\right) \right]$$

$$= \lambda \left[t - \frac{x}{v}\left(1 - \frac{1}{\lambda^2}\right) \right] \qquad [\text{as } \lambda = \lambda']$$

$$= \lambda \left[t - \frac{x}{v}\left(1 - \frac{c^2 - v^2}{c^2}\right) \right] \qquad \left[\because \lambda^2 = \frac{c^2}{(c^2 - v^2)} \right]$$

$$= \lambda \left[t - \frac{v^2 x}{vc^2} \right] = \lambda \left[t - \frac{vx}{c^2} \right]$$

$$\Rightarrow \qquad t = \frac{t - \frac{vx}{c^2}}{\sqrt{\left(1 - \frac{v^2}{c^2}\right)}} \quad \text{[using Eq. (1.18)]} \tag{1.19}$$

Thus on combining all equations, we have

$$x' = \frac{x - vt}{\sqrt{\left(1 - \frac{v^2}{c^2}\right)}} \tag{1.20a}$$

$$y' = y \tag{1.20b}$$

$$z' = z \tag{1.20c}$$

$$t' = \frac{t - \frac{vx}{c^2}}{\sqrt{\left(1 - \frac{v^2}{c^2}\right)}} \tag{1.20d}$$

Equations [1.20(a–d)] are known as Lorentz Transformation Equations of space and time. It we use $\beta = v/c$, then Eq. (1.20) becomes

$$x' = \frac{x - vt}{\sqrt{\left(1 - \beta^2\right)}} \tag{1.21a}$$

$$y' = y \tag{1.21b}$$

$$z' = z \tag{1.21c}$$

$$t' = \frac{t - \frac{vx}{c^2}}{\sqrt{\left(1 - \beta^2\right)}} \tag{1.21d}$$

If we assume that system S is moving with velocity $-v$ along positive X-axis relative to S′, then the Lorentz transformation equations become

$$x' = \frac{x' + vt'}{\sqrt{\left(1 - \beta^2\right)}} \tag{1.22a}$$

$$y' = y \tag{1.22b}$$

$$z' = z \tag{1.22c}$$

$$t = \frac{t' + \frac{vx'}{c^2}}{\sqrt{\left(1 - \beta^2\right)}} \tag{1.22d}$$

Equations [1.22(a–d)] are said to be Inverse Lorentz Transformation Equations of space and time.

Important Features of Lorentz Transformation Equations

The following features should be considered while solving problems with Lorentz transformations:

(i) The measurements of position and time depend upon the frame of reference of the observer, e.g. if some event happens in frame of reference S, which is at rest and we are observing the same from moving frame of reference S', then we will have to use Lorentz transformation equations. On the other hand, if the event happens in moving frame of reference S' and the observations are being made from frame of reference S, which is at rest, then only inverse Lorentz transformation will have to be applied to solve the problem.

(ii) Lorentz transformation equations will reduce to classical Galilean equation when velocity v is extremly less than the velocity of light c, i.e., $v \ll c$. In this case $\beta \ll 1$ and the quantity $\left[1/\sqrt{1-(v^2/c^2)} \right]$ will approach 1. Therefore, the Lorentz transformation equations become $x' = x - vt, y' = y, z' = z$ and $t' = t - (vx/c^2)$ which are Galilean transformations.

EXAMPLE 1.1   As measured by O, a flash bulb goes off at $x = 100$ km, $y = 10$ km, $z = 1$ km at $t = 5 \times 10^{-4}$ s. What are the co-ordinates x', y', z' and t' of this event as determined by a second observer, O', having relative to O at $\div 0.8c$ along the common X–X' axis?

Solution   Given $x = 100$ km, $y = 1$ km, $z = 1$ km and $t = 5 \times 10^{-4}$ s, $v = -0.8c = -0.8 \times 3 \times 10^5$ km/s

From Lorentz transformation equations,

$$x' = \frac{x - vt}{\sqrt{\left(1 - \dfrac{v^2}{c^2}\right)}} = \frac{100 \text{ km} - \left(-0.8 \times 3 \times 10^5 \text{ km/s}\right)\left(5 \times 10^4 \text{s}\right)}{\sqrt{1-(0.8)^2}} = 367 \text{ km}$$

$$y' = y = 10 \text{ km}; \quad z' = z = 1 \text{ km}$$

and
$$t' = \frac{t - \dfrac{vx}{c^2}}{\sqrt{\left(1 - \dfrac{v^2}{c^2}\right)}} = \frac{5 \times 10^{-4}\text{s} - \dfrac{(0.8)(100 \text{ km})}{\left(3 \times 10^5\right)^2 \text{ km/s}}}{\sqrt{1-(0.8)^2}} = 12.8 \times 10^{-4}\text{s}.$$

1.8   CONCEPT OF SIMULTANEITY IN RELATIVITY

Let two events E_1 and E_2 occur in a frame S. Let E_1 be specified by the co-ordinates (x_1, y_1, z_1, t_1) and E_2 by the co-ordinate (x_1, y_1, z_1, t_1). Let S' be another frame of reference moving with uniform velocity v with respect to S along the common X-axis of two frames. Let the co-ordinates of E_1 and E_2 as seen by the observer at origin O' of S' be (x_1', y_1', z_1', t_1') and (x_2', y_2', z_2', t_2'). Then from Lorentz transformation for time, we get

$$t_1' = \frac{t_1 - (vx_1/c^2)}{\sqrt{\left(1 - \dfrac{v^2}{c^2}\right)}} \quad \text{and} \quad t_2' = \frac{t_2 - (vx_2/c^2)}{\sqrt{\left(1 - \dfrac{v^2}{c^2}\right)}}$$

Subtracting one from the other, we get

$$t_2' - t_1' = \frac{t_2 - t_1}{\sqrt{\left(1 - \dfrac{v^2}{c^2}\right)}} + \frac{v(x_1 - x_2)/c^2}{\sqrt{\left(1 - \dfrac{v^2}{c^2}\right)}} \tag{1.23}$$

Suppose for the observer in S, the two events are simultaneous, i.e., $t_1 = t_2$. Then

$$t_2' - t_1' = \frac{v(x_1 - x_2)/c^2}{\sqrt{\left(1 - \dfrac{v^2}{c^2}\right)}} \tag{1.24}$$

We see that t_2' is not equal to t_1'. Thus two events which are simultaneous in frame S are not simultaneous with respect to frame S'. The principle of simultaneity of two events is, therefore, not an absolute concept. The simultaneity of two events depends on the observer and the frame of reference.

The effect discussed above is not due to time-dilation. Time-dilation results from the first term in the Lorentz transformation $t' = \dfrac{t - \dfrac{vx}{c^2}}{\sqrt{\left(1 - \dfrac{v^2}{c^2}\right)}}$

The effects of simultaneity arise from the second term. The two events, which are simultaneous in one frame of reference, are not simultaneous in another frame which is in relative motion with respect to first. The only exception is when the two events occur at the same point in space, then $x_1 = x_2$ and $t_1' = t_2'$.

Hence two events that are simultaneous in one inertial frame of reference are in general not simultaneous in another reference frame moving with respect to first.

1.9 TIME-DILATION

The time interval between two events those occur at the same place in an observer's frame of reference is called the proper time of the interval between the events.

A clock moving with respect to an observer appears to tick less rapidly than it does when at rest with respect to him. Let a clock be placed at the point x' in the moving frame S'. An observer in S' finds that the clock gives two ticks at time t_1' and t_2'.

After a time interval of t_0, the observer in the moving system finds that the time is now t_2' according to his clock, i.e.,

$$t_0 = t_2' - t_1'$$

$$t_1 = \frac{t_1' + \frac{vx'}{c^2}}{\sqrt{\left(1 - \frac{v^2}{c^2}\right)}}; \quad t_2 = \frac{t_2' + \frac{vx'}{c^2}}{\sqrt{\left(1 - \frac{v^2}{c^2}\right)}}$$

So the time internal as judged from S system is:

$$t = t_1 - t_2$$

$$t = \frac{t_2' + \frac{vx'}{c^2} - t_1' - \frac{vx'}{v^2}}{\sqrt{\left(1 - \frac{v^2}{c^2}\right)}}$$

$$= \frac{t_2' - t_1'}{\sqrt{\left(1 - \frac{v^2}{c^2}\right)}}$$

$$t = \frac{t_0}{\left(1 - \frac{v^2}{c^2}\right)} \tag{1.25}$$

Equation (1.25) shows that for the stationary observer in S, the time interval appears to be lengthened by a factor $[t/\sqrt{\{1 - (v^2/c^2)\}}]$. Thus the moving clock appears to be slowed down in a stationary observer. This effect is known as time-dilation (to dilate is to become larger).

There are following cases:

(i) If $v = c$, than $(v^2/c^2) = 1$. This means $t = \infty$, i.e., the clock moving with the speed of light will appear to be completely stopped to a stationary observer.

(ii) If $v \ll c$, then $\dfrac{1}{\sqrt{\left(1 - \dfrac{v^2}{c^2}\right)}} = \left(1 - \dfrac{v^2}{c^2}\right)^{-1/2} = 1 + \dfrac{v^2}{2c^2}$

In this v^2/c^2 is very small and can be neglected.

$\therefore$ $t = t_0$, i.e., if the clock is moving with the speed extremely smaller than the speed of light, then the time interval will remain same for moving observer.

1.9.1 Time-dilation is Real Effect

All clocks will run slow for an observer in relative motion. Biological clocks are also subjected to this time-dilation. Even the growth and decay of living objects are slowed down by this effect. Consider, for example, the birth and decay of the elementary particle muon. The muon is an unstable particle. It decays into electron and neutrinos. In the frame of reference fixed with respect to a muon, its birth and decay to place at the same point in this frame, say the laboratory, is formed to be about 2 μs. Muons are also formed in upper atmosphere, when high energy cosmic rays collide with atmosphere molecules. The muons thus created rush towards the earth at speed nearly equal to that of light in free space. The distance that muon could then be expected to travel at such a speed in its life time would be about $(2 \times 10^{-6}\ \text{s}) \times (3 \times 10^{8}\ \text{m/s})$ = 600 m. The height at which the muons are formed in several kilometres above the atmosphere, we should not expect any muon to reach the surface of the earth. But muons do reach the surface of the earth in large numbers. This is explained by the fact that, viewed from our frame of reference, the life time of muon is much dilated time, the muon can travel distances much larger than 600 km. They are able to reach the earth. Frequency is reciprocal of period. Hence the apparent frequency v and the proper frequency v_0 are related by the equation:

$$v = v_0 \sqrt{1 - \frac{v^2}{c^2}} \tag{1.26}$$

1.9.2 Experimental Verification of Time-dilation

Time-dilation is verified in experiments on a nuclear particle, called meson. There are two types of mesons, i.e., π(pi) and μ(mu) mesons. The π^+ meson decay (break into fragments) in such a way that in every 1.8×10^{-8} s, half of them die out, i.e., their flux decreases to 2^{-1} in every 1.8×10^{-8} s.

Now, in an experiment, in the laboratory, π^+ mesons were produced with speed $0.99c$ and their flux was measured at two places separated by 30 m. The laboratory time interval Δt for travelling this distance was given by

$$\Delta t = \frac{30\ \text{m}}{0.99c} \cong \frac{30}{3 \times 10^8} \approx 10 \times 10^{-8}\ \text{s}$$

This is about 5.6 times of 1.8×10^{-8} s. Hence, the flux of π^+ mesons should decrease to $2^{-5.6}$ or less than 2% of the original flux in travelling 30 m. But the actual flux at the second place was nearly 60% of that at first place.

This discrepancy is explained by computing the proper time (t_0) given by relation:

$$t = t_0 \left(1 - \frac{v^2}{c^2}\right)^{1/2} = 10 \times 10^{-8}[1 - (0.99)^2]^{1/2} = 1.4 \times 10^{-8}\ \text{s}$$

This is 0.78 times 1.8×10^{-8} s. Hence in this time, the flux should fall to $2^{-0.78}$ (nearly 60%) of the original flux. This is exactly what is observed.

It accounts to this: In laboratory measurements, the elapsed time for 30 m travel is 10×10^{-8} s, while the π^+ mesons themselves measure the time as 1.4×10^{-8} s only. A seven-fold dilation has occurred in this case.

Alternately, for an observer moving with mesons, the apparent travel is contracted by a factor $\sqrt{\left(1 - \dfrac{v^2}{c^2}\right)} = \sqrt{1 - (0.99)^2} = 0.14$ and the same result follows.

That is, a journey of 80 m for the mesons, as measured by an observer on the earth, will simply correspond to a journey of 30×0.14 m, as measured by an observer moving with the meson.

The journey will be performed in

$$\frac{30 \times 0.14 \text{ m}}{0.99c} \cong \frac{30 \times 0.14 \text{ m}}{3 \times 10^8 \text{ m/s}} \cong 1.4 \times 10^{-8} \text{ s}$$

which is 0.78 times of 1.8×10^{-8}. Hence, in this time, the flux should fall to $2^{-0.78}$ (nearly 60%) of the original flux.

EXAMPLE 1.2   A beam of particle of half life 2.0×10^{-8} s, travel in the laboratory with the speed of 0.96 times the speed of light. How much distance does the beam travel before the flux of fall to ½ times of the initial flux.

Solution   Given $t_0 = 2 \times 10^{-8}$ s and $v = 0.96c = 0.96 \times 3 \times 10^8$ m/s
Then, from Lorentz transformation equations,

$$t_0 = t\sqrt{\left(1 - \frac{v^2}{c^2}\right)} \quad \Rightarrow \quad t = \frac{t_0}{\sqrt{\left(1 - \dfrac{v^2}{c^2}\right)}} = \frac{2.0 \times 10^{-8}}{\sqrt{(1 - 0.16)}} = 7.1 \times 10^{-8} \text{ s}$$

The distance travelled by the beam in this time in the laboratory frame is:

$$= 0.96c \times 7.1 \times 10^{-8} = 20.45 \text{ m}$$

1.10   LENGTH CONTRACTION OR LORENTZ–FIZERALD CONTRACTION

Lorentz–Fitzerald observed that the length of a moving body (moving with a velocity comparable with velocity of light) measured with respect to an observer does not remain constant but gets discussed due to its relative motion. This decrease in length in the direction of motion is called length contraction.

According to Lorentz–Fitzerald, "when an object moves with a velocity v (comparable with velocity of light) relative to a stationary observer, its measured length appears to be contracted in the direction of its motion by a factor $\sqrt{1 - \dfrac{v^2}{c^2}}$, whereas its other dimensions perpendicular to the direction of motion remains unaffected".

To derive the expression for length contraction, let us consider a frame reference S', moving along +X-direction with a constant velocity v relative to S. Let (x, y, z, t) be the space and time co-ordinates of an event for an observer in frame S, and (x', y', z', t') be the co-ordinates of the same event for an observer in frame S'.

Let us consider a rod lying the X'-axis in the moving frame S' as shown in Figure 1.10.

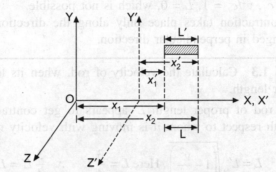

Figure 1.10   Illustration for length contraction.

An observer in this frame determines the co-ordinates of its ends to be x_1' and x_2'.

Now,
$$x_1' - x_2' = L'$$

The length L' of the rod in its rest frame is called its proper length. Here rest frame means observer's frame of reference.
$$x_1 - x_2 = L$$

From Lorentz transformation,

$$x_1' = \frac{x_1 - vt}{\sqrt{\left(1 - \dfrac{v^2}{c^2}\right)}} \quad \text{and} \quad x_2' = \frac{x_2 - vt}{\sqrt{\left(1 - \dfrac{v^2}{c^2}\right)}}$$

$$\therefore \qquad L' = x_2' - x_1' = \frac{x_2 - vt}{\sqrt{\left(1 - \dfrac{v^2}{c^2}\right)}} - \frac{x_1 - vt}{\sqrt{\left(1 - \dfrac{v^2}{c^2}\right)}} = \frac{x_2 - x_1}{\sqrt{\left(1 - \dfrac{v^2}{c^2}\right)}}$$

$$\Rightarrow \qquad L' = \frac{L}{\sqrt{\left(1 - \dfrac{v^2}{c^2}\right)}}$$

Hence
$$L = L'\sqrt{\left(1 - \frac{v^2}{c^2}\right)} \qquad\qquad (1.27)$$

As
$$L < L'$$

Hence length is appeared to be contracted, by a factor $\sqrt{\left(1-\dfrac{v^2}{c^2}\right)}$. This is known as Lorentz–Fitzerald contraction.

(i) If $v \ll c$, then v^2/c^2 is negligible. $\therefore L = L'$

(ii) $v \approx c$, $L < L'$

(iii) If $v = c$, $v^2/c^2 = 1$, $L = 0$, which is not possible.

(iv) The contraction takes place only along the direction of motion and remains unchanged in perpendicular direction.

EXAMPLE 1.3   Calculate the velocity of rod, when its length appears three-fourth of its proper length.

Solution A rod of proper length L' appears to get contracted to a length L for an observer with respect to whom it is moving with velocity v, such that

$$L = L'\sqrt{\left(1-\frac{v^2}{c^2}\right)}, \text{ Here } L = \frac{3}{4}L' \quad \therefore \quad \frac{3}{4}L' = L'\sqrt{\left(1-\frac{v^2}{c^2}\right)}$$

This gives

$$\left(1-\frac{v^2}{c^2}\right) = \frac{9}{16} \Rightarrow \frac{v^2}{c^2} = \frac{7}{16}$$

$$\Rightarrow \qquad \frac{v}{c} = \sqrt{\frac{7}{16}} = 0.66 \Rightarrow v = 0.66c$$

The rod moves with a velocity $0.66c$.

1.11   ADDITION OF VELOCITIES

Consider two systems S and S'. S' is moving with a velocity v relative to S along time X-axis. Let us express the velocity of a body in them. Suppose a body moves a distance dx in time dt in system S and through a distance dx' in time dt' in system S', then

$$\frac{dx}{dt} = u, \quad \text{and} \quad \frac{dx'}{dt'} = u' \tag{1.28}$$

We know from Lorentz transformation equations,

$$x = \phi(x' + vt') \tag{1.29}$$

Here

$$\phi = \frac{1}{\sqrt{\left(1-\dfrac{v^2}{c^2}\right)}} \quad \text{and} \quad t = \phi\left(t' + \frac{vx'}{c^2}\right) \tag{1.30}$$

Differentiating Eqs. (1.29) and (1.30), we have

$$dx = \phi(dx' + vdt') \tag{1.31}$$

$$dt = \phi\left(dt' + v\frac{dx'}{c^2}\right) \tag{1.32}$$

Dividing Eq. (1.31) by Eq. (1.32), we have

$$\frac{dx}{dt} = \frac{dx' + vdt'}{dt' + \dfrac{vdx'}{c^2}} = \frac{\dfrac{dx'}{dt'} + v}{1 + \dfrac{v}{c^2}\dfrac{dx'}{dt'}} = \frac{u' + v}{1 + \dfrac{u'v}{c^2}} \Rightarrow u = \frac{u' + v}{1 + \dfrac{u'v}{c^2}} \tag{1.33}$$

Equation (1.33) represents the relativistic law of addition of velocities, whereas in classical mechanics, it is simply $u = u' + v$.

Following points are drawn:

(i) When u' and v are smaller as compared to c, $u'v/c^2$ can be neglected in comparison to unity, then

$$u = u' + v \tag{1.34}$$

which is classical formula.

(ii) When v or $u' \cong c$, then

$$u = \frac{u' + c}{1 + \dfrac{u'c}{c^2}} = \frac{u' + c}{c + u'} \times c = c$$

i.e., if one object moves with velocity c with respect to other, then their relative velocity is always c, whatever may be velocity of the other.

(iii) When $u' = c = v$, then $u = u = \dfrac{c + c}{1 + \dfrac{c^2}{c^2}} = \dfrac{2c}{2} = c$

This shows addition of velocity of light to the velocity of light merely produced velocity of light.

EXAMPLE 1.4   A spaceship moving away from the earth with velocity $0.6c$ fires a rocket whose velocity relative to the spaceship is $0.9c$ (i) away from the earth. (ii) towards the earth. What will be the velocity of the rocket as observed from the earth in the two cases?

Solution   Let S and S′ be the frame of reference of the earth and spaceship respectively. We are given the velocity of rocket in the S′ frame, and we have to determine its velocity in the S frame, when the velocity of S′ relative to S is $0.6c$ away from the earth.

Now, we know that if a particle has velocity u'_x in S′ along the direction of relative motion between S and S′, then the velocity u_x in system S is given by

$$u_x = \frac{u'_x + v}{1 + \dfrac{u'_x v}{c^2}}$$

where v is the velocity of S′ relative to S.

Case 1 $u' = 0.9c$ and $v = 0.6c$

$$\therefore \qquad u = \frac{0.9c + 0.6c}{1 + 0.9c \times \dfrac{0.6c}{c^2}} = u = \frac{1.5c}{1.54} = 0.97c \quad \text{away from the earth}$$

Case 2 $u' = -0.9c$, $v = 0.6c$

$$\therefore \qquad u = \frac{-0.9c + 0.6c}{1 + 0.6c \dfrac{0.9c}{c^2}}$$

$$\therefore \qquad u = \frac{0.3c}{0.46} = 0.65c$$

$$u = 0.65c \quad \text{towards the earth}$$

1.12   VARIATION OF MASS WITH VELOCITY

According to the theory of relativity, like length and time, mass of a body is also not an absolute quantity and varies with velocity. This results in contradiction with classical mechanics, according to which mass is invariant and does not depend upon velocity.

To prove this, suppose an inertial frame S′ is associated with the moving mass. Let S′ frame be moving with velocity v along the positive X-direction of another inertial frame S. Now consider that further in the S′ frame, there are two masses m_1 and m_2 which move in opposite directions with velocities v and $-v$ relative to the origin of the S′ frame. Suppose they collide inelastically and come to rest in S′ frame. When viewed from S frame, m_2 will appear to be initially at rest, while m_1 will have some velocity (say u) before collision. For S frame after collision, the two masses will attain a common velocity v as shown in Figure 1.11.

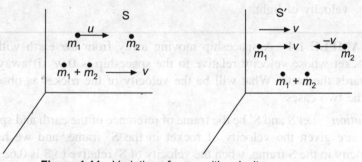

Figure 1.11   Variation of mass with velocity.

We relate the velocity of mass m_1 in S frame with its velocity in S′ frame using the velocity transformation relation:

$$u = \frac{u' + v}{1 + \dfrac{u'v}{c^2}}$$

Here we put $u = u$ and $u' = v$, we have

$$u = \frac{v + v}{1 + \frac{v^2}{c^2}} = \frac{2v}{1 + \frac{v^2}{c^2}} \qquad (1.35)$$

Since the law of conservation of momentum holds in all the inertial frames, we have in S frame

$$m_1 u + m_2.0 = (m_1 + m_2)\, v$$

or

$$v = \frac{m_1}{m_1 + m_2} \times u \qquad (1.36)$$

Putting this value of v in Eq. (1.35), we obtain

or

$$u = \frac{\dfrac{2m_1}{m_1 + m_2} u}{1 + \left(\dfrac{m_1}{m_1 + m_2}\right)^2 \left(\dfrac{u^2}{c^2}\right)} \qquad \text{or} \qquad 1 + \left(\frac{m_1}{m_1 + m_2}\right)^2 \frac{u^2}{c^2} = \frac{2m_1}{m_1 + m_2}$$

Multiplying both sides by $(m_1 + m_2)^2 \, c^2$, we obtain

$$c^2 (m_1 + m_2)^2 + m_1^2 u^2 = 2m_1 (m_1 + m_2) c^2$$

or

$$m_1^2 (c^2 - u^2) = m_2^2 \, 2c^2 \Rightarrow m_1^2 = \frac{m_2^2 c^2}{c^2 - u^2} \qquad \text{or} \qquad m_1^2 = \frac{m_2^2}{1 - \dfrac{u^2}{c^2}}$$

or

$$m_1 = \frac{m_2}{\sqrt{1 - \dfrac{u^2}{c^2}}} \qquad (1.37)$$

We take the two masses to be identical since m_2 is at rest and m_1 is moving with velocity u (the viewpoint of S frame), we refer to m_2 as m_0 and m_1 as m and put these values in Eq. (1.37). We get

$$m = \frac{m_0}{\sqrt{\left(1 - \dfrac{v^2}{c^2}\right)}} \qquad (1.38)$$

Equation (1.38) represents the variation of mass with velocity and shows how the mass increases with velocity. Mass m_0 is called the rest mass and m the relativistic mass or moving mass of the same body while in motion.

Now we will consider the following points:

(i) If v is very small in comparison to c, i.e., $v \ll c$, then $m = m_0$, the mass of the moving body is the same as its rest mass.

(ii) If v is comparable to c, i.e., $v \cong c$, then $m > m_0$, the mass of the moving object appears to be greater than at rest.

(iii) If $v = c$, then $m \rightarrow \infty$, which is impossible and implies no material body can attain the velocity of light.

In agreement with mass formula, the relativistic expression for the linear momentum is:

$$p = mv = \frac{m_0 v}{\sqrt{\left(1 - \dfrac{v^2}{c^2}\right)}} \qquad (1.39)$$

and the conservation of linear momentum for an isolated system is:

$$\sum_{i=1}^{n} m_i . v_i = \sum_{i=1}^{n} \frac{m_0 v_i}{\sqrt{\left(1 - \dfrac{(v_i)^2}{c^2}\right)}} = \text{Constant}$$

Relativistically, the force acting on the particle cannot be just the product of its mass and the acceleration. Since, the time rate of change of momentum defines force, we have

$$F = \frac{dp}{dt} = \frac{d}{dt}\left[\frac{m_0 v}{\sqrt{1 - \dfrac{v^2}{c^2}}}\right] = m_0 \frac{dv}{dt}\left(1 - \frac{v^2}{c^2}\right)^{-1/2} \neq ma \qquad (1.40)$$

EXAMPLE 1.5 Calculate the velocity of a particle at which its mass will be 5 times the mass at rest.

Solution Given $m = 5m_0$

We know variation of mass with velocity is given as:

$$m = \frac{m_0}{\sqrt{\left(1 - \dfrac{v^2}{c^2}\right)}} \Rightarrow 5m_0 = \frac{m_0}{\sqrt{\left(1 - \dfrac{v^2}{c^2}\right)}} \Rightarrow 1 - \frac{v^2}{c^2} = \frac{1}{25} \Rightarrow v^2 = \frac{24}{25}c^2 \Rightarrow v = 0.98c$$

1.13 EQUIVALENCE OF MASS AND ENERGY

The variation of mass with velocity has modified the idea of energy so that a relationship can be established between mass and energy. The relationship can be derived directly from the definition of kinetic energy of a moving body, i.e., work done in bringing the body from rest to its present state of motion.

$\therefore$ Work done $W = Fx$

and Kinetic energy K.E. $= Fx$

$$\text{K.E.} = \int\limits_{0}^{x} F dx \qquad (1.41)$$

In non-relativistic mechanics, the kinetic energy is given by

$$\text{K.E.} = \frac{1}{2} m_0 v^2$$

In relativistic mechanics, the kinetic energy can be calculated as:

Increase in K.E.,

$$dE_k = F dx$$

$$= \frac{d}{dt} (mv)x$$

$$= v \, d(mv) \left[\text{as } \frac{dx}{dt} = v \right] \quad \text{(where } dx \text{ is the small change in displacement.)}$$

$$dE_k = v^2 dm + mv dv \qquad (1.42)$$

We know the mass variation with velocity as:

$$m = \frac{m_0}{\sqrt{\left(1 - \dfrac{v^2}{c^2}\right)}}$$

$$\therefore \qquad m^2 \left(1 - \frac{v^2}{c^2}\right) = m_0^2 \Rightarrow m^2(c^2 - v^2) = m_0^2 c^2$$

or $\qquad m^2 c^2 - m^2 v^2 = m_0^2 c^2 \Rightarrow m^2 c^2 = m^2 v^2 + m_0^2 c^2 \qquad (1.43)$

Differentiating Eq. (1.43), we have

$$2mc^2 dm = 2mv^2 dm + 2vm^2 dv \quad \text{or} \quad c^2 dm = v^2 dm + mv dv \qquad (1.44)$$

Comparing Eqs. (1.42) and (1.44), we have

$$dE_k = c^2 dm \quad \text{or} \quad E_k = \int\limits_{0}^{E_k} dE_k = c^2 \int\limits_{m_0}^{m} dm \qquad [E_k = 0, \text{ if } v = 0]$$

$$E_k = c^2(m - m_0)$$

$$\text{Kinetic energy K.E.} = mc^2 - m_0 c^2 = c^2(m - m_0) \qquad (1.45)$$

This is the increase in kinetic energy due to increase in mass.

$$\text{Total energy} = \text{Kinetic energy} + \text{Rest energy} = \text{K.E.} + m_0 c^2$$

$$= c^2(m - m_0) + m_0 c^2$$

$$E = mc^2 \qquad (1.46)$$

Equation (1.46) gives the universal equivalence between mass and energy.

From Eq. (1.45), we obtain

$$\text{K.E.} = mc^2 - m_0c^2 = \frac{m_0c^2}{\sqrt{\left(1 - \dfrac{v^2}{c^2}\right)}} - m_0c^2 \Rightarrow m_0c^2\left(1 - \frac{v^2}{c^2}\right)^{-1/2} - m_0c^2 \qquad (1.47)$$

If $v \ll c$, $v^2/c^2 < \ll 1$, so using the binomial theorem,

$$\text{K.E.} = m_0c^2\left(1 + \frac{v^2}{2c^2}\right) - m_0c^2 = \frac{1}{2}mv^2$$

which is the classical expression for kinetic energy.

Figure 1.12 shows variation in relative energy $(mc^2 - m_0c^2)/m_0c^2$ versus v/c. For comparison, classical kinetic energy is also plotted.

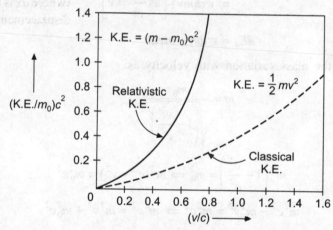

Figure 1.12 Variation in relativistic and classical kinetic energy with v/c.

It is seen that up to speed $0.3c$ there is not much difference between classical and relativistic energies. At $v = 0.5c$, the relativistic energy exceeds classical energy by about 19%. Thus the relativistic formula is needed when velocities are quite high.

1.13.1 Experimental Evidence of Einstein's Mass–Energy Relation

Blackett and Anderson found some cloud chamber photographs double tracks of equal intensity and curvature starting from some point in mid air in opposite directions. The curvature of the one of these tracks was identified as that due to an electron. As the other track was of equal intensity and opposite in direction, it was concluded that this second track must be due to a position. The two tracks are taken to be those due to an electron and a positron produced by annihilation of a γ-ray in the strong field near the nucleus of an atom as shown in Figure 1.13.

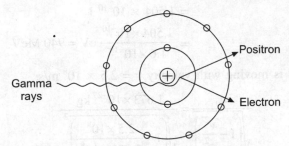

Figure 1.13 Experimental evidence of mass–energy relation.

The reverse process, i.e., the fusion of an electron and a positron to form a γ-ray photon is called annihilation of matter. The radiations resulting from the annihilation of matter is called annihilation radiation. It will appear as photons when positrons from radioactive nuclides are absorbed by matter. The reaction is:

$$_{-1}e^0 + {}_1e^0 \rightarrow 2h\nu \tag{1.48}$$

The photon travels in opposite directions. It is known by experiment that positron captured by matter is most probable at low velocities. The energy of each photon produced will, therefore, be nearly equal to the rest mass energy of the electron. Thus if the frequency of the photon is ν, then

$$h\nu = m_0 c^2 \tag{1.49}$$

where m_0 is the rest mass of the electron, and c is the velocity of light.

Substituting the known values of h, m_0 and c, we get

$$\text{Wavelength of photon, } \lambda = \frac{c}{\nu}$$

$$= \frac{h}{m_0 c}$$

$$= 2.4 \times 10^{-12} \text{m}$$

Dumand had verified experimentally that positron emission by radioactive elements is always accompanied by γ- rays of this wavelength.

EXAMPLE 1.6   A proton, which is the nucleus of a hydrogen atom, has a rest mass 1.673×10^{-27} kg. Find the total energy, when

  (i) It is at rest.
 (ii) It is moving with a velocity of 2.50×10^8 m/s
        (Take $c = 2.998 \times 10^8$ m/s).

Solution   Given $m_0 = 1.673 \times 10^{-27}$ kg, $c = 2.998 \times 10^8$ m/s, $v = 2.5 \times 10^8$ m/s

  (i) At rest, total energy $E_0 = m_0 c^2$

$$\Rightarrow \qquad\qquad E_0 = (1.673 \times 10^{-27} \text{ kg}) \times (2.998 \times 10^8 \text{ m/s})^2$$

$$= 1.504 \times 10^{-10} \text{ J}$$

$$= \frac{1.504 \times 10^{-10}}{1.6 \times 10^{-19}} \text{ eV} = 940 \text{ MeV}$$

(ii) When it is moving with velocity $v = 2.5 \times 10^8$ m/s

$$\frac{m_0}{\sqrt{\left(1 - \dfrac{v^2}{c^2}\right)}} = \frac{1.673 \times 10^{-27} \text{ kg}}{\sqrt{\left(1 - \left(\dfrac{2.5 \times 10^8}{2.998 \times 10^8}\right)^2\right)}}$$

$$\frac{1.673 \times 10^{-27} \text{ kg}}{\sqrt{1 - 0.6952}} = 3.074 \, 10^{-27} \text{ kg}$$

$$\text{Total energy} = mc^2 = \frac{3.074 \times 10^{-27} \times 2.998 \times 10^8}{1.6 \times 10^{-19} \times 10^6} \text{ MeV} = 1730 \text{ MeV}$$

1.13.2 Relation between Total Energy and Momentum

Since, we know that

$$E = mc^2 = \frac{m_0 c^2}{\sqrt{\left(1 - \dfrac{v^2}{c^2}\right)}} \tag{1.50}$$

and

$$p = mv^2 = \frac{m_0 v^2}{\sqrt{\left(1 - \dfrac{v^2}{c^2}\right)}} \tag{1.51}$$

$$\therefore \qquad E^2 = \frac{m_0^2 c^4}{\left(1 - \dfrac{v^2}{c^2}\right)} \text{ and } p^2 = \frac{m_0^2 v^2}{\left(1 - \dfrac{v^2}{c^2}\right)}, \text{ then}$$

$$p^2 c^2 = \frac{m_0^2 v^2 c^2}{\left(1 - \dfrac{v^2}{c^2}\right)}$$

Subtracting $p^2 c^2$ from E^2 yields

$$E^2 - p^2 c^2 = \frac{m_0^2 c^4 - m_0^2 v^2 c^2}{1 - \dfrac{v^2}{c^2}} = \frac{m_0^2 c^4 \left(1 - \dfrac{v^2}{c^2}\right)}{1 - \dfrac{v^2}{c^2}} = \frac{m_0^2 c^4 (c^2 - v^2)}{(c^2 - v^2)} = m_0^2 c^4$$

$$\Rightarrow \qquad\qquad E^2 - p^2c^2 = m_0^2c^4$$

$$\Rightarrow \qquad\qquad E^2 = p^2c^2 + m_0^2c^4 \qquad\qquad (1.52)$$

Following results may be drawn on the basis of Eq. (1.52):

(i) $E^2 - p^2c^2$ is invariant under Lorentz transformation because m_0 and c both are constant.

(ii) For $v << c$, $E = \dfrac{p^2}{2m_0}$

(iii) For massless particle (like photon),

$$m_0 = 0 \qquad \therefore \qquad E = pc$$

▌1.14 TRANSFORMATION EQUATIONS FOR MOMENTUM AND ENERGY

Let there be two inertial systems S and S′ having a velocity v relative to S along the positive direction of X-axis. Suppose that at the instant when the origins of the two systems coincide, a light signal starts from here. Let this instant be taken as zero, i.e., $t = t′ = 0$. The light signal will spread out a sphere in both the system having equations [Figure 1.14].

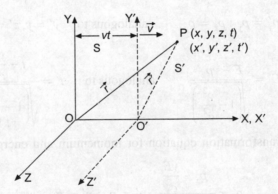

Figure 1.14 Illustration for momentum and energy transformation.

$$\left. \begin{array}{ll} x^2 + y^2 + z^2 = c^2t^2 & \text{(in systems)} \\ x'^2 + y'^2 + z'^2 = c^2t'^2 & \text{(in systems)} \end{array} \right\} \qquad (1.53)$$

If we consider a light photon be situated on this light sphere, it will have energy and momentum, which are related by

$$p^2 = \frac{E^2}{c^2} \qquad\qquad (1.54)$$

Since, for a photon of frequency v, we have

$$E = hv \text{ and } p = \frac{hv}{c}, \qquad \text{So that } p = \frac{E}{c}$$

Now, if the components of momentum along the three axes are p_x, p_y and p_z respectively, then

$$p^2 = p^2{}_x + p^2{}_y + p^2{}_z$$

From Eq. (1.54), we have

$$p_x^2 + p_y^2 + p_z^2 = \frac{E^2}{c^2} \qquad (1.55a)$$

Similarly

$$p_x'^2 + p_y'^2 + p_z'^2 = \frac{E'^2}{c^2} \qquad (1.55b)$$

Equations [1.55(a) and (b)] are exactly same as Eq. (1.53) and can be written by replacing (x, y, z, t) by $(p_x, p_y, p_z, E/c^2)$.

Further, since Eq. (1.55) is true for all inertial frames (systems) just like Eq. (1.53), it follows by symmetry that the quantities $(p_x, p_y, p_z, E/c^2)$ must be transform according to Lorentz transformation equations for (x, y, z, t). Hence

$$p_x' = \frac{p_x - \dfrac{vE}{c^2}}{\sqrt{\left(1 - \dfrac{v^2}{c^2}\right)}} \qquad \text{analogous to} \qquad x' = \frac{x - vt}{\sqrt{\left(1 - \dfrac{v^2}{c^2}\right)}}$$

$$p_y' = p_y : p_z' = p_z \qquad \text{analogous to} \qquad y' = y, z' = z$$

$$E' = \frac{E - vp_x}{\sqrt{\left(1 - \dfrac{v^2}{c^2}\right)}} \qquad \text{analogous to} \qquad t' = \frac{t - \dfrac{vx}{c^2}}{\sqrt{\left(1 - \dfrac{v^2}{c^2}\right)}}$$

Thus, the transformation equation for momentum and energy are:

$$p_x' = \frac{p_x - \dfrac{vE}{c^2}}{\sqrt{\left(1 - \dfrac{v^2}{c^2}\right)}}, p_y' = p_y : p_z' = p_z \quad \text{and} \quad E' = \frac{E - vp_x}{\sqrt{\left(1 - \dfrac{v^2}{c^2}\right)}} \qquad (1.56)$$

SOLVED NUMERICAL PROBLEMS

PROBLEM 1.1 Calculate the energy in MeV, which is equivalent of a proton of mass 1.67×10^{-27} kg.

Solution Given $\Delta m = 1.67 \times 10^{-27}$ kg, $c = 3 \times 10^8$ m/s

By mass–energy relation, we have

$$\Delta E = \Delta mc^2$$

$$\therefore \qquad \Delta E = (1.67 \times 10^{-27}) \times (3 \times 10^8)^2 = 1.503 \times 10^{-11} \text{ s}$$

$$= \frac{1.503 \times 10^{-11}}{1.6 \times 10^{-13}} \text{ MeV} = 9.39 \text{ MeV}$$

PROBLEM 1.2 Calculate the velocity of a particle at which its mass will become eight times of the rest mass.

Solution   Given $m = 8m_0$

Using mass variation with velocity,

$$m = \frac{m_0}{\sqrt{\left(1 - \dfrac{v^2}{c^2}\right)}}$$

$$\Rightarrow \qquad 8m_0 = \frac{m_0}{\sqrt{\left(1 - \dfrac{v^2}{c^2}\right)}}$$

or $\qquad 1 - \dfrac{v^2}{c^2} = \dfrac{1}{64}$ or $\dfrac{v^2}{c^2} = 1 - \dfrac{1}{64} \Rightarrow v^2 = \dfrac{63}{64}c^2 \Rightarrow v = 0.99c$

PROBLEM 1.3 A spaceship 50 m long passes the earth at a speed of 2.8×10^8 m/s. What will be its apparent length?

Solution   Given $\qquad L' = 50$ m, $v = 2.8 \times 10^8$ m/s

From length contraction formula,

$$L = L' \sqrt{\left(1 - \frac{v^2}{c^2}\right)}$$

$$= 50 \sqrt{\left(1 - \frac{(2.8)^2}{3^2}\right)} = 18 \text{ m}$$

PROBLEM 1.4 With what velocity should a rocket move, so that every year spent on it corresponds to 3 years on earth?

Solution   The time of one year spent on the rocket is the proper time for the observer sitting on it, and the corresponding time on the earth is the observed time $t = 4$ years. From time-dilation formula,

$$t = \frac{t_0}{\sqrt{\left(1 - \dfrac{v^2}{c^2}\right)}}$$

where v is the velocity of the rocket.

$$3 = \frac{1}{\sqrt{\left(1 - \dfrac{v^2}{c^2}\right)}} \Rightarrow \sqrt{\left(1 - \frac{v^2}{c^2}\right)} = \frac{1}{3}$$

$$v = \frac{\sqrt{8}}{9}c = 0.943c$$

PROBLEM 1.5 Two photons approach each other, what is their relative velocity?

Solution Velocity of photons $= c$ and $-c$ in laboratory frame S.

Keeping a frame S′ attached to one photon $v = c$, but the velocity of other photon be u'.

$$\therefore \qquad u' = \frac{u - v}{1 - \dfrac{uv}{c^2}} = \frac{-c - c}{1 - \dfrac{c^2}{c^2}} = \frac{-2c}{2} = -c$$

Hence the relative velocity of two photons approaching each other is equal to the velocity of light.

PROBLEM 1.6 Prove that the particle having rest mass zero always moves with velocity of light.

Solution Given $m_0 = 0$

From the relativistic momentum–energy relation,

$$E = \sqrt{p^2 c^2 + m_0^2 c^4} \Rightarrow E = pc \tag{i}$$

From the definition of momentum $p = mv$, and from relativistic mass–energy relation,

$$E = mc^2 \tag{ii}$$

From Eqs. (i) and (ii), we have

$$\therefore \qquad\qquad mc^2 = mvc$$

$$\therefore \qquad\qquad v = c$$

PROBLEM 1.7 Show that if (x_1, y_1, z_1, t_1) and (x_2, y_2, z_2, t_2) be the co-ordinates of one event in S_1 frame and the corresponding event in S_2 frame respectively, then expression $dS_1^2 = dx_1^2 + dy_1^2 + dz_1^2 - c^2 dt_1^2$ is invariant under a Lorentz transformation co-ordinates.

Solution The inverse Lorentz transformation equations are:

$$\left.\begin{array}{l} x_1 = \dfrac{x_2 + vt_2}{\sqrt{\left(1 - \dfrac{v^2}{c^2}\right)}}, y_1 = y_2, z_1 = z_2 \quad \text{and} \quad t_1 = \dfrac{t_2 + \dfrac{vx_2}{c^2}}{\sqrt{\left(1 - \dfrac{v^2}{c^2}\right)}} \end{array}\right\} \tag{i}$$

Differentiating Eq. (i), we obtain,

$$dx_1 = \frac{dx_2 + vdt_2}{\sqrt{1 - \dfrac{v^2}{c^2}}}, dy_1 = dy_2, dz_1 = dz_2 \quad \text{and} \quad dt_1 = \left.\frac{dt_2 + \dfrac{vdx_2}{c^2}}{\sqrt{1 - \dfrac{v^2}{c^2}}}\right\}$$ (ii)

$$dS_1^2 = \left[\frac{dx_2 + vdt_2}{\sqrt{\left(1 - \dfrac{v^2}{c^2}\right)}}\right]^2 + dy_2^2 + dz_2^2 - c^2\left[\frac{dt_2 + \dfrac{v}{c}dx_2}{\sqrt{1 - \dfrac{v^2}{c^2}}}\right]^2$$

On simplication, it gives

$$dS_1^2 = dx_2^2 + dy_2^2 + dz_2^2 - c^2dt_2^2$$
$$= dS_2^2$$

Hence proved.

PROBLEM 1.8 A stationary body exploded into two fragments each of mass 1.0 kg that move apart at speed of 0.6c relative to the original body. Find the mass of the original body.

Solution Given $m_1 = m_2 = 1.0$ kg, $v = 0.6c = 0.6 \times 3 \times 10^8$ m/s

The rest energy of the original body must equal the sum of the total energy of the fragments. Hence

$$E_0 = mc^2$$
$$= \phi m_1 c^2 + \phi m_2 c^2$$
$$= \frac{m_1 c^2}{\sqrt{\left(1 - \dfrac{v^2}{c^2}\right)}} + \frac{m_2 c^2}{\sqrt{\left(1 - \dfrac{v^2}{c^2}\right)}} \quad \text{and} \quad m = \frac{E_0}{c^2} = \frac{(m_1 + m_2)c^2}{c^2\sqrt{\left(1 - \dfrac{v^2}{c^2}\right)}} = \frac{2 \times 1.0 \text{ kg}}{\sqrt{1 - (0.6)^2}} = 2.5 \text{ kg}$$

PROBLEM 1.9 Find the velocity that an electron must be given so that its momentum is 10 times its rest mass the speed of light. What is the energy at this speed?

Solution Given $p = m_0 c$, $m_0 = 9.1 \times 10^{-31}$ kg
We know that

$$m = \frac{m_0}{\sqrt{\left(1 - \dfrac{v^2}{c^2}\right)}}$$

and
$$\text{Momentum } p = mv = \frac{m_0 v}{\sqrt{\left(1 - \dfrac{v^2}{c^2}\right)}}$$

According to problem,

$$10\, m_0 c = \frac{m_0 v}{\sqrt{\left(1 - \dfrac{v^2}{c^2}\right)}}$$

or
$$\frac{v}{c} = 10 \sqrt{\left(1 - \frac{v^2}{c^2}\right)} \qquad \text{(i)}$$

Squaring Eq. (i),

$$\frac{v^2}{c^2} = 100 - 100\frac{v^2}{c^2} \implies 101\frac{v^2}{c^2} = 100 \text{ or } \frac{v}{c} = \frac{100}{101}$$

Now, mass of electron,

$$m = \frac{m_0}{\sqrt{\left(1 - \dfrac{v^2}{c^2}\right)}} = \frac{9.1 \times 10^{-31}}{\sqrt{1 - \left(\dfrac{100}{101}\right)^2}} = 9.15 \times 10^{-30} \text{ kg}$$

PROBLEM 1.10 Show by analysing a collision between a photon and a free electron (using relativistic mechanics), that it is impossible for a photon to give all of its energy to free electron. In other terms, the photoelectric effect cannot oocur for completely free electrons; the electrons must be bound in a solid or in an atom.

Solution Suppose a photon of energy $h\nu$ is completely absorbed by a free electron. Then the photoelectron must be ejected in forward direction in order to conserve momentum. Conservation energy gives

$$h\nu + m_0 c^2 = \sqrt{p^2 c^2 + m_0^2 c^4} \qquad \text{(i)}$$

The conservation of momentum gives

$$\frac{h\nu}{c} = p \implies pc = h\nu \qquad \text{(ii)}$$

Using Eqs. (i) and (ii), we get

$$h\nu + m_0 c^2 = \sqrt{h^2 \nu^2 + m_0^2 c^4} \qquad \text{(iii)}$$

Squaring both sides and simplifying $2h\nu m_0 c^2 = 0$, which is absurd, since $h\nu \neq 0$ and $m_0 c^2 \neq 0$.

So it is impossible for a photon to give all of its energy to free electron.

EXERCISES

THEORETICAL QUESTIONS

1.1 Describe Michelson–Morley experiment. Discuss its aims and results achieved.

1.2 State the postulates of special theory of relativity and deduce Lorentz transformations.

1.3 Define an inertial frame of reference and derive the Lorentz transformations.

1.4 Show that Lorentz transformation reduces to Galilean transformation for velocities much smaller than the velocity of light in free space. What is relativity of simultaneity?

1.5 Deduce the relativistic law of addition of velocities. Prove that the velocity of a particle moving with velocity of light is same in all inertial frame of reference.

1.6 What is meant by length contraction and proper length? Explain the null result of Michelson–Morley experiment.

1.7 Explain (i) length contraction, (ii) time-dilation in special theory of relativity. What are proper length and proper time interval? How is the time-dilation experimentally verified?

1.8 What are time-dilation and proper time? Show that two simultaneous events at different positions in reference frame S are not in general simultaneous in another frame S′ moving with constant velocity relative to frame S.

1.9 Stating with Einstein's velocity addition formula show that it is in conformity with the principle of special theory of relativity.

1.10 Derive the formula for relativistic variation of mass with velocity. Hence prove that it is not possible for a material particle to have velocity equal or greater than the velocity of light.

1.11 Derive Einstein's mass–energy relation and explain its importance.

1.12 Find relativistic expression for the energy of a particle moving with momentum p. Is it true for a photon?

1.13 Derive mass–energy relation. Show that total energy E, and momentum p are related to $E^2 = p^2c^2 + m_0{}^2c^4$ where m_0 is the rest mass and c is the speed of light.

NUMERICAL PROBLEMS

P1.1 Calculate the percentage contraction of rod moving with velocity $0.6c$ in a direction inclined at $45°$ to its own length. **(Ans. 9.45%)**

P1.2 Calculate the velocity of a clock with respect to an observer, so that it apparently loses 48 s in a day. **(Ans. 10^7m/s)**

P1.3 A cosmic ray μ-meson has a proper half life 2.2×10^{-6} s. What will be its half life as measured by an observer on the earth, if it approaches the earth at a speed of $0.999c$. **(Ans. 49.2×10^{-6} s)**

P1.4 Spaceship A is travelling with a speed of $0.9c$ with respect to earth. Spaceship B travelling in the same direction passes A with a relative speed of $0.5c$. What is the speed of B with respect to earth? **(Ans. $0.9655c$)**

P1.5 An observer is at rest with respect to a cube of copper. The density of cube at rest is 8.9×10^3 kg/m^3. What would an observer moving parallel to one edge with a velocity $0.6c$ relative to the cube observe the density? **(Ans.** 13.90×10^3 kg/m^3)

P1.6 A certain young lady decides on her twenty fifth birthday that it is time to slenderize. She weighs 100 kg. She has heard that she moves fast enough, she will appear thinner to her stationary friends.
    (i) How fast must she move to appear slenderized by a factor 50%?
    (ii) At this speed, what will her mass appear to her stationary friend.
    (iii) If she maintained her speed until her twenty ninth birthday, how old will her stationary friends claim; she is according to their measurement?
                                    (Ans. (i) $0.866c$, (ii) 200 kg (iii) 33 years old)

P1.7 How fast must a body be travelling if its correct relativistic momentum is 1% greater than the classical $p = mv$. **(Ans.** $v = 0.14c$)

P1.8 A nuclear particle has mass 3 GeV/c^2 and momentum 4 GeV/c.
    (i) What is its energy?
    (ii) What is its speed? **(Ans.** (i) $E = 5$ GeV, (ii) $u = 0.8c$)

P1.9 A particle with a rest energy of 2400 MeV has energy of 15 GeV. Find the time (in Earth's frame of reference) necessary for this particle to travel from earth to a star 4 light years distant. **(Ans.** 4.05 light years)

P1.10 At what speed would a body's relativistic energy E be twice its rest energy mc^2?
                                             (Ans. $v = 0.87c$)

Multiple Choice Questions

MCQ 1.1 Inertial frame are those frames of reference, in which a free particle moves
    (a) Along a straight line with a constant speed
    (b) With a variable speed along a straight line
    (c) With a variable speed of a curved path
    (d) With constant speed on a curved path

MCQ 1.2 If S is an inertial frame of reference, then another frame accelerated with constant acceleration with respect to it, will be
    (a) Inertial
                     (b) Non-inertial
    (c) Inertial and non-inertial
    (d) Neither inertial nor non-inertial

MCQ 1.3 The frame attached to any point on earth's surface is
    (a) Inertial
                     (b) Non-inertial
    (c) Neither inertial nor non-inertial
    (d) Newton's frame of reference

MCQ 1.4 Correct transformation equations connecting to inertial frames of reference are
    (a) Galilean
                     (b) Newtonian
    (c) Lorentz
                     (d) None of these

MCQ 1.5 A particle can have velocity
    (a) Equal to the velocity of light
    (b) More than the velocity of light
    (c) Slightly less than the velocity of light
    (d) Equal to the velocity of sound

MCQ 1.6 A clock is moving with velocity $c/3$. In one hour, the clock appears to be slow by
    (a) 3 minutes
    (b) 3.4 minutes
    (c) 3.7 sec
    (d) Not at all

MCQ 1.7 The relativistic mass of a particle
    (a) Increases with velocity
    (b) Decreases with velocity
    (c) Decreases with velocity finally becomes zero
    (d) Decreases or increases with velocity and finally becomes zero

MCQ 1.8 Correct relation is

    (a) $L = \dfrac{L_0}{\sqrt{\left(1 - \dfrac{v^2}{c^2}\right)}}$
    (b) $t = \dfrac{t_0}{\sqrt{\left(1 + \dfrac{v^2}{c^2}\right)}}$

    (c) $m = \dfrac{m_0}{\sqrt{\left(1 - \dfrac{v^2}{c^2}\right)}}$
    (d) Not at all

MCQ 1.9 Gain in kinetic energy is equal to

    (a) $(m_0 - m)c^2$
    (b) $\dfrac{(m_0 - m)}{c^2}$

    (c) $\dfrac{(m - m_0)}{c^2}$
    (d) $(m_0 - m)c^2$

MCQ 1.10 On the basis of theory of relativity, momentum–energy relation is
    (a) $E^2 = p^2 c^2 + m_0^2 c^4$
    (b) $E^2 = p^2 c^2$
    (c) $E^2 = p^2 c^2 - m^2 c^4$
    (d) $E^2 = p^2 c^2 - m_0^2 c^4$

Answers

1.1 (a)	1.2 (b)	1.3 (b)	1.4 (c)	1.5 (c)	1.6 (b)	1.7 (a)
1.8 (c)	1.9 (d)	1.10 (a)				

CHAPTER 2

Quantum Mechanics I

CHAPTER OUTLINES

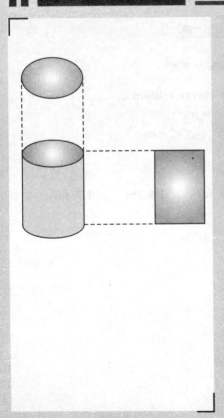

- Introduction
- Black Body Radiation
- Planck's Theory of Radiation
- Photoelectric Effect
- The Compton Effect
- De Broglie Hypothesis and Matter Waves
- Wave-Particle Duality
- Phase and Group Velocities
- Heisenberg's Uncertainty Principle

2.1 INTRODUCTION

The physics prior to 1900 had been enriched by many giant workers and wide variety of natural phenomenon was explained on basis of theories and experiments. The physics developed during this era is termed as *classical physics* or *Newtonian physics*. The classical physics was very successful in describing macroscopic phenomenon. The motion of mechanical objects was successfully discussed in terms of Newton's equation of motion on both celestial and terrestrial scales. Application of this theory to molecular motions produced useful results in kinetic theory of gases and the discovery of electron.

During the first quarter of twentieth century, many unresolved problems of physics compelled physics community to think over problems in a new way. When world of investigation changes, the physics must also change. It means that the line of thinking about problems of terrestrial bodies will not work at atomic scale. In other words, the transition from the macroscopic system to the microscopic system must be accompanied by the transition from classical theory to quantum theory.

In 1900 Planck explained the spectral distribution of black body radiation by assuming that emission and absorption of electromagnetic radiation occurs in from of discrete quanta. The discrete quanta responsible for the process contains an amount of energy E equal to multiplication of frequency of radiation v and universal constant h known as Planck's constant. Later on this quantum idea was used by Einstein in explaining photoelectric effect which could not be explained by using the wave theory of light. In this way, the concept of dual nature (wave and particle) of electromagnetic radiation emerged. Bohr's theory of hydrogen atom revealed the discreteness in atomic systems and Compton's experiment established the light quanta as particle of negligible rest mass. Infact the relativity and quantum theory remodelled the classical physics to bring it to modern era.

2.2 BLACK BODY RADIATION

A piece of hot metal gives off visible light whose colour varies with the temperature of the metal. At room temperature most of the radiation is in the infrared region of electromagnetic spectrum and, therefore, is invisible. Any body capable of radiating energy has ability to absorb radiations also. A black body is an ideal body which absorbs all radiations incident upon it irrespective of wavelength. Its radiation emitting properties are proved to be independent of nature of material of the body and vary in a simple way with temperature. Therefore, consideration on such a body makes the study convenient and easy. In practice, a black body can be approximated as a hollow cavity with a small hole. Any radiation entering to cavity through hole is trapped inside the cavity. The walls of cavity are continuously emitting and absorbing radiation. Such a black body is also known as cavity radiator. The theoretical study of the observed features of black body radiation in 1900 by the German physicist, Max Planck laid the foundation of modern quantum physics.

The spectral distribution of black body radiation is shown in Figure 2.1. The higher the temperature, the greater the amount of radiation and higher the frequency at which maximum emission occurs. From general thermodynamic arguments, it had been found that this spectral distribution depends only on temperature T and independent of material of walls and shape and size of black body.

Lord Rayleigh and James Jeans examined the black body spectrum. They assumed the walls of black body made of simple harmonic oscillators with average energy kT (k is Boltzmann's constant). The energy density in the cavity in frequency interval ν and $\nu + d\nu$ is given by Rayleigh–Jeans formula:

$$u(\nu)d\nu = \frac{8\pi}{c^3}\nu^2 d\nu\, kT \tag{2.1}$$

This formula predicts that as frequency of radiation increases, the energy density also increases, but the observation of Figure 2.1 finds this untrue.

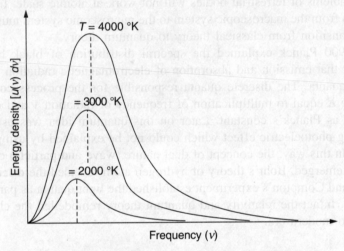

Figure 2.1 Spectral distribution of black body radiation.

The Rayleigh–Jeans formula is based on classical thoughts. Many other physicists tried to account the observed results, but they have limited success. Infact black body radiation problem was outstanding unsolved problem in physics during those days.

2.3 PLANCK'S THEORY OF RADIATION

Planck developed a theoretical model to explain black body radiation on the basis of atomic processes taking place at the cavity walls. He assumed that atoms of walls of cavity (black body) behave like electromagnetic oscillators. The oscillators may emit and absorb electromagnetic energy. Planck's theory was based on the following two assumptions:

1. An oscillator can have only energies given by

$$E_n = nh\nu \tag{2.2}$$

where v is frequency of oscillator, h is a universal constant termed as Planck's constant and n is a number such that $n = 0, 1, 2, 3, \ldots$. This assumption dictates that energy is not continuous as in classical physics, but it is quantized.

2. The oscillators radiate energy only when it changes its quantized energy states. In a system of such oscillators at temperature T, the number of oscillators of energy nhv will be given by the Boltzmann factor:

$$N_n = Ae^{-E_n/kT} \tag{2.3}$$

The average energy of the oscillator ε may be obtained by summing energies E_nN_n (total energy of cavity) over all n from 0 to ∞ and divide it by the total number of oscillators over all n. Thus average energy of oscillators is given by

$$\varepsilon = \frac{\sum\limits_{n=0}^{\infty} E_n Ae^{-\frac{E_n}{kT}}}{\sum\limits_{n=0}^{\infty} Ae^{-\frac{E_n}{kT}}}$$

or

$$\varepsilon = \frac{\sum\limits_{n=0}^{\infty} nhve^{-\frac{nhv}{kT}}}{\sum\limits_{n=0}^{\infty} e^{-\frac{nhv}{kT}}}$$

The result of summation may be given as:

$$\varepsilon = hv(e^{hv/kT} - 1) \tag{2.4}$$

Planck replaced kT by the value of ε and given his formula for energy density in cavity as:

$$u(v)dv = \frac{8\pi v^2}{c^3} dv \frac{hv}{(e^{hv/kT} - 1)} \tag{2.5}$$

Planck's energy density formula is in excellent agreement with the experimental findings. Clearly by introducing the idea of quantization of electromagnetic energy, Planck successfully explained the spectral distribution of black body radiation. It was this idea which transforms classical physics into quantum physics. Later on the quantization concept was used by A. Einstein to explain the photoelectric effect.

2.4 PHOTOELECTRIC EFFECT

In 1887, Hertz observed that when light rays are allowed to fall on the negative plate of an electric discharge tube, the electric discharge becomes easier. This experiment was also confirmed by Hallwachs, but he could not explain this phenomenon.

J.J. Thomson established that when light falls on a metal surface, electrons are emitted from the surface. They are attracted towards positive plate so the circuit is completed and current flows in circuit. The phenomenon of emission of electrons from metals under the effect of light rays is called 'photoelectric effect'. The emitted electrons are known as 'photoelectrons' and resulting current is termed as 'photoelectric current'. Photoemission is more effective for high frequency light.

2.4.1   Experimental Results

Lenard and Millikan performed many experiments on photoelectric effect. Figure 2.2 illustrates the apparatus employed to observe photoelectric effect. An evacuated tube contains two electrodes A and C which are connected by external circuit. Lenard and Millikan used different metals and illuminated them with light of different frequencies and intensities. Photoelectric current was measured in each case, and they obtained many results on the basis of this experiment.

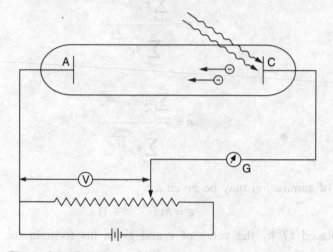

Figure 2.2   Experimental observation of photoelectric current.

Dependence on the intensity of light

When light of definite intensity and definite frequency falls on metallic plate, the emitted electrons are attracted towards relatively positive plate. When they reach the plate, current flows in the circuit. On increasing the positive potential, there is no increase in the photoelectric current because all emitted electrons have already reached the plate. However, if the intensity of light falling on the plate is increased, the current increases. Thus, rate of emission of photoelectrons depends directly on the intensity of incident light as shown in Figure 2.3.

If now potential of plate A is made negative relative to plate C, the current immediately decreases, but does not become zero, which proves that the emitted electrons have finite velocity (or kinetic energy). Some electrons will reach plate A inspite of the fact that the electric field opposes their motion. However, if the negative potential at A is made large enough, a value V_0, known as stopping potential is reached at which the photoelectric current drops to zero. Since the stopping potential

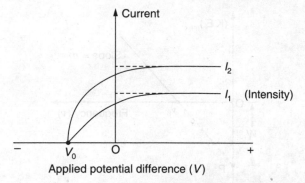

Figure 2.3   Photocurrent is proportional to light intensity for all retarding potentials.

stops the motion of electrons, even they have maximum obtainable kinetic energy, it is a measure of kinetic energy (K.E.) of electrons.

$$(K.E.)_{max} = eV_0 \qquad (2.6)$$

where e is electronic charge. Since at stopping potential, the photoelectric current cannot be increased by increasing the intensity of the incident light, therefore, stopping potential or maximum kinetic energy of electrons does not depend on the intensity of light.

Dependence on the frequency of light

If at the stopping potential, the light of higher frequency is made to fall on plate C, then photoelectric current is re-established. On increasing further the stopping potential (a negative potential) the flow of current again stops. It shows that higher the frequency of incident light, higher will be the stopping potential or higher will be the kinetic energy of emitted electrons. This is shown in Figure 2.4.

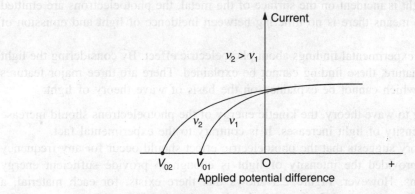

Figure 2.4   Dependence of stopping potential on frequency of incident light.

If a graph is plotted between maximum kinetic energy $(K.E.)_{max}$ of electron and frequency v of incident light, a straight line is obtained. Figure 2.5 shows that maximum kinetic energy of electrons increases as the frequency of incident radiation increases.

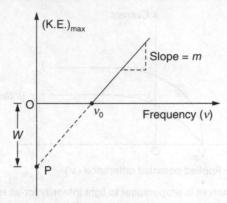

Figure 2.5   A plot of maximum kinetic energy $(K.E.)_{max}$ with respect to frequency of incident radiation.

This graph intersects the frequency axis at v_0 and energy axis in negative direction at P such that $(K.E.)_{max} = -W$. If we take slope of the graph as m, then

$$(K.E.)_{max} = mv + (-W) \qquad (2.7)$$

This equation is the result of experiment. It is clear from the graph that for emission of photoelectrons, the frequency of incident light must be greater than v_0. The frequency v_0 is known as threshold frequency or cut-off frequency of given material.

For $\qquad\qquad\qquad v = v_0, \qquad (K.E.)_{max} = 0$

and, therefore $\qquad\qquad W = mv_0 \qquad\qquad\qquad\qquad (2.8)$

Here W represents kinetic energy for minimum frequency v_0. This minimum energy W needed for emission of photoelectrons for a given material is known as work function of that material. Normally the order of work function is of a few electron volts. As soon as the light is incident on the surface of the metal, the photoelectrons are emitted instantly. This means there is no time lag between incidence of light and emission of electrons.

These are experimental findings about photoelectric effect. By considering the light having wave nature, these finding cannot be explained. There are three major features of this effect which cannot be explained on the basis of wave theory of light:

1. According to wave theory, the kinetic energy of the photoelectrons should increase as the intensity of light increases. It is contrary to the experimental fact.
2. Wave theory suggests that the photoelectric effect should occur for any frequency of light, provided the intensity of light is enough to provide sufficient energy to electrons. However, Figure 2.5 shows that there exists, for each material, a threshold frequency v_0 below which the photoelectric effect disappears even for very intense illumination.
3. The energy transferred by the light waves will not go to a particular electron, but it will be distributed among all the electrons present in the illuminated surface. Hence electrons will take sometime in accumulating energy required for emission. But according to experiment, emission of electrons takes place immediately after the light is incident on surface.

These objections against wave theory of photoelectric effect are resolved by Einstein using quantum model of light in 1905.

2.4.2   Einstein's Photon Theory

Einstein explained the photoelectric effect on the basis of Planck's theory. He assumed that light travels in form of small energy packets termed as *photons*. The energy of each photon is $h\nu$, where h is Planck's constant and ν is the frequency of light. The intensity of light depends on the number of these photons.

When a photon falls on a metal surface, it transfers whole of its energy $h\nu$ to any one of the electrons present in the metal and its own existence is vanished. A part of this energy is used in ejecting the electrons from metal (as work function W) and the rest is given to electron as kinetic energy which will be maximum kinetic energy (K.E.)$_{max}$ for that electron. Photoelectric effect is inelastic because incident photon disappears.

Applying the photon concept to the photoelectric effect, Einstein wrote

$$h\nu = W + (K.E.)_{max} \tag{2.9}$$

The comparison of Eq. (2.9) with Eq. (2.7) gives that m is actually Planck's constant and W is work function of the metal. Clearly if energy of photon absorbed is less than the work function of the metal, no electron will be emitted. If v_{max} is maximum velocity of ejected electron and threshold frequency is ν_0 then Eq. (2.9) may be written as:

$$\frac{1}{2}mv_{max}^2 = h(\nu - \nu_0) \tag{2.10}$$

or

$$eV_0 = h(\nu - \nu_0) \tag{2.11}$$

where m is mass of electron and $W = h\nu_0$.

Einstein removes all three objections raised against wave theory and gives interpretation of photoelectric effect by using his photon theory. The increment in light intensity increases the number of photons hence current increases. It does not change the energy $h\nu$ of the photons. This explanation meets out the first objection. If maximum kinetic energy (K.E.)$_{max}$ equals to zero, we have

$$h\nu = h\nu_0 = W \tag{2.12}$$

which means photon has energy to eject the electron and no energy remains for providing kinetic energy to electron. Now, if ν is reduced below ν_0, no electron is emitted even for intense illumination (larger number of photons). This confirms existence of threshold frequency ν_0 and removes the second objection.

As soon as the first photon falls on metal, one electron in the metal absorbs it as such and ejected out. It spreads out over a large area, as in the wave theory, thus there is no time lag between incidence of photon and emission of electron. This explanation removes the last objection against the wave theory of photoelectric effect.

Later on Einstein's theory was tested experimentally by Millikan and found correct. Although photon theory was found correct, it seems to be in direct conflict with the wave theory of light. The modern view of the nature of light is that it has dual character. In some circumstances, it behaves like wave and behaves like a particle or photon under other circumstances.

2.5   THE COMPTON EFFECT

The Compton scattering is a type of scattering that X-rays and gamma rays undergo in matter. In the scattering process, the emitted photons have reduced energy. This phenomenon is called Compton effect and was observed by Arthur Holly Compton in 1923 for which he received noble prize in 1927.

The Compton effect is important because it demonstrates that light cannot be explained purely as a wave phenomenon. This effect confirmed the concept of photon as a packet of energy. Compton convinced physicists that light can behave as a stream of particles of energy $h\nu$, known as photon.

2.5.1   Experiment

A.H. Compton allowed a beam of X-rays of sharply defined wavelength λ to fall on a graphite block, as shown in Figure 2.6. Although the incident beam consists of a single wavelength λ, the scattered X-rays have two wavelengths. One of them is the same as the incident wavelength λ and other λ' being larger by an amount $\Delta\lambda$, termed as *Compton shift*. The Compton shift varies with the angle of scattering.

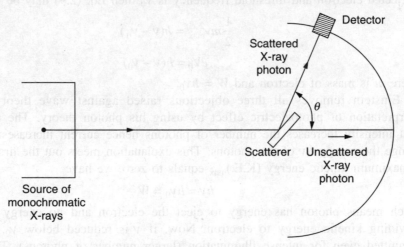

Figure 2.6   Schematic of Compton's experimental arrangement.

According to the electromagnetic wave theory, the incident wave of wavelength λ causes electrons of scatterer to oscillate with wavelength λ'. These oscillations radiate electromagnetic waves again with same wavelength λ as the incident wave. Clearly, the presence of scattered wave of wavelength λ' cannot be understood on the basis of wave theory.

2.5.2 Scattering Formula

This effect was explained by Compton on the basis of quantum theory of radiation. Compton postulated that the incoming X-ray beam was not a wave but an assembly of photons of energy $h\nu$. The whole process is treated as a particle collision event between X-ray photon and a loosely bound electron of the scatter. In this process, energy and momentum are conserved, and thus interaction of photon and electron is elastic. A portion of photon energy is transferred to electron and, therefore, scattered photon is of higher wavelength. Such a scattering is termed as *incoherent scattering*. Figure 2.7 illustrates the collision of photon with electron.

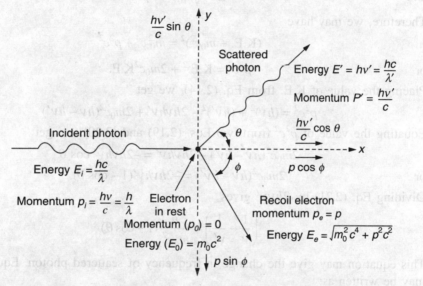

Figure 2.7 Collision of photon with electron initially at rest.

Let us consider an incident photon of energy $h\nu$ and momentum $h\nu/c$ collide with an electron initially at rest. The initial momentum of electron is zero, and rest energy is m_0c^2. The scattered photon of momentum $h\nu'/c$ and energy $h\nu'$ makes an angle θ from X-axis. Again the electron recoils along a direction making angle ϕ from X-axis with momentum p and energy mc^2.

The energy conservation principle gives

$$h\nu + m_0c^2 = h\nu' + mc^2 \qquad (2.13)$$

and

$$h\nu - h\nu' = \text{K.E.} \qquad (2.14)$$

where K.E. is kinetic energy gained by electron. Since the momentum is a vector quantity and is conserved, the x any y components must obey the following equations:

$$\frac{h\nu}{c} + 0 = \frac{h\nu'}{c}\cos\theta + p\cos\phi \qquad (2.15)$$

and

$$0 = \frac{h\nu'}{c}\sin\theta - p\sin\phi \qquad (2.16)$$

Now, multiplication of Eqs. (2.15) and (2.16) by c gives

$$pc \cos \phi = h\nu - h\nu' \cos \theta \qquad (2.17)$$

and

$$pc \sin \phi = h\nu' \sin \theta \qquad (2.18)$$

Squaring each of these equations and adding them, we have

$$p^2c^2 = (h\nu)^2 + (h\nu')^2 - 2h\nu h\nu' \cos \phi \qquad (2.19)$$

The total energy of a particle may be written in two ways:

$$E = \text{K.E.} + m_0 c^2$$

and

$$E = \sqrt{m_0^2 c^4 + p^2 c^2}$$

Therefore, we may have

$$(\text{K.E.} + m_0 c^2)^2 = m_0^2 c^4 + p^2 c^2$$

or

$$p^2 c^2 = \text{K.E.}^2 + 2m_0 c^2 \text{K.E.}$$

Placing the value of K.E. from Eq. (2.14), we get

$$p^2 c^2 = (h\nu)^2 + (h\nu')^2 - 2h\nu h\nu' + 2m_0 c^2 (h\nu - h\nu') \qquad (2.20)$$

Equating the value of $p^2 c^2$ from two Eqs. (2.19) and (2.20), we get

$$2m_0 c^2 (h\nu - h\nu') - 2h\nu h\nu' = -2h\nu h\nu' \cos \theta$$

or

$$2m_0 c^2 (h\nu - h\nu') = -2h\nu h\nu' (1 - \cos \theta) \qquad (2.21)$$

Dividing Eq. (2.21) by $2h\nu\nu'$ gives

$$\left(\frac{1}{\nu} - \frac{1}{\nu'} \right) = \frac{h}{m_0 c^2} (1 - \cos \theta) \qquad (2.22)$$

This equation may give the change in frequency of scattered photon. Equation (2.22) may be written as:

$$\left(\frac{c}{\nu'} - \frac{c}{\nu} \right) = \frac{h}{m_0 c} (1 - \cos \theta)$$

or

$$\lambda' - \lambda = \frac{h}{m_0 c} (1 - \cos \theta) = \Delta\lambda \qquad (2.23)$$

Equation (2.23) gives shift in wavelength of photon scattered at an angle θ. The quantity $\dfrac{h}{m_0 c}$ is known as *Compton wavelength* of the scattering particle. For electron, its value comes out as 0.024 Å. The change in wavelength $\Delta\lambda$ is independent of the wavelength of incident radiation and nature of scattering substance and depends only on scattering angle.

For $\theta = 0$, scattering occurs in direction of incident photon, and there is no change in wavelength of scattered photon, i.e., $\Delta\lambda = 0$.

When scattering occurs at right angle ($\theta = 90°$), the shift in wavelength $\Delta\lambda = 0.024$ Å is termed as *Compton wavelength*. However, we have maximum value of $\Delta\lambda$ (= 0.048 Å) when scattering occurs in the direction opposite to the incident photon.

Compton recoil electron

The recoil phenomenon predicted by Compton was verified by Wilson and Bothe. The value of kinetic energy of recoil electron is in agreement with theory, as confirmed by Bless in 1927.

Dividing Eq. (2.18) by Eq. (2.17), we get

$$\tan \phi = \frac{h\nu' \sin \theta}{h\nu - h\nu' \cos \theta} \tag{2.24}$$

and kinetic energy of recoil electron is given by

$$\text{K.E.} = h\nu - h\nu'$$

Equation (2.24) gives the direction of recoil electron.

2.5.3 Applications

Compton scattering is of prime importance in radiobiology because it is most probable interaction of gamma rays and X-rays with atoms of living organism. In material science and technology, Compton scattering can be used to probe the wave function of electron in momentum representation.

EXAMPLE 2.1   X-rays of wavelength 10.0 pm are scattered from a target at an angle of 45°. Find the wavelength of scattered X-rays.

Solution $\lambda' - \lambda = \dfrac{h}{m_0 c}(1 - \cos \theta)$

$$= 0.0242(1 - \cos 45°)\text{Å}$$

$$\lambda' - \lambda = 0.00708 \text{ Å}$$

or $\lambda' = \lambda + 0.00708°$

$$= 0.10 + 0.00708 \text{ Å}$$

Wavelength of scattered X-rays $\lambda' = 0.10708$ Å

EXAMPLE 2.2   A beam of radiation of wavelength 0.0242 Å is scattered from a target at an angle of 90°. Find the energy and direction of recoil electron.

Solution $\lambda' - \lambda = \dfrac{h}{m_0 c}(1 - \cos \theta)$

$$\lambda' - \lambda = 0.0242(1 - \cos 90°)$$

$$= 0.0242 \text{ Å}$$

$$\lambda' = \lambda + 0.0242 \text{ Å}$$

$$= 0.0242 + 0.0242$$

$$\lambda' = 0.0484 \text{ Å}$$

Energy of recoil electron is obtained as:

$$\text{K.E.} = h\nu - h\nu'$$

$$= hc\left(\frac{1}{\lambda} - \frac{1}{\lambda'}\right)$$

$$= hc\left(\frac{\lambda' - \lambda}{\lambda\lambda'}\right)$$

$$= \frac{6.64 \times 10^{-34} \times 3 \times 10^8 \times 0.0242 \times 10^{-10}}{0.0484 \times 10^{-10} \times 0.0242 \times 10^{-10}}$$

Energy of recoil electron K.E. = 4.07×10^{-14} J

The direction of recoil electron is given by

$$\tan\phi = \frac{h\nu'\sin\theta}{h\nu - h\nu'\cos\theta}$$

$$= \frac{\lambda\sin\theta}{\lambda' - \lambda\cos\theta}$$

$$= \frac{0.0242 \times 10^{-10} \times \sin 90^0}{0.0484 \times 10^{-10} - 0.0242 \times 10^{-10} \times \cos 90^0} = 0.5$$

Direction of recoil electron $\phi = \tan^{-1}(0.5) = 26.5°$

2.6 DE BROGLIE HYPOTHESIS AND MATTER WAVES

The explanation of spectral distribution of black body radiation by Planck results in the concept of energy quanta which was later known as photon. Einstein explained photoelectric effect by considering that the light wave is consist of photons. In 1922, the discovery of Compton effect confirmed the behaviour of photons as particle of negligible rest mass. These facts established the particle nature of the wave.

In 1925, Louis de Broglie reasoned that (a) nature loves symmetry, and it is symmetrical in many ways, (b) the universe is composed of light and matter, (c) if light, a wave, shows particle character, then the particles of material may show wave nature. Clearly one may expect the wave nature of electrons, protons and other material particles. De Broglie made the theoretical suggestion that the momentum p of a particle can be associated with a wave (matter wave) of wavelength λ such that

$$\lambda = \frac{h}{p} \tag{2.25}$$

The momentum of particle of mass m and velocity v is:

$$p = \gamma m v$$

where γ is relativistic factor given as:

$$\gamma = \frac{1}{\sqrt{1 - \left(\dfrac{v}{c}\right)^2}}$$

The de Broglie wavelength of such particle is accordingly

$$\lambda = \frac{h}{\gamma m v} \tag{2.26}$$

For particle velocity $v \ll c$, the relativistic factor γ may be taken as unity, therefore,

$$\lambda = \frac{h}{mv} \tag{2.27}$$

Now, if the particle has charge q and accelerated through a potential difference V, it acquires kinetic energy K.E. $= qV$. The wavelength of such a particle is:

$$\lambda = \frac{h}{\sqrt{2m\text{K.E.}}} \qquad \left(\because \frac{1}{2}mv^2 = \text{K.E.} \right) \tag{2.28}$$

In other form, it may be

$$\lambda = \frac{h}{\sqrt{2mqV}} \tag{2.29}$$

Therefore, de Broglie wavelength of an electron may be written as

$$\lambda = \frac{h}{\sqrt{2meV}} \tag{2.30}$$

de Broglie had no direct experimental evidence to support his conjecture. However, he shows that his theory accounts energy quantization in natural way. The quantity whose variation makes up matter waves or de Broglie wave, is wave function ψ similar to variation of electric and magnetic fields in light waves. The value of wave function is related to probability of finding the particle at a time and position. Therefore, matter waves are also termed as *waves of probability*.

The idea of matter waves was tested by C.J. Davisson and L.H. Germer in U.S.A. and G.P. Thomson in England independently. They confirmed de Broglie hypothesis experimentally. Thus idea of wave nature of material body was established.

In our everyday life wave nature of material body cannot be observed. When momentum of everyday objects divides the Planck's constant ($= 6.63 \times 10^{-34}$ Js), the wavelength comes out too small to be observed by any measurement tools. Therefore, we do not observe wave nature in everyday life. On the other hand, small particles like electron, having very small momentum gives observable wave like characteristics. Electron microscopes utilises wave nature of electrons instead of light to see very small objects.

Since electrons have typically more momentum than photon, their de Broglie wavelength will be smaller, resulting in better spatial resolution.

EXAMPLE 2.3 An electron is accelerated through a potential difference of 100 V. Calculate the de Broglie wavelength of the electron.

Solution The de Broglie wavelength of electron is given by

$$\lambda = \frac{h}{\sqrt{2meV}}$$

$$= \frac{6.64 \times 10^{-34}}{\sqrt{2 \times 91 \times 10^{-31} \times 1.6 \times 10^{-19} \times 100}}$$

∴ de Broglie wavelength $\lambda = 1.22$ Å

2.7 WAVE–PARTICLE DUALITY

The light is an electromagnetic wave, and it consists of concentrated energy packets known as photon, whose behaviour is like a particle. Thus light exhibits both natures—wave and particle. This phenomenon is known as the wave–particle duality. It is generally accepted that the wave–particle duality is two different visions of a single object. To understand this, we may consider a cylinder shown in Figure 2.8. We observe either a rectangle or a circle depending on where we stand. This metaphor is very interesting, but it does not explain anything scientifically and logically.

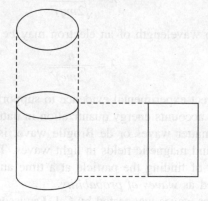

Figure 2.8 An example of cylinder to understand the duality.

The dual nature of light shows up strikingly in Eq. (2.25) and in equation $E = h\nu$. Each equation contains both wave concept (ν and λ) and particle concept (E and p). The particle nature of photon was confirmed by Compton. This confirmation establishes the concept of particle nature of wave.

The conjecture of de Broglie about matter waves got confirmation through the experiment by Davisson and Germer for electron. This theory was also tested for many other elementary particles and found correct. These facts lead to establishment of concept of wave nature of particles. Thus we find that wave–particle theory is well established and classical theory cannot deal with these ideas. An electron is neither a particle nor a wave; it is an entity which may manifest one of these features (wave or particle) depending on the type of the interaction with the external world.

2.7.1   Experimental Study of De Broglie Waves—(Davisson–Germer Experiment)

In 1926, Elsasser pointed out that the wave nature of matter might be tested in the same way the wave nature of X-rays was first tested. This idea was utilised by C.J. Davisson and L.H. Germer in 1927 and G.P. Thomson in 1928 independently. They observed diffraction effects with beam of electrons scattered by crystals. In this way, the de Broglie hypothesis of matter waves is confirmed experimentally. Here the experiment of Davisson and Germer, shown in Figure 2.9, is considered because of its more direct interpretation.

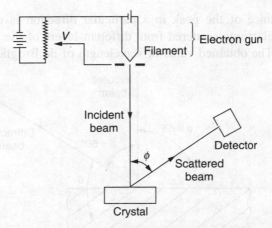

Figure 2.9   The schematic design of Davisson–Germer experiment.

The experimental arrangement consists of an electron gun which comprises tungsten filament and a heating system. Electrons emitted by the filament are accelerated by a variable potential difference V and, therefore, they acquire kinetic energy eV. This electron beam is allowed to fall on surface of a nickel crystal. The electrons are scattered in all directions by the atoms of the crystal. The intensity of the electron beam scattered in a given direction is measured by the electron detector. The detector records current proportional to intensity of electron. By rotating the detector on the circular scale at different positions, the intensity of the scattered electron beam is measured for different values of ϕ. It was found that at $\phi = 50°$ for $V = 54$ volt a strong maxima occurs as shown in Figure 2.10.

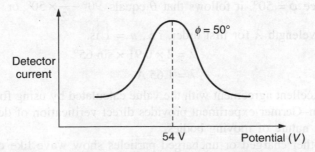

Figure 2.10   Detector current is plotted as function of potential showing diffraction maxima at $\phi = 50°$

The strong reflected beam may be accounted for by assuming that electron beam has a wavelength given by de Broglie relation:

$$\lambda = \frac{h}{p}$$

or $\qquad\qquad\qquad \lambda = \dfrac{h}{\sqrt{2meV}}$ (ignoring relativistic considerations)

This equation gives de Broglie wavelength λ associated with electron as $\lambda = 1.66$ Å for $V = 54$ V.

The appearance of the peak in a particular direction is due to the constructive interference of electrons scattered from different layers of the regularly spaced atoms of the crystals. The obtained value of wavelength of de Broglie waves can be verified

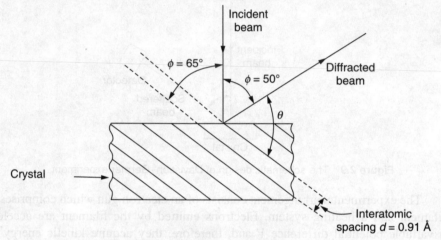

Figure 2.11   The strong diffraction from crystal planes at $\phi = 50°$ and $V = 54$ V. In this situation, Bragg angle $\theta = 65°$.

by using Bragg's law of X-ray diffraction. Figure 2.11 shows such a Bragg reflection, obeying the Bragg's law.

$$n\lambda = 2d \sin \theta \qquad\qquad (2.31)$$

The effective interplanar spacing d can be shown by X-ray diffraction analysis to be 0.91 Å. Since $\phi = 50°$, it follows that θ equals $90° - \dfrac{1}{2} \times 50°$ or $65°$.

The wavelength λ for first order, i.e., $n = 1$ is:

$$\lambda = 2 \times 0.91 \times \sin 65°$$

or $\qquad\qquad\qquad \lambda = 1.65$ Å

This is in excellent agreement with the value calculated by using formula $\lambda = h/p$. Thus the Davisson–Germer experiment provides direct verification of de Broglie hypothesis of the wave nature of moving bodies.

Many other charged or uncharged particles show wave like characteristics. Slow neutrons beam is often used to investigate the atomic structure of solids. Clearly like light wave, the matter also shows the existence of dual character.

2.8   PHASE AND GROUP VELOCITIES

There are points of constant phase in all the electromagnetic waves whether they are plane waves or any other type of waves. For plane waves, these constant phase points form a surface, termed as *wavefront*. As a monochromatic light wave propagates, these points of constant phase travel with a velocity known as phase velocity v_p.

In quantum mechanics, particles behave as wave and a de Broglie wave can be associated with a moving body. In terms of angular frequency ω and phase propagation constant k, the de Broglie phase velocity (or wave velocity) v_p may be given as:

$$v_p = \frac{\omega}{k} \tag{2.32}$$

Using relations $E = \hbar\omega$ and $p = \hbar k$, we have

$$v_p = \frac{E}{p}$$

The use of relativistic relations for energy and momentum give another form of de Broglie phase velocity,

$$v_p = \frac{\gamma mc^2}{\gamma mv}, \gamma = \frac{1}{\sqrt{1 - \left(\dfrac{v}{c}\right)^2}}$$

or

$$v_p = \frac{c^2}{v} \tag{2.33}$$

Here γ is the relativistic factor and v is the velocity of particle. Clearly, the de Broglie phase velocity is not same as the particle velocity v. Since v will always be less than the velocity of light, the de Broglie waves always travel with a speed greater than light. Infact, this concept of phase velocity is purely theoretical, and it does not indicate any information of energy transfer. To understand this unexpected result, we must look into somewhat deeper.

Having adopted the wave and particle character on the basis of de Broglie hypothesis, it is reasonable to associate particles with wave groups or wave packets as shown in Figure 2.12. The simple sine or cosine wave with fixed wavelength does not allow to localise or associate the particle definitely. A wave group may be described in terms of superposition of individual sine or cosine waves of different wavelengths whose interference results in wave groups.

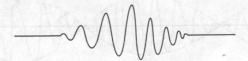

Figure 2.12   A wave group.

Consider two waves that have same amplitudes A_0, angular frequencies ω and $(\omega + \Delta\omega)$ and wave numbers k and $(k + \Delta k)$. These waves are:

$$\Psi_1 = A_0 \cos(\omega t - kx)$$

and

$$\Psi_2 = A_0 \cos[(\omega + \Delta\omega)t - (k + \Delta k)x]$$

Addition of Ψ_1 and Ψ_2 gives resultant displacement.

$$\Psi = A_0 \cos(\omega t - kx) + A_0 \cos[(\omega + \Delta\omega)t - (k + \Delta k)x] \tag{2.34}$$

With the help of trigonometric identities,

$$\cos A + \cos B = 2\cos\frac{(A+B)}{2}\cos\frac{(A-B)}{2}$$

and

$$\cos(-\theta) = \cos\theta$$

Eq. (2.34) is reduced to

$$\Psi = 2A_0 \cos\frac{1}{2}\,[(2\omega + \Delta\omega)t - (2k + \Delta k)x]\cos\frac{1}{2}\,(\Delta\omega t - \Delta k x) \tag{2.35}$$

Here $\Delta\omega \ll 2\omega$ and $\Delta k \ll 2k$, so we may have

$$\Psi = 2A_0\,(\omega t - kx)\cos\left(\frac{\Delta\omega}{2}t - \frac{\Delta k}{2}x\right)$$

or

$$\Psi = 2A_0\,\cos\left(\frac{\Delta\omega}{2}t - \frac{\Delta k}{2}x\right)\cos(\omega t - kx) \tag{2.36}$$

This is a simple cosine wave with the angular frequency ω, wave number k, and the modulated amplitude $2A_0\cos\left(\dfrac{\Delta\omega}{2}t - \dfrac{\Delta k}{2}x\right)$. In other words, the amplitude of the wave is itself a wave. The effect of the modulations is thus to produce successive wave groups as shown in Figure 2.13.

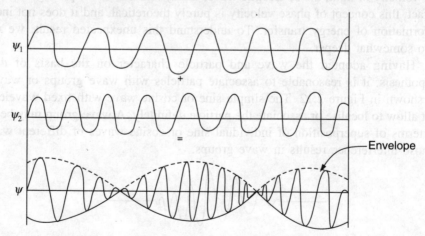

Figure 2.13 Production of wave groups through superposition of two waves with different frequencies.

Generally, some modulation of the frequency or amplitude of a wave is required in order to convey information, and it is this modulation that represents the signal content. The actual speed of the content in the situation described above is $\Delta\omega/\Delta k$. This is phase velocity of the amplitude wave, but since each amplitude wave contains a group of internal waves, this speed is usually called *group velocity*. The group velocity v_g of a wave is velocity with which the overall shape of the wave's amplitudes, known as modulation or envelope of the wave, propagates. It is given as:

$$v_g = \frac{\Delta\omega}{\Delta k} \qquad (2.37)$$

if the ω and k have continuous spread, Eq. (2.37) can be written as:

$$v_g = \frac{d\omega}{dk} \qquad (2.38)$$

and phase velocity of wave represented by Eq. (2.36) is given as:

$$v_p = \frac{\omega}{k} \qquad (2.39)$$

The group velocity is often thought of as the velocity at which energy or information is conveyed along a wave. The group velocity may be less or greater than phase velocities of its member waves depending on the variation of phase velocity with wave number. If the phase velocity is same for all waves constituting wave group, the group and phase velocities are the same. This is the case for a light wave propagating in a free space. The de Broglie group velocity v_g of particle with mass m and velocity v is:

$$v_g = \frac{d\omega}{dk}$$

$$v_g = \frac{d\left(\dfrac{E}{\hbar}\right)}{d\left(\dfrac{p}{\hbar}\right)} = \frac{dE}{dp}$$

For a free non-relativistic particle,

$$v_g = \frac{d}{dp}\left(\frac{p^2}{2m}\right) = \frac{p}{m}$$

so $\qquad\qquad\qquad\qquad v_g = v \qquad\qquad\qquad\qquad\qquad (2.40)$

The group velocity of a non-relativistic free particle is equal to velocity of the particle.

Further, for a relativistic particle,

$$E^2 = p^2c^2 + m^2c^4$$

Differentiating w.r.t. to p, we get

$$2E\frac{dE}{dp} = 2pc^2$$

and, therefore,
$$\frac{dE}{dp} = \frac{pc^2}{F}$$

But
$$E = \gamma mc^2 \quad \text{and} \quad p = \gamma mv$$

Giving
$$\frac{pc^2}{E} = v$$

Thus
$$v_g = v \tag{2.41}$$

Equations (2.40) and (2.41) tell us that de Broglie group velocity of the wave packet is to be associated with the velocity of the particle.

The phase velocity v_p is given by Eq. (2.32). It represents the velocity of propagation of an infinitely long monochromatic plane wave. The group velocity v_g and phase velocity v_p are equal only when ω is proportional to k. In general, the group velocity of packet is of physical significance.

Relation between v_g and v_p

The wave number is defined as:

$$k = \frac{2\pi}{\lambda} \tag{2.42}$$

$\therefore$
$$\frac{dk}{d\lambda} = -\frac{2\pi}{\lambda^2} \tag{2.43}$$

Also
$$\omega = 2\pi v = 2\pi \frac{v_p}{\lambda}$$

For the case when v_p is function of λ, we have

$$\frac{d\omega}{d\lambda} = 2\pi \left[-\frac{v_p}{\lambda^2} + \frac{1}{\lambda} \frac{dv_p}{d\lambda} \right]$$

or
$$\frac{d\omega}{d\lambda} = -\frac{2\pi}{\lambda^2} \left[v_p - \lambda \frac{dv_p}{d\lambda} \right] \tag{2.44}$$

Now, we can get group velocity v_g from Eqs. (2.43) and (2.44) as:

$$v_g = \frac{d\omega}{dk} = \frac{d\omega/d\lambda}{dk/d\lambda}$$

So
$$v_g = v_p - \lambda \frac{dv_p}{d\lambda} \tag{2.45}$$

For a dispersive medium, v_p is function of λ and usually $dv_p/d\lambda$ is positive for normal dispersion,

$$v_g < v_p \tag{2.46}$$

This is the case with de Broglie waves. For non-dispersive medium, v_p is not a function of λ and, therefore, $dv_p/d\lambda = 0$. We have

$$v_g = v_p \tag{2.47}$$

This is the case with electromagnetic waves propagating in a free space.

▌2.9 HEISENBERG'S UNCERTAINTY PRINCIPLE

The best known consequence of the wave–particle duality is the uncertainty principle given by Werner Heisenberg in 1927 for which he was awarded Nobel Prize in 1932. In some sense, the uncertainly principle protects wave particle duality. This principle states that "it is impossible to measure simultaneously the position and momentum of a particle with unlimited accuracy".

Louis de Broglie's proposed that a moving particle can be regarded as a wave group. This suggests that the particle properties like position and momentum can be measured with limited accuracy. Narrower the wave group, lesser the uncertainty in position of particle, but at the same time the wavelength of group cannot be found correctly, and hence the momentum. Figure 2.14 shows wave groups that result from interference of waves having same amplitudes, but different frequencies.

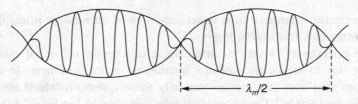

Figure 2.14 Wave groups.

It is reasonable to suppose that the particle is somewhere in the wave packet. The width of the packet is evidently equal to half the wavelength λm of the modulation. Then the uncertainty in position Δx of the particle corresponds to width of the group, that is,

$$\Delta x \approx \frac{\lambda_m}{2} \tag{2.48}$$

From equation of resultant wave Ψ,

$$k_m = \frac{1}{2}\Delta k$$

and, therefore,

$$\lambda_m = \frac{2\pi}{\frac{1}{2}\Delta k}$$

Equation (2.48) results in

$$\Delta x \approx \frac{2\pi}{\Delta k} \tag{2.49}$$

The uncertainty in propagation constant Δk may be associated with uncertainty in momentum Δp as:

$$\Delta p = \frac{h\Delta k}{2\lambda} \qquad \left(p = \frac{hk}{2\pi} \right) \tag{2.50}$$

Equations (2.49) and (2.50) give

$$\Delta x\, \Delta p \approx h \tag{2.51}$$

Since wave groups may have different shapes, the values of Δx and Δp are minimum. In actual experiment, we get greater values of these quantities and, therefore, product of Δx and Δp will always be greater than or equal to Planck's constant.

$$\Delta x \Delta p \geq h \tag{2.52}$$

Equation (2.52) is uncertainty principle obtained by Heisenberg. In terms of components, we may write

$$\Delta x \Delta p_x \geq h$$
$$\Delta y \Delta p_y \geq h$$

and

$$\Delta z \Delta p_z \geq h \tag{2.53}$$

A more realistic and exact calculation changes Eq. (2.52) to

$$\Delta x \Delta p \geq \frac{\hbar}{2} \tag{2.54}$$

This is sometimes referred to as exact statement of uncertainty principle. One can write it in components form accordingly.

The uncertainty is due to imprecise character in nature of quantities involved. It is not due to inefficiency of any instrument or measurement. In this regard, once Max Born said that, "we can answer only those questions which are compatible with uncertainty principle." The uncertainty principle obtained in Eq. (2.54) is from the point of view of the wave properties of particles. It can also be derived from the point of view of the particle properties of waves.

If an electron has to be observed using a light of wavelength λ, at least one photon must come back to eye from the electron. Consider the momentum of electron is p. When photon of momentum h/λ incident on electron for its observation, the change in momentum Δp will be of order of h/λ. In fact, by this amount the momentum of electron will be uncertainty. Hence

$$\Delta p \approx \frac{h}{\lambda} \tag{2.55}$$

The minimum uncertainty in position of electron may be estimated as equal to wavelength of light, that is,

$$\Delta x \approx \lambda \tag{2.56}$$

Hence

$$\Delta x \Delta p \approx h$$

which is similar to Eq. (2.51).

Equations (2.55) and (2.56) tell us that to reduce uncertainty in position, λ has to be reduced which in turn increases uncertainty in momentum. Therefore, both quantities cannot be controlled simultaneously. The uncertainty relation is an order relation. The exact value of product of Δx and Δp cannot be obtained.

The Heisenberg uncertainty relationship applies to any pair of variables which are canonically conjugate in the Hamiltonian formulation of mechanics. This relationship may be written for other quantities, for example, between energy E and time t and between angular momentum J_z and angle ϕ. The time–energy uncertainty relation is:

$$\Delta E \Delta t \geq \frac{\hbar}{2} \tag{2.57}$$

This relation implies that an energy determination that has an accuracy ΔE must occupy at least a time interval:

$$\Delta t \approx \frac{\hbar}{2\Delta E}$$

For angle variables, the uncertainty relation is given as:

$$\Delta J_z \; \Delta \phi \geq \frac{\hbar}{2} \tag{2.58}$$

where J_z represents the z-component of angular momentum and ϕ is the corresponding angle.

EXAMPLE 2.4   Find out the minimum uncertainty in the energy state of an atom if an electron remains in the state for $10^{-8}s$.

Solution   The uncertainty relation is:

$$\Delta E \; \Delta t \geq \frac{\hbar}{2}$$

$$= \frac{h}{4\pi}$$

or

$$\Delta E \approx \frac{h}{4\pi\Delta t}$$

$$= \frac{6.64 \times 10^{-34}}{4 \times 3.14 \times 10^{-8} \times 1.6 \times 10^{-19}} \, \text{eV}$$

Uncertainty in energy $\Delta E \approx 0.33 \times 10^{-7}$ eV.

2.9.1   Applications of Uncertainty Relation

The uncertainty principle imposes its limitations at atomic scale because of extremely small value of Planck's constant, i.e., (6.63×10^{-34}) Js. There are many phenomena that can be understood by applying uncertainty principle.

Existence of Electron in Nuclei

Typically nuclei are less than 10^{-14} m in radius. If an electron exists inside the nucleus, the uncertainty in position of electron may not exceed 10^{-14} m. So,

$$\Delta x \approx 10^{-14} \, \text{m}$$

and hence

$$\Delta p \geq \frac{\hbar}{2\Delta x}$$

or

$$\Delta p \geq 0.53 \times 10^{-20} \, \text{kg m/s}$$

Uncertainty in momentum of such amount implies that momentum itself must be at least of this value. An electron with momentum 0.53×10^{-20} kg m/s has kinetic energy (K.E.) much greater than rest mass energy $m_0 c^2$. According to relativistic idea,

$$\text{K.E.} = pc$$

Substituting the value of p and c, we get

$$K.E. = 0.53 \times 10^{-20} \times 3 \times 10^8$$

or

$$K.E. = 1.59 \times 10^{-12} \text{ J}$$

$$= 1.59 \times 10^{-12}/1.6 \times 10^{-19} \text{ eV}$$

or

$$K.E. \approx 10 \text{ MeV}$$

This calculation shows that if an electron exists in atomic nuclei, it must posses kinetic energy more than 10 MeV. Experiments revealed that maximum energy of β-particles emitted is only 2 MeV to 3 MeV. Hence we conclude that electrons are not present inside nuclei of atom.

Zero-point Energy of Harmonic Oscillator

Consider the problem of one-dimensional linear harmonic oscillator. The total energy of the oscillator is given by

$$E = \frac{p_x^2}{2m} + \frac{1}{2}kx^2$$

or

$$E = \frac{p_x^2}{2m} + \frac{1}{2}m\omega^2 x^2$$

If the particle is confined to the region of dimension a, then uncertainty in position $\Delta x \approx a$ and, therefore,

$$\Delta p_x \approx \frac{\hbar}{2a}$$

The uncertainties in position and momentum implies that position and momentum must be at least of this value, i.e.,

$$x \approx a \quad \text{and} \quad p_x \approx \frac{\hbar}{2a}$$

Thus total energy becomes

$$E = \frac{\hbar^2}{8ma^2} + \frac{1}{2}m\omega^2 a^2$$

The ground state energy or zero point energy of oscillator corresponds to minimum value of E for which $dE/da = 0$. Therefore,

$$\frac{dE}{da} = \frac{\hbar^2}{4ma^3} + m\omega^2 a$$

giving

$$a \approx \left(\frac{\hbar}{2m\omega}\right)^{1/2}$$

Now, minimum energy becomes

$$E = \frac{\hbar^2}{8m}\left(\frac{2m\omega}{\hbar}\right) + \frac{1}{2}m\omega^2\left(\frac{\hbar}{2m\omega}\right)$$

or

$$E = \frac{1}{2}\hbar\omega$$

which is zero point energy of harmonic oscillator, and it is derived using uncertainty principle.

SOLVED NUMERICAL PROBLEMS

PROBLEM 2.1 Which of the metals sodium and copper will be suitable for a photoelectric cell using light of wavelength 4000 Å. The work functions of sodium and copper are 2.0 eV and 4.0 eV respectively. (Planck's constant $h = 6.6 \times 10^{-34}$ Js, speed of light $c = 3 \times 10^8$ m/s, 1 eV $= 1.6 \times 10^{-19}$ J).

Solution If v_0 is threshold frequency, then the minimum energy required for photoelectric emission is $h v_0$, which is work function W of the material, i.e.,

$$W = h v_0$$

Then   Threshold wavelength $\lambda_0 = \dfrac{hc}{W}$

This is the largest wavelength for ejection of photoelectron.
For sodium,

$$\lambda_0 = \frac{6.6 \times 10^{-34} \times 3 \times 10^8}{2.0 \times 1.6 \times 10^{-19}}$$

or $\lambda_0 = 6188$ Å

For copper,

$$\lambda_0 = \frac{6.6 \times 10^{-34} \times 3 \times 10^8}{4.00 \times 1.6 \times 10^{-19}}$$

or $\lambda_0 = 3094$ Å

Thus, for light of wavelength 4000 Å, sodium is suitable for photoelectric emission.

PROBLEM 2.2 A reverse voltage 2.5 V is required to reduce the photo-current zero when light of wavelength 4000 Å strikes a certain metal. Find the work function of the metal.

Solution Total energy $h v$ = Work function + Kinetic energy

or $h v = W + e V_0$

or   Work function $W = h v - e V_0$

So $W = \dfrac{hc}{\lambda} - e V_0$

Thus
$$W = \frac{6.6 \times 10^{-34} \times 3 \times 10^8}{4000 \times 10^{-10}} - 1.6 \times 10^{-19} \times 2.5$$

$$W = 1 \times 10^{-19} \text{ J}$$

∴ Work function $W = 0.63$ eV

PROBLEM 2.3 The work function of a metal plate is 2 eV. A light of wavelength 1800 Å is able to eject photoelectron from the metal surface. Find the maximum velocity of photoelectrons.

Solution
$$h\nu = W + \text{K.E.}$$

$$\text{K.E.} = h\nu - W$$

Now,
$$\text{K.E.}_{\text{max}} = \frac{hc}{\lambda} - W$$

$$\text{K.E.}_{\text{max}} = \frac{6.6 \times 10^{-34} \times 3 \times 10^8}{1800 \times 10^{10}} - 2 \times 1.6 \times 10^{-19}$$

$$\text{K.E.}_{\text{max}} = 7.8 \times 10^{-19} \text{ J}$$

Maximum kinetic energy of electron of mass m may be given as:

$$\text{K.E.}_{\text{max}} = \frac{1}{2} m v_{\text{max}}^2$$

So
$$v_{\text{max}} = \sqrt{\frac{2 \text{ K.E.}_{\text{max}}}{m}}$$

$$v_{\text{max}} = \sqrt{\frac{2 \times 7.8 \times 10^{-19}}{9.1 \times 10^{-31}}} = 1.3 \times 10^6$$

Thus maximum velocity of photoelectrons is 1.3×10^6 m/s.

PROBLEM 2.4 Find the kinetic energy of recoil electron when a photon of wavelength 0.50 Å is scattered by a free electron through 90°.

Solution Kinetic energy of the recoil electron is the difference in energies of incident and scattered photon, i.e.,

$$\text{K.E.} = h\nu - h\nu'$$

K.E. of recoil electron $= hc\left(\frac{1}{\lambda} - \frac{1}{\lambda'}\right)$

Now, from Compton's formula,

$$\lambda' - \lambda = \frac{h}{m_0 c}(1 - \cos\theta)$$

So
$$\lambda' = \lambda + \frac{h}{m_0 c}(1 - \cos\theta)$$

$$= 0.5 \times 10^{-10} + \frac{6.6 \times 10^{-34}}{9.1 \times 10^{-31} \times 3 \times 10^8} (1 - \cos 90°)$$

$$\lambda' = 0.52 \text{ Å}$$

Therefore, $\text{K.E.} = 6.6 \times 10^{-34} \times 3 \times 10^8 \left(\frac{1}{0.5 \times 10^{-10}} - \frac{1}{0.52 \times 10^{-10}} \right)$

Kinetic energy of recoil is 1.8×10^{-16} J.

PROBLEM 2.5 X-rays of wavelength 1 Å are scattered by a free electron such that the recoil electron gets maximum kinetic energy. Calculate the wavelength of scattered ray.

Solution The maximum Compton shift corresponds to maximum kinetic energy of recoil electron. In Compton effect, maximum shift in wavelength is obtained at scattering angle of 180°, which is,

$$\Delta\lambda = 0.048 \text{ Å}$$

or $\lambda' - \lambda = 0.048$

or $\lambda' = \lambda + 0.048$

$$= 1 \text{ Å} + 0.048 \text{ Å}$$

Thus wavelength of scattered ray is 1.048 Å.

PROBLEM 2.6 Calculate the de Broglie wavelength for a beam of electrons of kinetic energy 1.6×10^{-17} J.

Solution Kinetic energy of electron $\text{K.E.} = \frac{1}{2} mv^2$

So $v = \sqrt{\dfrac{2k}{m}} = \sqrt{\dfrac{2 \times 1.6 \times 10^{-17}}{9.1 \times 10^{-31}}}$

$$v = 6 \times 10^6 \text{ m/s}$$

The de Broglie wavelength of electron is given by

$$\lambda = \frac{h}{p} = \frac{h}{mv}$$

$$= \frac{6.6 \times 10^{-34}}{9.1 \times 10^{-31} \times 3 \times 10^6}$$

$$= 0.12 \text{ Å}$$

The de Broglie wavelength of electron is 0.12 Å.

PROBLEM 2.7 What will be the de Broglie wavelength associated with a 92-g ball moving with velocity 72 m/s? Give your comment about wave character of the ball.

Solution   The velocity of the ball is much less than the speed of light, so

$$m = \text{Rest mass of ball} = 92 \text{ g}$$
$$= 0.092 \text{ kg}$$

The de Broglie wavelength of the ball is:

$$\lambda = \frac{h}{mv} = \frac{6.6 \times 10^{-34}}{0.09 \times 72}$$

or
$$\lambda = 1.0 \times 10^{-34} \text{ m}$$

The wavelength associated with the moving ball is quite less in comparison to dimension of the ball and, therefore, wave character of ball cannot be observed.

PROBLEM 2.8   A beam of thermal neutrons at temperature 27 °C is made to fall on a crystal at incidence angle 30°. Find out the wavelength of neutrons and interplanar spacing of the crystal for first order reflection (mass of neutron = 1.67×10^{-27} kg, Boltzmann constant $k = 1.38 \times 10^{-23}$ N/m²).

Solution   The kinetic energy K.E. of thermal neutrons is given by

$$\text{K.E.} = kT$$
$$\text{K.E.} = 1.38 \times 10^{-23} \times 300$$
$$= 4.14 \times 10^{-21} \text{ J}$$

The wavelength of thermal neutrons is obtained as:

$$\lambda = \frac{h}{\sqrt{2m\,K.E.}} = \frac{6.6 \times 10^{-34}}{\sqrt{(2 \times 1.67 \times 10^{-27} \times 4.14 \times 10^{-21})}}$$

$$\lambda = 1.8 \text{ Å}$$

To calculate interplanar spacing, we may use Bragg's law:

$$2d \sin \theta = n\lambda$$

or
$$2d \sin 30° = 1 \times \lambda$$

or
$$\frac{2d.1}{2} = \lambda$$

or
$$d = \lambda$$

So interplanar spacing is about 1.8 Å.

PROBLEM 2.9   The speed of a bullet of mass 50 g is 400 m/s with accuracy 0.01%. Find out the accuracy in position of the bullet.

Solution   The momentum of the bullet is:

$$p = mv$$
$$= 50 \times 10^{-3} \times 400$$
$$p = 20 \text{ kg m/s}$$

Thus uncertainty in momentum is:

$$\Delta p = \frac{0.01}{100} \times 20$$

$$= 20 \times 10^{-4} \text{ kg m/s}$$

According to uncertainty principle,

$$\Delta x \, \Delta p \geq \frac{\hbar}{2}$$

So

$$\Delta x \approx \frac{\hbar}{2} \Delta p = \frac{h}{4\pi \, \Delta p}$$

$$= \frac{6.6 \times 10^{-34}}{(4 \times 3.14 \times 20 \times 10^{-4})}$$

$$\Delta x = 2.6 \times 10^{-32} \text{ m}$$

This is uncertainty in position of bullet.

PROBLEM 2.10   An electron of mass 9.1×10^{-31} kg is moving with a velocity which is uncertain by 3×10^9 cm/s. Find out uncertainty in position of electron.

Solution   The uncertainty principle states that

$$\Delta x \, \Delta p \geq \frac{\hbar}{2}$$

So

$$\Delta x \approx \frac{\hbar}{2\Delta p}$$

$$= \frac{h}{4\pi m \, \Delta v}$$

$$= \frac{6.6 \times 10^{-34}}{(4 \times 3.14 \times 9.1 \times 10^{-31} \times 3 \times 10^9 \times 10^{-2})}$$

$$= 1.9 \times 10^{-10} \text{ cm}$$

Thus the uncertainty in position of electron is 1.9×10^{-10} cm.

PROBLEM 2.11   Find the inherent uncertainty in the frequency of a photon which is emitted by an excited atom during transition to lower state. The mean life time of state is 10^{-8} s.

Solution   The uncertainty in frequency of the photon is given by

$$\Delta v \approx \frac{2\Delta E}{\hbar} \qquad \left(\because \Delta E \Delta t \approx \frac{\hbar}{2} \right)$$

The energy of photon $E = hv = \dfrac{h}{t}$,   (t = Life time of the state)

$$= \frac{6.6 \times 10^{-34}}{10^{-8}}$$

$$E = 6.6 \times 10^{-26} \text{ J}$$

So energy is uncertain by

$$\Delta E \approx 6.6 \times 10^{-26} \text{ J}$$

Therefore,

$$\Delta \nu \approx \frac{2 \Delta E}{\hbar} = \frac{4\pi \, \Delta E}{h}$$

or

$$\Delta \nu = \frac{4 \times 3.14 \times 6.6 \times 10^{-2}}{6.6 \times 10^{-34}}$$
$$= 12.56 \times 10^{-8} \text{ Hz}$$

The uncertainty in frequency is 12.56×10^{-8} Hz.

PROBLEM 2.12   The angular position of an electron initially in the ground state of hydrogen atom is determined by an experiment to an accuracy of 1 s. Calculate the uncertainty in angular momentum.

Solution   Uncertainty in angular position is:

$$\Delta \phi = 1 \text{ s}$$
$$= \frac{\pi}{60 \times 60 \times 180} \text{ radian}$$

Uncertainty in angular momentum is

$$\Delta J \approx \frac{\hbar}{2 \Delta \phi}$$
$$= 1.08 \times 10^{-28} \text{ Js}$$

Uncertainty in angular momentum is 1.08×10^{-29} Js.

EXERCISES

THEORETICAL QUESTIONS

2.1 Explain photoelectric effect on the basis of quantum theory, and hence derive Einstein's relation for photoelectric emission.

2.2 What is photoelectric effect? How does the emission of photons depend on the intensity and frequency of the incident radiation?

2.3 On the basis of Einstein's relation, explain that the velocity of electrons ejected due to photoelectric effect depends on the frequency and not on the intensity of the incident light.

2.4 Prove that in Compton scattering, the angle at which electron recoils is given by relation:

$$\cos \phi = \frac{(h\nu)^2 - (h\nu')^2 + p^2 c^2}{2h\nu pc}$$

2.5 Explain in brief the Compton effect and photoelectric effect. How they differ from each other?

2.6 Show that in collision of a photon and a free electron, it is impossible for a photon to give all of its energy to the free electron.

2.7 Explain the consequence of Compton effect, and prove that Compton shift is independent of wavelength of incident radiation.

2.8 Develop an expression for de Broglie wavelength of a particle in terms of its kinetic energy and rest mass energy.

2.9 What do you understand by the term matter waves? Give experimental proof for existence of matter waves.

2.10 Explain group velocity and phase velocity. Find out the relation between them.

2.11 Show that if the uncertainty in the location of particle is equal to its de Broglie wavelength, the uncertainty in its velocity is equal to its velocity.

2.12 Explain uncertainty principle. Relate uncertainty in angular momentum with uncertainty in angular position.

2.13 Discuss the dual nature of matter and radiation.

2.14 Why wave nature of material bodies is not apparent in our daily life? Explain.

NUMERICAL PROBLEMS

P2.1 The work function for the surface of aluminium is 4.2 eV. How much potential difference will be required to stop the emission of maximum energy electrons emitted by light of 2000 Å wavelength? **(Ans. 1.98 V)**

P2.2 Light of wavelength 5000 Å falls on a sensitive surface. If the surface has received 10^{-7} J of energy, then how many photons have fallen on the surface? **(Ans. 2.5×10^{11})**

P2.3 Calculate the frequency of the light which can eject photoelectron from a metal surface. The retarding potential for these electrons is 3 V, and photoelectric effect in this metal occurs at frequency 6×10^{14} Hz. Also find the work function for given metal.
(Ans. 1.32×10^{15} Hz, 3.97×10^{-19} J)

P2.4 Calculate the wavelength of incident X-rays which gives scattered X-rays of wavelength 0.22 nm when scattered by a free electron at an angle 45°. **(Ans. 0.15 nm)**

P2.5 An energetic photon is scattered at an angle 30° after collision with a free electron. Find out the energy of incident photon if electron recoils at an angle 30°. **(Ans. 4.33×10^{-13} J)**

P2.6 Find the energy of a photon scattered at an angle 90° when a photon of energy 1.6×10^{-12} J is incident on a free electron. **(Ans. 0.48 MeV)**

P2.7 What will be the wavelength of a helium atom when it is heated at temperature 400 °K. (mass of helium atom = 6.7×10^{-27} kg). **(Ans. 6.6 nm)**

P2.8 A proton is moving with wavelength 1×10^{-15} m. What will be its total energy?
(Ans. 1556 MeV)

P2.9 The de Broglie wavelength of an electron is 1.5 nm. Find out the kinetic energy of electron. Also compute the phase and group velocities of de Broglie waves.

P2.10 Compare the uncertainties in the velocities of an electron and a proton confined to a 1.0 nm box.

P2.11 To determine the mass of a particle, number of experiments are carried out. Experiments show variation of $\pm$ 20 m_e, where m_e is mass of electron. Find out uncertainty in mean life of the particle.   (**Ans.** 3×10^{-23} s)

P2.12 A proton is confined to a nucleus of radius 5×10^{-15} m. Calculate the uncertainty in its momentum.   (**Ans.** 0.52×10^{-20} kg m/s)

P2.13 How accurately can the position of a proton with $v \ll c$ be determined without giving it more than 1.0 k eV of kinetic energy?   (**Ans.** 1.44×10^{-13} m)

MULTIPLE CHOICE QUESTIONS

MCQ2.1 The unit of Planck's constant is that of:
  (a) Work   (b) Energy
  (c) Linear momentum   (d) Angular momentum

MCQ2.2 In a photoelectric experiment, the stopping potential for the incident light of wavelength 4000 Å is 2 V. If the wavelength be changed to 3000 Å, the stopping potential will be:
  (a) 2 V   (b) Less than 2 V
  (c) Zero   (d) More than 2 V

MCQ2.3 Energy of photon $E = h\nu$, momentum of photon $p = h/\lambda$, then velocity of photon will be:
  (a) $\dfrac{E}{p}$ (b) Ep

  (c) $\left(\dfrac{E}{p}\right)^2$ (d) 3×10^8 m/s

MCQ2.4 The work function of aluminium is 4.2 eV. If two photons, each of energy 2.5 eV, strike an electron of aluminium then:
  (a) Photon emission occurs
  (b) No photo emission
  (c) Photo emission occurs in certain specific conditions
  (d) Photo emission occurs when photons transfer energy jointly

MCQ2.5 A photon recoils back after striking an electron at rest. The shift in wavelength is:
  (a) 0.0048 nm   (b) 0.024 Å
  (c) Zero   (d) Between 0.024 Å and 0.0048 Å

MCQ2.6 In Compton's scattering effect, the shift in wavelength is:
  (a) Independent of scatterer   (b) Dependent on scatterer
  (c) Sometimes dependent   (d) Independent in few situations

MCQ2.7 The Compton effect is:
  (a) Elastic and coherent   (b) Elastic and incoherent
  (c) Inelastic and incoherent   (d) Inelastic and coherent

MCQ 2.8 In Compton's scattering, the Compton wavelength is obtained at an angle of:

(a) 270°   (b) 280°

(c) 90°   (d) 60°

MCQ 2.9 The de Broglie wavelength is expressed as:

(a) $\dfrac{h}{p}$ (b) $\dfrac{hc}{E}$

(c) $\dfrac{p}{h}$ (d) ph

MCQ 2.10 The Davisson–Germer experiment confirms

(a) Wave nature of particles   (b) Particle nature of waves

(c) (a) and (b) both   (d) None of these

MCQ 2.11 The exact expression for uncertainty in position and momentum is:

(a) $\Delta x \Delta p \geq h$ (b) $\Delta x \Delta p \geq \hbar$

(c) $\Delta x \Delta p \geq \dfrac{\hbar}{2}$ (d) $\Delta x \Delta p \geq \dfrac{h}{2}$

MCQ 2.12 The statement of uncertainty principle is:

(a) Order statement   (b) Exact statement

(c) Neither (a) nor (b)   (d) False statement

MCQ 2.13 The uncertainty principle is true for:

(a) Wave nature   (b) Particle nature

(c) Wave and particle nature both   (d) Neither of them

MCQ 2.14 The de Broglie phase velocity v_p and particle velocity v are related as:

(a) $v_p = cv$ (b) $v_p = \dfrac{c^2}{v}$

(c) $v = \dfrac{v_p}{c^2}$ (d) $v_p = c$

Answers

2.1 (d)	2.2 (d)	2.3 (a)	2.4 (b)	2.5 (a)	2.6 (a)	2.7 (b)
2.8 (c)	2.9 (a)	2.10 (a)	2.11 (c)	2.12 (a)	2.13 (c)	2.14 (b)

Quantum Mechanics II

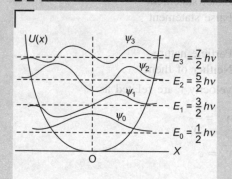

3.1   INTRODUCTION

Niels Bohr, working in Rutherford's laboratory in 1913, succeeded in combining Rutherford's nuclear atom with Planck's quanta and Einstein's photons to predict the observed spectral lines of hydrogen atom. In spite of several limitations, the Bohr's theory explained many aspects of atomic system. The bright side of Bohr's contribution is that it transforms scientific thought. Now, it was clear that to understand the atomic phenomena, a more general approach is needed. This led to the development of quantum mechanics by Max Born, Erwin Schrödinger, Werner Heisenberg, Paul Dirac, and many other physicists.

In a lecture delivered in 1900, Lord Kelvin began with words, "The beauty and clearness of the dynamical theory, which asserts light and heat to be modes of motion, is at present obscured by two clouds". The first cloud was the question of 'how the earth moves through the luminiferous ether', and the second was the failure of the 'Maxwell–Boltzmann's theory regarding the equipartition of energy' to predict results consistent with experiments in all cases.

The classical way of thinking was unable to remove these both clouds and a new approach was needed. The first cloud was removed by Einstein's theory of relativity and the removal of second cloud needed Planck's energy quantum hv and a new mechanics known as quantum mechanics.

Instead of asserting on certain value (as in classical mechanics) the quantum mechanics explores probabilities. For example, as per Bohr's theory, the radius of hydrogen atom in ground state is 5.3×10^{-11} m. The quantum mechanics says that it is the most probable radius. It means that if we go on measuring the radius, we get different values, either larger or smaller than 5.3×10^{-11} m. But the value most likely to be found will be 5.3×10^{-11} m. The certainties of classical (or Newtonian) mechanics are actually illusory, and their apparent agreement with experiment occurs because ordinary objects are made of so many atoms that deviations from average behaviour are unnoticeable.

3.2   WAVE FUNCTION

The wave function ψ (Greek letter psi) is an important quantity in quantum mechanics. It measures the wave variation of matter waves. For string waves, the wave variation may be measured by a transverse displacement y, for sound waves, it may be measured by a pressure variation p, and for electromagnetic waves, variation may be measured by the variations in electric and magnetic fields. The value of the wave function associated with a moving body at a particular point and time in space is related to the probability of finding the body there.

The wave function ψ cannot be interpreted in terms of an experiment and, therefore, has no direct physical significance. Max Born first suggested that the quantity $|\psi|^2$ at any particular point is a measure of the probability that the particle will be near that point. If a volume element dV is constructed around that point, the probability that the particle will be found in the volume element at a given time is $|\psi|^2 dV$. This interpretation of ψ provides a statistical connection between the wave and the associated particle. This tells us where the particle is likely to be, not where it is. A large value of $|\psi|^2 \, dV$ means the strong possibility of the body's presence, while small value means the slight possibility of its presence.

The problem of quantum mechanics is to determine the wave function ψ for a body under action of external forces. Usually wave functions are complex quantity. However, a probability must be positive and real quantity. The probability density $|\psi|^2$ for a complex ψ is taken as product of ψ and its complex conjugate $\psi*$. A complex function ψ can be written as:

$$\psi = a + ib \qquad (a \text{ and } b \text{ are real})$$

and
$$\psi* = a - ib$$

Thus
$$|\psi|^2 = a^2 + b^2$$

which is a real quantity.

3.3 CHARACTERISTICS OF WAVE FUNCTION

A wave function is a mathematical tool in quantum mechanics describing the quantum state of a particle or system of particles. Mathematically, it is a function from a space that maps the possible states of the system. The laws of quantum mechanics describe how the wave function evolves over the time.

In quantum mechanics, probability amplitude is a complex number whose modulus squared represents a probability density. The principle use of probability amplitudes is as the physical meaning of the wave function, a link first proposed by Max Born and a pillar of the Copenhagen interpretation of quantum mechanics. Born was awarded the 1954 Nobel Prize in physics for this understanding. The probability thus calculated is sometimes called the *Born Probability*, and the relationship used to calculate probability from the wave function is sometimes called the *Born rule*.

Before going through actual calculation of ψ, it must satisfy certain requirements. In quantum mechanics, wave function describing real particle must be normalisable, i.e., the integral of $|\psi|^2$ over all space must be finite–after all the body is located somewhere. Mathematically, it is expressed as:

$$\int_{-\infty}^{\infty} \psi* \, \psi \, dV = 1$$

or
$$\int_{-\infty}^{\infty} |\psi|^2 \, dV = 1 \qquad (3.1)$$

A wave function that obeys Eq. (3.1) is said to be normalised. The normalisation of wave function is done to obtain the physically applicable wave function or probability amplitudes. All wave function representing real particles must be normalisable. This helps in discarding solutions of Schrödinger's equation which do not have a finite integral in a given interval.

If integral of Eq. (3.1) gives zero value, i.e.,

$$\int_{-\infty}^{\infty} |\psi|^2 \, dV = 0$$

the particle does not exist. This integral cannot be infinite. The way through which $|\psi|^2$ is defined restricts it to be negative or complex quantity. Therefore, integral must be a real quantity greater than or equal to zero if ψ has to describe a real body properly. To understand normalisation procedure, consider a particle that is restricted to one-dimensional region between $x = 0$ and $x = l$, and its wave function is given as:

$$\psi(x, t) = \begin{cases} Ae^{i(kx - \omega t)}, & 0 \leq x \leq l \\ 0, & \text{elsewhere} \end{cases}$$

To normalise the given wave function, we need to calculate the value of arbitrary constant A. Hence we have to solve the equation:

$$\int_{-\infty}^{\infty} |\psi|^2 dV = 1$$

Substituting ψ, we get $|\psi|^2$ as:

$$|\psi|^2 = A^2 e^{i(kx - \omega t)} e^{-i(kx - \omega t)} = A^2$$

So

$$\int_{-\infty}^{\infty} |\psi|^2 dx = \int_{-\infty}^{0} |\psi|^2 dx + \int_{0}^{l} |\psi|^2 dx + \int_{l}^{\infty} |\psi|^2 dx$$

$$= \int_{-\infty}^{0} 0 \, dx + \int_{0}^{l} A^2 dx + \int_{l}^{\infty} 0 \, dx$$

or

$$\int_{-\infty}^{\infty} |\psi|^2 dx = A^2 l$$

Normalisation condition requires that

$$\int_{-\infty}^{\infty} |\psi|^2 dx = 1$$

Therefore,

$$A^2 l = 1$$

$$A = \frac{1}{\sqrt{l}}$$

or

Hence normalised wave function is:

$$\psi(x, t) = \begin{cases} \dfrac{1}{\sqrt{l}} e^{i(kx - \omega t)}, & 0 \leq x \leq l \\ 0, & \text{elsewhere} \end{cases}$$

The properties associated with wave function must not be altered while normalising the wave function, otherwise the process becomes meaningless. Each and every information contained in unnormalised wave function must be associated with the normalised wave function also. Since properties of particles like probability distribution, momentum, energy, etc. are calculated by using Schrödinger's equation, therefore, it must be invariant under normalisation.

In addition to normalisable characteristic of wave function ψ, it must satisfy certain constraints to represent a physically observable system. Wave function ψ must be the solution of Schrödinger's equation. It must be single-valued at every point in space, since probability of finding the particle can have only one value at a particular place and time. The wave function must also be continuous. Figure 3.1(a) shows a function $f(x)$ that is discontinuous at $x = x_1$ and Figure 3.2(b) presents a function $f(x)$ which is not single valued at $x = x_1$.

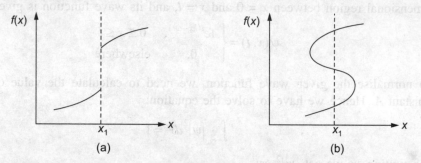

Figure 3.1   A function $f(x)$ (a) that is not continuous and (b) not single valued.

The momentum considerations require that the partial derivatives of ψ be finite, continuous and single-valued. The wave function satisfying these properties, known as 'well-behaved wave functions', can yield physically meaningful results when used in calculations. These constraints on wave function are summarised as follows:

1. ψ must be the solution of the Schrödinger's equation.
2. ψ must be normalisable. This implies that the wave function approaches zero as:

$$x \to \pm\infty, y \to \pm\infty, z \to \pm\infty$$

3. Wave function ψ must be continuous and single valued everywhere.
4. The slope of the wave function, i.e., $\partial\psi/\partial x$, $\partial\psi/\partial y$, $\partial\psi/\partial z$ must also be continuous and single-valued everywhere.

EXAMPLE 3.1   The wave function of a particle is $\psi = A \cos^2 x$ for interval $-\pi/2$ to $\pi/2$. Find the value of A.

Solution $\psi = A \cos^2 x$ for $-\pi/2$ to $\pi/2$

Then

$$\int_{-\pi/2}^{\pi/2} |\psi|^2 \, dx = 1 \qquad \text{(for normalisation)}$$

or

$$\int_{-\pi/2}^{\pi/2} A^2 \cos^4 x \, dx = 1$$

or

$$2A^2 \int_0^{\pi/2} \cos^4 x \, dx = 1$$

or

$$2A^2 \, \frac{3\pi}{16} = 1$$

Hence

$$A = \sqrt{\frac{8}{3\pi}}$$

3.4   SCHRÖDINGER'S WAVE EQUATION

In the beginning of the twentieth century, experimental evidences suggested that atomic particles were also wave-like in nature. For example, electrons were found to give diffraction patterns when passed through a double slit in a similar way to light waves. Therefore, it was thought that there should be a wave equation to explain the behaviour of atomic particles. Erwin Schrödinger presented such wave equation. The eigenvalues of the wave equation were shown to be equal to the energy levels of the quantum mechanical systems. Infact Schrödinger's wave equation is the fundamental equation of quantum mechanics in the same sense the second law of motion is the fundamental equation of Newtonian (or classical) mechanics.

The usual basis of the mathematical theory of the wave motion is given by a differential equation. An electromagnetic wave moving in the X-direction can be described by

$$\frac{\partial^2 E_y(x,t)}{\partial x^2} = \frac{1}{c^2} \frac{\partial^2 E_y(x,t)}{\partial t^2}$$

where E_y is y-component of electric vector and c is velocity of light.

The logical development from first principle to find the differential equation for describing matter waves cannot be carried out. It is a matter of guess keeping in mind properties of waves. Such a wave equation may be verified by comparison of experimental data with predicted results from this equation. Consider a free particle of completely undetermined position travelling in the positive X-direction with precisely known momentum p_x and total energy E. The wave function for the free particle can be expressed as a sine or cosine wave, [sin $2\pi i$ ($vt -x/\lambda$), cos $2\pi i(vt -x/\lambda)$] or for convenience combining the two in exponential form as:

$$\psi(x, t) = e^{-2\pi i(vt-x/\lambda)} \tag{3.2}$$

where frequency v and wavelength λ of the particle can be expressed in terms of energy E and momentum p_x as:

$$E = hv = 2\pi \hbar v$$

and

$$\lambda = \frac{h}{p_x} = \frac{2\pi \hbar}{p_x}$$

we have

$$\psi(x, t) = e^{-\frac{i}{\hbar}(Et-p_x x)} \tag{3.3}$$

Equation (3.3) is mathematical representation of the wave equivalent of a free (unrestricted) particle of total energy E and momentum p_x moving in positive X-direction.

The non-relativistic energy of a free particle of mass m moving in the X-direction with momentum p_x is given by:

$$E = \frac{p_x^2}{2m}$$

If external forces are also acting on the particle so that it acquires potential energy $U(x)$, the total energy E becomes

$$E = \frac{p_x^2}{2m} + U(x) \tag{3.4}$$

Multiplying both sides of Eq. (3.4) by the wave function ψ, we have

$$\frac{p_x^2}{2m} \psi + U\psi \tag{3.5}$$

To obtain fundamental differential equation for Ψ, differentiate Eq. (3.3) twice with respect to x and once with respect to t,

$$\frac{\partial^2 \psi}{\partial x^2} = -\frac{p_x^2}{\hbar^2} \psi \tag{3.6}$$

and

$$\frac{\partial^2 \psi}{\partial t^2} = -i \frac{E}{\hbar} \psi \tag{3.7}$$

Equations (3.6) and (3.7) may be written as:

$$p_x^2 \psi = -\hbar \frac{\partial^2 \psi}{\partial n^2} \tag{3.8}$$

or

$$E\psi = i\hbar \frac{\partial \psi}{\partial t} \tag{3.9}$$

Substituting these expressions for $E\psi$ and $p_x^2 \psi$ into Eq. (3.5), we get

$$i\hbar \frac{\partial \psi}{\partial t} = -\frac{\hbar}{2m} \frac{\partial^2 \psi}{\partial x^2} + U\psi \tag{3.10}$$

Equation (3.10) is time-dependent form of Schrödinger's wave equation in one dimension. The agreement of results deduced from this equation justifies the belief that the equation is valid for matter waves as long as relativistic effects can be neglected.

In three dimensions, the time-dependent form of Schrödinger's equation can be written as:

$$i\hbar \frac{\partial \psi}{dt} = -\frac{\hbar^2}{2m} \left(\frac{\partial^2 \psi}{\partial x^2} + \frac{\partial^2 \psi}{\partial y^2} + \frac{\partial^2 \psi}{\partial z^2} \right) + U\psi$$

$$i\hbar \frac{\partial \psi}{\partial t} = -\frac{\hbar^2}{2m} \nabla^2 \psi + U\psi \tag{3.11}$$

Here ψ and U are functions of x, y, z, and t, and Laplacian operator ∇^2 is defined as:

$$\nabla^2 \equiv \frac{\partial^2}{\partial x^2} + \frac{\partial}{\partial y^2} + \frac{\partial}{\partial z^2}$$

Equation (3.11) is Schrödinger's wave equation in three dimensions. This equation holds good in all co-ordinate systems provided that Laplacian operator is suitably defined. This equation is extension of Eq. (3.10) which is derived for a free particle (U = constant). But Eq. (3.11) is a general equation applicable to particle subjected to arbitrary forces that vary in time and space, i.e., their potential energy U is function of x, y, z and t. There is no way to prove that extension is correct except the justification of predicted results by Eq. (3.11) to experimental findings. In practice, Schrödinger's equation has been found correct in predicting the results of experiments. Infact Schrödinger's wave equation cannot be derived from first principle, but it represents a first principle itself.

If we are dealing with static potentials, i.e., the potential energy U is not function of time, and it varies with position of particle only, the Schrödinger's equation may be simplified by removing time dependence. The wave function of a free particle in one dimension is written as:

$$\psi = e^{-\frac{i}{\hbar}(Et - p_x x)}$$

$$= e^{\frac{i}{\hbar}p_x x} e^{-\frac{i}{\hbar}Et}$$

$$\psi = \psi_0 e^{-\frac{i}{\hbar}Et} \tag{3.12}$$

Here $\psi_0 = e^{\frac{ip_x x}{\hbar}}$. It is position-dependent function, while $\psi_0 = e^{\frac{iEt}{\hbar}}$ is a time-dependent function. Substituting the value of ψ from Eq. (3.12) into time-dependent form of Schrödinger's wave equation, we get

$$i\hbar\psi_0 \left(-\frac{i}{\hbar}E\right)e^{-\frac{i}{\hbar}Et} = -\frac{\hbar^2}{2m}e^{-\frac{i}{\hbar}Et}\frac{\partial^2\psi_0}{\partial x^2} + U\psi_0 e^{-\frac{i}{\hbar}Et}$$

Dividing by $e^{-\frac{i}{\hbar}Et}$, we get

$$E\psi_0 = -\frac{\hbar^2}{2m}\frac{\partial^2\psi_0}{\partial x^2} + U\psi_0$$

or

$$\frac{\partial^2\psi_0}{\partial x^2} + \frac{2m}{\hbar^2}(E - U)\psi_0 = 0 \tag{3.13}$$

Equation (3.13) is the steady-state form of one-dimensional Schrödinger's equation. In three dimensions, it may be expressed as

$$\nabla^2\psi_0 + \frac{2m}{\hbar^2}(E - U)\psi_0 = 0 \tag{3.14}$$

Schrödinger's wave equation is used in physics to deal with problems related to the atomic structure of matter. It is an extremely powerful mathematical tool and basis of wave mechanics. The time-dependent equation is used for describing progressive wave and motion of free particles. The time-independent form of this equation is often used to describe standing waves. The time-dependent equation is of the first order in time and of the second order with respect to the position, therefore, it is not consistent with relativity. The Schrödinger's equation can be used only for non-relativistic problems. For speed of particle comparable to speed of light, more elaborate formulation is needed.

3.5 EXPECTATION VALUES AND OPERATORS

Let us consider a particle confined to the X-axis and described by wave function $\psi(x, t)$. The expectation value of position of particle can be obtained if the positions of many particles described by $\psi(x, t)$ is determined experimentally and then averaged out. Consider particles are distributed along the X-axis in such a way that N_1 particles at x_1, N_2 particles at x_2 and so on. The average position $\bar{x}$ of particles is given as:

$$\bar{x} = \frac{N_1 x_1 + N_2 x_2 + \cdots}{N_1 + N_2 + \cdots}$$

or

$$\bar{x} = \frac{\sum_i N_i x_i}{\sum_i N_i}$$

As per probability theory, expectation value is a mathematical expectation. It is average of the results of a large number of measurements on systems. For a single particle, N_i may be replaced by the probability P_i that the particle be found in an interval dx at x_i, i.e.,

$$p_i = |\psi|^2 \, dx$$

Making this substitution and changing the summation by integral, we have

$$\bar{x} = \frac{\int_{-\infty}^{\infty} x |\psi|^2 \, dx}{\int_{-\infty}^{\infty} |\psi|^2 \, dx} \tag{3.15}$$

If wave function ψ is taken as normalised wave function, then

$$\int_{-\infty}^{\infty} |\psi|^2 \, dx = 1$$

and we have

$$\bar{x} = \int_{-\infty}^{\infty} x |\psi|^2 \, dx$$

or equivalently

$$\bar{x} = \int_{-\infty}^{\infty} \psi^* x \psi \, dx \tag{3.16}$$

For a wave function $\Psi(r, t)$ depending on x, y, z and t, the expectation value of r is written as:

$$\bar{r} = \int \psi^*(r, t) \, r \, \psi(r, t) dV \tag{3.17}$$

The expectation value of any other quantity can be found in similar way if it depends only on particle co-ordinates r.

We have seen the process of finding the expectation value for any quantity at any time which is function of position of particle described by wave function ψ. The expectation value of other dynamical quantities like momentum and total energy of particle cannot be obtained in same way because to evaluate the integral, momentum p and energy E must be expressed as functions of x and t both [since $\psi = f(x, t)$]. But the uncertainty principle implies that functions like $p(x, t)$ and $E(x, t)$ cannot exist. If x and t are specified precisely, the relationship:

$$\Delta p \Delta x \geq \hbar$$

and

$$\Delta E \Delta t \geq \hbar$$

implies that in principle p and E cannot be determined exactly. Therefore, to evaluate the expectation value of momentum p and energy E, a different way has to be adopted. The free particle wave function is:

$$\psi = e^{-\frac{i}{\hbar}(Et - p_x x)}$$

Differentiating with respect to x and t, we find that

$$p_x \psi = -i\hbar \frac{\partial \psi}{\partial x} \tag{3.18}$$

and

$$E\psi = i\hbar \frac{\partial \psi}{\partial t} \tag{3.19}$$

Hence we see that the dynamical quantity p_x corresponds to operator $-i\hbar \partial/\partial x$ and dynamical quantity E corresponds to operator $i\hbar \partial/\partial t$. Clearly an operator tells about the operation to be carried out on the quantity that follows it. Operators are often denoted by placing a cap on usual symbol of concerned quantities. The momentum and energy operators may be given as:

$$\hat{p}_x = -i\hbar \frac{\partial}{\partial t} \tag{3.20}$$

and

$$\hat{E} = -i\hbar \frac{\partial}{\partial t} \tag{3.21}$$

Similarly kinetic energy $(p_x^2/2m)$ operator is:

$$\hat{T} = -\frac{\hbar^2}{2m} \frac{\partial^2}{\partial x^2} \tag{3.22}$$

The expectation values of quantities involving operators are expressed as:

$$\overline{p}_x = \int_{-\infty}^{\infty} \psi^* \hat{p}_x \psi \, dx$$

or

$$p_x = \int_{-\infty}^{\infty} \psi^* (-i\hbar) \frac{\partial \psi}{\partial x} \, dx \tag{3.23}$$

and

$$\overline{E} = \int_{-\infty}^{\infty} \psi^* i\hbar \frac{\partial \psi}{\partial t} \, dx \tag{3.24}$$

In general, expectation value of a quantity $O(x, t)$ is expressed as:

$$\bar{O}(x, t) = \int_{-\infty}^{\infty} \psi^* \hat{O} \psi \, dx \tag{3.25}$$

3.6   EIGENVALUE EQUATION

The operator equation that corresponds to total energy $E = $ K.E. $+$ P.E. when E is constant is:

$$E = -\frac{\hbar^2}{2m}\frac{\partial^2}{\partial x^2} + U$$

Multiplying by ψ, we have

$$E\psi = -\frac{\hbar^2}{2m}\frac{\partial^2 \psi}{\partial x^2} + U\psi$$

or

$$\left(-\frac{\hbar^2}{2m}\frac{\partial^2}{\partial x^2} + U\right)\Psi = E\Psi \tag{3.26}$$

Obviously, the term

$$-\frac{\hbar^2}{2m}\frac{\partial^2}{\partial x^2} + U$$

is a total energy operator. This operator is known as Hamiltonian operator H. Equation (3.26) may be written as:

$$H\psi = E\psi \tag{3.27}$$

Equation (3.27) is termed as *eigenvalue equation*. The ψ is said to be an eigen function of the operator H that appears on left. The multiplying constant E that appears on the right is called the corresponding eigenvalue. An energy eigen function ψ is said to represent a stationary state of the particle provided $|\psi|^2$ is constant in time.

EXAMPLE 3.2   An operator $\dfrac{d^2}{dx^2}$ has eigen function $\psi = e^{\alpha x}$. Calculate the eigenvalue.

Solution   The operator and eigenvalue relation may be written as:

$$\hat{O}\psi = O\psi$$

where $\hat{O}$ is operator and O is eigenvalue.

Here

$$\hat{O} = \frac{d^2}{dx^2} \quad \text{and} \quad \psi = e^{\alpha x}$$

$$\hat{O}\psi = \frac{d^2}{dx^2}e^{\alpha x}$$

$$\hat{O}\Psi = \alpha^2 e^{\alpha x}$$

We can write this as:

$$\hat{O}\Psi = \alpha^2\Psi \quad \text{such that} \quad O = \alpha^2$$

Clearly α^2 is eigenvalue.

3.7 POSTULATES OF QUANTUM MECHANICS

In the development of quantum mechanics, there are certain basic postulates which are of fundamental importance.

1. A wave function may be associated with any particle moving in a conservative field of force, and it determines everything that can be known about the system in consistence of uncertainty principle.

2. The wave function or state function of a system evolves in time according to the time-dependent Schrödinger's equation:

$$\hat{H}\psi(x, t) = i\hbar\frac{\partial\psi}{\partial t}$$

3. The total wave function must be antisymmetric with respect to the interchange of all co-ordinates of one fermion with those of another. The Pauli's exclusion principle is a direct result of this antisymmetry principle.

4. Corresponding to every observable in classical mechanics there is a linear operator in quantum mechanics. The operator may be simply a multiplication operator such as $\hat{r}$ for position, or it may be a differential operator such as $-i\hbar\nabla$ for the momentum. Table 3.1 summarises the quantum operators for some physical observables.

Table 3.1 Observables and Corresponding Operators

Observable	Observable symbol	Operator symbol	Operator
Position	r	$\hat{r}$	multiply by r
Momentum	p	$\hat{p}$	$-i\hbar\nabla$
Kinetic energy	T	$\hat{T}$	$-\dfrac{\hbar^2}{2m}\nabla^2$
Potential energy	$U(r)$	$\hat{U}(r)$	multiply by $U(r)$
Total energy	E	$\hat{E}$	$-\dfrac{\hbar^2}{2m}\nabla^2 + U(r)$
Angular momentum	L	$\hat{L}$	$\hat{r}\times(-i\hbar\nabla)$

5. In any measurement of the observable associated with operator $\hat{A}$, the observed values are eigenvalues a satisfying the equation:

$$\hat{A}\psi = a\psi$$

This postulate captures the central point of quantum mechanics. Although measurements must always yield an eigenvalue, the state does not have to be an eigen state of $\hat{A}$ initially. An arbitrary state can be expanded in the complete set of eigen vectors of $\hat{A}$:

$$\psi = \sum_{i}^{n} C_i \psi_i$$

such that

$$\hat{A}\psi_i = a_i \psi_i$$

6. If a system is in a state described by a normalised wave function ψ, then the average value of the observable corresponding to $\hat{A}$ is given by

$$<a> = \int_{-\infty}^{\infty} \psi^* \hat{A}\psi \, dV$$

3.8 APPLICATIONS OF SCHRÖDINGER'S EQUATION

The Schrödinger's wave equation can be used to solve many problems. The solution of such problems helps in understanding of modern physics. Here some simple problems are taken which need limited mathematical background.

The one-dimensional free particle ($U = 0$) wave equation:

$$i\hbar \frac{\partial \psi}{\partial t} = -\frac{\hbar^2}{2m} \frac{\partial^2 \psi}{\partial x^2}$$

has to be changed to include the effects of external force that may act on the particle. These forces may be electrostatic, gravitational and nuclear force. They can be combined into single force F such that

$$F = -\frac{\partial U}{\partial x}$$

Including effects of external force one-dimensional Schrödinger's wave equation for a particle of mass m may be given as:

$$i\hbar \frac{\partial \psi}{\partial t} = -\frac{\hbar^2}{2m} \frac{\partial^2 \psi}{\partial x^2} + U(x)\psi$$

3.8.1 Particle in a Box

The particle in a box problem, also known as *infinite square well potential problem* or *infinite potential well problem*, describes a particle free to move in a narrow space surrounded by impenetrable barriers. The particle in a box problem can be solved without approximations. Such problems are rare in quantum mechanics. It is a very simple model to understand quantum effects. This model is a hypothetical example to compare the predictions of quantum mechanics with those of classical (Newtonian) mechanics.

Consider a particle of mass m in a box such that its motion is restricted along X-axis between $x = 0$ and $x = L$. The particle may only move forward and backward along a straight line with impenetrable barriers at either end. The potential energy U of the particle is infinite on both sides of the box, while U is a constant, say, zero for convenience, on the inside as shown in Figure 3.2.

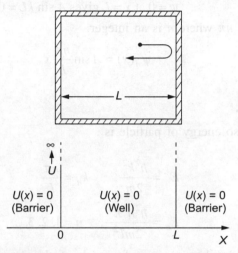

Figure 3.2 A particle confined to a box of width L.

The zero potential energy inside the box means that no forces act upon the particle inside the box and it can move freely in that region. However, infinitely large force (due to infinite P.E.) repel the particle as it touches the walls of the box, preventing it from escaping. Further, the particle cannot have infinite amount of energy so it cannot exist outside the box and, therefore, its wave function Ψ is 0 for $x \leq 0$ and $x \geq L$. The potential energy of this system may be given as:

$$U(x) = \begin{cases} 0, & 0 < x < L \\ \infty, & 0 \geq x \geq L \end{cases}$$

The behaviour of particle and measurable properties of the particle (position, momentum and energy) may be obtained from wave function. The wave function ψ within the box can be found by solving Schrödinger's equation:

$$\frac{d^2\psi}{dx^2} + \frac{2m}{\hbar^2}[E - U(x)]\psi = 0 \qquad (3.28)$$

Inside the box $U(x) = 0$. Here total derivative is used because ψ is function of x only.

$$\frac{d^2\psi}{dx^2} + \frac{2m}{\hbar^2}E\psi = 0 \qquad (3.29)$$

or $\qquad \dfrac{d^2\psi}{dx^2} + k^2\psi = 0 \qquad$ such that $\qquad k^2 = \dfrac{2mE}{\hbar^2} \qquad (3.30)$

The general solution of this equation is:

$$\psi = A \sin kx + B \cos kx \tag{3.31}$$

The boundary conditions can be used to evaluate the constants A and B,

$$\psi = 0 \text{ at } x = 0 \text{ and hence } B = 0$$

$$\psi = 0 \text{ at } x = L \text{ gives } A \sin kL = 0$$

Since $A \neq 0$, $kL = n\pi$ where n is an integer.

Thus

$$\psi_n(x) = A \sin \frac{n\pi}{L} x \tag{3.32}$$

Energy Levels

Since $k^2 = \dfrac{2mE}{\hbar^2}$ so energy of particle is

$$E_n = \frac{\hbar^2 k_n^2}{2m}, \qquad k_n = \frac{n\pi}{L} \tag{3.33}$$

or

$$E_n = \frac{\hbar^2 n^2 \pi^2}{2mL^2}, \qquad n = 1, 2, 3, \ldots \tag{3.34}$$

Equation (3.33) describes energies for each of the permitted wave number. It is clear from Eq. (3.34) that energy of particle can have only certain values (discrete). The integer n is known as *quantum number* for energy level or eigenvalue E_n, and corresponding wave function or eigen function ψ_n is given by Eq. (3.32). Clearly, particle in a box cannot have any energy value. It can have only certain discrete energy values.

Since integer n is positive and cannot be zero, therefore, particle cannot have zero or negative energy inside the box. The zero energy of the particle inside the box corresponds zero wave function, which means particle is not present inside the box. Such a characteristic cannot be found in classical mechanics where all energies are possible. The nonzero value of energy inside the box can be confirmed through uncertainty principle.

The energy levels increase with n^2, meaning that the high energy levels are separated from each other by a greater amount than low energy levels. The lowest possible energy for $n = 1$ is known as zero-point energy. The energy of a particle in a box and of a free particle both depend on wave number in same way. First one shows discrete energy levels while second gives continuous nature as shown in Figure 3.3.

Wave Function

The eigen function corresponding to eigenvalue E_n is given by

$$\psi_n = A \sin(k_n x)$$

or

$$\psi_n = A \sin\left(\frac{n\pi x}{L}\right)$$

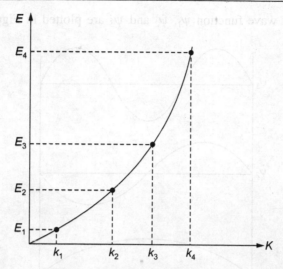

Figure 3.3 Energy of a particle as a function of wave number. Black circles show energy of particle in a box, while continuous line is that of free particle.

This eigen function meets all necessary requirements for a wave function: for each n, ψ_n is single-valued function of x, and ψ_n and $\partial \psi_n / \partial x$ are continuous. Further the integral $|\psi|^2$ over all space is finite. Since particle is confined between $x = 0$ and $x = L$, so

$$\int_{-\infty}^{\infty} |\psi_n|^2 \, dx = \int_0^L |\psi_n|^2 \, dx$$

$$= A^2 \int_0^L \sin^2 \left(\frac{n\pi x}{L} \right)$$

$$= A^2 \frac{L}{2} \tag{3.35}$$

To normalise ψ, the value of A should be so as to make

$$\int_{-\infty}^{\infty} |\psi_n|^2 \, dx = 1 \tag{3.36}$$

Comparing Eqs. (3.35) and (3.36), we get for normalisation of wave function of a particle in box

$$A = \sqrt{\frac{2}{L}} \tag{3.37}$$

The normalised wave functions of the particle are:

$$\psi_n = \sqrt{\frac{2}{L}} \sin \left(\frac{n\pi x}{L} \right) \tag{3.38}$$

The normalised wave function ψ_1, ψ_2 and ψ_3 are plotted in Figure 3.4.

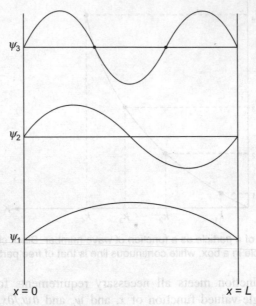

Figure 3.4 Wave functions of a particle inside the box with rigid walls of infinite potential energy.

Spatial Location

In classical physics, the particle can be detected anywhere in the box with equal probability. However, in quantum mechanics, the probability densities $|\psi_n|^2$ for finding the particle at a given position are obtained from the wave function ψ_n. The probability densities are given by

$$P_n(x) = \begin{cases} \dfrac{2}{L}\sin^2\left(\dfrac{n\pi x}{L}\right); & 0 < x < L \\ 0; & \text{otherwise} \end{cases} \tag{3.39}$$

Obviously, for any value of n (> 1), there are regions within the box for which $P_n(x) = 0$, i.e., there exist spatial nodes at which the particle cannot be found. The probability densities are plotted in Figure 3.5. In each case $|\psi_n|^2 = 0$ at $x = 0$ and $x = L$. The $|\psi_n|^2$ has maximum value at middle ($x = L/2$) of the box. A particle in lowest energy level ($n = 1$) is most likely to be found in the middle of the box. However, a particle in the next higher state ($n = 2$) will never be there as $|\psi_2|^2 = 0$ for this case. Therefore, probability of finding particle, in quantum mechanics, is different for different energies (or quantum numbers).

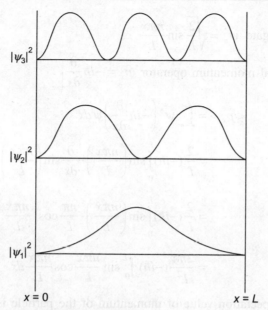

$x = 0$ $x = L$

Figure 3.5   Probability densities of particle confined to a box of rigid walls.

EXAMPLE 3.3   Calculate the energy of ground state and first excited state of an electron trapped in a box of length 10^{-8} cm.

Solution   For one-dimensional box,

$$E_n = \frac{n^2 h^2}{8mL^2}$$

For ground state $n = 1$,

$$E_1 = \frac{(6.6 \times 10^{-34})^2}{8 \times 9.1 \times 10^{-31} \times (10^{-10})^2} \text{ J}$$

Energy of ground state $E_1 = 37.7$ eV
For first excited state $n = 2$, and hence

$$E_2 = 4 \frac{h^2}{8mL^2} = 4 \times 37.7 \text{ eV}$$

Energy of first excited state $E_2 = 150.8$ eV

EXAMPLE 3.4   Calculate the expectation value $< p_x >$ of the momentum of a particle trapped inside a one-dimensional box.

Solution   For one-dimensional box of length L, the wave function may be given as:

$$\psi_n = \sqrt{\frac{2}{L}} \sin \frac{n\pi x}{L}$$

Complex conjugate $\psi_n^* = \sqrt{\dfrac{2}{L}} \sin \dfrac{n\pi x}{L}$

One-dimensional momentum operator $\hat{p}_x = -i\hbar \dfrac{\partial}{\partial x}$

so
$$< p_x > = \int_{-\infty}^{\infty} \psi^* \left(-i\hbar \dfrac{\partial}{\partial x} \right) \psi \, dx$$

$$= \dfrac{2}{L}(-i\hbar) \int_0^L \sin\left(\dfrac{n\pi x}{L} \right) \cdot \dfrac{\partial}{\partial x} \sin\left(\dfrac{n\pi x}{L} \right) dx$$

$$= \dfrac{2}{L}(-i\hbar) \int_0^L \sin\left(\dfrac{n\pi x}{L} \right) \cdot \dfrac{n\pi}{L} \cos\left(\dfrac{n\pi x}{L} \right) dx$$

$$= \dfrac{2n\pi}{L^2}(-i\hbar) \int_0^L \sin \dfrac{n\pi x}{L} \cos \dfrac{n\pi x}{L} \, dx$$

Clearly, the expectation value of momentum of the particle is $< p_x > = 0$.

3.8.2   One-dimensional Step Potential

The one-dimensional step potential is a model system in quantum mechanics and scattering theory. Such a step barrier does not exist in nature. This problem is reasonably approximate to the instantaneous potential difference between the dees of a cyclotron, normal metal-superconductor interfaces and the potential barrier at the surface of a metal. Therefore, understanding of step potential problem is also applicable to these problems upto certain extent.

Consider a uniform beam of electrons of energy E is incident on barrier from left as shown in Figure 3.6. Let the potential energy $U(x)$ be 0 to the left of the origin and U_0 to the right. Such a potential field may be defined by

$$U(x) = 0 \quad \text{for } x < 0$$
$$= U_0 \quad \text{for } x \geq 0 \tag{3.40}$$

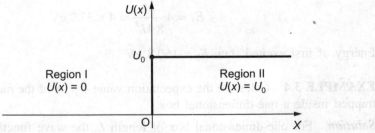

Figure 3.6   Typical potential step.

The time-independent one-dimensional Schrödinger's equation is:

$$\frac{d^2\psi}{dx^2} + \frac{2m}{\hbar^2}[E - U(x)]\psi = 0$$

For first region, $U(x) = 0$ and $\psi = \psi_1$. Therefore, the Schrödinger's equation takes the form:

$$\frac{d^2\psi_1}{dx^2} + \frac{2m}{\hbar^2}E\psi_1 = 0 \qquad (3.41)$$

and the solution of Eq. (3.41) is given as:

$$\psi_1 = Ae^{i\alpha x} + Be^{-i\alpha x} \qquad (3.42)$$

where A and B are constants and $\alpha = \sqrt{\dfrac{2mE}{\hbar^2}}$. The first term represents the incident particles and second term that of reflected particles.

For second region, $\psi = \psi_2$ and $U(x) = U_0$. The Schrödinger's wave equation for this region is:

$$\frac{d^2\psi_2}{dx^2} + \frac{2m}{\hbar^2}[E - U(x)]\psi_2 = 0 \qquad (3.43)$$

and the solution of Eq. (3.43) is:

$$\psi_2 = Ce^{i\beta x} + De^{-i\beta x} \qquad (3.44)$$

where C and D are constants to be determined and $\beta = \sqrt{\dfrac{2m(E - U_0)}{\hbar^2}}$. In Eq. (3.44), the first term represents the transmitted wave, and the second term represents a wave coming from $+\infty$, i.e., a reflected wave in second region. Since there is no boundary in second region, therefore, for $x > 0$, no particle can flow to the left. Hence Eq. (3.44) becomes

$$\psi_2 = Ce^{i\beta x} \qquad (3.45)$$

Now, we place two requirements on ψ and its derivatives in order that the probability be finite and single-valued so that wave function represents a specific physical situation. The continuity of ψ at $x = 0$, needs

$$\psi_1 = \psi_2 \text{ at } x = 0$$

giving
$$A + B = C \qquad (3.46)$$

and continuity of derivatives of ψ at $x = 0$ means

$$\frac{d\psi_1}{dx} = \frac{d\psi_2}{dx} \qquad \text{at} \qquad x = 0$$

which gives
$$\alpha(A - B) = \beta C \qquad (3.47)$$

Solving Eqs. (3.46) and (3.47), we get

$$\frac{B}{A} = \frac{\alpha - \beta}{\alpha + \beta} \qquad (3.48)$$

and

$$\frac{C}{A} = \frac{2\alpha}{\alpha + \beta} \qquad (3.49)$$

In region on right of $x = 0$, two possibilities must be considered: $E > U_0$ and $E < U_0$ as shown in Figure 3.7.

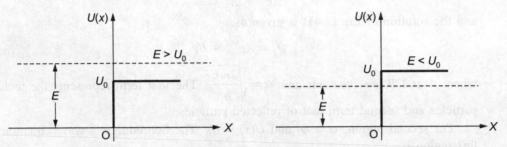

Figure 3.7   Schematic of step potential for (a) $E > U_0$ and (b) $E < U_0$.

CASE 1: $E > U_0$ (Energy of particles is greater than barrier height):   In this case, $\beta = \sqrt{\dfrac{2m(E - U_0)}{\hbar^2}}$ is real. To study the transmission and reflection characteristics, we have to find out the current densities in both regions. The probability current density is defined as:

$$J = \frac{i\hbar}{2m}(\psi \nabla \psi^* - \psi^* \nabla \psi) \qquad (3.50)$$

where ψ^* is complex conjugate of ψ. For the first region, this equation takes the form:

$$J_1 = \frac{i\hbar}{2m}\left(\psi_1 \frac{d\psi_1^*}{dx} - \psi_1^* \frac{d\psi_1}{dx}\right) \qquad (3.51)$$

$$= \frac{i\hbar}{2m}[\{(Ae^{i\alpha x} + Be^{-i\alpha x})(-i\alpha)(A^* e^{-i\alpha x} - B^* e^{i\alpha x})\}$$

$$- \{(A^* e^{-i\alpha x} + B^* e^{i\alpha x})(i\alpha)(A^* e^{-i\alpha x} - B^* e^{i\alpha x})\}]$$

$$= \frac{\alpha \hbar}{m}(AA^* - B^* B) \qquad (\alpha^* = \alpha \text{ because } \alpha \text{ is real})$$

or

$$J_1 = \frac{\alpha \hbar}{m}(|A|^2 - |B|^2) \qquad (3.52)$$

The probability current density for second region is:

$$J_2 = \frac{i\hbar}{2m}\left(\psi_2 \frac{d\psi_2^*}{dx} - \psi_2^* \frac{d\psi_2}{dx}\right) \qquad (3.53)$$

$$= \frac{i\hbar}{2m}[Ce^{i\beta x}(-i\beta)C^*e^{-i\beta x} - C^*e^{-i\beta x}(i\beta)e^{i\beta x}]$$

$$= \frac{\beta\hbar}{m}(CC^*) \quad (\beta^* = \beta \text{ because it is real})$$

Thus $\qquad J_2 = \frac{\beta\hbar}{m}|C|^2$ \hfill (3.54)

Equation (3.52) consists of two terms. First term represents incident current density (incident flux) and second term gives reflected flux. The transmitted current density is given by Eq. (3.54). These may be defined as:

$$J_{1i} = \frac{\alpha\hbar}{m}|A|^2$$

$$J_{1r} = \frac{\alpha\hbar}{m}|B|^2$$

$$J_{2t} = \frac{\beta\hbar}{m}|C|^2 \hfill (3.55)$$

Here nonzero B (nonzero J_{1r}) gives the non-classical result of the possibility of the particle being reflected by the barrier although it has enough energy ($E > U_0$) to pass over it. A classical particle with energy larger than barrier height does not feel the barrier at all.

The reflected fraction (of particles) known as reflection coefficient R is:

$$R = \frac{|J_{1r}|}{|J_{1i}|} \hfill (3.56)$$

or

$$R = \frac{|B|^2}{|A|^2}$$

From Eq. (3.42), we get

$$R = \frac{|\alpha - \beta|^2}{|\alpha + \beta|^2} \hfill (3.57)$$

And the transmitted fraction known as transmission coefficient T is given as:

$$T = \frac{|J_{2t}|}{|J_{1i}|} \hfill (3.58)$$

or

$$T = \frac{|C|^2}{|A|^2} \cdot \frac{\beta}{\alpha}$$

From Eq. (3.49), we can have transmission coefficient:

$$T = \frac{4\alpha\beta}{(\alpha + \beta)^2} \hfill (3.59)$$

The reflection and transmission coefficients are such that

$$R + T = 1$$

which verifies the results obtained.

CASE 2: $E < U_0$ **(Energy of particles E is less than the step height U_0):**   When $E < U_0$, the term $\beta = \sqrt{\dfrac{2m(E - U_0)}{\hbar^2}}$ becomes imaginary. Therefore, we have

$$\beta = \frac{\sqrt{2m(U_0 - E)}}{\hbar} = i\gamma$$

Such that

$$\gamma = \frac{\sqrt{2m(U_0 - E)}}{\hbar}$$

and

$$\beta^* = -\frac{i\sqrt{2m(U_0 - E)}}{\hbar} = -\beta \tag{3.60}$$

From Eq. (3.45), the wave function

$$\psi_2 = Ce^{-\gamma x} \quad (\beta \rightarrow i\gamma)$$

The probability current density for second region is:

$$J_2 = \frac{i\hbar}{2m}\left(\psi_2 \frac{d\psi_2^*}{dx} - \psi_2^* \frac{d\psi_2}{dx} \right)$$

$$= \frac{i\hbar}{2m}[Ce^{-\gamma x}(-\gamma^*)C^* e^{-\gamma^* x} - C^* e^{-\gamma^* x}(-\gamma)Ce^{-\gamma x}]$$

$$J_2 = 0 \qquad \text{(for } \gamma^* = -\gamma) \tag{3.61}$$

This represents transmitted probability current density. Its zero value gives zero transmission coefficients:

$$T = 0 \tag{3.62}$$

and by definition $R + T = 1$, so we get

$$R = 1 \tag{3.63}$$

Clearly particles are reflected back completely in this case. Same result is expected from classical physics also. Therefore, for $E < U_0$, the classical and quantum approaches give the same results.

3.8.3   The Barrier Penetration (Tunnel Effect)

The possibility in quantum physics that a particle may penetrate into a region where its total energy E is less than the potential energy $U(x)$ of barrier leads to tunneling of particles through thin barriers. Although, the probability of transmission (penetration) is not great, but it is always nonzero. Such a phenomenon is unheard in classical physics. According to classical mechanics, the particle must be reflected. In quantum mechanics, the de Broglie waves corresponding to particle are partly reflected and partly transmitted.

Consider a beam of identical particles of energy E is incident from left on a barrier of height U_0 and width L as shown in Figure 3.8. On both sides of the barrier potential, energy is zero which means that no forces act on the particles there. This barrier is described by

$$U(x) = 0, -\infty < x \le 0$$
$$= U_0, 0 \le x \le L$$
$$= 0, L \le x < \infty \tag{3.64}$$

Let us assume the wave functions ψ_1, ψ_2 and ψ_3 for region I, region II and region III respectively.

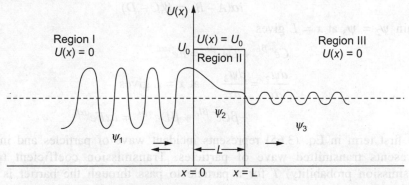

Figure 3.8   A one-dimensional potential barrier of height U_0 and thickness L.

Since $U(x) = 0$ in regions I and II, the Schrodinger's wave equation for both regions is Eq. (3.41) and its solution is like Eq. (3.42). The general solution for these regions is:

$$\psi_1 = Ae^{i\alpha x} + Be^{-i\alpha x} \tag{3.65}$$

$$\psi_3 = Fe^{i\alpha x} + Ge^{-i\alpha x} \tag{3.66}$$

where $\alpha = \sqrt{\dfrac{2mE}{\hbar^2}}$. The Schrodinger's equation for region II is given by

$$\frac{d^2\psi_2}{dx^2} + \frac{2m}{\hbar^2}(E - U_0)\psi_2 = 0$$

Since $E < U_0$, we have

$$\frac{d^2\psi_2}{dx^2} + \frac{2m}{\hbar^2}(U_0 - E)\psi_2 = 0$$

or

$$\frac{d^2\psi_2}{dx^2} + (i\beta)^2\psi_2 = 0, \quad \beta = \sqrt{\frac{2m(U_0 - E)}{\hbar^2}} \tag{3.67}$$

The solution of Eq. (3.67) may be written as:

$$\psi_2 = Ce^{-\beta x} + De^{\beta x} \tag{3.68}$$

There is no boundary in third medium, therefore no reflection of particles occur. The second term of Eq. (3.66) may be neglected. Thus,

$$\psi_3 = Fe^{i\alpha x} \tag{3.69}$$

Applying the conditions that wave function ψ and its derivatives must be continuous at $x = 0$ and at $x = L$ leads to the following equations,

$$\psi_1 = \psi_2 \quad \text{at } x = 0 \text{ gives}$$

$$A + B = C + D \tag{3.70}$$

and

$$\frac{d\psi_1}{dx} = \frac{d\psi_2}{dx} \quad \text{at } x = 0$$

$$i\alpha(A - B) = \beta(C - D) \tag{3.71}$$

Again $\psi_2 = \psi_3$ at $x = L$ gives

$$Ce^{-\beta L} + De^{+\beta L} = Fe^{i\alpha L} \tag{3.72}$$

and

$$\frac{d\psi_2}{dx} = \frac{d\psi_3}{dx} \quad \text{at } x = L \text{ gives}$$

$$-\beta Ce^{-\beta L} + \beta De^{+\beta L} = i\alpha Fe^{i\alpha L} \tag{3.73}$$

The first term in Eq. (3.65) represents incident wave of particles and in Eq. (3.69) represents transmitted wave of particles. Transmission coefficient (or may be transmission probability) T for a particle to pass through the barrier is fraction of incident beam that gets through the barrier, and it is defined as:

$$T = \frac{|F|^2}{|A|^2} \tag{3.74}$$

The solution of Eqs. (3.70), (3.71), (3.72) and (3.73) for A in terms of F yields

$$A = Fe^{i\alpha L} \left[\cosh \beta L + \frac{i}{2} \left(\frac{\beta}{\alpha} - \frac{\alpha}{\beta} \right) \sinh \beta L \right]$$

or

$$\frac{|A|^2}{|F|^2} = 1 + \frac{1}{4} \left(\frac{\beta}{\alpha} + \frac{\alpha}{\beta} \right)^2 \sinh^2 \beta L$$

$$= \left[1 + \frac{U_0^2 \sin^2 h\beta L}{4E(U_0 - E)} \right]$$

The transmission coefficient is:

$$T = \left[1 + \frac{U_0 \sin^2 h\beta L}{4E(U_0 - E)} \right]^{-1} \tag{3.75}$$

When $\beta L \gg 1$, the transmission coefficient T becomes very small and is given approximately

$$T = \left[\frac{16E(U_0 - E)}{U_0^2}\right] e^{-2\beta L} \tag{3.76}$$

In this formula, the quantity in bracket varies much less in comparison to the exponential term with E and U_0. A reasonable approximation for T is, therefore,

$$T = e^{-2\beta L}$$

or

$$T = \exp\left(-2\frac{\sqrt{2m(U_0 - E)}}{\hbar} \cdot L\right) \tag{3.77}$$

Clearly T depends on the height of barrier as well as on the width of the barrier. The transmission coefficient T of a barrier is plotted as a function of particle energy in Figure 3.9.

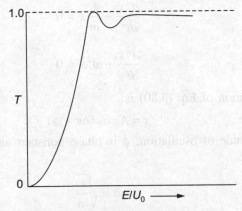

Figure 3.9 Transmission coefficient T as a function of particle energy.

The tunnel effect was used by Gamow in 1928 to explain the emission of alpha particles by certain radioactive nuclei. An α-particle, whose energy is only a few MeV, escapes from a nucleus whose potential wall is perhaps 25 MeV high, although probability of escape is extremely small. This effect also occurs in the operation of certain semiconductor diodes (e.g. Tunnel diode). The ability of electrons to tunnel through a potential barrier is used in the Scanning Tunneling Microscope (STM) to study surfaces on an atomic scale size.

3.8.4 Linear Harmonic Oscillator

The one-dimensional harmonic oscillator is a mass m vibrating back and forth on a line around an equilibrium position. In quantum mechanics, the one-dimensional harmonic oscillator is one of the few systems for whom the Schrodinger's equation can be solved analytically. The study of harmonic oscillator is important because some complicated systems can always be analysed in terms of harmonic oscillators.

Whenever one studies the behaviour of a physical system in the neighborhood of a stable equilibrium position, one arrives at equations, in the limit of small oscillations, are those of a harmonic oscillator. The examples of such systems are the vibrations of individual atoms in molecules and in crystals. It also provides the key to quantum theory of the electromagnetic field. The energy of electromagnetic waves in a cavity can be taken as the energy of a large set of harmonic oscillators.

In case of simple harmonic motion, the restoring force F on a particle of mass m is proportional to the particles displacement from its equilibrium position, so that

$$F = -kx \tag{3.78}$$

where k is force constant. According to second law of motion,

$$F = m\frac{d^2x}{dt^2} \tag{3.79}$$

From Eqs. (3.78) and (3.79), we have

$$\frac{d^2x}{dt^2} + \frac{k}{m}x = 0$$

or

$$\frac{d^2x}{dt^2} + \omega^2 x = 0 \tag{3.80}$$

A convenient solution of Eq. (3.80) is:

$$x = A\cos(\omega t + \phi) \tag{3.81}$$

where A is amplitude of oscillation, ϕ is phase constant and ω is angular frequency which is given as:

$$\omega = \sqrt{\frac{k}{m}} \tag{3.82}$$

The potential energy $U(x)$ that corresponds to the Hooke's law $F = -kx$ may be given by

$$U(x) = \frac{1}{2}kx^2 \tag{3.83}$$

and thus the Schrodinger's equation for such a case is:

$$\frac{d^2x}{dt^2} + \frac{2m}{\hbar^2}\left(E - \frac{1}{2}kx^2\right)\psi = 0 \tag{3.84}$$

To deal such an equation, it is convenient to express this equation in dimensionless form. We introduce a dimensionless independent variable $\xi = \alpha x$ and a dimensionless eigenvalue λ to put Eq. (3.84) in the form:

$$\frac{d^2x}{d\xi^2} + (\lambda - \xi^2)\psi = 0 \tag{3.85}$$

such that

$$\alpha^4 = \frac{mk}{\hbar^2} \text{ and } \lambda = \frac{2E}{\hbar\omega} \tag{3.86}$$

For sufficiently large ξ, $\psi(\xi) = \xi^n \exp\left(\pm\dfrac{\xi^2}{2}\right)$ satisfies Eq. (3.85) when n has any finite value. The boundary condition on wave function allows only negative sign in exponential term. Therefore, an exact solution of Eq. (3.85) may be

$$\psi(\xi) = H(\xi)e^{-\xi^2/2} \qquad (3.87)$$

Here $H(\xi)$ is a polynomial of finite order in ξ. With Eq. (3.87), Eq. (3.85) takes the form:

$$\frac{d^2H}{d\xi^2} - 2\xi\frac{dH}{d\xi} + (\lambda - 1)H = 0 \qquad (3.88)$$

The acceptable solutions of Eq. (3.86) are those for which $\psi \to 0$ as $\xi \to \infty$ such that

$$\int_{-\infty}^{\infty} |\psi|^2 \, d\xi = 1$$

Only in this situation the wave function can represent an actual particle. This condition can be fulfilled by having

$$\lambda = 2n + 1 \qquad n = 0, 1, 2, 3, \ldots \qquad (3.89)$$

Now, Eq. (3.88) can be written as:

$$\frac{d^2H}{d\xi^2} - 2\xi\frac{dH}{d\xi} + 2nH = 0 \qquad (3.90)$$

This is a standard mathematical equation known as Hermite's equation. The solutions of this equation are Hermite polynomials, given by

$$H_n(\xi) = (-1)^n \exp(\xi^2)\frac{d^n e^{-\xi^2}}{d\xi^n} \qquad (3.91)$$

where n is the order of polynomial. The general formula for the nth wave function of harmonic oscillator is:

$$\psi_n(\xi) = N_n H_n(\xi)e^{-(\xi^2/2)} \qquad (3.92)$$

Here N_n is normalisation constant.

Now, from Eqs. (3.86) and (3.89), we have

$$\frac{2E}{\hbar\omega} = 2n + 1$$

or

$$\frac{2E}{h\nu} = 2n + 1$$

The permitted eigenvalues (energy) of harmonic oscillator are given as:

$$E_n = \left(\frac{n+1}{2}\right)h\nu \qquad \text{such that } n = 0, 1, 2, 3, \ldots \qquad (3.93)$$

This gives an infinite set of discrete (quantised) energy states in steps of hv. The ground state ($n = 0$) of the oscillator has energy $E_0 = (1/2)hv$, called as zero-point energy. The energy levels are evenly spaced unlike the energy levels of a particle in a box. The existence of zero-point energy is related to uncertainty principle. Classically this energy would be zero. The potential energy of a simple harmonic oscillator as a function of displacement (heavy curve) and the first four energy levels together with the associated wave functions are shown in Figure 3.10.

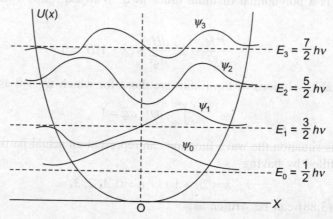

Figure 3.10   Potential energy and first four energy levels together with associated wave functions of one-dimensional harmonic oscillator.

SOLVED NUMERICAL PROBLEMS

PROBLEM 3.1   Calculate the ground state energy of an electron confined to a box 1 Å wide.

Solution   We have $m = 9.1 \times 10^{-31}$ kg, $L = 1 \times 10^{-10}$ m

Since energy of electron in nth state is given by

$$E_n = \frac{n^2 \pi^2 \hbar^2}{2mL^2}$$

$$= \frac{n^2 h^2}{8mL^2}$$

For ground state,

$$E_1 = \frac{h^2}{8mL^2}$$

$$= \frac{(6.625 \times 10^{-34})^2}{8 \times 9.1 \times 10^{-31} \times (1 \times 10^{-10})^2}$$

$$E_1 = 6.016 \times 10^{-18} \text{ J}$$

PROBLEM 3.2 An electron is confined to move between two rigid walls separated by 10^{-9} m. Find the de Broglie wavelengths representing the first three allowed energy states of the electron and the corresponding energies (Given: electronic mass is 9.1×10^{-31} kg and $h = 6.63 \times 10^{-34}$ Js).

Solution An electron moving back and forth between rigid walls will form a stationary wave pattern with nodes at the walls. For this, the distance between the walls must be an integral multiple of the de Broglie half wavelengths.

Thus,
$$L = n\frac{\lambda}{2} \qquad \text{where } n = 1, 2, 3, \dots$$

or
$$\lambda = \frac{2L}{n}$$

Here
$$L = 10^{-9} \text{ m} = 10 \text{ Å}$$

$$\lambda = \frac{2 \times 10}{n} \text{ Å}$$

$$\lambda = \frac{2 \times 10 \text{ Å}}{n} \ (n = 1, 2, 3, \dots)$$

$$= 20 \text{ Å}, 10 \text{ Å}, 6.7 \text{ Å} \dots$$

The corresponding energies are:
$$E_n = \frac{n^2 h^2}{8mL^2}$$

Here $m = 9.1 \times 10^{-31}$ kg, $L = 10^{-9}$ m

$$E_n = \frac{n^2 (6.63 \times 10^{-34})^2}{8 \times 9.1 \times 10^{-31} \times (10^{-9})^2} \text{ J}$$

$$= 6.04 \times 10^{-20} n^2 \text{ J}$$

$$= \frac{6.04 \times 10^{-20} n^2}{1.6 \times 10^{-19}} \text{ eV} = 0.38 n^2 \text{ eV}$$

For $n = 1, 2, 3, \dots$, we have
$$E_1 = 0.38 \text{ eV}, E_2 = 1.52 \text{ eV}, E_3 = 608 \text{ eV}$$

PROBLEM 3.3 An electron is confined to a one-dimensional box of side 1 Å. Obtain the first four eigenvalues in eV of the electron.

Solution
$$E_n = \frac{n^2 h^2}{8mL^2}$$

$$E_n = 38 n^2 \text{ eV}, L = 1 \text{ Å}$$

$$E_1 = 38 \text{ eV}, E_2 = 152 \text{ eV}, E_3 = 342 \text{ eV}, E_4 = 608 \text{ eV}$$

PROBLEM 3.4   A particle of mass m is moving in an infinitesimal deep square well potential expending from $x = 0$ to $x = L$. Show that $\langle E \rangle = \dfrac{n^2 h^2}{8ma^2}$

or

Prove that the value of energy obtained for a particle of mass m moving in a one-dimensional box can also be obtained with the help of relation used to obtain expectation value.

Solution   The normalised wave function for a particle of mass m in one-dimensional box is given by

$$\psi_n(x) = \sqrt{\frac{2}{L}} \sin \frac{n\pi x}{L}$$

Now, Energy operator $E = -\dfrac{\hbar^2}{2m} \dfrac{d^2}{dx^2}$

So, that $\displaystyle \langle E \rangle = \int_0^L \psi_n^*(x) \hat{E} \psi_n(x)\,dx$

$$= \frac{\hbar^2}{2m} \int_0^L \sqrt{\frac{2}{L}} \sin \frac{n\pi x}{L} \times \frac{d^2}{dx^2} \left(\sqrt{\frac{2}{L}} \sin \frac{n\pi x}{L} \right) dx$$

$$= \frac{\hbar^2}{2m} \times \frac{2}{L} \int_0^L \sin \frac{n\pi x}{L} \left(-\frac{n^2 \pi^2}{L^2} \right) \sin \frac{n\pi x}{L}\,dx$$

$$= \frac{n^2 \pi^2 \hbar^2}{mL^3} \int_0^L \sin^2 \frac{n\pi x}{L}\,dx$$

$$= \frac{n^2 \pi^2 \hbar^2}{2mL^3} \int_0^L \left[1 - \cos \frac{2n\pi x}{L} \right] dx$$

$$= \frac{n^2 \pi^2 \hbar^2}{2mL^3} \times L = \frac{n^2 \pi^2 \hbar^2}{2mL^2} = \frac{n^2 h^2}{8mL^2}$$

PROBLEM 3.5   A proton or a neutron in a nucleus can be roughly regarded as a particle in a box; the surface of the nucleus plays the role of the wall of the box, and the proton moves more or less freely between these walls. What is the energy released when a proton makes a transition from first excited state to the ground state of a box of nuclear size, say, 10×10^{-14} m?

Solution $L = 10 \times 10^{-14}$ m and $m_p = 1.67 \times 10^{-14}$ kg.
   The energy of the ground state (for $n = 1$) is:

$$E_1 = \frac{\hbar^2 \pi^2}{2m_p L^2} = 3.3 \times 10^{-13}\,\text{J}$$

The energy of the first excited state is four times larger than the energy of ground state.

$$E_2 = 4E_1 = 13.2 \times 10^{-13} \text{ J}$$

$$(E_2 - E_1) = 13.2 \times 10^{-13} \text{ J} - 3.3 \times 10^{-13} \text{ J} = 9.9 \times 10^{-13} \text{ J} = 6.2 \text{ MeV}$$

PROBLEM 3.6   A particle is moving in a one-dimensional box (of infinite height) of width 10 Å. Calculate the probability of finding the particle within an interval of 1 Å at the centre the box, when it is in its state of least energy.

Solution $\psi_1 = \sqrt{\left(\dfrac{2}{L}\right)} \sin \dfrac{\pi x}{L}$

The probability of finding the particle in unit interval at the centre of the box $\left(x = \dfrac{L}{2}\right)$ is:

$$P = |\psi_1|^2 \, dx = \left(\frac{2}{L}\right) dx \text{ at } x = \frac{a}{2}$$

With $L = 10$ Å and $dx = 1$ Å

$$P = \frac{2}{10} \times 1 = 0.2$$

$$P = 0.2$$

PROBLEM 3.7   A particle is moving in a one-dimensional potential box of infinite height. What is the probability of finding the particle in a small interval Δx at the centre of the box when it is in the energy state, next to the least energy state.

Solution   The wave function of the particle in the first excited state ($n = 2$) is:

$$\psi_2 = \sqrt{\frac{2}{L}} \sin \frac{2\pi x}{L}$$

At the centre of box, $x = L/2$, then $\psi_2 = 0$ and probability $P = |\psi_2|^2 \, dx = 0$

EXERCISES

THEORETICAL QUESTIONS

3.1 Discus the physical meaning of wave function.

3.2 What are the basic postulates of quantum mechanics? Explain in detail.

3.3 Derive time-independent Schrödinger's equation.

3.4 Explain the terms 'eigenvalue' and 'eigen function'. What are the requirements for a suitable wave function?

3.5 What do you understand by normalisation of a wave function? Explain with the help of a suitable example.

3.6 If ψ_1 and ψ_2 are wave function of a system then show that the linear combination $a_1\psi_1 + a_2\psi_2$ is also wave function of that system. Here a_1 and a_2 are constants.

3.7 Explain the terms 'expectation value' and 'operator' in detail.

3.8 Prove that the normalisation process of a wave function does not change associated properties of the wave function.

3.9 What is tunnel effect? Write down Schrödinger's equation for potential barrier problem and steps to find out the transmission coefficient of a particle having less energy than the height of potential barrier.

3.10 Explain, what do you understand by the term 'potential well' and 'potential barrier'? How does a particle with energy lower than the barrier height tunnels through its quantum mechanically? Give one example.

Numerical Problems

P3.1 An electron is confined by electrical forces between two rigid walls separated by 1.0×10^{-9} m. Find the energy value for first state.   (**Ans.** 0.38 eV)

P3.2 Calculate the probability of finding a particle within distance $l/3$ for $n = 1$. Here l is separation of rigid walls.   (**Ans.** 0.20)

P3.3 Find the expectation values of position and momentum of a particle whose wave function is $\psi(x) = A \, \exp\left(-\dfrac{x^2}{2L^2} + ikx\right)$ (**Ans.** 0, $\hbar k$)

P3.4 A beam of electrons of energy 15 eV is incident at the boundary of step potential of height 5 eV. Find the fraction of the beam that is reflected and transmitted.

(**Ans.** $R = 1\%$, $T = 99\%$)

P3.5 Calculate the probability of transmission that an electron with energy 1 eV will penetrate a potential barrier of 4 eV, when barrier width is 4 Å.   (**Ans.** 0.024)

P3.6 Calculate the ground state energy of an electron confined to a box 1 Å wide.

(**Ans.** 6.016×10^{-18} J)

P3.7 Determine the expectation value of position of a particle in one-dimensional box.

(**Ans.** $L/2$).

P3.8 A particle limited to the X-axis has a wave function $\psi = ax$ between $x = 0$ and $x = 1$, $\psi = 0$ elsewhere. Find (a) the probability that particle can be found between $x = 0.45$ and $x = 0.55$, (b) expectation value $<x>$ of the particle's position.

(**Ans.** (a) 0.0251 a^2, (b) $a^2/4$)

P3.9 The wavelength of laser is 632.8 nm. Assuming this light is due to transition from $n = 4$ state to $n = 3$ state of an electron in a box, determine width of the box.   (**Ans.** 1.16 nm)

P3.10 Consider an electron, mass $m = 9.0 \times 10^{-31}$ kg in an infinite wall that is 2 cm wide. For what value of n will the electron have an energy of 1 eV?   (**Ans.** 3×10^7)

MULTIPLE CHOICE QUESTIONS

MCQ3.1 The wave function associated with a free electron is:

(a) $\psi = A \exp\left(-\dfrac{iEt}{h}\right)$

(b) $\psi = A \exp\left(-\dfrac{ip_x}{h}\right)$

(c) $\psi = A \exp\left[\dfrac{-i}{\hbar}(Et - px)\right]$

(d) None of these

MCQ3.2 If E_1 is the energy of the electron in the ground state of a one-dimensional potential box of length a, E_2 is the energy of the electron in the second energy state of a one-dimentional potential box of length $a/3$, then $E_1 - E_2$ is:

(a) $-\dfrac{h^2}{ma^2}$

(b) $-\dfrac{10h^2}{ma^2}$

(c) $\dfrac{h^2}{ma^2}$

(d) None of these

MCQ3.3 The wave function for the motion of a particle in a one-dimensional potential box of length a is given by $\psi_n = A \sin\left(\dfrac{n\pi x}{a}\right)$, where A is the normalisation constant. Its value is:

(a) $\left(\dfrac{2}{a}\right)^{1/2}$

(b) $\dfrac{2}{a}$

(c) a

(d) $\dfrac{1}{a}$

MCQ3.4 The expression for the discrete energy values of the electrons in a cubical box of lattice parameter a is $\dfrac{n^2h^2}{8ma^2}$, then n will be

(a) Sum of the square of three integers

(b) An integer

(c) Square root of an integer

(d) None of these

MCQ3.5 The energy of the lowest state in one-dimensional potential box of length a is:

(a) Zero

(b) $\dfrac{2h^2}{8ma^2}$

(c) $\dfrac{h^2}{8ma^2}$

(d) $\dfrac{h}{8ma^2}$

MCQ3.6 If E_1 is the energy of the electron in the lowest state of a one-dimensional potential box of length a and E_2 is the energy of the electron in the first and second energy state respectively, then $\dfrac{\psi_1}{\psi_2}$ is

(a) 0

(b) ∞

(c) 1

(d) 2

MCQ 3.7 In a one-dimensional box of length L, the probability of finding a particle will be maximum at

(a) $\dfrac{L}{2}$

(b) L

(c) Zero

(d) $\dfrac{L}{4}$

MCQ 3.8 Electrons with energy E are incident on a potential step of height V such that $E > V$. Which of the following is wave mechanically true?
(a) Electrons are transmitted
(b) Electrons will be reflected
(c) Partial reflection and partial transmission will take place
(d) None of these

Answers

3.1 (d)   **3.2** (a)   **3.3** (a)   **3.4** (a)   **3.5** (d)   **3.6** (b)   **3.7** (a)
3.8 (c)

CHAPTER 4

Atomic Physics

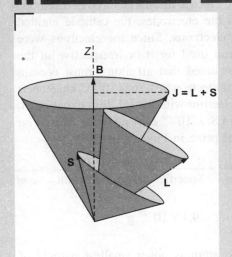

4.1 INTRODUCTION

In this chapter we will study about atom and atomic spectra based on different atomic model. Initially we will go through the basics of charges on atomic subparticles. Later on we will study various atomic models and their drawbacks. The spectra of hydrogen atom, helium atom and alkali elements will be understood with the help of atomic model. After that we will explain the different phenomenon like *Normal Zeeman effect, Anomalous Zeeman effect, Paschen back effect and Stark effect.*

4.2 CONCEPT OF CHARGE

The electric charge is the property that measures the electrification of material. In other words, it is the property due to which materials get electrified. According to the Standard Model, The electric charge is a fundamental conserved property of certain subatomic particles that determines their electromagnetic interactions. Electrically charged particles are influenced by electromagnetic fields. Moreover, they can also able to create electromagnetic field around themselves.

According to Dalton, the fundamental particle of matter is atom and is electrically neutral. An English physicist, J.J. Thomson studied electric discharge through a cathode ray tube. When high voltage was applied across the electrodes, the cathode emitted a stream of negatively charged particles, called **electrons.** Since the electrons were released from the cathode irrespective of the metal used for it or irrespective of the gas filled in the cathode ray tube, Thomson concluded that all atoms must contain electrons. He found that the charge per unit mass of electron is 1.76×10^{11} C/kg. Robert Millikan (1868–1953) determined the charge on electron with his oil drop experiment. He obtained that *an electron has* 1.6×10^{-19} C $(4.8 \times 10^{-10}$ esu) *negative charge on it.* For this invention, Millikan received the Nobel prize in Physics in 1923.

$$\text{Mass of electron} = \frac{\text{Charge on electron}}{\text{Charge per unit mass of electron}} = \frac{\text{Charge on electron}}{\text{Specific charge of electron}} = \frac{e}{e/m}$$

$$m = \frac{1.6 \times 10^{-19}\,\text{C}}{1.76 \times 10^{11}\,\text{C/kg}} = 9.1 \times 10^{-31}\,\text{kg} = 9.1 \times 10^{-28}\,\text{g}$$

The discovery of the electron concludes that the atom is not a smallest particle of matter, but it is also divisible. Hence the idea of indivisibility of atom as suggested by Dalton was proved incorrect. If the atom was divisible, what were are its *constituents*?

In 1886, Eugen Goldstein observed that rays flowing in a direction opposite to that of the cathode rays were positively charged. Such rays were named as **canal rays** because they passed through the holes or the canals present in the perforated cathode. In 1898, Wilhelm Wien, a German physicist, measured *e/m* for canal rays. It was found that the particles constituting the canal rays are much heavier than electrons. Also unlike cathode rays, the nature and the type of these particles varied depending upon the gas present in the cathode ray tube. The canal rays had positive charges which were whole number multiples of the amount of charge present on the electron. When the cathode ray tube contained hydrogen gas, the particles of the canal rays obtained were the lightest and their *e/m* ratio was the highest. Rutherford showed that these particles were identical to the hydrogen ion. These particles were named as **protons** *which has* 1.6×10^{-19} C *of positive charge and* 1.67×10^{-27} kg *mass*. According to Rutherford atomic model, protons exist inside the nucleus.

After the discovery of nucleus, several particles are discovered that exist inside it. The second basic particle of nucleus is **neutron** *that is electrically neutral*. The deep studies of these nuclear particles show that *the nucleus particles are made up of* **quarks**. The charges on up quark *u*, up anti-quark $\bar{u}$, down quark *d* and down anti-quark $\bar{d}$ are equal to $+2e/3$, $-2e/3$, $+ e/3$ and $-e/3$ respectively. These nucleus particles and its constituent particles can be studied in nuclear and particle physics.

4.3 ATOMIC MODELS

According to Dalton, the matter is made up of extremely small particles called atoms. After it, Prout suggested that all elements are made up of atoms of hydrogen. This suggestion was suitable because many of the elements were found to have atomic weights that were not exact multiples of that of hydrogen. Several atomic models came into effect after the discovery of radioactivity and electron. These models are: Thomson model, Rutherford model, Bohr model, Sommerfeld model, Sommerfeld relativistic model, vector atom model, and Dirac model.

4.4 THOMSON MODEL

The study of discharge of electricity through gases confirms that an atom consists of positive and negative charges. The physicist, J.J. Thomson explained the arrangement of these charges inside the atom. The atomic model given by him is called *Thomson model*.

According to this model, an atom is a sphere of radius of the order of 10^{-10} m in which positive charges are uniformly distributed, while the electrons are embedded in it (Figure 4.1). The atom is electrically neutral because total positive charge inside the atom is equal to the total negative charge. This model is also called *plum pudding model* or *raisin pudding model* because the electrons resembled the raisins dispersed in a pudding.

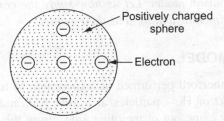

Figure 4.1 Distribution of charges in the atom according to Thomson model.

Thomson also gave an idea about the distribution of electrons inside the uniformly positively charge distributed sphere. He suggested that the electrons of an atom are positioned in a symmetrical pattern with respect to the centre of the sphere. In case of a single electron in the atom, the electron resides at the centre of the sphere. In case of atom having two electrons, the electrons are situated on the opposite sides of the centre symmetrically, i.e., at equal distance of $r/2$, where r is the radius of the positive sphere. For the case of three-electron system of the atom, the electrons exist at the corners of a symmetrically placed equilateral triangle of side r (Figure 4.2).

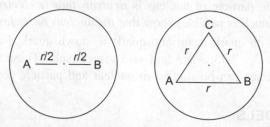

Figure 4.2   Arrangement of two or three electrons in sphere of positive charge.

The simple harmonic vibration of these electrons on both sides of their mean positions results the spectral radiations. Yet this model represents a clear explanation about the stability of the atom but it has some drawbacks.

Drawbacks

(i) This atomic model is unable to explain the simplest hydrogen atom spectrum which has five series of spectral lines. The vibrating electron radiates energy whose frequency is same as the electron. In the case of hydrogen atom, this model found a spectral line about 1300 Å, which contradicts the five series spectrum of hydrogen atom.

(ii) This model is also unable to explain the scattering of α-particles.

During this period, the phenomenon of radioactivity was also being studied by the scientists. This phenomenon of spontaneous emission of rays from atoms of certain elements also proved that the atom was divisible and it contained sub-atomic particles. Ernest Rutherford and his coworkers were also carrying out experiments which revealed that the radiation could be of three types: alpha, beta and gamma. In 1910–11, Rutherford and his co-workers performed an experiment which led to the downfall of the Thomson model. Let us now study the contribution of Rutherford and his coworker experimental finding and their model.

4.5   RUTHERFORD MODEL

In 1911, Ernest Rutherford performed an experiment to test the plum pudding model. He fired energetic α or He^{2+} particles at a gold foil, and measured the deflection of the particles as they came out of the other side. From this he deduced the information about the structure of the foil.

Experimental Arrangement

It consists of source of α–particles, two slits, thin gold foil and a fluorescent screen coated with zinc sulfide (scintillation screen). The experimental arrangement is shown in Figure 4.3. The radioactive materials like radium or radon are the good source of α–particles. When these materials are placed in a lead box with narrow opening, then a beam of α-particles emerges from the lead box opening. The α-particles emitted except the direction of opening are absorbed in the lead box itself. This beam is allowed to incident on a thin gold foil after passing through the diaphragms (slits) D_1 and D_2, which results in scattering of α-particles at different angles. The scattered α-particles strike a fluorescent screen and produces tiny flashes of light. The observations of scattered particles are made with the help of a low power microscope. The different possible scattering of α-particles through gold foil is shown in Figure 4.4.

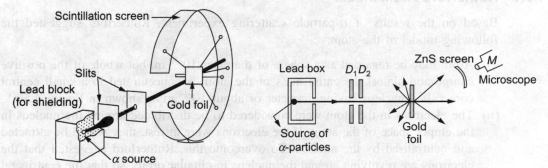

Figure 4.3 Experimental set-up for the α-particle bombardment on thin gold foil.

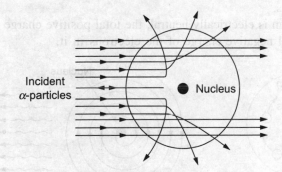

Figure 4.4 Scattering of α-particle through gold foil or gold atom.

From the experiment, Rutherford observed that:

- Most of the alpha particles pass straight through the gold foil.
- Some of the alpha particles get deflected by very small amounts.
- A very few get deflected greatly.
- Even fewer get bounced back after striking on foil.

Discard of Thomson Model

If the positively charged particles are uniformly distributed through nucleus then no alpha particles will pass through the foil. But in the experiment it was not observed, so this experiment discards the Thomson model.

On the basis of experimental observations, Rutherford concluded that:

- The maximum space of an atom is empty because most of the α-particles pass the gold foil without scattering.
- Scattering and back to the left of α-particles through the foil indicates that central part of atom is positively charged and have maximum mass of atom. This central part is termed as *nucleus*.
- The nucleus is approximately 1,00,000 times smaller than the atom.

4.5.1 Rutherford Atom Model

Based on the results of α-particle scattering experiment, Rutherford suggested the following model of the atom:

(i) Atom may be regarded as a sphere of diameter 10^{-10} m, but whole of the positive charge and almost the entire mass of the atom is concentrated in a small central core called nucleus having diameter of about 10^{-14} m as shown in Figure 4.5(a).

(ii) The electrons in the atom were considered to be distributed around the nucleus in the empty space of the atom. If the electrons were at rest, they would be attracted and neutralised by the nucleus. To overcome this, Rutherford suggested that the electrons are revolving around the nucleus in circular orbits, so that the centripetal force is provided by the electrostatic force of attraction between the electron and the nucleus.

(iii) As the atom is electrically neutral, the total positive charge of the nucleus is equal to the total negative charge of the electrons in it.

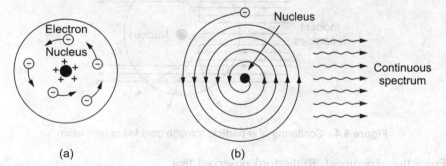

(a)   (b)

Figure 4.5   (a) Rutherford atomic model (b) motion of electron in spiral path.

Drawbacks

Following are the two drawbacks of Rutherford's model:

(i) **Stability of atom:**   According to the electromagnetic theory, an accelerated electric charge radiates energy in the form of electromagnetic waves. Since the circulating electron possesses an acceleration thus it radiates energy. If the accelerated electron loses energy by radiation, the energy of the electron must decrease continuously. So, it should be spiral down into the nucleus, as shown in Figure 4.5(b). Hence, the atom cannot be stable. But, it is well known that most of the atoms are stable.

(ii) **Provision of only continuous spectrum:** The frequency of radiation emitted by circularly accelerated charge is proportional to its the angular velocity. If the electron spirals towards the nucleus, the angular velocity tends to become infinity, and hence the frequency of the emitted energy will tend to infinity. This will result in a continuous spectrum with all possible wavelengths. But experiments indicate only line spectra of fixed wavelength for the atoms.

Impact Parameter, Distance of Closest Approach and Scattering Angle

When the α-particle approaches to the nucleus, it gets deflected/scattered due to repulsive electrostatic interaction between the particle and nucleus. The *impact parameter b* is the minimum distance to which the α-particle would approach the nucleus if there were no force between them.

The head-on collision occurs when impact parameter becomes zero. In case of head-on collision, when α-particle approaches to nucleus, then at a certain distance the repulsive force of nucleus stops the approaching α-particle momentarily. In this position the distance between α-particle and nucleus is termed as *distance of closest approach D*. Furthermore, at this point, all of its kinetic energy is transformed into potential energy. From this point, the particle get back to reverse direction. Suppose the α-particle of mass m approaches to the nucleus with velocity v. If Ze and $2e$ is charge on nucleus and α-particle, then at the distance of closest approach,

Kinetic energy of α-particle = Potential energy between nucleus and α-particle

$$\text{K.E.} = \frac{1}{4\pi\varepsilon_0} \frac{Ze\,2e}{D} = \frac{1}{4\pi\varepsilon_0} \frac{2Ze^2}{D}$$

$$D = \frac{1}{4\pi\varepsilon_0} \frac{2Ze^2}{\text{K.E.}} = \frac{1}{4\pi\varepsilon_0} \frac{Ze^2}{mv^2} \tag{4.1}$$

The *scattering angle θ* is the angle between the asymptotic direction of approach of the α-particle and the asymptotic direction in which it recedes. The pictorial presentation of impact parameter, distance of closest approach and scattering angle is shown in Figure 4.6.

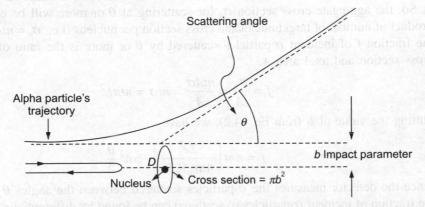

Figure 4.6 Illustration of impact of α-particle closer to nucleus.

The relation between impact parameter and scattering angle is given by the following equation:

$$b = \frac{1}{4\pi\varepsilon_0} \frac{Ze^2}{\text{K.E.}} \cot\frac{\theta}{2} = \frac{D}{2}\cot\frac{\theta}{2} \qquad (4.2)$$

The variation of b with θ is shown in Figure 4.7. Equation (4.2) indicates that the scattering angle increases with decrease in impact parameter, i.e., all the particles that approach to the nucleus with an impact parameter $\leq b$ will be scattered by an angle $\geq \theta$. Let b_1 and b_2 are the two impact parameters and their corresponding scattering angles are θ_1 and θ_2. If $b_1 < b_2$, then $\theta_1 > \theta_2$ [Figure 4.7(a)]. The area around each nucleus with radius equal to an impact parameter is known as **cross section of interaction** σ which is equal to πb^2. The scattering angle increases with decrease in cross section of interaction [Figure 4.7(b)].

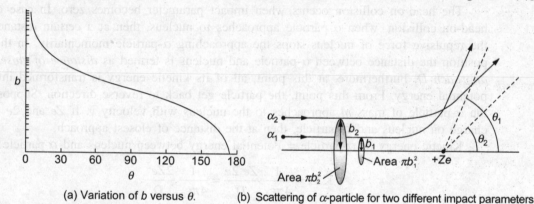

(a) Variation of b versus θ. (b) Scattering of α-particle for two different impact parameters

Figure 4.7 Variation of impact parameter with respect to scattering angle.

4.5.2 Rutherford's Scattering Formula

Consider the foil thickness is t that contains n atoms per unit volume. Then the number of target nuclei per unit area will be nt. If an α-particle particle beam is incident upon an area A, then it encounters ntA nuclei. The α-particle, which is directed within the area of cross section of interaction $\sigma(= \pi b^2)$ around a nucleus, is scattered at an angle θ. So, the aggregate cross section σ_a for scattering at θ or more will be equal to the product of number of target nuclei and cross section per nucleus (i.e., $\sigma_a = ntA\sigma$). Hence the fraction f of incident α-particles scattered by θ or more is the ratio of aggregate cross section and total area A.

$$f = \frac{\sigma_a}{A} = \frac{ntA\sigma}{A} = nt\sigma = nt\pi b^2 \qquad (4.3)$$

Putting the value of b from Eq. (4.2), we have

$$f = \pi nt \left(\frac{Ze^2}{4\pi\varepsilon_0 \text{K.E.}}\right)^2 \cot^2\frac{\theta}{2} \qquad (4.4)$$

Since the detector measures the α-particles scattered between the angles θ to $\theta + d\theta$, the fraction of incident α-particles so scattered can be found by differentiating Eq. (4.4).

Differentiating Eq. (4.4), we have

$$df = -\pi nt \left(\frac{Ze^2}{4\pi\varepsilon_0 \text{K.E.}} \right)^2 \cot\frac{\theta}{2} \text{cosec}^2\frac{\theta}{2} \, d\theta \tag{4.5}$$

Here minus sign indicates that f decreases with increase in scattering angle. Suppose r is the distance between foil and florescent screen, and α-particles scatter between the angles θ to $\theta + d\theta$. The scattered particles reach to the screen in a zone of sphere of radius r whose width is $rd\theta$ (Figure 4.8). So, the area ds of screen (detector/flourescent screen) on which these particles strike is product of perimeter of zone and its width, i.e.,

$$ds = (2\pi r \sin\theta)(rd\theta) = 2\pi r^2 \sin\theta d\theta = 4\pi r^2 \sin\left(\frac{\theta}{2}\right)\cos\left(\frac{\theta}{2}\right) d\theta \tag{4.6}$$

If the total of N_i number of α-particles strikes the foil during the experiment, then the number of scattered in range $d\theta$ at θ is $N_i df$. Suppose $N(\theta)$ is number of α-particles per unit area that reach at the screen at a scattering angle θ. Then

$$N(\theta) = \frac{N_i |df|}{ds}$$

$$N(\theta) = \frac{N_i \pi nt \left(\dfrac{Ze^2}{4\pi\varepsilon_0 \text{K.E.}} \right)^2 \cot\dfrac{\theta}{2} \text{cosec}^2\dfrac{\theta}{2} \, d\theta}{4\pi r^2 \sin(\theta/2)\cos(\theta/2)d\theta}$$

$$N(\theta) = \frac{N_i nt Z^2 e^4}{(8\pi\varepsilon_0)^2 r^2 (\text{K.E.})^2 \sin^4(\theta/2)} \tag{4.7}$$

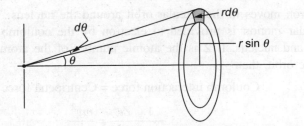

Figure 4.8 Areal projection of striking α-particles on the screen at an angle θ to $\theta + d\theta$.

Equation (4.7) is called *Rutherford scattering formula*. This formula provides a theoretical justification to the findings of Geiger and Marsden measurement which was performed on the suggestion of Rutherford. Since $N(\theta)$ is inversely proportional to $\sin^4(\theta/2)$, the variation of $N(\theta)$ with θ is highly pronounced. This shows that a fraction (≈ 0.14) per cent of the incident particle are scattered by more than $1°$, i.e., there exists a tiny volume inside the atom where its maximum mass and total positive charge is concentrated. Thus it gives a theoretical justification for the existence of nucleus. Since the physicist Rutherford exposed first time the nucleus, so he is called *father of nuclear physics*.

Size of Nucleus

An α-particle have its smallest distance when it approaches to nucleus during head-on collision, which is followed by 180° scattering. At the distance of closest approach, the kinetic energy of α-particles is converted to electric potential energy. Thus the distance of closest approach D can be written as:

$$D = \frac{1}{4\pi\varepsilon_0} \frac{2Ze^2}{\text{K.E.}}$$

Since Rutherford assumed that the size of target nucleus is small as compared to the distance of closest approach, D provides the upper limit to nuclear dimension. The maximum kinetic energy of α-particles of natural origin is 7.7 MeV or 1.2×10^{-12} J. So, the upper limit to nuclear dimension becomes

$$D = 9 \times 10^9 \frac{2 \times Z \times (1.6 \times 10^{-19})^2}{1.2 \times 10^{-12}} = 3.8 \times 10^{-16} \ Zm$$

Since for gold (Au), $Z = 79$, $D(\text{Au}) \sim 3.0 \times 10^{-14}$ m. Thus the radius of the gold nucleus is less than 3.0×10^{-14} m. Later on the actual radius of gold nucleus is found to be about one fifth of D(Au).

4.6 BOHR ATOMIC MODEL

In 1913, Bohr proposed an atomic model with refinement in Rutherford atomic model and using the Planck's law of radiation for the explanation of hydrogen or hydrogen like spectra. The atomic model given by him is called *Bohr atomic model* which have following postulates:

(i) An electron moves in the circular orbit around the nucleus. The centripetal force for circular motion is provided to electron by the coulomb interaction between electron and nucleus. If Z is the atomic number of the atom, and r is the radius of atomic orbit, then

Coulomb interaction force = Centripetal force

$$\frac{1}{4\pi\varepsilon_0} \frac{Ze^2}{r^2} = \frac{mv^2}{r} \tag{4.8}$$

Here, m and v are the mass and orbital velocity of electron.

(ii) The electron cannot revolve in all circular orbits. It revolves only in those circular orbits for which the angular momentum is integral multiple of $\hbar$ ($\hbar$ = Planck constant divided by $2\pi = 1.055 \times 10^{-34}$ Js).

$$mvr = n\hbar = \frac{nh}{2\pi}; \quad n = 1, 2, 3, \ldots \tag{4.9}$$

The allowed circular orbits for electron motion are called *stationary orbits/states*. n is a positive integer, called *principal quantum number*. Equation (4.9) is termed as *Bohr quantum condition*.

The electromagnetic radiation is not emitted by moving electron in an orbit. The emission and absorption of radiation occur by the electronic transition between stationary states of atom, i.e., a jump of electron from higher to lower state causes the emission, while the jump from lower to higher state causes the absorption of radiation. The energy quanta of radiation emitted/absorbed by an atom are equal to energy difference between the stationary states. If ΔE is energy difference, E_i and E_f are the energies of stationary states and v is the frequency of radiation, then

$$\left.\begin{array}{l} \Delta E = hv = \dfrac{hc}{\lambda} \\[2mm] E_i - E_f = hv \qquad \text{(for emission)} \\[2mm] E_f - E_i = hv \qquad \text{(for absorption)} \end{array}\right\} \qquad (4.10)$$

E_i and E_f are the energies of initial and final stationary states. Equation (4.10) is called *Bohr frequency condition*.

4.6.1 Bohr Radius (Radius of Stationary Orbit/State)

On the basis of postulates, radius of stationary states can be determined. Suppose the atom has Z atomic number, then charge on nucleus will be Ze. If the electron having mass m is moving around the nucleus circular orbit of radius r_n with orbital velocity v_n (Figure 4.9), then from Bohr's first postulate,

$$\frac{1}{4\pi\varepsilon_0} \frac{Ze^2}{r_n^2} = \frac{mv_n^2}{r_n} \qquad (4.11)$$

and from Bohr's quantum condition of angular momentum, we can write

$$mv_nr_n = n\hbar = \frac{nh}{2\pi}$$

or

$$v_n = \frac{nh}{2\pi}\left(\frac{1}{mr_n}\right) \qquad (4.12)$$

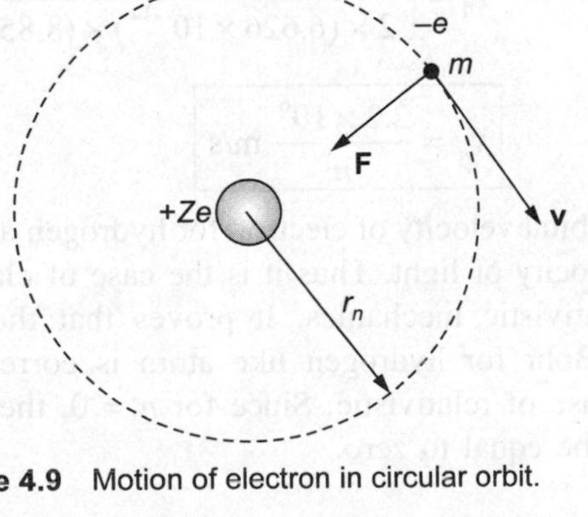

Figure 4.9 Motion of electron in circular orbit.

Putting the value of v_n in Eq. (4.11), we have

$$\frac{1}{4\pi\varepsilon_0}\frac{Ze^2}{r_n^2} = \frac{m}{r_n}\left[\frac{nh}{2\pi}\left(\frac{1}{mr_n}\right)\right]^2$$

$$\boxed{r_n = \left(\frac{h^2\varepsilon_0}{\pi me^2}\right)\frac{n^2}{Z}} \Rightarrow r_n \propto \frac{n^2}{Z} \qquad (4.13)$$

Thus the radii of different stationary orbits/states are proportional to quantity (n^2/Z). For hydrogen atom $Z = 1$, r_n is proportional to the square of principal quantum number, i.e.,

$$r_n = \left(\frac{h^2\varepsilon_0}{\pi me^2}\right)n^2$$

$$r_n = \left(\frac{(6.626\times10^{-34})^2(8.85\times10^{-12})}{3.14\times(9.1\times10^{-31})\times(1.6\times10^{-19})^2}\right)n^2$$

$$r_n = 0.529\times10^{-10}n^2 \text{ m}$$

$$\boxed{r_n = 0.53\,n^2 \text{ Å}} \qquad (4.14)$$

So, we can say that the radius of smallest orbit for hydrogen atom is 0.53 Å which is found to be in good agreement with data estimated from the kinetic theory. Now, from Eqs. (4.13) and (4.14), we can find

$$\boxed{v_n = \left(\frac{e^2}{2h\varepsilon_0}\right)\frac{Z}{n}} \Rightarrow v_n \propto \frac{Z}{n} \qquad (4.15)$$

For the hydrogen atom $Z = 1$, so

$$v_n = \left(\frac{e^2}{2h\varepsilon_0}\right)\frac{1}{n}$$

$$v_n = \left(\frac{(1.6\times10^{-19})^2}{2\times(6.626\times10^{-34})\times(8.85\times10^{-12})}\right)\frac{1}{n}$$

$$\boxed{v_n = \frac{2.2\times10^6}{n} \text{ m/s}} \qquad (4.16)$$

The smallest orbital velocity of electron for hydrogen atom is 2.2×10^8 cm/s. This is less than 1% of velocity of light. Thus it is the case of classical/non-relativistic mechanics instead of relativistic mechanics. It proves that the non-relativistic approximation employed by Bohr for hydrogen like atom is correct. But for large value of Z, it becomes the case of relativistic. Since for $n = 0$, the velocity becomes infinite, Bohr restricted n to be equal to zero.

4.6.2 Energy of Stationary Orbits/States

As we know that the total energy of a system is sum of its potential energy and kinetic energy.

$$E = \text{P.E.} + \text{K.E.} \tag{4.17}$$

Since the potential energy of an electron in an orbit is equal to the work done in taking the electron from r_n distance to infinity against the electrostatic attraction between nucleus and electron. So

$$\text{P.E} = -\int_{r_n}^{\infty} F\, dr$$

$$\because \qquad F = \frac{1}{4\pi\varepsilon_0}\frac{Ze^2}{r^2}$$

$$\therefore \qquad \text{P.E.} = -\int_{r_n}^{\infty} \frac{1}{4\pi\varepsilon_0}\frac{Ze^2}{r^2}\, dr = -\frac{1}{4\pi\varepsilon_0}\frac{Ze^2}{r_n} \tag{4.18}$$

and

$$\text{K.E.} = \frac{1}{2}mv_n^2$$

Using Eq. (4.11), we can write

$$\text{K.E.} = \frac{1}{8\pi\varepsilon_0}\frac{Ze^2}{r_n} \tag{4.19}$$

Putting the value of potential energy and kinetic energy in Eq. (4.17), we can find the total energy of nth state:

$$E_n = -\frac{1}{4\pi\varepsilon_0}\frac{Ze^2}{r_n} + \frac{1}{8\pi\varepsilon_0}\frac{Ze^2}{r_n}$$

$$E_n = -\frac{1}{8\pi\varepsilon_0}\frac{Ze^2}{r_n}$$

$$E_n = -\frac{Ze^2}{8\pi\varepsilon_0}\left(\frac{\pi m e^2}{h^2\varepsilon_0}\right)\frac{Z}{n^2} \qquad \text{[Using Eq. (4.13)]}$$

$$\boxed{E_n = -\left(\frac{me^4}{8\varepsilon_0^2 h^2}\right)\frac{Z^2}{n^2}} \Rightarrow E_n \propto -\frac{Z^2}{n^2} \tag{4.20}$$

The quantity $\left(\dfrac{mZ^2 e^4}{8\varepsilon_0^2 h^3 c}\right)$ for $Z = 1$ is called *Rydberg constant* (R_∞ or R_H). The value of *Rydberg constant* is 1.09737×10^7 m^{-1}. In terms of it, Eq. (4.20) takes the form:

$$\boxed{E_n = -\frac{R_\infty hcZ^2}{n^2}} \tag{4.21}$$

The negative value of total energy signifies that the electron is bound to the nucleus. This expression also indicates that the energy is also quantised and is inversely proportion to n^2. So, the energy of state decreases with negative sign with increase in n, i.e., E increases with n and it approaches to zero value for large n. Thus the electron is no longer bound to the nucleus. The lowest energy occurs for the smallest quantum number $n = 1$, that corresponds to the innermost orbit. Sometimes the energy of state is also defined in terms of *term value* ($T_n = -E_n/hc$). So in terms of wave number, the energy of state becomes

$$T_n = \frac{R_\infty Z^2}{n^2} \qquad (4.22)$$

The lowest and most stable energy state has the largest term value. For the Hydrogen atom ($Z = 1$), the Eq. (4.20) takes the following form:

$$E_n = -\left(\frac{me^4}{8\varepsilon_0^2 h^2}\right)\frac{1}{n^2}$$

$$E_n = -\left(\frac{(9.1\times10^{-31})(1.6\times10^{-19})^4}{8\times(8.85\times10^{-12})^2\times(6.626\times10^{-34})^2}\right)\frac{1}{n^2}$$

$$E_n = -\frac{21.76\times10^{-19}}{n^2}\text{ J}$$

$$E_n = -\frac{13.6}{n^2}\text{ eV}$$

If E_1, E_2, E_3 and E_4 are the energies of stationary states $n = 1, 2, 3$ and 4, then

$$E_1 = -\frac{13.6}{1^2}\text{ eV} = -13.6\text{ eV}$$

$$E_2 = -\frac{13.6}{2^2}\text{ eV} = -3.4\text{ eV}$$

$$E_3 = -\frac{13.6}{3^2}\text{ eV} = -1.51\text{ eV}$$

$$E_4 = -\frac{13.6}{4^2}\text{ eV} = -0.85\text{ eV}$$

Here E_1 is called *ground or normal state energy*, while E_2, E_3 and E_4 are termed as *excited state energies*. The ground state energy of hydrogen atom is also energy unit, called *Rydberg* (ry).

In M.K.S. system, $1\text{ ry} = \left(\frac{me^4}{8\varepsilon_0^2 h^2}\right) = 21.76\times10^{-19}\text{ J} = 13.6\text{ eV}$

In C.G.S. system, $1\text{ ry} = \left(\frac{2\pi^2 me^4}{h^3}\right) = 21.76\times10^{-12}\text{ erg} = 13.6\text{ eV}$

A typical presentation of energy levels for hydrogen atom is shown in Figure 4.10. The value of $\dfrac{1}{4\pi\varepsilon_0}$ is equal to unity in the C.G.S. system. Thus the radius of stationary orbit, r_n, orbital velocity of electron, v_n, the energy of states, E_n and *Rydberg constant* (R_∞ or R_H) in the C.G.S. system have the following forms:

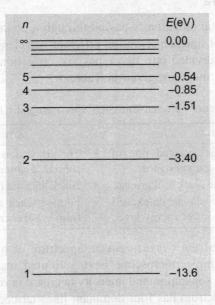

n	$E(eV)$
∞	0.00
5	−0.54
4	−0.85
3	−1.51
2	−3.40
1	−13.6

Figure 4.10 Energy states of hydrogen atom.

$$r_n = \left(\frac{h^2}{4\pi^2 me^2}\right)\frac{n^2}{Z} = \left(\frac{n^2 h^2}{4\pi^2 mZe^2}\right);\ v_n = \left(\frac{2\pi e^2}{h}\right)\frac{Z}{n} = \left(\frac{2\pi Ze^2}{nh}\right) = \frac{Ze^2}{n\hbar};$$

$$E_n = -\left(\frac{2\pi^2 me^4}{h^2}\right)\frac{Z^2}{n^2} = -\left(\frac{2\pi^2 mZ^2 e^4}{n^2 h^2}\right) \text{ and } R_\infty = \left(\frac{2\pi^2 me^4}{h^3 c}\right)$$

For the hydrogen atom ($Z = 1$), quantity v_1/c ($= 2\pi e^2/hc$) is a pure number and is called *fine structure constant* α. Its value is about 1/137.

EXAMPLE 4.1 Prove that the potential energy of an electron existing in any orbit of hydrogen is twice of its kinetic energy in magnitude.

Solution Let r_n is the radius of nth orbit of hydrogen then $r_n = 0.53n^2$ Å

Potential energy P.E. $= -\dfrac{1}{4\pi\varepsilon_0}\dfrac{e^2}{r_n} = -9 \times 10^9 \dfrac{(1.6 \times 10^{-19})^2}{0.53 \times 10^{-10} n^2}$ J

Potential energy $= -9 \times 10^9 \dfrac{(1.6 \times 10^{-19})}{0.53 \times 10^{-10} n^2}$ eV $= -\dfrac{27.2}{n^2}$ eV

Kinetic energy = E – Potential energy = $-\dfrac{13.6}{n^2} + \dfrac{27.2}{n^2} = \dfrac{13.6}{n^2}$

Thus the potential energy is twice of kinetic energy in magnitude.

4.6.3 Hydrogen Spectra

When electromagnetic radiation is passed through a prism or grating, it is split up and forms a collection of lines representing different wavelengths. This is called spectrum. The spectra can be divided into two types viz., emission and absorption spectra. The differences between them are given in Table 4.1.

Table 4.1   Difference between Emission Spectrum and Absorption Spectrum

Emission spectrum	*Absorption spectrum*
• The emission spectrum is obtained due to emission of radiation from the substances.	• The absorption spectrum is obtained when the substance absorbs the radiation.
• White lines are formed on the black background.	• Black lines are formed on the white background.
• Formed when atoms or molecules are de-excited from higher energy level to lower energy level.	• Formed when atoms or molecules are excited from lower energy level to higher energy levels.

Initially by *Angstrom's measurement*, spectrum of hydrogen atom is obtained in visible region by simple microscope having limited resolving power. It consists of series of lines whose separation and intensity decrease in regular pattern towards shorter wavelength. The spectrum has four dominant lines termed as H_α, H_β, H_γ and H_δ lines, while the series converges to 364.6 nm at violet end (as shown in Figure 4.11).

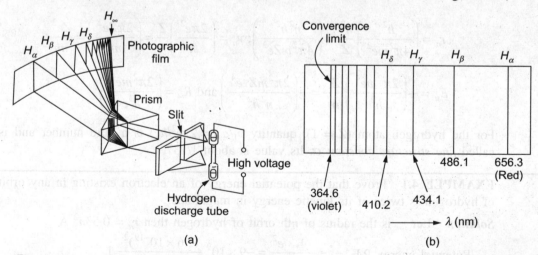

Figure 4.11   (a) Experimental arrangement to achieve hydrogen spectra, (b) Balmer series spectra obtained from hydrogen discharge tube.

In 1885, Balmer gave an empirical formula for the wavelength of spectral lines of hydrogen obtained by *Angstrom's measurement*. Balmer noticed that a single number had a relation to every line in the hydrogen spectrum that was in the visible light region. That number was 364.6 nm. His formula can be expressed in the following form:

$$\lambda = B\left(\frac{n^2}{n^2 - 4}\right); B = 3646 \text{ Å} \tag{4.23}$$

where λ is the wavelength of the absorbed/emitted light, and n is the positive integer that can have value greater than 2. Using the Balmer equation, the wavelength of the emission spectral lines in visible region for the hydrogen can be evaluated. In 1888, the physicist Johannes Rydberg generalised the Balmer equation for all transitions of hydrogen.

For Balmer lines, $\quad \dfrac{1}{\lambda} = \dfrac{4}{B}\left(\dfrac{1}{2^2} - \dfrac{1}{n^2}\right) = R_H\left(\dfrac{1}{2^2} - \dfrac{1}{n^2}\right); n = 3, 4, 5, \ldots \quad$ (4.24)

General formula, $\quad \dfrac{1}{\lambda} = \dfrac{4}{B}\left(\dfrac{1}{m^2} - \dfrac{1}{n^2}\right) = R_H\left(\dfrac{1}{m^2} - \dfrac{1}{n^2}\right) \quad$ (4.25)

where $n > m$ and $R_H = 4/B = 1.09709 \times 10^7$ m^{-1} is the *Rydberg constant* and is the same for all the different hydrogen series lines. Hydrogen has five series of lines. These series came to be named after the experimentalists that were first to see them. But at that time, the formula is only used for the evaluation of wavelength of Balmer series of lines.

Bohr Explanation of Hydrogen Spectra

Yet the empirical formulas given by Balmer or Rydberg explain the hydrogen spectral lines, but they have no theoretical justification. It was solved when the Bohr atomic model came into effect. The Bohr theory was a marvelous success in explaining the spectrum of the hydrogen atom. His calculated wavelengths agreed perfectly with the experimentally measured wavelengths of the spectral lines.

According to Bohr postulate, when an electron jumps from higher energy state to lower then emission of light occurs. While, absorption of radiation occurs by the electron transition from lower to higher state. The energy of emitted/absorbed photon is equal to energy difference between the states. Suppose an electron jumps from higher energy state n_i to lower energy state n_f. Due to this, v frequency of radiation is emitted. Then from Bohr postulate, we can write

$$hv = \frac{hc}{\lambda} = \Delta E = E_i - E_f \tag{4.26}$$

From Eq. (4.21),

$$E_n = -\frac{R_\infty hcZ^2}{n^2}$$

Thus expression of E_i and E_f become

$$E_i = -\frac{R_\infty hcZ^2}{n_i^2} \text{ and } E_f = -\frac{R_\infty hcZ^2}{n_f^2} \tag{4.27}$$

Using Eqs. (4.26) and (4.27), we have

$$\frac{hc}{\lambda} = \left(-\frac{R_\infty hcZ^2}{n_i^2} \right) - \left(-\frac{R_\infty hcZ^2}{n_f^2} \right)$$

$$\frac{hc}{\lambda} = R_\infty hcZ^2 \left(\frac{1}{n_f^2} - \frac{1}{n_i^2} \right)$$

$$\boxed{\frac{1}{\lambda} = \bar{v} = R_\infty Z^2 \left(\frac{1}{n_f^2} - \frac{1}{n_i^2} \right)} \qquad (4.28)$$

where *Rydberg constant* $R_\infty = \left(\dfrac{me^4}{8\varepsilon_0^2 h^3 c} \right) = 1.09678 \times 10^7 \text{ m}^{-1}$. The calculated value of

Rydberg constant with Bohr theory is very close to R_H used in *Rydberg empirical formula*. By Eq. (4.28), the wavelength of hydrogen spectral lines can be calculated. This expression for $z = 1$ is same as *Rydberg empirical formula*. Hence, the Bohr theory presents the theoretical justification of *Rydberg fomula* or provides theoretical explanation of Balmer line which was initially observed. The hydrogen spectrum has five series of lines depending upon the final position of transition (as shown in Figure 4.12).

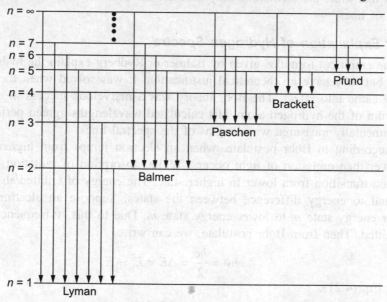

Figure 4.12 Electronic transition showing the hydrogen spectrum.

1. *Lyman series* is named after its discoverer, Theodore Lyman, who discovered the spectral lines in 1916. All the wavelengths in the Lyman series are in the ultraviolet region. For this series $n_f = 1$ and $n_i = 2, 3, 4, \ldots$.
2. *Balmer series* is named after Johann Balmer, who discovered the Balmer formula, an empirical equation to predict the Balmer series, in 1885. Balmer lines are

historically referred to as 'H-alpha', 'H-beta', 'H-gamma' and so on, where H is the element hydrogen. Four of the Balmer lines are in the technically 'visible' part of the spectrum, with wavelengths longer than 400 nm. Parts of the Balmer series can be seen in the solar spectrum. For this series, $n_f = 2$ and $n_i = 3, 4, 5, \ldots$.

3. *Paschen series* is named after the Austro–German physicist, Friedrich Paschen, who first observed them in 1908. The Paschen lines all lie in the near infrared region. For this series, $n_f = 3$ and $n_i = 4, 5, 6, \ldots$.

4. *Brackett series* is named after the American physicist, Frederick Sumner Brackett who first observed the spectral lines in 1922. The Brackett lines lie in the infrared region. For this series, $n_f = 4$ and $n_i = 5, 6, 7, \ldots$.

5. *Pfund series* was experimentally discovered in 1924 by August Herman Pfund which are observed in infrared region. For this series, $n_f = 5$ and $n_i = 6, 7, \ldots$.

Thus the expression for different series of line can be obtained by selecting different values of n_f. Hence

$$Lyman\ series \qquad \frac{1}{\lambda} = R_\infty \left(\frac{1}{1^2} - \frac{1}{n_i^2} \right); n_i = 2, 3, 4, 5, 6, 7, \ldots$$

$$Balmer\ series \qquad \frac{1}{\lambda} = R_\infty \left(\frac{1}{2^2} - \frac{1}{n_i^2} \right); n_i = 3, 4, 5, 6, 7, \ldots$$

$$Paschen\ series \qquad \frac{1}{\lambda} = R_\infty \left(\frac{1}{3^2} - \frac{1}{n_i^2} \right); n_i = 4, 5, 6, 7, \ldots$$

$$Brackett\ series \qquad \frac{1}{\lambda} = R_\infty \left(\frac{1}{4^2} - \frac{1}{n_i^2} \right); n_i = 5, 6, 7, \ldots$$

$$Pfund\ series \qquad \frac{1}{\lambda} = R_\infty \left(\frac{1}{5^2} - \frac{1}{n_i^2} \right); n_i = 6, 7, \ldots$$

The electronic transition from $n = \infty$ to $n = 1, 2, 3, 4, 5$ provides the series limit. The limit series wavelengths are $1/R_\infty$, $4/R_\infty$, $9/R_\infty$, $16/R_\infty$ and $25/R_\infty$ respectively. The calculated wavelengths and wave numbers for the different series lines are listed in Table 4.2.

Table 4.2 Wavelengths and Wave Number for Series

Lyman series ($n_f = 1$)		Balmer series ($n_f = 2$)		Paschen series ($n_f = 3$)		Brackett series ($n_f = 4$)		Pfund series ($n_f = 5$)	
n_i	λ (Å)	n_i	λ (Å)	n_i	λ (Å)	n_i	λ (Å)	n_i	λ (Å)
2	1216	3	6563	4	18756	4	40519	6	74596
3	1026	4	4861	5	12821	5	26261	7	46538
4	9725	5	4341	6	10939	6	21659	8	37406
5	9495	6	4102	7	10052	7	19451	9	32969
∞	9118	∞	3647	∞	8205	∞	14588	∞	22794

In the absorption spectra of hydrogen, only Lyman series of line are obtained because normally the electron/atom is found to be in the ground state. On the excitement by any reason (optical or electrical), it goes to any higher state by absorbing energy equivalent to energy difference between ground and excited states.

EXAMPLE 4.2 Calculate the wavelength of the 8th line of the Balmer series of hydrogen. ($R_H = 1.09 \times 10^5$ cm^{-1}).

Solution The 8th line of the Balmer series of hydrogen arises due to transition from $n_i = 10$ to $n_f = 2$. Thus

$$\frac{1}{\lambda} = R_H \left(\frac{1}{n_f^2} - \frac{1}{n_i^2} \right)$$

$$\frac{1}{\lambda} = R_H \left(\frac{1}{2^2} - \frac{1}{10^2} \right) = \frac{6R_H}{25}$$

$$\lambda = R_H \left(\frac{1}{2^2} - \frac{1}{10^2} \right) = \frac{25}{6R_H} = \frac{25}{6 \times 1.09 \times 10^5} = 3.82 \times 10^{-5} \text{ cm} = 3820 \text{ Å}$$

4.6.4 Critical Potential

We know that an atom possesses number of stationary states. The lowest energy state is called *ground state*, while above it, states are termed as *excited states*. The uppermost state ($n = \infty$) having energy zero is called ionization state. The minimum energy (in eV) required to excite the electron/atom from its ground state is called *critical potential*. The critical potential is of two types: excitation potential and ionization potential. The energy (in eV) required to raise the electron/atom from its ground state position to excited state position, is termed as *excitation potential*. Similarly, the amount of energy (in eV) needed to send the electron/atom from its ground state to ionization state position is called *ionization potential*. In other words, the *ionization potential* is the amount of energy in electron volts to remove the electron from a given orbit to an infinite distance from the nucleus.

Suppose E_g, E_{ex} and E_∞ are the energies of ground, excited and ionization states. The excitation and ionization potentials can be determined with the following expressions:

Excitation potential = $E_{ex} - E_g$
Ionization potential = $E_\infty - E_g$

For the hydrogen atom, the energy of stationary states $n = 1, 2, 3, 4, ..., \infty$ are –13.6 eV, –3.40 eV, –1.51 eV, –0.85 eV, ..., 0 eV respectively. Then the excitation potentials will be E_2–E_1, E_3–E_1, E_4–E_1 and so on, while the ionization potential will be E_∞–E_1.

Ist excitation potential = $E_2 - E_1 = (-3.40) - (-13.6) = -3.40 + 13.6 = 10.20$ eV

IInd excitation potential = $E_3 - E_1 = (-1.51) - (-13.6) = -1.51 + 13.6 = 12.09$ eV

IIIrd excitation potential = $E_4 - E_1 = (-0.85) - (-13.6) = -0.85 + 13.6 = 12.75$ eV

Ionization potential = $E_\infty - E_1 = (0.00) - (-13.6) = 0.00 + 13.6 = 13.60$ eV

Similarly, the excitation potentials for the mercury are 4.9 eV and 6.7 eV, while the ionization potential is 10.4 eV.

EXAMPLE 4.3 Calculate the wavelength of photon that would ionize a hydrogen atom in the ground state and gives the ejected electron of a kinetic energy 10.5 eV.

Solution Energy of photon = Ionization energy + Kinetic energy of electron

$$E = 13.6 + 10.5 = 24.1 \text{ eV}$$

So, $$\lambda = \frac{hc}{E} = \frac{6.626 \times 10^{-34} \times 3 \times 10^{8}}{24.1 \times 1.6 \times 10^{-19}} = 0.5155 \times 10^{-7} \text{ m} = 515.5 \text{ Å}$$

4.6.5 Frank–Hertz Experiment

In 1914, J. Frank and G. Hertz were performed an experiment that gave an experimental justification to the Bohr concept of discrete energy levels or critical potentials. They were awarded by Nobel Prize in 1925 for this work. They were taken Hg under study. The energies of ground and first excited states of Hg are –10.42 eV and –5.44 eV. So, the first excitation potential will be 4.88 ≈ 4.9 eV. The experiment verifies this excitation potential and existence of stationary states for the atoms. The description and analysis of the experiment can be understood in following ways.

The experimental setup has a Tube T filled with mercury at pressure 1 mm of mercury and temperature 150 °C. The tube also consists of filament, grid and collector (plate). The filament is connected to battery B for the thermal emission of electrons. The grid and collector are connected to positive (*V*: accelerating voltage) and negative (*V′*: retarding voltage) terminals of batteries (Figure 4.13). The variation in accelerating potential (0–60 V) of grid at constant collector voltage (0.5 V) provides a non-linear graph in accelerating voltage and collector current having number of dips at a potential difference of ≈5 V. Initial dips are obtained at 4.9 V and 9.8 V as shown in Figure 4.14.

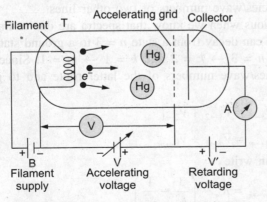

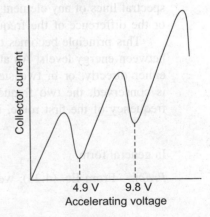

Figure 4.13   Frank–Hertz experimental setup.   Figure 4.14   Initial dips at 4.9 V and 9.8 V.

Explanation of Graph

When the accelerating grid potential increases, then the collector current increases because number of electron reaching to collector increases. The dip in current at 4.9 V suggests that most of the accelerated electrons loose its energy (4.9 eV) in inelastic collision with Hg atoms (in excitement of Hg) and unable to reach at collector. Before 4.9 V, the accelerated electrons do not loose its energy in collision (elastic collision) with medium (Hg) atoms and hence, current increases continuously before this voltage. This suggests that the excitation energy of Hg is 4.9 eV. Further continuous increase in collector current after 4.9 V is because of elastic collision of electrons with medium atoms. The second dip in current at 9.8 V indicates that the electron having energy 9.8 eV makes two consecutive inelastic collision with two Hg atoms and excites them. A similar concept can be understood for the later dips and continuous increase in current and voltage graphs.

During the process of inelastic collision, Hg atom is excited. Since the excited atom can reside in excited state only for a time 10^{-8} s, thus they come to ground state by the emission of photon. The emitted light through tube is found to be of wavelength of 2536 Å by spectroscopic technique. The energy ($E = hc/\lambda$) corresponding to this wavelength is equivalent to 4.9 eV. This proves excitation energy to be 4.9 eV for Hg. The analysis of graph shows that only definite amount of energy is required to excite the atom. Thus atom has discrete energy states, which is justification to Bohr concept atomic levels. By this experiment, one cannot distinguish whether it is excitement or ionization potential. This is the main cause of failure of the experiment. Later on, Davis and Goucher performed the same experiment with slight amendment (two-grid system arrangement). They found that the excitation potentials for the mercury are 4.9 eV and 6.7 eV, while the ionization potential is 10.4 eV.

4.6.6   Ritz Combination Principle

The Rydberg–Ritz combination principle is the theory proposed by Walter Ritz in 1908 to explain relationship of the spectral lines for all atoms. The principle states that the spectral lines of any element include frequencies/wave numbers that are either the sum or the difference of the frequencies/wave numbers of two other lines.

This principle becomes obvious when we know that spectra are due to transitions between energy levels. An atom can decay from a state $n = 3$ to a ground state $m = 1$ either directly, or in two steps $n = 3 \rightarrow k = 2$ and $k = 1 \rightarrow m = 1$. Since energy is conserved, the two frequencies/wave numbers of the latter route add to give the frequency of the first route, i.e.,

$$\overline{v}_{3 \rightarrow 1} = \overline{v}_{3 \rightarrow 2} + \overline{v}_{2 \rightarrow 1}$$

In general form,
$$\overline{v}_{n \rightarrow m} = \overline{v}_{n \rightarrow k} + \overline{v}_{k \rightarrow m}$$

Proof: From Eq. (4.28), we can write

$$\overline{v}_{n \rightarrow k} = R_\infty Z^2 \left(\frac{1}{k^2} - \frac{1}{n^2} \right)$$

$$\overline{v}_{k \rightarrow m} = R_\infty Z^2 \left(\frac{1}{m^2} - \frac{1}{k^2} \right)$$

So
$$\overline{v}_{n \to k} + \overline{v}_{k \to m} = R_\infty Z^2 \left(\frac{1}{k^2} - \frac{1}{n^2} \right) + R_\infty Z^2 \left(\frac{1}{m^2} - \frac{1}{k^2} \right)$$

$$\overline{v}_{n \to k} + \overline{v}_{k \to m} = R_\infty Z^2 \left(\frac{1}{m^2} - \frac{1}{n^2} \right)$$

$$\boxed{\overline{v}_{n \to k} + \overline{v}_{k \to m} = \overline{v}_{n \to m}} \qquad (4.29)$$

Example: Let $\overline{v}_\alpha$ and $\overline{v}_\beta$ are the wave numbers for the H_α and H_β spectral lines of hydrogen, then

$$\overline{v}_\alpha = \overline{v}_{3 \to 2} = R_\infty \left(\frac{1}{2^2} - \frac{1}{3^2} \right) \text{ and } \overline{v}_\beta = \overline{v}_{4 \to 2} = R_\infty \left(\frac{1}{2^2} - \frac{1}{3^2} \right)$$

So
$$\overline{v}_\beta - \overline{v}_\alpha = \overline{v}_{4 \to 2} - \overline{v}_{3 \to 2} = R_\infty \left(\frac{1}{2^2} - \frac{1}{4^2} \right) - R_\infty \left(\frac{1}{2^2} - \frac{1}{3^2} \right)$$

$$\overline{v}_\beta - \overline{v}_\alpha = \overline{v}_{4 \to 3} = R_\infty \left(\frac{1}{3^2} - \frac{1}{4^2} \right)$$

Thus, the difference of these wave numbers $\overline{v}_\alpha$ and $\overline{v}_\beta$ provides a new wave number which is the wave number of first line of *Paschen* series.

4.6.7 Inadequacy in Bohr's Theory

The Bohr atomic model clearly explains the emission and absorption spectra of hydrogen, but it has difficulty with, or else fails to explain the following facts.

1. The spectrum of hydrogen with high resolution spectroscope has five lines for the H_α spectral line, which is termed as *hydrogen fine structure*. Bohr atomic model was unable to explain the existence of *fine structure* and *hyperfine structure* in spectral lines.
2. The spectra of larger atoms were found to be unexplained by this model. The spectrum of some atoms has doublets and triplets spectral lines. Bohr's model does not provide the correct explanation for the presence of these lines.
3. This model predicts that multi-electron atoms do not have energy levels. Thus the theory does not work for multi-electron atoms or helium (neutral).
4. Bohr model assumes that only electron moves, while nucleus is stationary (having infinite mass), but actually both electron and nucleus (finite mass) move in closed orbit around the centre of mass.
5. This theory is unable to explain the Zeeman Effect (splitting of spectral line in presence of magnetic field) and Stark effect (splitting of spectral line in presence of electric field).
6. The model also violates the *uncertainty principle*. Because it considers that electrons have known orbits and definite radius which is opposite to *uncertainty principle*.

7. The model was fail to explain the rate at which electronic transitions occur. In other words, the selection rule for the electronic transitions was not given in this model.

8. The reason behind the variation in intensity of spectral lines was not explained by this model.

4.6.8 Effect of Nuclear Motion in Atomic Spectra or Correction in Bohr Model for the Finite Mass Nucleus

From the Bohr atomic model, we have seen that the wave number ($\bar{v} = 1/\lambda$) of emission/absorption spectral line for hydrogen like atom is given by the following formula:

$$\bar{v} = R_\infty Z^2 \left(\frac{1}{n_f^2} - \frac{1}{n_i^2} \right); \quad R_\infty = \left(\frac{me^4}{8\varepsilon_0^2 h^3 c} \right) \tag{4.30}$$

where n_i and $n_f (<n_i)$ are the initial and final states of electronic transition. If we consider the case of ionized helium (He$^+$), which have nuclear charge $+2e$ and atomic number $Z = 2$, then Eq. (4.30) for He$^+$ becomes

$$\bar{v} = R_\infty 2^2 \left(\frac{1}{n_f^2} - \frac{1}{n_i^2} \right) = 4R_\infty \left(\frac{1}{n_f^2} - \frac{1}{n_i^2} \right) \tag{4.31}$$

The transition from $n_i = 5, 6, 7, 8, \ldots$, to $n_f = 4$ for He$^+$ gives a series of spectral line known as *Pickering series*. The wave number of this series of lines can be written with the help of Eq. (4.31).

$$\bar{v} = 4R_\infty \left(\frac{1}{4^2} - \frac{1}{n_i^2} \right); \quad n_i = 5, 6, 7, 8, \ldots \tag{4.32}$$

$$\bar{v} = R_\infty \left(\frac{1}{4} - \frac{4}{n_i^2} \right) = R_\infty \left(\frac{1}{2^2} - \frac{1}{n^2} \right); \quad n = n_i/2 \qquad \text{for } n_i = 6, 8, \ldots \tag{4.33}$$

The wave number of Balmer series line has same expression as of Eq. (4.33). Hence alternate lines of *Pickering series* with $n_i = 6, 8, \ldots$ should have the same wave number/wavelength as the spectral lines of Balmer series of hydrogen. But experimentally, it is found that the lines of *Pickering series* have slightly higher wave number than the corresponding lines of Balmer series. This mismatch needs the correction in Bohr atomic model.

This dissimilarity is due to the fact that the nucleus was assumed to be stationary because of infinite mass in Bohr model. So, the correction in the model can be done if the nucleus is considered to be of finite mass.

Let the nucleus and electron have masses M and m respectively. Both nucleus and electron are rotating around the common axis (centre of mass) with angular velocity ω (Figure 4.15). The separation between them is r. If r_1 and r_2 are the distances of electron and nucleus from the axis, then we can write,

$$mr_1 = Mr_2$$

$$r_2 = \frac{m}{M} r_1$$

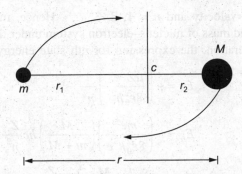

Figure 4.15 Motion of electron and nucleus around the common axis (C is the centre of mass.)

From Figure 4.15, we have

$$r = r_1 + r_2$$

So

$$r = r_1 + \frac{m}{M}r_1 = r_1\left(1 + \frac{m}{M}\right)$$

If L be the combined angular momentum of nucleus and electron, then

$$L = I_1\omega + I_2\omega = mr_1^2\omega + Mr_2^2\omega$$

$$L = mr_1^2\omega + M\left(\frac{m}{M}r_1\right)^2\omega \qquad \text{(using value of } r_2\text{)}$$

$$L = mr_1^2\omega + \frac{m}{M}m\,r_1^2\omega = mr_1^2\omega\left(1 + \frac{m}{M}\right)$$

$$L = m\frac{r^2}{(1 + m/M)^2}\,\omega\left(1 + \frac{m}{M}\right) \qquad \text{(using value of } r_1\text{)}$$

$$L = \frac{m}{(1 + m/M)}r^2\omega = \frac{mM}{(m + M)}r^2\omega$$

Suppose the quantity $\dfrac{m}{(1 + m/M)}$ or $\dfrac{mM}{(m + M)}$ is the reduced mass μ of the electron –nucleus system, then

$$L = \mu r^2\omega = I\omega \qquad (4.34)$$

where, the symbol I is moment of inertia. The expression of angular momentum indicates that rotating nucleus–electron system can be regarded as a single system/ particle having mass μ which is rotating with angular velocity ω on circular path of radius r (i.e., single particle of mass μ rotating around the stationary nucleus). In this condition, the Bohr quantum condition becomes

$$\mu r^2\omega = \mu v r = n\hbar \qquad (4.35)$$

where v is orbital velocity and $n = 1, 2, 3, \ldots$. Hence, mass of electron should be replaced by reduced mass of nucleus–electron system under the finite nucleus approach. Under this consideration, the expression for nth state energy becomes

$$E_n = -\left(\frac{\mu Z^2 e^4}{8\varepsilon_0^2 h^2}\right)\frac{1}{n^2}$$

$$E_n = -\left(\frac{me^4}{8\varepsilon_0^2 h^3 c}\right)\left(\frac{M}{m + M}\right)hc\frac{Z^2}{n^2}$$

$$E_n = -R_\infty\left(\frac{M}{m + M}\right)hc\frac{Z^2}{n^2}$$

Suppose
$$R_M = R_\infty\left(\frac{M}{m + M}\right) = \frac{R_\infty}{1 + m/M} \tag{4.36}$$

The *Rydberg constant* depends on mass of nucleus M. Since $M \gg m$, the difference in R_M and R_∞ will be very small. As M approaches to infinity, the R_M also tends to R_∞. The quantity R_∞ is called *Rydberg constant* for infinitely heavy nucleus. So

$$\boxed{E_n = -R_M hc\frac{Z^2}{n^2}} \tag{4.37}$$

If $\bar{v}$ is the wave number corresponding to electronic transition n_i to n_f, then

$$\bar{v} = \frac{1}{\lambda} = \frac{E_{n_i} - E_{n_f}}{hc} = R_M Z^2\left(\frac{1}{n_f^2} - \frac{1}{n_i^2}\right) = \frac{R_\infty}{1 + m/M}Z^2\left(\frac{1}{n_f^2} - \frac{1}{n_i^2}\right)$$

$$\bar{v} = \frac{R_\infty}{1 + m/M}Z^2\left(\frac{1}{n_f^2} - \frac{1}{n_i^2}\right) \tag{4.38}$$

Justification of Correction to Bohr Theory

1. The values of *Rydberg constant* for hydrogen R_H and helium R_{He} with Eq. (4.36) are 1.09677×10^7 m^{-1} and 1.09722×10^7 m^{-1} respectively. The R_H measured with spectroscopic technique justifies this result.

2. Another justification of R_M can be done by the evaluation of $m_e/m_p = m/M_H$ (ratio of electron and proton masses).

    Using Eq. (4.36), we can write

$$R_{He} = \frac{R_\infty}{1 + m/M_{He}} \text{ and } R_H = \frac{R_\infty}{1 + m/M_H}$$

So
$$\frac{R_{He}}{R_H} = \frac{1 + m/M_H}{1 + m/M_{He}}$$

$$\frac{R_{He}}{R_H} = \frac{1 + m/M_H}{1 + m/4M_H} \text{; as } M_{He} \approx 4M_H$$

$$R_{He} + \frac{R_{He}}{4}\frac{m}{M_H} = R_H + R_H\frac{m}{M_H}$$

$$R_{He} - R_H = \left(R_H - \frac{R_{He}}{4}\right)\frac{m}{M_H}$$

$$\frac{m}{M_H} = \frac{R_{He} - R_H}{(R_H - R_{He}/4)} \tag{4.39}$$

The calculated ratio of m and M_H with the help of spectroscopic R_H and R_{He} values is found to be approximately equal to 1/1836, which is found to be in good agreement with the results obtained by other methods.

3. The modified equation of wave number justifies the experimental values of *Pickering series*. Since $R_{He} > R_H$, the spectral line of the Pickering series of ionized helium have slightly higher wave numbers than the corresponding lines of the Balmer series of hydrogen.

Thus the modification made in Bohr theory based on nuclear motion has good evidences for its justification.

4.6.9 Discovery of Heavy Hydrogen

H.C. Urey and co-workers (1931) were observed the presence of faint companion lines on the short wavelength sides of the hydrogen spectral lines. On the basis of these faint lines, they had confirmed the existence of isotope of hydrogen. This isotope was termed as *deuterium*. By the measurement of wave number difference in faint and bright spectral lines, they predicted that atomic mass of deuterium is approximately twice of hydrogen ($M_D \approx 2 M_H$). The faintness of spectral line was explained on the ground that this isotope has low percentage in natural occurring hydrogen. Later on, it is found that the natural hydrogen has 99.985% $_1H^1$ (protium or hydrogen) and 0.015% $_1D^2$ (deuterium).

The wave number/wavelength difference between $_1H^1$ and $_1D^2$ can be theoretically determined on the basis of modified Bohr theory based on consideration of nuclear motion. According to this theory, the expression of wave number corresponding to an electronic transition can be written as:

$$\bar{v} = \frac{1}{\lambda} = R_M Z^2\left(\frac{1}{n_f^2} - \frac{1}{n_i^2}\right) = \frac{R_\infty}{1 + m/M}Z^2\left(\frac{1}{n_f^2} - \frac{1}{n_i^2}\right)$$

So, for hydrogen ($Z = 1$) and deuterium ($Z = 1$), the expression of wave number becomes

$$\bar{v}_H = \frac{1}{\lambda_H} = \frac{R_\infty}{1 + m/M_H}Z^2\left(\frac{1}{n_f^2} - \frac{1}{n_i^2}\right) \tag{4.40}$$

$$\bar{v}_D = \frac{1}{\lambda_D} = \frac{R_\infty}{1 + m/M_D}Z^2\left(\frac{1}{n_f^2} - \frac{1}{n_i^2}\right) \tag{4.41}$$

From Eqs. (4.40) and (4.41), we have

$$\Delta\lambda_{HD} = \lambda_H - \lambda_D = \frac{m/M_H - m/M_D}{R_\infty\left(\dfrac{1}{n_f^2} - \dfrac{1}{n_i^2}\right)} = \frac{m/M_H - m/2M_H}{\dfrac{1}{\lambda_H}}$$

$$\boxed{\Delta\lambda_{HD} = \lambda_H - \lambda_D = \lambda_H \frac{m}{2M_H} = \frac{\lambda_H}{3672}} \qquad (4.42)$$

For the α spectral line, $n_i = 3$, $n_f = 2$, $\lambda_H = 6563$ Å thus $\Delta\lambda_{HD} = 1.787$ Å and for β spectral line, $n_i = 4$, $n_f = 2$, $\lambda_H = 4681$ Å, thus $\Delta\lambda_{HD} = 1.274$ Å. A good concurrence between theoretical and experimental (by Urey and co-workers) $\Delta\lambda$ gives an evidence for the existence of deuterium.

Note: Hydrogen has three isotopes: protium ($_1H^1$), deuterium ($_1D^2$) and tritium ($_1T^3$). The difference of wavelength for the spectral lines of protium and tritium can be calculated with following expression:

$$\boxed{\Delta\lambda_{HD} = \lambda_H - \lambda_D = \lambda_H \frac{m}{2M_H} = \frac{\lambda_H}{3672}} \text{ and } \boxed{\Delta\lambda_{HT} = \lambda_H - \lambda_T = \lambda_H \frac{2m}{3M_H} = \frac{\lambda_H}{2754}}$$

For α and β spectral lines of Balmer series, $\Delta\lambda_{HT}$ are 2.383 Å and 1.274 Å respectively.

4.6.10 Bohr's Correspondence Principle

As we know that the new laws of physics under quantum approach (e.g. Wave theory of matter, quantum theory of radiation, relativistic electromagnetic theory, etc.) reduces to laws in classical approach when it is taken for a macroscopic system (a system of large number of particles). The similar thing is pointed out by Bohr in 1932 for the subparticle of atom (electron), which is known as Bohr correspondence principle. According to him, laws, of atomic physics describing the motion of electron transform to classical laws when they are considered macroscopically, i.e., the theory derived by Bohr switches into the classical theory of atomic physics for high quantum state (stationary state having large principle quantum number).

Classically, a revolving electron in circular path radiates electromagnetic radiation, whose frequency is equal to orbital frequency of electron. For the justification of his statement, he showed that the frequency of emitted radiation is same as orbital frequency of the electron, when n is large, i.e.,

$$\nu_{\text{transition}} = \nu_{\text{orbital}}$$

Proof: From Bohr's theory, the wave number of emitted radiation under transition from n_i to n_f is given by following expression:

$$\bar{\nu} = \frac{1}{\lambda} = R_\infty Z^2 \left(\frac{1}{n_f^2} - \frac{1}{n_i^2}\right)$$

$$\nu_{\text{transition}} = c\bar{\nu} = \frac{c}{\lambda} = R_\infty c Z^2 \left(\frac{1}{n_f^2} - \frac{1}{n_i^2}\right)$$

$$v_{\text{transition}} = R_\infty c Z^2 \left(\frac{n_i^2 - n_f^2}{n_f^2 n_i^2} \right)$$

$$v_{\text{transition}} = R_\infty c Z^2 \left(\frac{(n_i + n_f)(n_i - n_f)}{n_f^2 n_i^2} \right)$$

$$v_{\text{transition}} = R_\infty c Z^2 \left(\frac{(2n + 1)}{n^2 (n + 1)^2} \right); \; n_i = n + 1 \text{ and } n_f = n$$

$$v_{\text{transition}} = R_\infty c \left(\frac{2n}{n^4} \right) = \frac{2R_\infty c}{n^3}; \; n = \text{large and } Z = 1 \tag{4.43}$$

If electron is rotating in nth circular stationary state of radius r_n with velocity v_n, then

$$v_{\text{orbital}} = \frac{v_n}{2\pi r_n}$$

Since $\quad\quad r_n = \frac{n^2 h^2 \varepsilon_0}{\pi m e^2}$ and $v_n = \frac{e^2}{2 n h \varepsilon_0}$ $\quad$ [using Eqs. (4.13) and (4.15)]

So $\quad\quad v_{\text{orbital}} = \frac{1}{2\pi} \left(\frac{e^2}{2 n h \varepsilon_0} \right) \left(\frac{\pi m e^2}{n^2 h^2 \varepsilon_0} \right) = 2 \left(\frac{m e^4}{8 \varepsilon_0^2 h^3 c} \right) \frac{c}{n^3}$

Hence $\quad\quad v_{\text{orbital}} = \frac{2 R_\infty c}{n^3} \tag{4.44}$

From Eqs. (4.43) and (4.45), we obtain that

$$\boxed{v_{\text{transition}} = v_{\text{orbital}}} \tag{4.45}$$

Thus the proof justifies the Bohr's correspondence principle.

EXAMPLE 4.4 Show that for a linear harmonic oscillator the angular momentum (moment of linear momentum) is equal to product of n (an integer) and h (Planck constant).

Solution Consider a particle of mass m is oscillating on linear path harmonically. If x is its displacement at instant t, then its equation of motion can be written as:

$$\frac{d^2 x}{dt^2} + \omega^2 x = 0$$

where ω is angular frequency and is equal to $2\pi v$ (v is frequency of oscillation). The total energy of such system can be written as sum of its kinetic energy and potential energy.

$$E = \text{Kinetic energy} + \text{Potential energy}$$

$$E = \frac{1}{2} m \dot{x}^2 + \frac{1}{2} m \omega^2 x^2$$

$$E = \frac{p^2}{2m} + \frac{1}{2}m\omega^2 x^2; \quad \text{(because, } p = m\dot{x})$$

$$1 = \frac{p^2}{2mE} + \frac{x^2}{2E}m\omega^2$$

$$\frac{p^2}{2mE} + \frac{x^2}{2E/m\omega^2} = 1$$

This equation implies that the variation in p and x of the particle is an ellipse whose semi-major and minor axes are $a = \sqrt{2mE}$ and $b = \sqrt{2E/m\omega^2}$ respectively. As the particle completes one oscillation, the point on this ellipse also completes one revolution. The total area of this ellipse provides the angular momentum of particle for one complete oscillation.

$$J = \oint p\,dx = \pi ab = \pi\sqrt{2mE}\sqrt{\frac{2E}{m\omega^2}}$$

$$J = \oint p\,dx = \frac{2\pi E}{\omega} = \frac{2\pi E}{2\pi v} = \frac{E}{v} = \frac{nhv}{v} \quad \text{(as from Planck's law, } E = nhv)$$

$$\boxed{J = \oint p\,dx = nh}$$

Thus the total anular momentum of a particle for one complete oscillation is a constant and is equal to product of n and h.

4.7 SOMMERFELD'S ATOMIC MODEL

In 1916, Sommerfeld extended the Bohr atomic theory for the removal of its inadequacy. He suggested that the path of an electron around the nucleus is an ellipse with the nucleus at one foci and circular path is a special case of it. We know that the motion of planet around sun is elliptical due to being in motion under the gravitational attractive force $(f \propto 1/r^2)$. Similarly, an electron moves under the coulomb attractive force, which is also an inverse square force. So, path of electron should be elliptical.

When the electron moves on circular orbit, the angular co-ordinate is sufficient to describe its motion, but in the case of elliptical motion, position of an electron at any instant is described by two co-ordinates r (radial) and θ (angular). The position of a particle describing an elliptical path in a plane can be represented by polar co-ordinates (r, θ).

We know that for a linear harmonic oscillator, the angular momentum (moment of linear momentum) is equal to product of n (an integer) and h (Planck constant), i.e.,

$$\oint p\,dx = nh \tag{4.46}$$

Wilson and Sommerfeld generalised the above condition for any pair of periodically varying co-ordinate and its conjugate momentum. If q and p_q represent generalised co-ordinate and momentum respectively, then the generalise quantum condition becomes

$$\oint p_q\,dq = n_q h \tag{4.47}$$

Here n_q is an integer. For the case of elliptical motion, there exist two-pair conjugate variables, i.e., $(r, p_r = m\dot{r})$ and $(\theta, p_\theta = mr^2\dot{\theta})$. So for an elliptical motion, there exist following two quantum conditions:

$$\oint p_r dr = n_r h \tag{4.48}$$

$$\oint p_\theta\, d\theta = n_\theta h = kh \tag{4.49}$$

where n_r and n_θ are called as radial and azimuthal/orbital quantum numbers respectively. As a result, now we have two quantum numbers instead of one as in Bohr's theory. The sum of n_r and n_θ is also an integer quantity, say, n which is called principal quantum number.

Energy of Elliptical Orbits and Condition for n_r, n_θ and n

Let an electron of rest mass $m = m_0$ and charge $-e$ is rotating on an elliptical orbit around about nucleus of charge Ze which is located at one focus of the orbit (Figure 4.16). The total energy of such system can be written as sum of its kinetic energy K.E. and potential energy V.

$$E = \text{K.E.} + V$$

$$E = \frac{1}{2}mV^2 - \frac{Ze^2}{r}$$

$\because \qquad v^2 = \dot{r}^2 + r^2\dot{\theta}^2 \qquad \text{(for an elliptical orbit)}$

$\therefore \qquad E = \frac{1}{2}m(\dot{r}^2 + r^2\dot{\theta}^2) - \frac{Ze^2}{r}$

$$E = \frac{1}{2}m\dot{r}^2 + \frac{1}{2}mr^2\dot{\theta}^2 - \frac{Ze^2}{r}$$

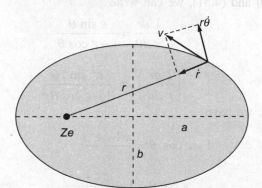

Figure 4.16 Elliptical motion of electron around nucleus.

Since $\dot{\theta} = \dfrac{p_\theta}{mr^2}$ and $\dot{r} = \dfrac{dr}{dt} = \dfrac{dr}{d\theta}\dfrac{d\theta}{dt} = \dfrac{dr}{d\theta}\dot{\theta} = \dfrac{dr}{d\theta}\dfrac{p_\theta}{mr^2}$

So
$$E = \frac{1}{2}m\left(\frac{dr}{d\theta}\frac{p_\theta}{mr^2}\right)^2 + \frac{1}{2}mr^2\left(\frac{p_\theta}{mr^2}\right)^2 - \frac{Ze^2}{r}$$

$$E = \frac{1}{2}m\left(\frac{dr}{d\theta}\frac{p_\theta}{mr^2}\right)^2 + \frac{1}{2}\frac{p_\theta^2}{mr^2} - \frac{Ze^2}{r}$$

By multiplying both sides by $\dfrac{2mr^2}{p_\theta^2}$, we have

$$\frac{2mr^2}{p_\theta^2}E = \frac{1}{2}m\left(\frac{dr}{d\theta}\frac{p_\theta}{mr^2}\right)^2\frac{2mr^2}{p_\theta^2} + 1 - \frac{Ze^2}{r}\frac{2mr^2}{p_\theta^2}$$

$$\frac{2mE}{p_\theta^2}r^2 = \left(\frac{1}{r}\frac{dr}{d\theta}\right)^2 + 1 - \frac{2mZe^2}{p_\theta^2}r$$

$$\left(\frac{1}{r}\frac{dr}{d\theta}\right)^2 = \frac{2mE}{p_\theta^2}r^2 - 1 + \frac{2mZe^2}{p_\theta^2}r \qquad (4.50)$$

Equation (4.50) is a differential equation of electron orbit moving around the nucleus. For the solution of this equation, let us consider the polar equation of ellipse,

$$\frac{1}{r} = \frac{1}{a}\frac{1-\varepsilon\cos\theta}{1-\varepsilon^2} \qquad (4.51)$$

where, quantity $\sqrt{1-\varepsilon^2}$ is equal to ratio of semi-minor and major axes (b/a). Differentiating Eq. (4.51) we have

$$-\frac{1}{r^2}\frac{dr}{d\theta} = \frac{1}{a}\frac{\varepsilon\sin\theta}{1-\varepsilon^2} \qquad (4.52)$$

From Eqs. (4.50) and (4.51), we can write

$$\frac{1}{r}\frac{dr}{d\theta} = \frac{\varepsilon\sin\theta}{1-\varepsilon\cos\theta}$$

$$\left(\frac{1}{r}\frac{dr}{d\theta}\right)^2 = \frac{\varepsilon^2\sin^2\theta}{(1-\varepsilon\cos\theta)^2} \qquad (4.53)$$

From Eq. (4.51), we have

$$1-\varepsilon\cos\theta = \frac{a(1-\varepsilon^2)}{r}$$

$$\varepsilon\cos\theta = 1 - \frac{a(1-\varepsilon^2)}{r}$$

So
$$\varepsilon^2\sin^2\theta = \varepsilon^2 - \varepsilon^2\cos^2\theta$$

$$\varepsilon^2\sin^2\theta = \varepsilon^2 - \left(1 - \frac{a(1-\varepsilon^2)}{r}\right)^2$$

$$\varepsilon^2 \sin^2 \theta = \varepsilon^2 - \left(1 + \frac{a^2(1 - \varepsilon^2)^2}{r^2} - \frac{2a(1 - \varepsilon^2)}{r} \right)$$

$$\varepsilon^2 \sin^2 \theta = -(1 - \varepsilon^2) - \frac{a^2(1 - \varepsilon^2)^2}{r^2} + \frac{2a(1 - \varepsilon^2)}{r}$$

Thus, $$\frac{\varepsilon^2 \sin^2 \theta}{(1 - \varepsilon \cos \theta)^2} = \frac{-(1 - \varepsilon^2) - \dfrac{a^2(1 - \varepsilon^2)^2}{r^2} + \dfrac{2a(1 - \varepsilon^2)}{r}}{\{a(1 - \varepsilon^2)/r\}^2}$$

$$\frac{\varepsilon^2 \sin^2 \theta}{(1 - \varepsilon \cos \theta)^2} = -\frac{r^2}{a^2(1 - \varepsilon^2)} - 1 + \frac{2r}{a(1 - \varepsilon^2)} \tag{4.54}$$

With help of Eqs. (4.53) and (4.54), we can write

$$\left(\frac{1}{r} \frac{dr}{d\theta} \right)^2 = -\frac{1}{a^2(1 - \varepsilon^2)} r^2 - 1 + \frac{2}{a(1 - \varepsilon^2)} r$$

$$\left(\frac{1}{r} \frac{dr}{d\theta} \right)^2 = -\frac{1}{b^2} r^2 - 1 + \frac{2a}{b^2} r \tag{4.55}$$

Equations (4.50) and (4.55) are identical to each other. Comparing the coefficients of r^2 and r, we get

$$\frac{2mE}{p_\theta^2} = -\frac{1}{b^2} \tag{4.56}$$

$$\frac{mZe^2}{p_\theta^2} = \frac{a}{b^2} \tag{4.57}$$

Dividing Eq. (4.56) with square of Eq. (4.57), we find that

$$\frac{2mE/p_\theta^2}{(mZe^2/p_\theta^2)^2} = -\frac{1/b^2}{(a/b^2)^2} = -\frac{b^2}{a^2}$$

$$\boxed{E = -mZ^2e^2 \frac{b^2/a^2}{p_\theta^2}} \tag{4.58}$$

And from Eq. (4.57), we can write

$$\boxed{a = \frac{mZe^2 b^2}{p_\theta^2}} \tag{4.59}$$

Equations (4.58) and (4.59) are the expressions of the energy and size of elliptical orbits in terms of angular momentum of electron. To find these expressions in terms of quantum numbers, we have to find the value of p_θ and b/a in terms of quantum numbers. For this, we have to solve the equations of quantum conditions.

From Eq. (4.49), we have

$$\int_0^{2\pi} p_\theta \, d\theta = kh$$

Since, according to classical mechanics the angular momentum of any isolated system is constant, thus

$$p_\theta \int_0^{2\pi} d\theta = kh$$

$$p_\theta 2\pi = kh$$

$$\Rightarrow \qquad p_\theta = kh/2\pi = k\hbar \qquad\qquad (4.60)$$

And from Eq. (4.48), we can write

$$\oint p_r dr = n_r h$$

$$\int_0^{2\pi} \left(m \frac{dr}{dt} \right) \left(\frac{dr}{d\theta} \, d\theta \right) = n_r h$$

$$\int_0^{2\pi} \left(m \frac{dr}{d\theta} \frac{d\theta}{dt} \right) \left(\frac{dr}{d\theta} \, d\theta \right) = n_r h$$

$$\int_0^{2\pi} \left(\frac{p_\theta}{r^2} \frac{dr}{d\theta} \right) \left(\frac{dr}{d\theta} \, d\theta \right) = n_r h \qquad \left(\text{Since } m\dot\theta = \frac{p_\theta}{r^2} \right)$$

$$p_\theta \int_0^{2\pi} \left(\frac{1}{r} \frac{dr}{d\theta} \right)^2 d\theta = n_r h$$

$$p_\theta \int_0^{2\pi} \frac{\varepsilon^2 \sin^2 \theta}{(1 - \varepsilon \cos \theta)^2} \, d\theta = n_r h \qquad \text{[From Eq. (4.54)]}$$

$$p_\theta 2\pi \left(\frac{1}{\sqrt{1 - \varepsilon^2}} - 1 \right) = n_r h$$

$$\frac{kh}{2\pi} 2\pi \left(\frac{1}{\sqrt{1 - \varepsilon^2}} - 1 \right) = n_r h$$

$$\left(\frac{1}{\sqrt{1 - \varepsilon^2}} \right) = \frac{n_r}{k} + 1$$

$$\sqrt{1 - \varepsilon^2} = \frac{k}{n_r + k}$$

Since $\sqrt{1 - \varepsilon^2} = \dfrac{b}{a}$ and $n_r + k = n$, thus

$$\frac{b}{a} = \frac{k}{n} \tag{4.61}$$

This equation indicates that electron can move only in those elliptical orbits for which ratio of minor to major axis is equal to ratio of two integers k and n. It also implies that azimuthal quantum number cannot have zero value because at this situation b becomes zero and ellipse becomes straight line which is not correct. So, azimuthal/orbital quantum number k can have values 1, 2, 3, 4, ... The radial quantum number, however, can take value zero for which the orbit is circular, i.e., n_r can have values 0, 1, 2, 3, 4, ... Therefore, $n(=n_r + k)$ can take values 1, 2, 3, 4, ... The path of electron is elliptical when $b < a$ or $k < n$ while path of electron becomes circular for $b = a$ or $k = n$. So, the azimuthal quantum number can take 1, 2, 3, ..., n values for a given principal quantum number. Thus, the theory predicts that for any given n value, there are n different quantised orbits for the electron corresponding to $k = 1, 2, 3, ..., n$.

Putting values of p_θ and b/a from Eqs. (4.60) and (4.61) in Eqs. (4.58) and (4.59), the expression of energy of elliptical orbit and semi-major axis takes the following form:

$$E = -\frac{2\pi^2 m Z^2 e^4}{n^2 h^2} \quad \text{(in C.G.S. system)} \tag{4.62a}$$

$$E_n = -\left(\frac{m Z^2 e^4}{8\varepsilon_0^2 n^2 h^2}\right) \quad \text{(in M.K.S. system)} \tag{4.62b}$$

and

$$a = \frac{n^2 h^2}{4\pi^2 m Z e^2 p_\theta^2} = a_0 \frac{n^2}{Z} \tag{4.63}$$

From Eq. (4.61), the expression of semi-minor axis can be written as:

$$b = a\frac{k}{n} = a_0 \frac{nk}{Z} \tag{4.64}$$

Equation (4.62) is the expression of energy of elliptical orbit under Sommerfeld's atomic model, which is exactly same as energy of Bohr's circular orbits. It means that energy of electron orbit depends only on the principal quantum number and is independent of azimuthal quantum number.

Equations (4.63) and (4.64) define the size and shape of elliptical orbits. These two equations define that there exists different quantised states/orbits for a given principal quantum number because azimuthal quantum number posses 1, 2, 3, ..., n values for a given principal quantum number n. The different orbits are assigned by s, p, d, f, g corresponding to $k = 1, 2, 3, 4, 5$ values. In case of hydrogen atom, the following types of orbits are found on the basis of shape–size expressions and conditions.

1. $n = 1$ $k = 1$ (s orbit) $a = a_0$ and $b = a_0$ (circular orbit)

2. $n = 2$ $k = 1$ (s orbit) $a = 4a_0$ and $b = 2a_0$ (elliptical orbit)
           $k = 2$ (p orbit) $a = 4a_0$ and $b = 4a_0$ (circular orbit)

3. $n = 3$ $k = 1$ (s orbit) $a = 9a_0$ and $b = 3a_0$ (elliptical orbit)
           $k = 2$ (p orbit) $a = 9a_0$ and $b = 6a_0$ (elliptical orbit)
           $k = 3$ (d orbit) $a = 9a_0$ and $b = 9a_0$ (circular orbit)

The pictorial presentations of these orbits are shown in Figure 4.17.

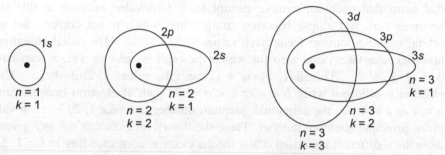

Figure 4.17   Change in shape of orbits for different combinations of n and k.

Yet the orbits/states for the electron are different, but the total energy for the given n is same, so these are called degenerate states. Furthermore, since the energy is same as predicted by Bohr, this means that Sommerfeld's elliptical model introduces no new energy levels and hence fails to explain the fine structure.

4.7.1   Sommerfeld's Relativistic Correction

For the removal of degeneracy in state and explanation of hydrogen fine structure, Sommerfeld introduced the concept of relativity in his atomic model. In the elliptical motion, the velocity of electron is large ($\approx c/137$) when it is close to nucleus (i.e., near perihelion), while it is relatively low when it is farther away from nucleus. In the relativistic approach ($v \to c$), the dynamic mass of electron, m is given by relation:
$m = \gamma m_0 = m_0/\sqrt{1 - \beta^2} = m_0/\sqrt{1 - (v/c)^2}$ (m_0 is rest mass, v is velocity of electron,

c is velocity of light, $\gamma = 1/\sqrt{1 - \beta^2}$). Thus the mass of electron is variable in elliptical motion and is large when it is close to the nucleus. The variation in mass of electron due to change in velocity at different points of its orbit causes a slow precession of the major axis in the plane of the ellipse about an axis through one of the foci. So, the trajectory of the electron becomes a rosette as shown in Figure 4.18.

Using the relativistic kinetic energy

$$(\text{K.E.} = E - m_0 c^2 = mc^2 - m_0 c^2 = m_0 c^2 \{(1/\sqrt{1 - \beta^2}) - 1\})$$

Sommerfeld obtained the following expression for the motion of electron in the case of variable velocity and mass (relativistic case) of electron.

$$\frac{1}{r} = \frac{1 + \varepsilon \cos \gamma \theta}{a(1 - \varepsilon^2)} \tag{4.65}$$

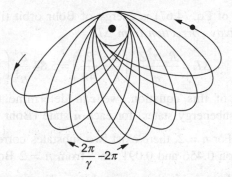

$$\frac{2\pi}{\gamma} - 2\pi$$

Figure 4.18   Motion of electron under Sommerfeld's relativistic model.

where γ is constant. For the non-relativistic case ($v \ll c$, $\gamma = 1$), r returns to the same value when θ changes by 2π. Hence the periodicity of r and θ are same for the non-relativistic case.

But in relativistic case ($\gamma < 1$), r does not return to the same value when θ changes by 2π, and it has different periodicity from θ. Since $\cos \gamma\{(2\pi/\gamma) + \theta\} = \cos \{2\pi + \gamma\theta\} = \cos \gamma\theta$, thus r becomes same only when θ changes by $2\pi/\gamma$ that is larger than 2π. Thus the axis of the ellipse advances through angle $2\pi/\gamma - 2\pi$ per revolution and perihelion of the orbit precesses in the same direction as rotation of electron. After the complex calculation under this approach, Sommerfeld obtained that the total energy of electronic state is not only function of principal quantum number n, but it also depends on azimuthal quantum number k. The expression of total energy is given by the following equation:

$$E_{n,k} = -\frac{2\pi^2 m_0 Z^2 e^4}{n^2 h^2}\left[1 + \frac{\alpha^2 Z^2}{n}\left(\frac{1}{k} - \frac{3}{4n}\right)\right]$$

$$E_{n,k} = -\frac{R_\infty hcZ^2}{n^2}\left[1 + \frac{\alpha^2 Z^2}{n}\left(\frac{1}{k} - \frac{3}{4n}\right)\right] \qquad (4.66)$$

where $R_\infty = \left(\frac{2\pi^2 m_0 e^4}{ch^3}\right)$ (in C.G.S.) $= \left(\frac{me^4}{8\varepsilon_0^2 h^3 c}\right)$ (in M.K.S.) $= 1.09678 \times 10^7$ m^{-1} and

fine structure constant $\alpha = 2\pi e^2/hc \approx 1/137$. Equation (4.66) indicates that there exists separate energy value for each set of n and k, so the problem of degeneracy has been removed under the Sommerfeld's relativistic approach. In the form of term value, the energy of state can be written as:

$$T = -\frac{E_{n,k}}{hc} = \frac{R_\infty Z^2}{n^2}\left[1 + \frac{\alpha^2 Z^2}{n}\left(\frac{1}{k} - \frac{3}{4n}\right)\right]$$

$$T = \frac{R_\infty Z^2}{n^2} + \frac{R_\infty \alpha^2 Z^4}{n^3}\left(\frac{1}{k} - \frac{3}{4n}\right) \qquad \text{(in cm}^{-1}\text{)} \qquad (4.67)$$

The first term of Eq. (4.67) is energy of Bohr orbit (in cm^{-1}), while the second term is called *relativity correction term* ΔT.

$$\Delta T = \frac{R_\infty \alpha^2 Z^4}{n^3}\left(\frac{1}{k} - \frac{3}{4n}\right) = 5.84 \frac{Z^4}{n^3}\left(\frac{1}{k} - \frac{3}{4n}\right) \text{cm}^{-1} \qquad (4.68)$$

On the basis of this equation, we can determine the shift of different k states (Sommerfeld subenergy states) for each n state (Bohr energy states).

Example 1: For $n = 2$, there exist two substates corresponding to $k = 1$ and 2 which are at separation 0.456 and 0.091 cm^{-1} from $n = 2$ Bohr orbit respectively.

Example 2: For $n = 3$, there exist three substates corresponding to $k = 1, 2$ and 3 which are at separation 0.162, 0.054 and 0.018 cm^{-1} from $n = 3$ Bohr orbit respectively.

4.7.2   Fine Structure of H_α Line under Sommerfeld's Relativistic Correction Model and its Shortcomings

The H_α spectral line in hydrogen spectrum is obtained when a transition takes place from $n = 3$ to $n = 2$. When this spectral line is seen by high resolution microscope, it is found to be composed of five lines which are named as Ia, IIa IIIa, IIb and IIIb. Out of five lines two components (Ia and IIb) are stronger and has separation of 0.3297 cm^{-1}, while the other three are weaker ones (as shown in Figure 4.19). Furthermore the IIb line is found to be stronger than the Ia line. This is called fine structure of H_α spectral line. At the first times it was obtained by Gerhard Hansen in 1925 {Annalen der Physik 78 (1925) 558} in his experiment.

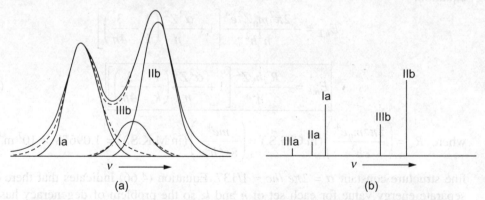

Figure 4.19   (a) Photometric curve and (b) fine structure of H_α line as measured by Gerhard Hansen in 1925.

Sommerfeld explained this fine structure in the following ways on the basis of his model. According to his model, there are two substates (orbitals) corresponding to $k = 1, 2$ for state $n = 2$ and there exist three substates corresponding to $k = 1, 2, 3$ for state $n = 3$. The pictorial presentation of these states along with their relative energy differences are shown in Figure 4.20.

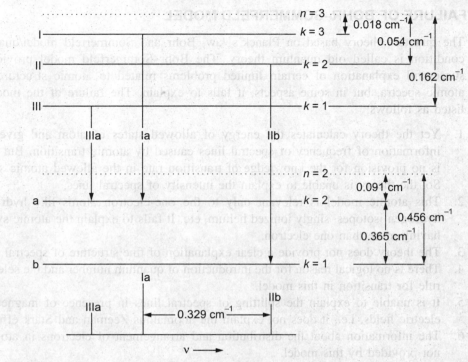

Figure 4.20 Electronic transitions among the Sommerfeld orbits for the fine structure of H_α-line.

The rule, under which the electronic transition can take place, is termed as *selection rule*. The selection rule for the transition from k levels of $n = 3$ to the k levels of $n = 2$ is $\Delta k = \pm 1$. Yet this selection rule was initially empirical, but justified later on the basis of correspondence principle and quantum mechanics. Under the selection rule, the three transitions take place that result three spectral lines (Figure 4.20).

1. $(n = 3, k = 3) \rightarrow (n = 2, k = 2)$; $\Delta k = -1$; Ia spectral line
2. $(n = 3, k = 2) \rightarrow (n = 2, k = 1)$; $\Delta k = -1$; IIb spectral line
3. $(n = 3, k = 1) \rightarrow (n = 2, k = 2)$; $\Delta k = +1$; IIIa spectral line

Sommerfeld found that the separation between two main components (Ia and IIb) is 0.329 cm^{-1}, which was in good agreement with the Hanson's experimental result. Except it, there are two main discrepancies.

1. There is no theoretical evidences for the lines IIa and IIIb because the transition $(n = 3, k = 2) \rightarrow (n = 2, k = 2)$ and $(n = 3, k = 1) \rightarrow (n = 2, k = 1)$ is forbidden under selection rule since $\Delta k = 0$ for these transitions.
2. The spectral line Ia is found to be stronger than IIb, which was reverse to the Hanson's experimental result.

For the removal of these discrepancies, the spin of electron and quantum mechanical relativistic approach was later on introduced.

4.8 FAILURE OF BOHR–SOMMERFELD MODEL

The quantum theory based on Planck's law, Bohr and Sommerfeld model/quantum condition is called old quantum theory. The Bohr–Sommerfeld model provides a successful explanation of certain limited problems related to atomic structure and atomic spectra, but in some aspects, it fails to explain. The failure of the model is listed as follows:

1. Yet the theory calculates the energy of allowed states of atom and gives the information of frequency of spectral lines caused by atomic transition. But there is no provision for the knowledge of transition rate in the allowed atomic states. So, the theory is unable to explain the intensity of spectral lines.
2. This atomic model is relevant only to the one-electron atoms like hydrogen, hydrogen isotopes, singly ionized helium, etc. It fails to explain the atomic system having more than one electron.
3. The theory does not provide a clear explanation of fine structure of spectral lines.
4. There is no logical reason for the introduction of quantum number and the selection rule for transition in this model.
5. It is unable to explain the splitting of spectral lines in presence of magnetic or electric fields, i.e., it does not explain the anomalous Zeeman and Stark effects.
6. The information about the distribution and arrangement of electrons in atoms is not provided by this model.

The solution of these problems came after the introduction of quantum mechanics in 1925 and onwards. In 1926, Schrödinger introduced his wave equation for the matter waves that presents a clear picture of atomic systems. When Schrödinger's wave equation is applied to the hydrogen atom, then three quantum numbers came into consideration which are known as principal quantum number n, azimuthal quantum number ℓ and magnetic quantum number m_ℓ. The azimuthal quantum number can have values 0, 1, 2, …, $n - 1$, while magnetic quantum number can have values 0, ± 1, ± 2, …, $\pm \ell$. The azimuthal quantum number is similar to orbital quantum number k introduced by Sommerfeld. This quantum number defines the quantisation of magnitude of orbital magnetic moment. The magnetic quantum number introduced a new concept, that not only the magnitude but also the orientation of orbital magnetic moment is quantised. This was termed as *space quantization*. On the basis of this concept, the problem of normal Zeeman Effect has been solved.

For the explanation of anomalous Zeeman Effect and alkali doublet spectra, the two Dutch Physicists, G.E. Uhlenbeck and S.A. Goudsmit proposed the *hypothesis of electron spin* that was confirmed by *Stern–Gerlach experiment* (1922). According to them, the electron has a spinning motion like top, i.e., the electron spins about its axis, while it also moves around the nucleus of atom in its orbit. On the basis of relativistic quantum mechanics, Dirac (1928) proved that an electron has intrinsic angular momentum and magnetic moment. The quantisation of spin angular momentum is called as *spin quantization*. This introduces a new quantum number known as spin quantum number s. Phipps and Taylor experiment in 1927, confirmed that s can have value 1/2. The quantisation of orbital and spin motion of electron both in magnitude and direction is called *spatial quantization*. The vector addition of both angular momentums provides the total angular momentum of atom, which is the basis of new atomic model known as **vector atom model**. This model presents a clear explanation of problems faced by old models.

4.9 MAGNETIC MOMENT OF ELECTRON

An electron has two types of motions: one is orbital and second is spin. Thus it has two types of angular momentums and magnetic moments.

Orbital Magnetic Moment

Let an electron of mass m is rotating in an orbit of radius r with speed v, and its time period of rotation is T. Since a rotating electron forms a current loop thus it has magnetic moment. The magnetic moment is defined as the product of current i and area A. So, the orbital magnetic moment μ_L can be written as:

$$|\boldsymbol{\mu_L}| = \mu_L = IA = \frac{e}{T} A = e\frac{1}{T} A = e\frac{v}{2\pi r}\pi r^2 = \frac{evr}{2} \qquad (4.69)$$

According to classical mechanics, the orbital angular momentum $|L|$ is defined as:

$$|\mathbf{L}| = L = mvr \qquad (4.70)$$

From Eqs. (4.69) and (4.70), we have

$$\frac{|\boldsymbol{\mu_L}|}{|\mathbf{L}|} = \frac{e}{2m} \qquad (4.71)$$

Equation (4.71) shows that the ratio of orbital magnetic moment and angular momentum is a constant which is known as orbital *gyromagnetic ratio* of electron. The vector form of Eq. (4.71) can be written as:

$$\boldsymbol{\mu_L} = -\frac{e}{2m}\mathbf{L} \qquad (4.72a)$$

$$\boldsymbol{\mu_L} = -g_\ell \frac{e}{2m}\mathbf{L}; \; g_\ell = 1 \qquad (4.72b)$$

The minus sign indicates that $\boldsymbol{\mu_L}$ is in opposite direction to $\mathbf{L}$. This is due the fact that electron is negatively charged. The term g_ℓ is called *orbital g factor*. In the quantum mechanics, the magnitude of $\mathbf{L}$ is equal to $\sqrt{\ell(\ell+1)}\,\hbar$. Here ℓ is *azimuthal quantum number* and $\hbar$ is Planck constant divided by 2π (i.e., $|\mathbf{L}| = \sqrt{\ell(\ell+1)}\,\hbar$). Using this, Eq. (4.71) can be written as:

$$|\boldsymbol{\mu_L}| = \frac{e}{2m}\sqrt{\ell(\ell+1)}\hbar = \sqrt{\ell(\ell+1)}\,\frac{e\hbar}{2m} = \sqrt{\ell(\ell+1)}\,\frac{eh}{4\pi m} \qquad (4.73)$$

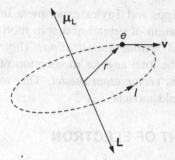

Figure 4.21 Direction of orbital angular momentum (L) and magnetic moment (μ_L) for the revolving electron in its orbit.

The quantity $eh/4\pi m$ is a constant quantity, called *Bohr magnetron* (μ_B). The numerical value of μ_B is equal to 0.928×10^{-23} amp-m² or 0.928×10^{-20} erg/gauss. It is the natural unit for the measurements of atomic magnetic moment. In the terms of μ_B, Eq. (4.73) becomes

$$|\mathbf{\mu_L}| = \mu_L = \sqrt{\ell(\ell+1)}\mu_B \qquad (4.74)$$

This is the expression of orbital magnetic moment.

Spin Magnetic Moment

Let $\mathbf{\mu_S}$ and $\mathbf{S}$ is spin magnetic moment and spin angular momentum respectively. Similar to Eq. (4.72b), the relation between these two quantities can be written as:

$$\mathbf{\mu_S} = -g_S \frac{e}{2m}\mathbf{S} \qquad (4.75)$$

Here g_s is called *spin g factor* which is equal to 2 because gyromagnetic ratio of electron spin is twice of orbital gyromagnetic ratio. The minus sign indicates that $\mathbf{\mu_S}$ acts in opposite direction to $\mathbf{S}$. From Eq. (4.75), we can write

$$|\mathbf{\mu_S}| = g_S \frac{e}{2m}|\mathbf{S}| \qquad (4.76)$$

In the quantum mechanics, the magnitude of $\mathbf{S}$ is equal to $\sqrt{s(s+1)}\hbar$ (i.e., $|\mathbf{S}| = \sqrt{s(s+1)}\hbar$). Here s is called *spin quantum number*. Using this, Eq. (4.76) can be re-written as:

$$|\mathbf{\mu_S}| = 2\frac{e}{2m}\sqrt{s(s+1)}\hbar = 2\sqrt{s(s+1)}\frac{eh}{4\pi m}$$

$$\boxed{|\mathbf{\mu_S}| = \mu_S = 2\sqrt{s(s+1)}\ \mu_B} \qquad (4.77)$$

This is the expression of spin magnetic moment. Since spin quantum number can have sole value equal to 1/2. So,

$$\mu_S = 2\sqrt{\frac{1}{2}\left(\frac{1}{2}+1\right)}\ \mu_B = \sqrt{3}\ \mu_B \qquad (4.78)$$

Thus the spin magnetic moment of electron is equal to $\sqrt{3}$ times of Bohr magnetron.

4.10 LARMOR PRECESSION

When an atom is subjected in external magnetic field, the electron orbit or orbital angular momentum precesses around about the field axis or direction. This precession is called *larmor precession*, and the frequency of this precession is called *Larmor frequency*.

Explanation

A rotating electron in an orbit behaves as current loop thus atom behaves as atomic dipole whose magnetic moment and angular momentum are μ_L and L respectively. Both μ_L and L operate in opposite direction perpendicular to plane of electron rotation as shown in Figure 4.22. When this atomic dipole is placed in magnetic field B, then a torque ($\tau = \mu_L \times B$) acts on it which is perpendicular to angular momentum. The acting torque causes a change in angular momentum because it is equal to rate of change of angular momentum (i.e., $\tau = dL/dt$; dL is change in L in time dt). The change in angular momentum dL acts perpendicular to L because it is in direction of τ which is perpendicular to L. Due to this reason, the direction of angular momentum changes with time, while its magnitude remains constant. As a result L traces a cone around B. This precession of angular momentum is called *Larmor precession*.

Suppose plane of electron orbit makes an angle θ with the direction of magnetic field, hence angle between L or μ_L and B will be also L. Let L precesses around B with angular velocity ω, and θ is the angle between μ_L and B. If L turns around with an angle $d\phi$ in time dt in plane perpendicular to B, then $d\phi = \omega dt$.

Since
$$\text{Angle} = \frac{\text{Arc}}{\text{Radius}}$$

Thus,
$$d\phi = \frac{|dL|}{|L|\sin\theta} = \frac{dL}{L\sin\theta}$$

$$\omega dt = \frac{dL}{L\sin\theta}$$

$$\omega = \frac{dL}{dt} \times \frac{1}{L\sin\theta}$$

$$\omega = \frac{\tau}{L\sin\theta} = \frac{|\mu_L \times B|}{L\sin\theta}$$

$$\omega = \frac{\mu_L B\sin\theta}{L\sin\theta} = \frac{\mu_L}{L}B$$

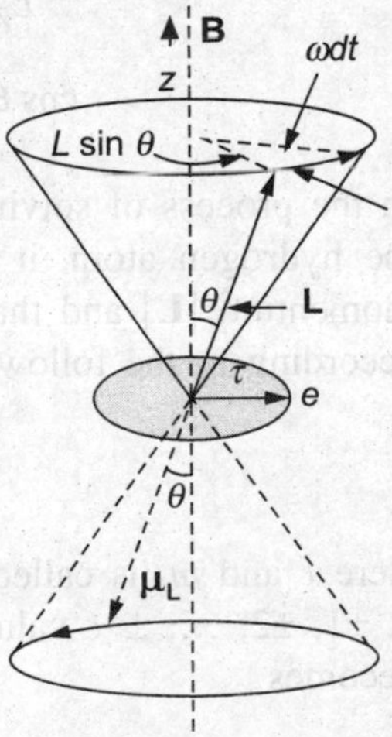

Using Eq. (4.71), we can write

Figure 4.22 Precession of orbital angular momentum in influence of magnetic field.

$$\boxed{\omega = \frac{e}{2m}B} \tag{4.79}$$

Here ω is also called angular frequency of Larmor precession. If ν is Larmor frequency, then $\omega = 2\pi\nu$. So from Eq. (4.79), we have

$$\nu = \frac{e}{4\pi m} B \qquad (4.80)$$

Equation (4.80) shows that the Larmor frequency is independent of the angle between $\mathbf{L}$ or $\mathbf{\mu_L}$ and $\mathbf{B}$.

4.11 SPACE QUANTIZATION

When the atom is placed in magnetic field, the orbital angular momentum can have number of orientation around the field direction. The discrete possible orientation of orbital angular momentum in magnetic field is called *space quantization.*

Explanation

The orbital angular momentum $\mathbf{L}$ of an atom precesses around the magnetic field $\mathbf{B}$ in the shape of cone such that an angle between $\mathbf{L}$ and $\mathbf{B}$ remains constant, when the atom is placed in external magnetic field. This precession is called *Larmor precession.* Let the magnetic field is along the Z-axis direction and $\mathbf{L}$ makes an angle θ with it. Let L_Z is component of $|\mathbf{L}|$ along the magnetic field direction, then from Figure 4.23, we can write.

$$L_Z = |\mathbf{L}| \cos\theta$$

$$\cos\theta = \frac{L_Z}{|\mathbf{L}|} \qquad (4.81)$$

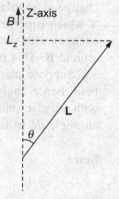

Figure 4.23 Orientation of **L** in magnetic field.

In the process of solving the Schrödinger's equation for the hydrogen atom, it is found that the orbital angular momentum $|\mathbf{L}|$ and that the z-component of the angular momentum L_Z is quantised according to the following relationship:

$$|\mathbf{L}| = \sqrt{\ell(\ell+1)}\, \hbar \qquad (4.82)$$

$$L_z = m_\ell \hbar \qquad (4.83)$$

Here ℓ and m_ℓ is called *azimuthal* and *magnetic quantum numbers.* The m_ℓ can have $0, \pm 1, \pm 2, \ldots, \pm \ell$ values. Putting the values from Eqs. (4.82) and (4.83), Eq. (4.81) becomes

$$\cos\theta = \frac{m_\ell}{\sqrt{\ell(\ell+1)}} \qquad (4.84)$$

Since m_ℓ can have $(2\ell + 1)$ values for a given ℓ, thus angle θ also have $(2\ell + 1)$ discrete values. Therefore, $\mathbf{L}$ have $(2\ell + 1)$ discrete orientation in the magnetic field. This orientation quantisation of $\mathbf{L}$ is termed as *space quantization.* Since $|m_\ell|$ is always smaller than $\sqrt{\ell(\ell+1)}$, thus **L** can never be aligned exactly parallel or anti-parallel to field direction, i.e., θ cannot have values $0°$ or $180°$.

Examples:

1. When $\ell = 1$, then $m_\ell = 0, \pm 1$

    So, $\cos \theta = +\dfrac{1}{\sqrt{2}}, 0, -\dfrac{1}{\sqrt{2}} \implies \theta = 45°, 90°, 135°$

2. When $\ell = 2$ then $m_\ell = 0, \pm 1, \pm 2$

    So, $\cos \theta = +\dfrac{2}{\sqrt{6}}, \dfrac{1}{\sqrt{6}} + 0, -\dfrac{1}{\sqrt{6}}, -\dfrac{2}{\sqrt{6}} \implies \theta = 35°, 66°, 90°, 114°, 145°$

Hence, for $\ell = 1$ and $\ell = 2$, there are three and five discrete orientations of **L** respectively (Figures 4.24 and 4.25).

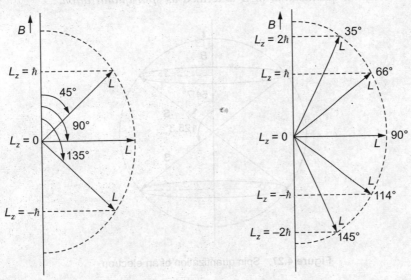

| **Figure 4.24** Space quantization for $\ell = 1$. | **Figure 4.25** Space quantization for $\ell = 2$. |

4.12 SPIN QUANTIZATION

The discrete possible orientation of spin angular momentum **S** in magnetic field is called *space quantization*. Let the magnetic field **B** is along the Z-axis direction, and **S** makes an angle θ with it. Let S_Z is component of $|\mathbf{S}|$ along the magnetic field direction, then from Figure 4.26 we can write

$$S_Z = |\mathbf{S}| \cos \theta$$

$$\cos \theta = \frac{S_Z}{|\mathbf{S}|} \qquad (4.85)$$

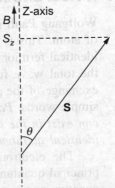

Figure 4.26 Orientation of **S** in magnetic field **B**.

Similar to the orbital angular momentum, the spin angular momentum $|S|$ and its Z-component S_Z is also quantised according to the following relationship:

$$|S| = \sqrt{s(s+1)}\,\hbar \tag{4.86}$$

$$S_z = m_s \hbar \tag{4.87}$$

Here s and m_s is called spin and spin magnetic quantum numbers. The s is equal to 1/2, while m_s can have $\pm 1/2$ values. Putting the values from Eqs. (4.82) and (4.83), the Eq. (4.85) becomes

$$\boxed{\cos\theta = \frac{m_s}{\sqrt{s(s+1)}}} \tag{4.88}$$

Since m_s have two values $\pm 1/2$, thus angle θ also have two discrete values (54.7° and 125.3°). Therefore, S have two discrete orientations in the magnetic field (Figure 4.27). This orientation quantisation of S is termed as *spin quantization*.

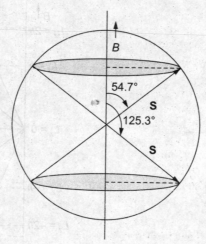

Figure 4.27   Spin quantization of an electron.

4.13 PAULI EXCLUSION PRINCIPLE

Wolfgang Pauli in 1925 gave a fundamental principle for the electronic configuration of atom. This is a quantum mechanical principle. According to this principle, no two identical fermions may occupy the same quantum state simultaneously. In other words, the total wave function for two identical fermions is anti-symmetric with respect to exchange of the particles. The electron is a half integral particle, so it is a fermion. In simple words, *Pauli Exclusion Principle states that no two electrons in a single atom can exist in the same quantum states, i.e. no two electrons in a single atom can have identical quantum numbers.*

The electron configuration is defined in terms of four quantum numbers as, principal quantum number n, azimuthal or orbital quantum number ℓ, orbital magnetic quantum number m_ℓ and spin magnetic quantum number m_s. Pauli's exclusion principle indicates that two electrons may have three same quantum numbers, but the fourth quantum number must be different. Thus, if n, l, and m_l are the same for the two electrons in an atom, m_s must be different such that the electrons have opposite spins (Table 4.3).

Table 4.3   Quantum Number for s Orbital System of Electrons

Electron	n	ℓ	m_ℓ	m_s
e_1	1	0	0	+1/2
e_2	1	0	0	−1/2

Since there are only two possible values of spin magnetic quantum number, therefore, third electron cannot be accommodated in s-orbital. This clearly points out that Pauli's exclusion principle can also be stated as: "An orbital cannot accommodate more than two electrons, i.e., maximum number of electrons that an orbital can accommodate is two."

Explanation

Let a system have two fermions. The fermions are those particles which have half integral spin and are identical. If particles 1 and 2 are in states a and b respectively, then wave function for the system can be written as:

$$\psi_{ab}(1, 2) = \psi_a(1)\,\psi_b(2) \tag{4.89}$$

If particles interchange the states (i.e., particle 1 goes to state b and particle 2 comes in state a), then the wave function of system would be

$$\psi_{ba}(1, 2) = \psi_b(1)\,\psi_a(2) \tag{4.90}$$

Since the particles are identical, they are indistinguishable. Due to being indistinguishable particles, one cannot say which wave function either $\psi_{ab}(1, 2)$ or $\psi_{ab}(2, 1)$ describes the system at any moment. For the complete information of system, the total wave function must be a linear combination of these two wave functions:

$$\psi_{ab}(1, 2) = N'[\psi_a(1)\,\psi_b(2) \pm \psi_b(1)\,\psi_a(2)] \tag{4.91}$$

Here N' is normalisation constant and has value $1/\sqrt{2}$. Since the fermions are anti-symmetrical with respect to the exchange of co-ordinates, the particles of half-integer spin must have anti-symmetric wave function. Therefore, for anti-symmetric wave function, Eq. (4.91) must be in following form:

$$\psi(1, 2) = \frac{1}{\sqrt{2}}[\psi_a(1)\,\psi_b(2) - \psi_b(1)\,\psi_a(2)] \tag{4.92}$$

or
$$\psi(1, 2) = \frac{1}{\sqrt{2}}\begin{vmatrix} \psi_a(1) & \psi_a(2) \\ \psi_b(1) & \psi_b(2) \end{vmatrix} \tag{4.93}$$

Equation (4.92) clearly explains that, if both particles exist in same quantum state, then $\psi(1, 2)$ will be equal to zero. Thus, two non-interacting Fermi particles cannot be in the same quantum state simultaneously, i.e., they both cannot be described by the same set of quantum numbers. Let a system contain n states that occupies the N

number of Fermi particles. The wave function for this system can be written in the following form with the help of Eq. (4.93):

$$\psi(1, 2) = \frac{1}{\sqrt{N}} \begin{vmatrix} \psi_a(1) & \psi_a(2) & \cdots & \psi_a(N) \\ \psi_b(1) & \psi_b(2) & \cdots & \psi_b(N) \\ \vdots & \vdots & \vdots & \vdots \\ \psi_n(1) & \psi_n(2) & \cdots & \psi_n(N) \end{vmatrix} \qquad (4.94)$$

The determinant in Eq. (4.94) is called Slater determinant. This determinant vanishes if any two rows are identical. This also indicates that more than one Fermi particles cannot be in the same quantum state simultaneously.

4.14 NUMBER OF ELECTRONS IN SHELL/ORBIT AND SUB-SHELL/ORBITAL

The quantum state defined by quantum number ℓ has $(2\ell + 1)$ orbital magnetic quantum number, hence it has $(2\ell + 1)$ sub-shell. According to Pauli's exclusion principle, an orbital or sub-shell can have maximum two electrons because there are two possible values of spin quantum number. Therefore, a quantum state defined by quantum number ℓ can have maximum $2(2\ell + 1)$ electrons. The sub-shells corresponding to $\ell = 0, 1, 2, 3, 4, \ldots$ are denoted by $s, p, d, f, g, \ldots$ respectively. So, sub-shells $s, p, d, f, g, \ldots$ can have maximum 2, 6, 10, 14, 18, ... number of electrons respectively (Table 4.4).

Table 4.4 Distribution of Electron in Subshell

Sub-shell	ℓ	m_ℓ	m_s	spin	Maximum electron $2(2\ell + 1)$	Notation of state
s	0	0	$\pm 1/2$	↑↓	2	s^2
		1	$\pm 1/2$	↑↓		
p	1	0	$\pm 1/2$	↑↓	6	p^6
		-1	$\pm 1/2$	↑↓		
		2	$\pm 1/2$	↑↓		
		1	$\pm 1/2$	↑↓		
d	2	0	$\pm 1/2$	↑↓	10	d^{10}
		-1	$\pm 1/2$	↑↓		
		-2	$\pm 1/2$	↑↓		
		3	$\pm 1/2$	↑↓		
		2	$\pm 1/2$	↑↓		
		1	$\pm 1/2$	↑↓		
f	3	0	$\pm 1/2$	↑↓	14	f^{14}
		-1	$\pm 1/2$	↑↓		
		-2	$\pm 1/2$	↑↓		
		-3	$\pm 1/2$	↑↓		

The quantum state defined by principal quantum number n has n subquantum states described by orbital quantum number $\ell = 0, 1, 2, 3, 4, \ldots (n - 1)$. Since each subquantum state defined by quantum number ℓ have maximum $2(2\ell + 1)$ electrons. Therefore, maximum number of electrons (M_n) in a state defined by quantum number n is equal to the sum of electrons in the constituent n subquantum states defined by quantum number ℓ.

$$M_n = 2 \sum_{\ell=0}^{\ell=n-1} (2l+1) = 2 \times [1 + 3 + 5 + 7 + \cdots + \{2(n-1)+1\}]$$

$$M_n = 2 \times [1 + 3 + 5 + 7 + \cdots + (2n+1)] = 2 \times \frac{n}{2}[1 + (2n-1)] = 2n^2 \quad (4.95)$$

The *principal quantum number* defines the shell or orbit of an atom. The atomic shells are denoted by K, L, M, N, O corresponding to $n = 1, 2, 3, 4, 5$ respectively. According to Eq. (4.95), the shells $K, L, M, N, \ldots$ can have 2, 8, 18, 32, … maximum number of electrons respectively Table 4.5.

Table 4.5 Distribution of Maximum Number of Electrons in Shell

Shell	n	ℓ	Notation of state	Maximum electron $2n^2$
K	1	0	$1s^2$	2
L	2	0	$2s^2$	8
		1	$2p^6$	
M	3	0	$3s^2$	18
		1	$3p^6$	
		2	$3d^{10}$	
O	4	0	$4s^2$	32
		1	$4p^6$	
		2	$4d^{10}$	
		3	$4f^{14}$	

The *Hund rule* states that electrons are placed in individual orbitals before they are paired up except s orbital. On the basis of above results and Hund's rule, the electronic configuration (distribution of electrons in various sub-shells around the nucleus of atom) of some elements can be written in following manner (Table 4.6).

Table 4.6 Electronic Configuration of Elements

Atom	Z	Electronic configuration	Orbitals filling with the electrons
H	1	$1s^1$	↑ 1s
He	2	$1s^2$	↓↑ 1s
Li	3	$1s^2, 2s^1$	↓↑ 1s ↑ 2s

(Contd.)

Table 4.6 Electronic Configuration of Elements (*Contd.*)

Atom	Z	Electronic configuration	Orbitals filling with the electrons
Be	4	$1s^2, 2s^2$	[↓↑]$_{1s}$ [↓↑]$_{2s}$
B	5	$1s^2, 2s^2, 2p^1$	[↓↑]$_{1s}$ [↓↑]$_{2s}$ [↑][ ][ ]$_{2p}$
C	6	$1s^2, 2s^2, 2p^2$	[↓↑]$_{1s}$ [↓↑]$_{2s}$ [↑][↑][ ]$_{2p}$
N	7	$1s^2, 2s^2, 2p^3$	[↓↑]$_{1s}$ [↓↑]$_{2s}$ [↑][↑][↑]$_{2p}$
O	8	$1s^2, 2s^2, 2p^4$	[↓↑]$_{1s}$ [↓↑]$_{2s}$ [↓↑][↑][↑]$_{2p}$
F	9	$1s^2, 2s^2, 2p^5$	[↓↑]$_{1s}$ [↓↑]$_{2s}$ [↓↑][↓↑][↑]$_{2p}$
Ne	10	$1s^2, 2s^2, 2p^6$	[↓↑]$_{1s}$ [↓↑]$_{2s}$ [↓↑][↑↑][↓↑]$_{2p}$
Na	11	$1s^2, 2s^2, 2p^6, 3s^1$	[↓↑]$_{1s}$ [↓↑]$_{2s}$ [↓↑][↓↑][↓↑]$_{2p}$ [↑]$_{3s}$
Mg	12	$1s^2, 2s^2, 2p^6, 3s^2$	[↓↑]$_{1s}$ [↓↑]$_{2s}$ [↓↑][↓↑][↓↑]$_{2p}$ [↓↑]$_{3s}$
Al	13	$1s^2, 2s^2, 2p^6, 3s^2, 3p^1$	[↓↑]$_{1s}$ [↓↑]$_{2s}$ [↓↑][↓↑][↓↑]$_{2p}$ [↓↑]$_{3s}$ [↑][ ][ ]$_{3p}$
Si	14	$1s^2, 2s^2, 2p^6, 3s^2, 3p^2$	[↓↑]$_{1s}$ [↓↑]$_{2s}$ [↓↑][↓↑][↓↑]$_{2p}$ [↓↑]$_{3s}$ [↑][↑][ ]$_{3p}$
P	15	$1s^2, 2s^2, 2p^6, 3s^2, 3p^3$	[↓↑]$_{1s}$ [↓↑]$_{2s}$ [↓↑][↓↑][↓↑]$_{2p}$ [↓↑]$_{3s}$ [↑][↑][↑]$_{3p}$

Similar to above, the electronic configuration of all the known atoms can be written.

4.15 VECTOR ATOM MODEL

An electron possesses two types of angular momentum: orbital and spin angular momentum ($\mathbf{L}$ and $\mathbf{S}$). These angular momentums act opposite to their corresponding magnetic moments ($\boldsymbol{\mu_L}$ and $\boldsymbol{\mu_S}$), i.e.,

$$\boldsymbol{\mu_L} = -g_L \frac{e}{2m} \mathbf{L}; \; g_L = 1 \tag{4.96}$$

$$\boldsymbol{\mu_S} = -g_S \frac{e}{2m} \mathbf{S}; \; g_S = 2 \tag{4.97}$$

Quantum mechanically, the magnitude of these two momentum vectors and their z-components are defined by following expressions:

$$|\mathbf{L}| = \sqrt{\ell(\ell+1)}\hbar \text{ and } L_Z = m_\ell \hbar \qquad (4.98)$$

$$|\mathbf{S}| = \sqrt{s(s+1)}\hbar \text{ and } S_Z = m_s \hbar \qquad (4.99)$$

where ℓ and m_ℓ are called orbital and magnetic quantum numbers. The m_ℓ can have 0, ±1, ±2, ..., ± ℓ values. The s and m_s are called spin and spin magnetic quantum numbers. The s is equal to 1/2, while m_s can have ±1/2 values.

Lande proposed a new atomic model, known as *Lande's vector atom model*. According to this model, *the total angular momentum of an atom* $\mathbf{J}$ *is a combination of orbital and spin angular momentum of its electrons*. Since total angular momentum $\mathbf{J}$ is a vector sum of $\mathbf{L}$ and $\mathbf{S}$ (Figure 4.28), it is termed as *vector atom model*.

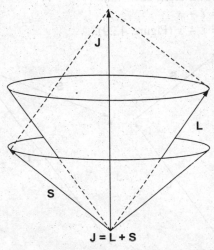

Figure 4.28 Vector representation of Lande's vector atom model.

Reason of coupling: The orbital motion of an electron produces an internal magnetic field in which spinning electron is present with a magnetic moment. In the similar way, an electron having orbital magnetic moment is also present in the internal magnetic field produced by spin motion. Thus, they exert torque on each other or interact magnetically. Such interaction is called spin–orbit interaction. These internal torques do not change the magnitude of momentum vectors, but it causes them to precess uniformly around their resultant vector $\mathbf{J}$. Hence the spin–orbit interaction causes the coupling of $\mathbf{L}$ and $\mathbf{S}$ to give the resultant angular momentum of atom. The vector $\mathbf{J}$ is conserved in magnitude and direction in absence of external torque.

According to vector atom model, we have

$$\mathbf{J} = \mathbf{L} + \mathbf{S} \qquad (4.100)$$

Similar to Eqs. (4.98) and (4.99), we can write the magnitude of $\mathbf{J}$ and its z-component in following form:

$$|\mathbf{J}| = \sqrt{j(j+1)}\ \hbar \text{ and } J_Z = m_j \hbar \qquad (4.101)$$

Here j is called total angular momentum/inner quantum number and m_j is termed as corresponding magnetic quantum number. The m_j can have 0, ±1, ±2, ..., ±j values.

Since (a) coupling results the formation of **J** and hence J_z, (b) ℓ , s, j, m_ℓ, m_s and m_j are scalar, and (c) S_z and m_s can positive or negative, thus we may write the following expressions:

$$J_Z = L_Z + S_Z \qquad (4.102)$$

and

$$m_j = m_\ell \pm m_s \qquad (4.103)$$

Since the maximum values of m_ℓ, m_s and m_j are ℓ, s and j respectively, Eq. (4.103) takes the following form:

$$j = \ell \pm s \qquad (4.104)$$

Equation (4.104) indicates that there are only two relative orientations of **J** which is possible for one-electron atom such that

(a) $|\mathbf{J}| > |\mathbf{L}|$ for $j = \ell + s$
(b) $|\mathbf{J}| < |\mathbf{L}|$ for $j = \ell - s$ (Figure 4.29).

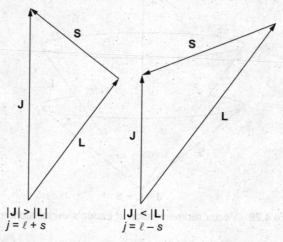

Figure 4.29 Vector representation of spin-orbit interaction.

Therefore, by spin–orbit interaction, each energy level of a given ℓ is splitted into two sublevels corresponding to $j = \ell + (1/2)$ and $j = \ell - (1/2)$ except $\ell = 0$. For $\ell = 0$, only one level is possible corresponding to $j = 1/2$ because $j = -1/2$ is not possible. The number of different possible values of j is known as *multiplicity*, which is equal to $(2s + 1)$. On the basis of it, new notations are defined for an energy level corresponding to an electron system. The energy levels of an electron is called *term*. For one-electron system, the multiplicity is 2, known as *doublet term*. The terms are denoted by following notation:

Energy level notation (spectroscopic term) = $^{2s+1}X_j$

where X is equal to S, P, D, F, ... corresponding to $\ell = 0, 1, 2, 3, ...$ respectively. The complete notation of energy level is expressed by writing principal quantum number before the term, i.e., $n\ ^{2s+1}X_j$. For the one-electron system, the terms are shown in Table 4.7.

Table 4.7 Spectroscopic Term (Energy Level Notation) for Different Orbitals

Term	ℓ	j	Energy level notation
S	0	1/2	$^2S_{1/2}$
P	1	1/2, 3/2	$^2P_{1/2}, {}^2P_{3/2},$
D	2	3/2, 5/2	$^2D_{3/2}, {}^2D_{5/2},$
F	3	5/2, 7/2	$^2F_{5/2}, {}^2F_{7/2},$

The notation of energy level for a system of more than one electron and selection rule for electron transition are defined under the *L-S* and *J-J* coupling scheme. On the basis of these new energy levels and selection rules, the fine structure spectra and alkali doublet spectra can be well explained.

Needed quantum number in vector atom model: On the basis of above discussion, we can say that the complete information of energy states under the vector atom model can be found on the knowledge of following quantum numbers:

1. Principal quantum number (n)
2. Orbital quantum number (ℓ)
3. Spin quantum number (s)
4. Orbital magnetic quantum number (m_ℓ)
5. Spin magnetic quantum number (m_s)
6. Total angular momentum or inner quantum number (j)
7. Total angular momentum magnetic quantum number (m_j)

4.16   TOTAL MAGNETIC MOMENT OF AN ATOM AND LANDE *g*-FACTOR

According to vector atom model, the spin and orbit angular momentum vector precess around a common axis due to spin–orbit interaction. The combination of these two vectors provides total angular momentum of an atom, which acts along the direction of common axis.

When the atom is placed in weak external magnetic field **B** then total angular momentum vector **J** precesses around the field direction, while **L** and **S** continue their precession around **J** (Figure 4.30). The vectors **L** and **S** do not precess around **B** because interaction between **L** and **B** and **S** and **B** are smaller than the interaction of **L** and **S** with **J**.

If ω is frequency of precession and $\mathbf{\mu_j}$ is total magnetic moment, then from Larmor's theorem, we have

$$\omega = \frac{|\mathbf{\mu_j}|}{|\mathbf{J}|}B = \frac{\mu_j}{|\mathbf{J}|}B \qquad (4.105)$$

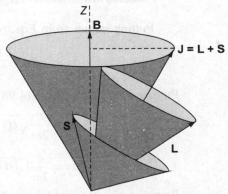

Figure 4.30 Precession of **J** around magnetic field **B**.

Suppose $\boldsymbol{\mu_L}$ and $\boldsymbol{\mu_S}$ are the orbital and spin magnetic moment of electron, which act opposite to **L** and **S** respectively. The resultant of $\boldsymbol{\mu_L}$ and $\boldsymbol{\mu_S}$ vectors gives the total magnetic moment (i.e., $\boldsymbol{\mu_j} = \boldsymbol{\mu_L} + \boldsymbol{\mu_S}$) that acts opposite to **J** (Figure 4.31). From this figure, it is clear that the angle between $\boldsymbol{\mu_L}$ and $\boldsymbol{\mu_j}$ is equal to the angle between **L** and **J**. Similarly, the angle between $\boldsymbol{\mu_S}$ and $\boldsymbol{\mu_j}$ is equal to the angle between **S** and **J**. If $\cos \theta_L$ and $\cos \theta_S$ are the angles between **L** and **J** and **S** and **J** respectively, then the total magnetic moment can be written as:

$$\mu_j = |\boldsymbol{\mu_j}| = \text{Component of } \boldsymbol{\mu_L} \text{ along } \boldsymbol{\mu_j} + \text{component of } \boldsymbol{\mu_S} \text{ along } \boldsymbol{\mu_j}$$

$$\mu_j = |\boldsymbol{\mu_L}| \cos \theta_L + |\boldsymbol{\mu_S}| \cos \theta_S$$

$$\mu_j = \mu_L \cos \theta_L + \mu_S \cos \theta_S$$

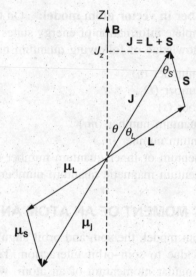

Figure 4.31   Vector representation of angular momentums and corresponding magnetic moments for an electron in magnetic field **B**.

Putting values from Eqs. (4.71) and (4.76),

$$\mu_j = \frac{e}{2m} |\mathbf{L}| \cos \theta_L + 2\frac{e}{2m} |\mathbf{S}| \cos \theta_S$$

Putting values from Eqs. (4.98) and (4.99),

$$\mu_j = \frac{e}{2m} \sqrt{\ell(\ell+1)}\hbar \cos \theta_L + 2\frac{e}{2m} \sqrt{s(s+1)} \, \hbar \cos \theta_S$$

$$\mu_j = \frac{e}{2m} [\sqrt{\ell(\ell+1)} \cos \theta_L + 2\sqrt{s(s+1)} \cos \theta_S] \, \hbar \qquad (4.106)$$

From vector atom model, we have

$$\mathbf{J} = \mathbf{L} + \mathbf{S}$$

$$\mathbf{J} - \mathbf{L} = +\mathbf{S}$$

$$(\mathbf{J} - \mathbf{L})(\mathbf{J} - \mathbf{L}) = \mathbf{SS}$$

$$|\mathbf{J}|^2 + |\mathbf{L}|^2 - 2\mathbf{L}.\mathbf{J} = |\mathbf{S}|^2$$

$$|\mathbf{J}|^2 + |\mathbf{L}|^2 - 2|\mathbf{L}||\mathbf{J}| \cos\theta_L = |\mathbf{S}|^2$$

$$2|\mathbf{L}||\mathbf{J}| \cos\theta_L = |\mathbf{J}|^2 + |\mathbf{L}|^2 - |\mathbf{S}|^2$$

$$\cos\theta_L = \frac{|\mathbf{J}|^2 + |\mathbf{L}|^2 - |\mathbf{S}|^2}{2|\mathbf{L}||\mathbf{J}|}$$

$$\cos\theta_L = \frac{j(j+1)\hbar^2 + \ell(\ell+1)\hbar^2 - s(s+1)\hbar^2}{2\sqrt{\ell(\ell+1)}\hbar \sqrt{j(j+1)}\,\hbar}$$

$$\cos\theta_L = \frac{j(j+1) + \ell(\ell+1) - s(s+1)}{2\sqrt{\ell(\ell+1)}\sqrt{j(j+1)}} \tag{4.107}$$

Similarly, the expression of $\cos\theta_S$ can be written as:

$$\cos\theta_S = \frac{j(j+1) + s(s+1) - \ell(\ell+1)}{2\sqrt{s(s+1)}\,j(j+1)} \tag{4.108}$$

Putting values of $\cos\theta_L$ and $\cos\theta_S$ in Eq. (4.97), we get

$$\mu_j = \frac{e}{2m}\left[\sqrt{\ell(\ell+1)}\,\frac{j(j+1) + \ell(\ell+1) - s(s+1)}{2\sqrt{\ell(\ell+1)}\sqrt{j(j+1)}} + 2\sqrt{s(s+1)}\,\frac{j(j+1) + s(s+1) - \ell(\ell+1)}{2\sqrt{s(s+1)}\sqrt{j(j+1)}}\right]\hbar$$

$$\mu_j = \frac{e}{2m}\left[\frac{j(j+1) + l(l+1) - s(s+1)}{2\sqrt{j(j+1)}} + \frac{j(j+1) + s(s+1) - l(l+1)}{\sqrt{j(j+1)}}\right]\hbar$$

$$\mu_j = \frac{e}{2m}\left[\frac{j(j+1) + \ell(\ell+1) - s(s+1)}{2j(j+1)} + \frac{j(j+1) + s(s+1) - \ell(\ell+1)}{j(j+1)}\right]\sqrt{j(j+1)}\,\hbar$$

$$\mu_j = \frac{e}{2m}\left[\frac{j(j+1) + \ell(\ell+1) - s(s+1) + 2j(j+1) + 2s(s+1) - 2\ell(\ell+1)}{2\,j(j+1)}\right]\sqrt{j(j+1)}\,\hbar$$

$$\mu_j = \frac{e}{2m}\left[\frac{2j(j+1) + j(j+1) + s(s+1) - \ell(\ell+1)}{2j(j+1)}\right]\sqrt{j(j+1)}\,\hbar$$

$$\mu_j = \frac{e}{2m}\left[1 + \frac{j(j+1) + s(s+1) - \ell(\ell+1)}{2j(j+1)}\right]|\mathbf{J}| \tag{4.109}$$

The quantity in bracket of Eq. (4.109) is called Lande g-factor, denoted by g, i.e.,

$$g = g_j = 1 + \frac{j(j+1) + s(s+1) - \ell(\ell+1)}{2j(j+1)} \tag{4.110}$$

Thus, the expression of μ_j becomes

$$\boxed{\mu_j = |\boldsymbol{\mu_j}| = g\frac{e}{2m}|\mathbf{J}|} \quad \text{or} \quad \frac{\mu_j}{|\mathbf{J}|} = g\frac{e}{2m} \tag{4.111}$$

Equation (4.111) is expression of total magnetic moment of electron. In vector form, it can be written as:

$$\boldsymbol{\mu_j} = -g\frac{e}{2m}\mathbf{J} \tag{4.112}$$

From Eqs. (4.105) and (4.111), we have

$$\boxed{\omega = g\frac{e}{2m}B} \tag{4.113}$$

If θ is orientation of $\mathbf{J}$ around $\mathbf{B}$, then from Figure 4.31, we can write

$$\cos\theta = \frac{J_z}{|\mathbf{J}|}$$

$$\cos\theta = \frac{m_j\hbar}{\sqrt{j(j+1)}\,\hbar} \tag{4.114}$$

$$\cos\theta = \frac{m_j}{\sqrt{j(j+1)}} \tag{4.115}$$

Since m_j can have $(2j+1)$ values for a given j, thus angle θ also have $(2j+1)$ discrete values. Therefore, $\mathbf{J}$ have $(2j+1)$ discrete orientation in the magnetic field, which corresponds to slightly different energy levels. On the basis of these levels, anomalous Zeeman effect is well explained.

EXAMPLE 4.5   Find the Lande g-factor for (a) pure orbital angular momentum (b) pure spin angular momentum (c) $^2S_{1/2}$ (d) $^2P_{1/2}$ (e) $^2P_{3/2}$ and (g) $^3D_2$.

Solution $g = g_j = 1 + \dfrac{j(j+1) + s(s+1) - \ell(\ell+1)}{2j(j+1)}$

(a) $s = 0$; $j = \ell$; $g = 1 + \dfrac{\ell(\ell+1+0-\ell(\ell+1))}{2\ell(\ell+1)} = 1$

(b) $\ell = 0$; $j = s$; $g = 1 + \dfrac{s(s+1)+s(s+1)-0}{2s(s+1)} = 1+1 = 2$

(c) For state $^2S_{1/2}$, $\ell = 0$; $2s+1=2 \Rightarrow s = 1/2$, $j = 1/2$

$g = 1 + \dfrac{(1/2)\{1+(1/2)\} + (1/2)\{1+(1/2)\} - 0}{2(1/2)\{1+(1/2)\}} = 1 + \dfrac{(3/4)+(3/4)-0}{3/2} = 1+1 = 2$

(d) For state $^2P_{1/2}$, $\ell = 1$; $2s+1 = 2 \Rightarrow s = 1/2$, $j = 1/2$

$g = 1 + \dfrac{(1/2)\{1+(1/2)\} + (1/2)\{1+(1/2)\} - (1+1)}{2(1/2)\{1+(1/2)\}} = 1 + \dfrac{(3/4)+(3/4)-2}{3/2} = 1 - \dfrac{1}{3} = \dfrac{2}{3}$

(e) For state $^2P_{3/2}$, $\ell = 1$; $2s+1 = 2 \Rightarrow s = 1/2$, $j = 3/2$

$g = 1 + \dfrac{(3/2)\{1+(3/2)\} + (1/2)\{1+(1/2)\} - (1+1)}{2(3/2)\{1+(3/2)\}} = 1 + \dfrac{(15/4)+(3/4)-2}{15/2} = 1 + \dfrac{1}{3} = \dfrac{4}{3}$

EXAMPLE 4.6 Find the magnetic moment for the state $^3P_1$ in terms of Bohr magnetron.

Solution For state $^3P_1$, $\ell = 1$; $2s + 1 = 3 \Rightarrow s = 1$, $j = 1$

$$g = 1 + \frac{(1+1) + (1+1) - (1+1)}{2(1+1)} = 1 + \frac{1}{2} = \frac{3}{2}$$

$$\mu = \sqrt{j(j+1)}\, g\mu_B = \sqrt{(1+1)}\, \frac{3}{2} \mu_B = \frac{3}{\sqrt{2}} \mu_B$$

4.17 COUPLING SCHEMES

The total angular momentum of an atom having two or more than two electrons can be found by two coupling schemes. The scheme of coupling depends on amount or intensity of interaction between spin and orbit angular momentums. Since the core electrons do not contribute to the angular momentum of the atom, only optical electrons (electrons outside the completed sub-shells) are considered under coupling schemes. There are two types of coupling methods for the determination of total momentum of atom:

(a) *L-S* coupling or Russell–Saunders coupling
(b) *J-J* coupling

4.17.1 *L-S* Coupling

The spin-orbit coupling, also known as *spin-pairing* that describes a weak magnetic interaction or coupling of the particle spin and the orbital motion of this particle, e.g. the electron spin and its motion around an atomic nucleus. One of its effects is to separate the energy of internal states of the atom, e.g. spin-aligned and spin-anti-aligned that would otherwise be identical in energy. This interaction is responsible for many of the details of atomic structure. In the macroscopic world of orbital mechanics, the term spin-orbit coupling is sometimes used in the same sense as spin-orbital resonance.

In light atoms (generally Z < 30), electron spins interact among themselves so they combine to form a total spin angular momentum **S**. *The same happens with orbital angular momenta of electrons, forming a total orbital angular momentum* **L**. *Then* **S** *and* **L** *couple together and form a total angular momentum* **J**. *The interaction between* **S** *and* **L** *is called Russell–Saunders coupling or L-S coupling.* Thus the *L-S* coupling completes in the following three processes or mechanisms:

Spin-spin interaction: In this, the individual spin angular momentum $\mathbf{S_i}$ of the optical electrons are strongly coupled with one another to form a resultant spin angular momentum **S**.

$$\mathbf{S} = \sum_i \mathbf{S_i}$$

The vector **S** has magnitude equal to $\sqrt{s(s+1)}\,\hbar$. The s is called *spin quantum number*. If s_1 and s_2 are the two spins of the individual electrons, then s **has values from**

$(|s_1 - s_2|)$ **to** $(|s_1 + s_2|)$ **in increment of unity.** In case of three spins, the resultant of two spins further combines with the spin of third electron in the same way to give the value of s. Similar treatment is followed for a large number of electron systems. The different s values correspond to different energy states, thus the unperturbed energy state is splitted into number of well separated levels. The different splitted levels are defined by multiplicity, which is equal to $(2s + 1)$. The high multiplicity states lie deeper because the energy level corresponding to lowest s has highest energy. The terms (states) corresponding to 1, 2, 3 and 4 values of multiplicity is called *singlet, doublet, triplet terms and quartets*.

Example: A single electron system has spin 1/2, so its multiplicity will be 2 and term will be doublet. The resultant values of s, multiplicity and terms for the two-and three-electron system can be found in the following manner:

(i) *Two-electron system* $(s_1 = 1/2, s_2 = 1/2)$:
$s = (|s_1 - s_2|), (|s_1 - s_2|) + 1, ..., (|s_1 + s_2|) = |s_1 - s_2|$ to $|s_1 + s_2| = 0, 1$
For $s = 0$, multiplicity = 1 and state: Singlet
For $s = 1$, multiplicity = 3 and state: Triplet

(ii) *Three-electron system* $(s_1 = 1/2, s_2 = 1/2, s_3 = 1/2)$:
$s' = $ combination of s_1 and $s_2 = 0, 1$
$s = $ combination of s' and s_3 (gives two combinations)
$s = $ combination of 0 and $1/2 = \left| 0 - \dfrac{1}{2} \right|$ to $\left| 0 + \dfrac{1}{2} \right| = \dfrac{1}{2}$

thus multiplicity: 2 and state: Doublet

$s = $ combination of 1 and $1/2 = \left| 1 - \dfrac{1}{2} \right|$ to $\left| 1 + \dfrac{1}{2} \right| = \dfrac{1}{2}, \dfrac{3}{2}$

thus multiplicity: 2 and 4 and states: Doublet and Quartets

Residual electrostatic interaction: In this, the individual orbital angular momentum (L_i) of the optical electrons are strongly coupled with one another to form a resultant spin angular momentum L.

$$L = \sum_i L_i$$

The vector L has magnitude equal to $\sqrt{l(l+1)}\hbar$. The l is called *orbital quantum number*. If ℓ_1 and ℓ_2 are the orbital quantum numbers of the two optical electrons, then l **has values from** $(|\ell_1 - \ell_2|)$ **to** $(|\ell_1 + \ell_2|)$ **in increment of unity.** In case of three optical electrons, the resultant of two orbital quantum numbers further combines with the third in the same way to give the value of l. The similar treatment is followed for a large number of electron systems. The different l values correspond to different energy states. Thus the energy level which is splitted by spin–spin coupling is further splitted to number of energy states due to residual electrostatic interaction. The states corresponding to $l = 0, 1, 2, 3, 4$ are denoted by S, P, D, F, G respectively. The state having high l lies lower due to having lower energy.

Example:

(i) *Two-electron system* ($4p$ $4d$: $\ell_1 = 1$, $\ell_2 = 2$):

$$l = |\ell_1 - \ell_2|, |\ell_1 - \ell_2| + 1, ..., |\ell_1 + \ell_2| = |\ell_1 - \ell_2| \text{ to } |\ell_1 + \ell_2| = |1 = 2| \text{ to } |1 + 2|$$

$l = 1, 2, 3$ thus states: *P, D, F*

(ii) *Three-electron system* ($3d$ $4s$ $5p$: $\ell_1 = 2$, $\ell_2 = 0$, $\ell_3 = 1$):

l' = combination of ℓ_1 and $\ell_2 = |2 - 0|$ to $|2 + 0| = 2$

l = combination of l' and $\ell_3 = |2 - 1|$ to $|2 + 1|$

$l = 1, 2, 3$ thus states: *P, D, F*

(iii) *Three-electron system* ($2p$ $3p$ $4d$: $\ell_1 = 1$, $\ell_2 = 1$, $\ell_3 = 2$):

l' = combination of ℓ_1 and $\ell_2 = |1 - 1|$ to $|1 + 1| = 0, 1, 2$

l = combination of l' and ℓ_3 (gives three set of combinations)

l = combination of 0 and 2 = $|0 - 2|$ to $|0 + 2| = 2$ thus states: *D*

l = combination of 1 and 2 = $|1 - 2|$ to $|1 + 2| = 1, 2, 3$ thus states: *P, D, F*

l = combination of 2 and 2 = $|2 - 2|$ to $|2 + 2| = 0, 1, 2, 3, 4$ thus states: *S, P, D, F, G*

Total states: one *S*, two *P*, three *D*, two *F* and one *G*

Spin-orbit interaction: In this, the resultant orbital and spin angular momentums (**L** and **S**) are weakly coupled with each another to form a resultant total angular momentum **J**. This coupling takes in such a way that the magnitude of **L**, **S** and **J** remains constant.

$$\mathbf{J} = \mathbf{L} + \mathbf{S} = \sum_i \mathbf{L_i} + \sum_i \mathbf{S_i}$$

The vector **J** has magnitude equal to $\sqrt{j(j+1)}\hbar$. The *j is called total or inner quantum number which has values from* $(|l - s|)$ **to** $(|l + s|)$ *in increment of unity.* The number of *j* values is $(2s + 1)$ when $l > s$ or $(2l + 1)$ when $l < s$. This indicates that the level characterised by *l* and *s* is further broken into $(2s + 1)$ or $(2l + 1)$ closer levels. The group of these *j* states forms a fine structure multiplet. The relative spacing of these *j*-levels within the multiplet is governed by *Lande interval rule*. According to which, *the energy interval between consecutive j and j + 1 levels of a fine structure multiplet is proportional to j + 1, i.e., to the larger of the two j-values involved.*

Note: This coupling scheme is good as long as external magnetic field is weak. In large magnetic fields, these two momenta decouple, giving rise to a different splitting pattern in the energy levels and the size of *L-S* coupling term becomes small.

Spectroscopic terms under L-S coupling scheme

The spectroscopic terms under *L-S* coupling scheme are denoted by $^{2S+1}X_j$. Here *X* is equal to *S, P, D, F,* ... corresponding $l = 0, 1, 2, 3,$ ... respectively. The complete notation of energy level is expressed by writing principal quantum number before the term, i.e., n $^{2S+1}X_J$.

The electronic configuration can be classified into three categories as: (i) single optical electron system, (ii) non-equivalent optical electron system, and (iii) equivalent optical electron system.

(i) *Single optical electron system:* The atom having single electron in the valence shell is called *single optical electron system*, e.g., hydrogen like atoms. In such cases, each orbital is divided into two states. The splitted terms/levels under the L-S coupling scheme are shown in Table 4.8.

Table 4.8 Levels under L-S Coupling of Single Optical Electron System

Electronic configuration of valance orbital	s	2s+1	l	$J = \lvert l - s \rvert$ to $\lvert l + s \rvert$ in increment of unity	Energy level notation
$1s^1; 2s^1; 3s^1; 4s^1$	1/2	2	0	1/2	$^2S_{1/2}$
$2p^1; 3p^1; 4p^1$	1/2	2	1	1/2, 3/2	$^2P_{1/2}, {}^2P_{3/2}$
$3d^1; 4d^1$	1/2	2	2	3/2, 5/2	$^2D_{3/2}, {}^2D_{5/2}$
$4f^1$	1/2	2	3	5/2, 7/2	$^2F_{5/2}, {}^2F_{7/2}$

(ii) *Non-equivalent optical electron system:* The atoms which have two or more electron optical electrons with non-equal principal n or orbital ℓ quantum numbers are called *non-equivalent optical electron system*. The splitted terms/levels under the L-S coupling scheme of such system are shown in Table 4.9.

Table 4.9 Levels under L-S Coupling of Non-equivalent Optical Electron System

Electronic configuration	s_1, s_2	s	2s + 1	ℓ_1, ℓ_2	l	j ($\lvert l - s \rvert$) to ($\lvert l + s \rvert$)	Terms
$n_1s^1; n_2s^1$ ($n_1 \neq n_2$)	1/2, 1/2	0	1	0, 0	0	For $s = 0, l = 0$: $j = 0$	$^1S_0$
		1	3			For $s = 1, l = 0$: $j = 1$	$^3S_1$
$n_1s^1; n_2p^1$	1/2, 1/2	0	1	0, 1	1	For $s = 0, l = 1$: $j = 1$	$^1P_1$
		1	3			For $s = 1, l = 1$: $j = 0, 1, 2$	$^3P_{0, 1, 2}$
$n_1s^1; n_2d^1$	1/2, 1/2	0	1	0, 2	2	For $s = 0, l = 2$: $j = 2$	$^1D_2$
		1	3			For $s = 1, l = 2$: $j = 1, 2, 3$	$^3D_{1, 2, 3}$
$n_1p^1; n_2p^1$ ($n_1 \neq n_2$)	1/2, 1/2	0	1	1,1	0, 1, 2	For $s = 0, l = 0$: $j = 0$	$^1S_0$
						For $s = 0, l = 1$: $j = 1$	$^1P_1$
						For $s = 0, l = 2$: $j = 2$	$^1D_2$
		1	3			For $s = 1, l = 0$: $j = 1$	$^3S_1$
						For $s = 1, l = 1$: $j = 0, 1, 2$	$^3P_{0, 1, 2}$
						For $s = 1, l = 2$: $j = 1, 2, 3$	$^3D_{1, 2, 3}$
$n_1p^1; n_2d^1$	1/2, 1/2	0	1	1, 2	1, 2, 3	For $s = 0, l = 1$: $j = 1$	$^1P_1$
						For $s = 0, l = 2$: $j = 2$	$^1D_2$
						For $s = 0, l = 3$: $j = 3$	$^1F_3$
		1	3			For $s = 1, l = 1$: $j = 0, 1, 2$	$^3P_{0, 1, 2}$
						For $s = 1, l = 2$: $j = 1, 2, 3$	$^3D_{1, 2, 3}$
						For $s = 1, l = 3$: $j = 2, 3, 4$	$^3F_{2, 3, 4}$

(iii) *Equivalent optical electron system:* The atoms which have two or more electron optical electrons with equal principal n and orbital ℓ quantum numbers are called *equivalent optical electron system.* According to *Pauli Exclusion Principle*, the two electrons cannot have all equal quantum numbers. If n and ℓ are same, then at least one in remaining quantum numbers (m_ℓ and m_s) must be different. So, for equivalent optical electron system, the values of m_ℓ and m_s must be different for both the electrons. Therefore, certain terms which are allowed in non-equivalent optical electron system are now not allowed. The following two extra rules are also followed in the determination of terms under *L-S* coupling scheme:

(a) A closed sub-shell like s^2, p^6, d^{10}, ... always forms a $^1S_0$ term only.

   Explanation: Since all the sub-shells are completely filled and each have anti-parallel pairs of electrons, the sum of orbital and spin magnetic quantum number will be zero, i.e.,

$$s^2: \quad \boxed{\downarrow\uparrow} \qquad \begin{array}{l} m_\ell = 0;\ 0 \\[6pt] m_s = \dfrac{1}{2};\ -\dfrac{1}{2} \end{array} \qquad \begin{array}{l} m_l = \sum m_\ell = 0 \\[6pt] m_s = \sum m_s = 0 \end{array}$$

$$p^6: \quad \boxed{\downarrow\uparrow}\ \boxed{\downarrow\uparrow}\ \boxed{\downarrow\uparrow} \qquad \begin{array}{l} m_\ell = 1, 0, -1;\ 1, 0, -1 \\[6pt] m_s = \dfrac{1}{2}, \dfrac{1}{2}, \dfrac{1}{2}; -\dfrac{1}{2}, -\dfrac{1}{2}, -\dfrac{1}{2} \end{array} \qquad \begin{array}{l} m_l = \sum m_\ell = 0 \\[6pt] m_s = \sum m_s = 0 \end{array}$$

   Due to $m_l = 0$ and $m_s = 0$ both the orbital and spin quantum numbers (l and s) of the whole system will be zero. For $l = s = 0$, the j will have zero value. Hence the spectroscopic term will be $^1S_0$.

(b) The spectroscopic terms of configurations p^X, d^Y and f^Z are equivalent to p^{6-X}, d^{10-Y} and f^{14-Z} respectively, i.e., spectroscopic terms of p^2 and d^2 are equivalent to p^4 and d^8 respectively. In such cases, the total quantum numbers are defined under G.Breit's scheme. In this scheme, all possible values of m_l which are formed by the combination of m_{ℓ_1} and m_{ℓ_2} of two electrons are written in the form of a table. The values of m_{ℓ_1} and m_{ℓ_2} are written in a row and column respectively, while values of m_l are written below m_{ℓ_1} and to the left of m_{ℓ_2}. The complete table is divided by L-shapes dotted lines. The values between the dotted lines provide the values of m_l. The diagonal values of m_l are not considered because they are formed by equal orbital magnetic quantum numbers. Similarly, the total spin magnetic quantum number m_s is determined. On the knowledge of m_l and m_s, the quantum numbers l, s, j are defined under Pauli Exclusion Principle and hence spectroscopic terms are determined.

Let us consider a system p^2 or p^4 for which the spectroscopic terms are to be determined. In the case of p^2 or p^4; $\ell_1 = 1$, $\ell_2 = 1$ and $s_1 = 1/2$, $s_2 = 1/2$. For this system, the table under Breit's scheme is shown in Figure 4.32.

From the table under Breit's scheme, it is clear that m_s has two set of values, while m_l has three set of values:

Set I: $m_s = 0$; \qquad\qquad $s = 0$

Set II: $m_s = 1, 0, -1$; \qquad $s = 1$

Set I: $m_l = 0$; $l = 0$
Set II: $m_l = 1, 0, -1$; $l = 1$
Set III: $m_l = 2, 1, 0, -1, -2$; $l = 2$

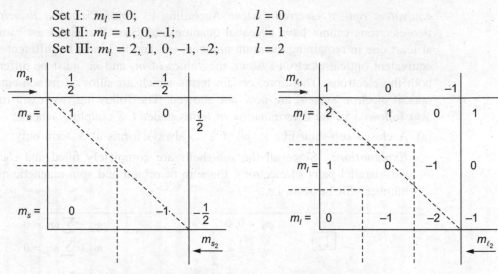

Figure 4.32 Evaluation of m_ℓ and m_s under Breit's scheme.

For $s = 1$ (triplet state), both the electrons have same spins, thus they must be different in orbital magnetic quantum numbers. For such case, those values of m_l are appropriate, which are on the both side of the diagonal of table. Only one side of m_l are taken because values of second side of diagonal are the mirror image of the first side. So, for $s = 1$, the correct values of l is 1 because the corresponding m_l is formed by different orbital magnetic quantum number of electrons.

For $s = 0$ (singlet state), both the electrons differ in their spins. Since, one quantum number is different, so any value of m_l is correct for this case. Furthermore Set II of m_l has been already used. So, for s = 0, the correct values of l is 0 and 2. On the basis of above analysis, the spectroscopic terms of p^2 or p^4 can be written with following combination of s and l as:

1. $s = 0, l = 0$; $2s + 1 = 3$; $j = 1$;   Term: $^1S_0$
2. $s = 0, l = 2$; $2s + 1 = 1$; $j = 1$;   Term: $^1D_2$
3. $s = 1, l = 1$; $2s + 1 = 3$; $j = 1, 2, 3$;   Term: $^3P_{0,1,2}$

The splitted terms/levels under the L-S coupling scheme of few equivalent optical electron system are shown in Table 4.10.

Table 4.10 Levels under L-S Coupling of Equivalent Optical Electron System

Electronic configuration	Terms
s^2, p^6, d^{10}	$^1S_0$
P^1, p^5	$^2P_{1/2}, {}^2P_{3/2}$
p^2, p^4	$^1S_0, {}^1D_2, 3P_{0,1,2}$
P^3	$^2P_{1/2}, {}^2P_{3/2}, {}^2D_{3/2}, {}^2D_{5/2}, {}^4S_{3/2}$
d^1, d^9	$^2D_{3/2}, {}^2D_{5/2}$
d^2, d^8	$^1S_0, {}^1D_2, {}^1G_4, {}^3P_{0,1,2}, {}^3F_{2,3,4}$

Selection rule for L-S coupling

The electronic transition in levels formed by *L-S* coupling scheme are governed by following rules:

1. $\Delta l = 0, \pm 1$, for one electron system, $\Delta l = 0$ is not allowed.
2. $\Delta s = 0$
3. $\Delta j = 0, \pm 1$ but $j = 0 \leftrightarrow j = 0$ is not allowed.

The rules for *l* and *s* hold only when the spin-orbit coupling is weak, i.e., when the splitting between the various fine structure levels of a multiplet is very small as compared to the separations between the various multiplets themselves. Therefore, the transition condition for *l* and *s* hold accurately for the light atoms.

EXAMPLE 4.7 Find term value for $4p\,4d$ system under *L-S* coupling scheme and draw clean diagram of energy level.

Solution The evaluation of term for $4p\,4d$ system under *L-S* coupling scheme is given in Table 4.11.

Table 4.11 Term for $4p\,4d$ System

System	s_1, s_2	s $(|s_1 - s_2|$ to $|s_1 + s_2|)$	$2s + 1$	ℓ_1, ℓ_2	l $(|\ell_1 - \ell_2|$ to $|\ell_1 + \ell_2|)$	j $(|l - s|$ to $|l + s|)$	Terms $^{2s+1}X_j$
$4p\,4d$	$1/2, 1/2$	0	1	$1, 2$	$1, 2, 3$	For $s = 0, l = 1: j = 1$	$^1P_1$
						For $s = 0, l = 2: j = 2$	$^1D_2$
						For $s = 0, l = 3: j = 3$	$^1F_3$
		1	3			For $s = 1, l = 1: j = 0, 1, 2$	$^3P_{0,1,2}$
						For $s = 1, l = 2: j = 1, 2, 3$	$^3D_{1,2,3}$
						For $s = 1, l = 3: j = 2, 3, 4$	$^3F_{2,3,4}$

The splitting of energy levels due to *L-S* coupling is shown in Figure 4.33.

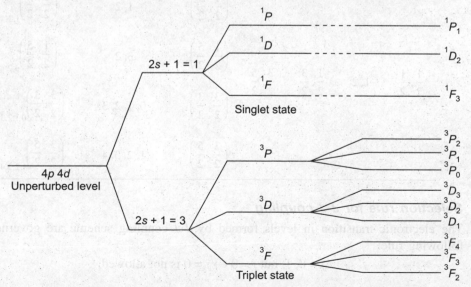

Figure 4.33 Splitting of energy level due to *L–S* coupling for $4p\,4d$ state.

4.17.2 *J-J* Coupling

The *J-J* coupling occurs in the heavy atoms. For these atoms, the interaction between the orbital and spin momenta of a single electron is much greater than the interaction between the orbital momenta of different electrons or between the spin momenta of different electrons. Therefore, in this coupling scheme, the orbital and spin angular momenta of single electron ($\mathbf{L_i}$ and $\mathbf{S_i}$) couple to form total angular momentum ($\mathbf{J_i}$). Later on, the total angular momentum ($\mathbf{J}$) of the atom is obtained by the coupling of total angular momenta of different electrons ($\mathbf{J_i}$), i.e.,

1. $\mathbf{J_i} = \mathbf{L_i} + \mathbf{S_i}$ and $|\mathbf{J_i}| = \sqrt{j_i(j_i+1)}\hbar$

   The total quantum number j_i of individual can have values from $|l - 1/2|$ to $|l + 1/2|$ in increment of unity. Hence, quantity j possesses a number of half integral values. This indicates that the unperturbed level is splitted into a number of well-spaced levels.

2. $\mathbf{J} = \sum_i \mathbf{J_i}$ and $|\mathbf{J}| = \sqrt{j(j+1)}\,\hbar$

   If there are two electrons, j can have values from $|j_1 - j_2|$ to $|j_1 + j_2|$ in increment of unity. Hence, quantity j possesses a number of half integral values. This indicates that the each splitted level is further splitted into a number of well-spaced levels.

The splitting terms of $4p\,4d$ system under *J-J* coupling scheme are listed in Table 4.12 and shown in Figure 4.34.

Table 4.12   Splitting Terms of $4p\,4d$ System under *J-J* Coupling

s_1, s_2	ℓ_1, ℓ_2	j_1	j_2	(j_1, j_2)	j $|j_1 - j_2|$ to $|j_1 + j_2|$	*Terms*
				$\left(\dfrac{1}{2}, \dfrac{3}{2}\right)$	1, 2	$\left(\dfrac{1}{2}, \dfrac{3}{2}\right)_{1,2}$
				$\left(\dfrac{1}{2}, \dfrac{5}{2}\right)$	3, 2	$\left(\dfrac{1}{2}, \dfrac{5}{2}\right)_{3,2}$
$\dfrac{1}{2}, \dfrac{1}{2}$	1, 2	$\dfrac{1}{2}, \dfrac{3}{2}$	$\dfrac{3}{2}, \dfrac{5}{2}$	$\left(\dfrac{3}{2}, \dfrac{3}{2}\right)$	0, 1, 2, 3	$\left(\dfrac{3}{2}, \dfrac{3}{2}\right)_{0,1,2,3}$
				$\left(\dfrac{3}{2}, \dfrac{5}{2}\right)$	1, 2, 3, 4	$\left(\dfrac{3}{2}, \dfrac{5}{2}\right)_{1,2,3,4}$

Selection rule for J-J coupling

The electronic transition in levels formed by *J-J* coupling scheme are governed by following rule:

$$\Delta j = 0, 1, \text{ but } j = 0 \leftrightarrow j = 0 \text{ is not allowed.}$$

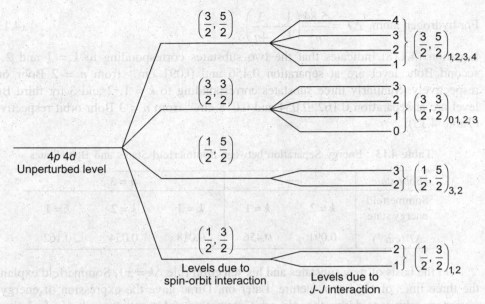

Figure 4.34 Splitting of energy level due to J-J coupling for 4p 4p state.

4.18 FINE STRUCTURE OF H_α LINE

Initially, Sommerfeld explained the fine structure of H_α line on the basis of his elliptical orbit model and relativistic correction to it. He gave the following expression of energy and term value of an atomic energy state.

$$E_{n,k} = -\frac{R_\infty hcZ^2}{n^2}\left[1 + \frac{\alpha^2 Z^2}{n}\left(\frac{1}{k} - \frac{3}{4n}\right)\right] \tag{4.116}$$

$$T = -\frac{E_{n,k}}{hc} = \frac{R_\infty Z^2}{n^2}\left[1 + \frac{\alpha^2 Z^2}{n}\left(\frac{1}{k} - \frac{3}{4n}\right)\right]$$

$$T = \frac{R_\infty Z^2}{n^2} + \frac{R_\infty \alpha^2 Z^4}{n^3}\left(\frac{1}{k} - \frac{3}{4n}\right) \tag{4.117}$$

where $R_\infty = \left(\dfrac{2\pi^2 m_0 e^4}{ch^3}\right)$ (in C.G.S.) = $\left(\dfrac{me^4}{8\varepsilon_0^2 h^3 c}\right)$ (in M.K.S.) $= 1.09678 \times 10^7 \text{m}^{-1}$ and

fine structure constant $\alpha = 2pe^2/hc \approx 1/137$. The quantities n and k are called principal and azimuthal quantum number respectively. The quantum number k can have 1, 2, ..., n values. The first term of Eq. (4.117) is energy of Bohr orbit (in cm^{-1}), while the second term is called *relativity correction term* ΔT.

$$\Delta T = \frac{R_\infty \alpha^2 Z^4}{n^3}\left(\frac{1}{k} - \frac{3}{4n}\right) = 5.84\frac{Z^4}{n^3}\left(\frac{1}{k} - \frac{3}{4n}\right) \text{cm}^{-1} \tag{4.118}$$

For hydrogen atom, $\Delta T = \dfrac{5.84}{n^3}\left(\dfrac{1}{k} - \dfrac{3}{4n}\right)$ (4.119)

Equation (4.118) indicates that the two substates corresponding to $k = 1$ and 2 for second Bohr level are at separation 0.456 and 0.091 cm^{-1} from $n = 2$ Bohr orbit respectively. Similarly three substates corresponding to $k = 1$, 2 and 3 for third Bohr level are at separation 0.162, 0.054 and 0.018 cm^{-1} from $n = 3$ Bohr orbit respectively (Table 4.13).

Table 4.13   Energy Separation between Sommerfeld States and Bohr States

Bohr state	\multicolumn{2}{c} *n = 2*		*n = 3*		
Sommerfeld energy state	$k = 2$	$k = 1$	$k = 3$	$k = 2$	$k = 1$
ΔT (cm^{-1})	0.091	0.456	0.018	0.054	0.162

On the basis of new substates and his selection rule $\Delta k = \pm 1$, Sommerfeld explained the three lines of H$_\alpha$ fine structure. Later on, Dirac gave the expression of energy of stationary orbit considering the spin-orbit interaction and quantum mechanical relativistic correction to it. He gave the following expression of energy and corresponding term value:

$$E_{n,j} = -\frac{R_\infty hcZ^2}{n^2}\left[1 + \frac{\alpha^2 Z^2}{n}\left(\frac{1}{j+\dfrac{1}{2}} - \frac{3}{4n}\right)\right]$$ (4.120)

$$T = -\frac{E_{n,j}}{hc} = \frac{R_\infty Z^2}{n^2}\left[1 + \frac{\alpha^2 Z^2}{n}\left(\frac{1}{j+\dfrac{1}{2}} - \frac{3}{4n}\right)\right]$$

$$T = \frac{R_\infty Z^2}{n^2} + \frac{R_\infty \alpha^2 Z^4}{n^3}\left(\frac{1}{j+\dfrac{1}{2}} - \frac{3}{4n}\right)$$ (4.121)

The second term of Eq. (4.120) represents the net term shit ($\Delta T'$) under the consideration of spin-orbit interaction and can be written as:

$$\Delta T' = \frac{R_\infty \alpha^2 Z^4}{n^3}\left(\frac{1}{j+\dfrac{1}{2}} - \frac{3}{4n}\right) = 5.84\frac{Z^4}{n^3}\left(\frac{1}{j+\dfrac{1}{2}} - \frac{3}{4n}\right)$$ (4.122)

For hydrogen atom, $\Delta T' = \dfrac{5.84}{n^3}\left(\dfrac{1}{j+\dfrac{1}{2}} - \dfrac{3}{4n}\right)$ (4.123)

Equation (4.123) is called the Dirac formula. This formula is similar to Eq. (4.119) [Sommerfeld formula; exception is only that there is j (total quantum number) in place of k (azimuthal quantum number)].

Under the *L-S* coupling scheme, the energy state of the hydrogen corresponding to $n = 2$ and $n = 3$ Bohr state are shown in Table 4.13. The evaluated value of shift $\Delta T'$ of these energy states (obtained from the *L-S* coupling scheme) from the Bohr energy states are also shown in Table 4.14.

Table 4.14 Spectroscopic Term Shift from Bohr Energy States

n	ℓ	j	Energy level notation	$\Delta T'$ (in cm^{-1})
2	1	3/2, 1/2	$2^2P_{3/2}, 2^2P_{1/2}$	0.091, 0.456
	0	1/2	$2^2S_{1/2}$	0.456
3	2	5/2, 3/2	$3^2D_{5/2}, 3^2D_{3/2}$	0.018, 0.054
	1	3/2, 1/2	$3^2P_{3/2}, 3^2P_{1/2}$	0.054, 0.162
	0	1/2	$3^2S_{1/2}$	0.162

The comparison of shifts calculated with Sommerfeld formula and Dirac formula indicates that following Dirac energy states exist corresponding to different combinations of n and k.

1. For $n = 2$, $k = 2$; energy state: $2^2P_{3/2}$
2. For $n = 2$, $k = 1$; energy state: $2^2S_{1/2}, 2^2P_{1/2}$
3. For $n = 3$, $k = 3$; energy state: $3^2D_{5/2}$
4. For $n = 3$, $k = 2$; energy state: $3^2P_{3/2}, 3^2D_{3/2}$
5. For $n = 3$, $k = 1$; energy state: $3^2S_{1/2}, 3^2P_{1/2}$

The selection rule for the transition in these new subenergy states of hydrogen follows the *L-S* coupling selection rule, which is:

1. $\Delta l = \pm 1$
2. $\Delta j = 0, \pm 1$, but j = 0 $\leftrightarrow$ j = 0 is not allowed.

These rules allow total seven transitions, i.e.,

(i) $3^2D_{5/2} \rightarrow 2^2P_{3/2}$ (ii) $3^2D_{3/2} \rightarrow 2^2P_{3/2}$

(iii) $3^2D_{3/2} \rightarrow 2^2P_{1/2}$ (iv) $3^2P_{3/2} \rightarrow 2^2S_{1/2}$

(v) $3^2P_{1/2} \rightarrow 2^2S_{1/2}$ (vi) $3^2S_{1/2} \rightarrow 2^2P_{3/2}$

(vii) $3^2S_{1/2} \rightarrow 2^2P_{1/2}$

The lines corresponding to transitions (iii) and (iv) coincide with each other because states $3^2P_{3/2}$ and $3^2D_{3/2}$ and $2^2S_{1/2}$ and $2^2P_{1/2}$ posses same energy. Similarly, lines corresponding to transitions (v) and (vi) coincide with each other. Therefore, transitions result five components of H$_\alpha$ line. In this way, Dirac theory justifies the existence of five lines in fine structure of H$_\alpha$ line. The energy states and transition among them for fine structure of H$_\alpha$ line are shown in Figure 4.35.

Yet the Dirac theory provides a good agreement with the Hanson's observation for fine structure of H$_\alpha$ line, but a slight deviation is found with the R.C. William

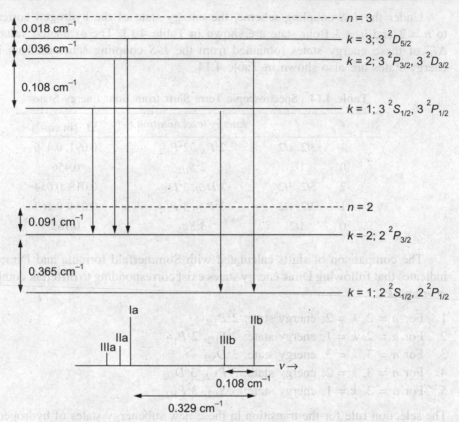

Figure 4.35   Energy level for *H*-atom under vector atom model and electronic transition for fine structure of H_α line.

observations for the same fine structure which was performed with the triple prism spectro-photograph and micrometer. The deviations are as follows.

1. The component Ia is found weaker than the component IIb, which is in contradiction with Dirac theory. In the same way, the component IIIb is found to be more intense than the intensity predicted theoretically.

2. The separation between Ia and IIb is observed to be 0.319 cm^{-1}, while theoretically it is found to be 0.329 cm^{-1}. Similarly the separation between IIb and IIIb is observed to be 0.134 cm^{-1}, while it is 0.108 cm^{-1} theoretically.

These two discrepancies are resolved later on the basis of unequal excitation of states and Lamb shift. The variation in intensity of fine structure lines is due to unequal excitation of $n = 3$ level of hydrogen atom and large transition for $3^2D_{3/2} \rightarrow 2^2P_{1/2}$ in comparison to $3^2D_{5/2} \rightarrow 2^2P_{3/2}$. Lamb and Rutherford predicted on the basis of their experiment that for Hydrogen like atoms, the states of a particular *n* value having terms with same *j* and different ℓ values (i.e. $2^2S_{1/2}$ and $2^2P_{1/2}$ or $3^2S_{1/2}$ and $3^2P_{1/2}$) are not degenerate, but are separated. Such separation is called *Lamb shift or Radiation shift*. It is found that the state $2^2S_{1/2}$ is higher than $2^2P_{1/2}$ by 0.0353 cm^{-1}. Similarly the state

$3^2S_{1/2}$ is above with $3^2P_{1/2}$ by 0.0105 cm^{-1}. Due to Lamb shift, the two separate lines are found in place of one line for the transitions $3^2P_{3/2} \rightarrow 2^2S_{1/2}$ and $3^2D_{3/2} \rightarrow 2^2P_{1/2}$. The intensity ratio between the lines corresponding to these transitions is 2.08 : 5.0, while the separation between the lines is 0.014 cm^{-1}. Therefore, the separation between the two main components Ia and IIb lines becomes equal to 0.315 cm^{-1} which is in good agreement with the separation observed in Williams measurement, i.e., 0.319 cm^{-1}. In this way, the Dirac theory with Lamb shift explains properly the hydrogen fine structure.

4.19 ALKALI SPECTRA

The atoms Li, Na, K, Rb, Cs and Fr are called *alkali atoms*. All these atoms have a single *s*-electron outside the closed shell. This *s*-electron is termed as *valence electron*. The inner electrons are called *core electrons*. *The valence electrons of charge $-(Z-1)e$ move in the field of the nucleus having charge +Ze.* The core electrons repel the valence electrons (electrons on the outer shell) to a certain degree. Thus the core electrons act as a shield between nucleus and valence electrons. Due to shielding or screening effect by the core electron on the attraction between valence electron and nucleus, the effective nuclear charge ($Z_{eff}\, e$) becomes less than +Ze. Considering the screening effect, the energy of atomic states is given by the following expression:

$$E_n = -\frac{Rhc\, Z_{eff}^2}{n^2} = -\frac{Rhc(Z-\sigma)^2}{n^2} \tag{4.124}$$

where R is Rydberg constant and σ is termed as screening constant which is different for the different ℓ-states of a given value of n. Due to screening effect, the energy of orbital increases with increases in ℓ for a given n. So, states s, p, d and f have increasing pattern of energy. Since the value of Z_{eff} is largest for 3s-state, it is much lower than the 3p-state. Furthermore, energy difference between states decreases with increase in n. Due to this, higher levels like 5d and 5f almost coincides. The core electrons do not take part in transition to give the spectrum. Only the valence electron and excited state electrons take part in transition to form the spectrum.

Emission spectra

The emission spectrum of such atom has four series of lines:

 (i) Principal series
 (ii) Sharp series
 (iii) Diffuse series and
 (iv) Fundamental or Bebergmann series.

The separation of lines and intensity of series lines decreases continuously. All the series overlap partially with one another. The lines of diffuse and sharp series appear diffuse and sharp as their names indicate. The fundamental series lies in infrared region.

The four series of lines indicate that alkali atom have four set of energy levels termed as $S(1s, 2s, 3s, \ldots)$, $P(2p, 3p, 4p, \ldots)$, $D(3d, 4d, 5d, \ldots)$ and $F(4f, 5f, 6f, \ldots)$.

The transition in these states follow the condition Δn = any integer and $\Delta \ell = \pm 1$. The transition from higher P-states to ground state of atom gives the Principal series and the transition from excited higher S-states to first excited P-state provides the sharp series. Similarly, the transition from D-states to first excited P-state and transition from F-states to first excited D-state provide the diffuse and fundamental series respectively. The electronic configuration of Li, Na and K are shown in Table 4.15.

Table 4.15 Electronic Configuration

Atom	Z	Electronic configuration	Orbitals filled with the electrons
Li	3	$1s^2, 2s^1$	[↓↑] $1s$ [↑] $2s$
Na	11	$1s^2, 2s^2, 2p^6, 3s^1$	[↓↑] $1s$ [↓↑] $2s$ [↓↑][↓↑][↓↑] $2p$ [↑] $3s$
K	19	$1s^2, 2s^2, 2p^6, 3s^2, 3p^6, 4s^1$	[↓↑] $1s$ [↓↑] $2s$ [↓↑][↓↑][↓↑] $2p$ [↓↑] $3s$ [↓↑][↓↑][↓↑] $3p$ [↑] $4s$

It is clear from Table 4.15 that except Li, the completed highest subshell is p subshell and the next s subshell is the outermost one. In the process of excitation, only this outermost electron participates and gives rise to optical spectra. For Li and Na, the transitions giving rise to four series of lines are presented in Table 4.16.

Table 4.16 Electronic Transitions for Four Series of Spectral Lines in Li and Na

Atom and transition → series ↓	Transition	
	Li	Na
Principal	$mp \rightarrow 1s$; $m = 2, 3, ..., \infty$	$mp \rightarrow 3s$; $m = 3, 4, ..., \infty$
Sharp	$ms \rightarrow 2p$; $m = 2, 3, ..., \infty$	$ms \rightarrow 3p$; $m = 4, 5, ..., \infty$
Diffuse	$md \rightarrow 2p$; $m = 3, 4, ..., \infty$	$md \rightarrow 3p$; $m = 3, 4, ..., \infty$
Fundamental	$mf \rightarrow 3d$; $m = 4, 5, ..., \infty$	$mf \rightarrow 3d$; $m = 4, 5, ..., \infty$

The wave number corresponding to the series lines of alkali spectra cannot be exactly represented by Balmer formula, but it is analogues to Balmer formula with modification. Rydberg gave the expressions for the wave number $\bar{v}$ of series lines, which are shown in Table 4.17.

In all expressions of wave number (Table 4.17), the first and second terms are called *fixed term* and *running term* respectively. The first term provides wave number corresponding to convergence limit of the series and the quantities s, p, d and f are called Rydberg corrections. The energy states and transition among them for Li and Na are shown in Figures 4.36 and 4.37.

Table 4.17 Wave Number of Series Lines

Atom and $\bar{\nu} \rightarrow$ series $\downarrow$	$\bar{\nu}$	
	Li	Na
Principal	$\bar{\nu}_P = \dfrac{R}{(1+s)^2} - \dfrac{R}{(m+p)^2}$; $m = 2, 3, ..., \infty$	$\bar{\nu}_P = \dfrac{R}{(3+s)^2} - \dfrac{R}{(m+p)^2}$; $m = 3, 4, ..., \infty$
Sharp	$\bar{\nu}_S = \dfrac{R}{(2+p)^2} - \dfrac{R}{(m+s)^2}$; $m = 2, 3, ..., \infty$	$\bar{\nu}_S = \dfrac{R}{(3+p)^2} - \dfrac{R}{(m+s)^2}$; $m = 4, 5, ..., \infty$
Diffuse	$\bar{\nu}_D = \dfrac{R}{(2+p)^2} - \dfrac{R}{(m+d)^2}$; $m = 3, 4, ..., \infty$	$\bar{\nu}_D = \dfrac{R}{(3+p)^2} - \dfrac{R}{(m+d)^2}$; $m = 3, 4, ..., \infty$
Fundamental	$\bar{\nu}_B = \dfrac{R}{(3+d)^2} - \dfrac{R}{(m+f)^2}$; $m = 4, 5, ..., \infty$	$\bar{\nu}_B = \dfrac{R}{(3+d)^2} - \dfrac{R}{(m+f)^2}$; $m = 4, 5, ..., \infty$

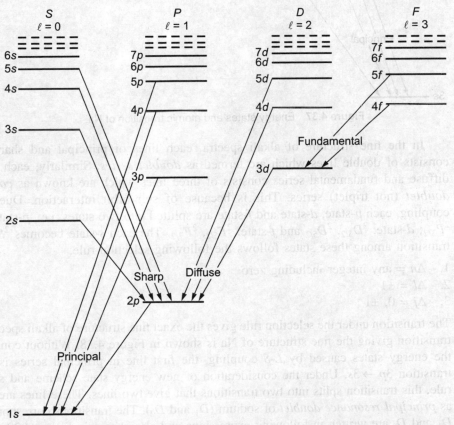

Figure 4.36 Energy states and atomic transition of Li.

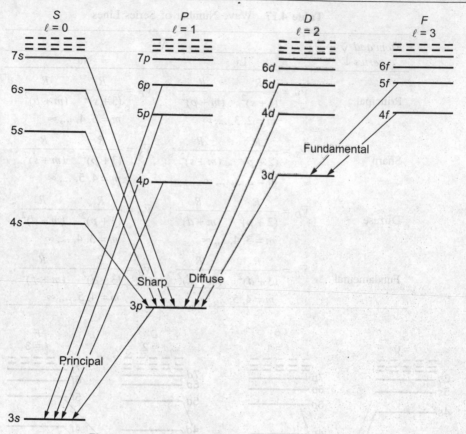

Figure 4.37 Energy states and atomic transition of Na.

In the fine structure of alkali spectra, each lines of principal and sharp series consists of double lines which are termed as *doublet series*. Similarly, each lines of diffuse and fundamental series consists of three lines which are known as *compound doublet* (not triplet) series. This is because of spin-orbit interaction. Due to *L-S* coupling, each *p*-state, *d*-state and *f*-state are splitted into two states, i.e., *p*-state: $^2P_{1/2}$, $^2P_{3/2}$; *d*-state: $^2D_{3/2}$, $^2D_{5/2}$ and *f*-state: $^2F_{5/2}$, $^2F_{7/2}$. The each *s*-state becomes $^2S_{1/2}$. The transition among these states follows the following selection rule:

1. Δn = any integer including zero
2. $\Delta l = \pm 1$
3. $\Delta j = 0, \pm 1$

The transition under the selection rule gives the exact fine structure of alkali spectra. The transition giving the fine structure of Na is shown in Figure 4.38. Without considering the energy states caused by *L-S* coupling, the first line in principal series is due to transition $3p \rightarrow 3s$. Under the consideration of new energy state scheme and selection rule, this transition splits into two transitions that give two lines. These lines are termed as *principal resonance doublet* of sodium (D_1 and D_2). The transitions corresponding to D_1 and D_2 are written in following expressions and also shown in Figure 4.39.

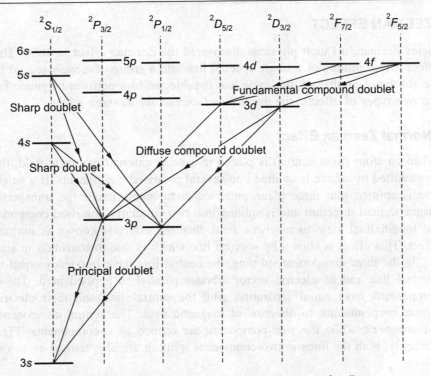

Figure 4.38 Atomic transition for the fine structure of sodium.

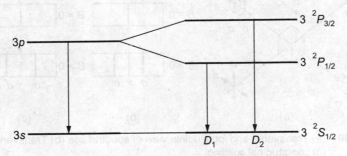

Figure 4.39 Transitions for the D_1 and D_2 of sodium.

D_1: $\lambda = 5896$ Å: caused by transition: $3^2P_{1/2} \rightarrow 3^2S_{1/2}$

D_2: $\lambda = 5890$ Å: caused by transition: $3^2P_{3/2} \rightarrow 3^2S_{1/2}$

Absorption spectra

The absorption spectra of alkali vapours consist of a series of lines similar to absorption spectra of hydrogen. The series of lines have a regular decrease in separation and intensity. The lines are due to atomic transition from valence s-state to p-state. The series of lines are called *Principal series*. The wave number of series lines can be determined with the expression written in Table 4.16.

4.20 ZEEMAN EFFECT

Pieter Zeeman, a Dutch physicist discovered the Zeeman effect in 1896. The Zeeman effect is the splitting of a single spectral line into a group of closely spaced lines when the substance producing the single line is subjected to a uniform magnetic field. There are two types of effects, the *normal and anomalous Zeeman effects*.

4.20.1 Normal Zeeman Effect

When an atom (light source) is placed in a weak external magnetic field, the spectral line emitted by source is splitted into several polarised components. If a single spectral line is splitted into three plane polarised components under the transverse view to magnetic field direction and is splitted into two circularly polarised components under the longitudinal view to magnetic field, then the effect is known as normal Zeeman effect. This effect is shown by spectral line which is due to transition in singlet state.

In the three-component splitting, the central line have frequency equal to original spectral line and its electric vector vibrates parallel to applied field. The two other components have equal separation with the central line and their electric vectors vibrate perpendicular to direction of magnetic field. The central component is called π-component, while the side component are termed as σ-components (Figure 4.40). Similarly, both the lines in two-component splitting are also termed as σ-components.

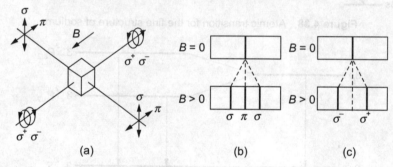

Figure 4.40 (a) Transverse and longitudinal view of spectral line (b) Transverse splitting (c) Longitudinal splitting.

Explanation

The normal Zeeman effect can be explained on the basis of space quantisation. When atom is placed in external magnetic field, then the orbital angular momentum of atom interacts with the magnetic field. Due to interaction, the orbital angular momentum vector **L** precesses around the direction of magnetic field, which is known as Larmour precession. If **L** makes an angle θ with the magnetic field and L_Z is component of $|\mathbf{L}|$ along the magnetic field direction.

$$L_Z = |\mathbf{L}| \cos \theta = \mathbf{L} \cos \theta$$

$$\cos \theta = \frac{L_Z}{L} = \frac{m_\ell \hbar}{L} \tag{4.125}$$

where m_ℓ is called magnetic quantum number. Let ΔE is the interaction energy of atom in the magnetic field, then

$$\Delta E = \mu_L B \cos \theta \qquad (4.126)$$

The orbital magnetic moment μ_L is given by the following expression:

$$\mu_L = g_\ell \frac{e}{2m} L; \, g_\ell = 1 \qquad (4.127)$$

Putting values in Eq. (4.126), we have

$$\Delta E = g_\ell \frac{e}{2m} LB \frac{m_\ell \hbar}{L}$$

$$\Delta E = g_\ell \, m_\ell \frac{e\hbar}{2m} B$$

$$\Delta E = g_\ell \, m_\ell \frac{eh}{4\pi m} B$$

$$\boxed{\Delta E = g_\ell \, m_\ell \, \mu_B B} \qquad (4.128)$$

Here μ_B is called Bohr magnetron. The m_ℓ can have $0, \pm 1, \pm 2, \ldots, \pm \ell$ values. In the terms of term value, the interaction energy can be given as:

$$-\Delta T = \frac{\Delta E}{hc} = g_\ell m_\ell \frac{eh}{4\pi m} B \frac{1}{hc} = g_\ell m_\ell \frac{eB}{4\pi mc}; \, g_\ell = 1$$

$$\boxed{-\Delta T = g_\ell m_\ell \frac{eB}{4\pi mc} = g_\ell m_\ell L' = m_\ell L'} \qquad (4.129)$$

Here L' is called Lorentz unit. So the shift of energy can also be determined in the unit of L'. If E_0 is energy of atomic state at $B = 0$ and E is energy of atomic state in magnetic field, then

$$E = E_0 + \Delta E = E_0 + g_\ell m_\ell \mu_B B \qquad (4.130)$$

Since m_ℓ have $(2\ell + 1)$ values for a given orbital quantum number ℓ, an energy state splits into $(2\ell + 1)$ states, i.e.,

1. When $\ell = 0$, $m_\ell = 0$, then $E = E_0$
2. When $\ell = 1$, $m_\ell = 0, \pm 1$, then $E = E_0, E_0 \pm \mu_B B$
3. When $\ell = 2$, $m_\ell = 0, \pm 1, \pm 2$, then $E = E_0, E_0 \pm \mu_B B, E_0 \pm 2\mu_B B$

Suppose a transition occurs between the initial state E^i and final state E^f. Using Eq. (4.130), we can write

$$E^f - E^i = (E_0^f + m_\ell^f \mu_B B) - (E_0^i + m_\ell^i \mu_B B)$$

$$E^f - E^i = (E_0^f - E_0^i) + (m_\ell^f - m_\ell^i) \mu_B B$$

$$h\nu = h\nu_0 + \Delta m_\ell \, \mu_B B$$

$$\boxed{\nu = \nu_0 + \Delta m_\ell \frac{eB}{4\pi m}} \qquad (4.131)$$

The selection rule for transition is $\Delta m_\ell = 0, \pm 1$, thus the spectral line in magnetic field can have three frequencies: $v_0 - \dfrac{eB}{4\pi m}$, v_0 and $v_0 + \dfrac{eB}{4\pi m}$. Therefore, a single line is splitted into three components in the magnetic field.

In absence of magnetic field, the transitions $^1D_2 \to {}^1P_1$, $^1P_1 \to {}^1S_0$ and $^1F_3 \to {}^1D_2$ give a single spectral line. Since in these transitions, all the terms have spin zero (singlet state), thus in the presence of magnetic field, the spectral lines by these transitions split into three components as shown in Figure 4.41.

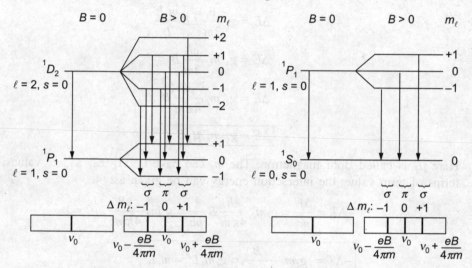

Figure 4.41 Zeeman splitting of spectral lines of $^1D_2 \to {}^1P_1$ and $^1P_1 \to {}^1S_0$ transitions.

EXAMPLE 4.8 A light source is placed in magnetic field of 0.3 W/m². Find the change in wavelength for the spectral line of wavelength 4500 Å.

Solution We know that,

$$\Delta v = \frac{eB}{4\pi m} = \frac{1.6 \times 10^{-19} \times 0.3}{4 \times 3.14 \times 9.1 \times 10^{-31}} = \frac{0.48 \times 10^{12}}{114.296} = 4.199 \times 10^9 \, \text{Hz}$$

$$\because \qquad v\lambda = c$$

$$\therefore \qquad v\Delta\lambda + \lambda\Delta v = 0$$

$$\because \qquad v\Delta\lambda = -\lambda\Delta v$$

$$\therefore \qquad \Delta\lambda = -\frac{\lambda}{v}\Delta v$$

$$\therefore \qquad |\Delta\lambda| = \frac{\lambda}{v}\Delta v = \frac{\lambda^2}{c}\Delta v = \frac{(4500 \times 10^{-10})^2}{3 \times 10^8} = 4.199 \times 10^9$$

$$\therefore \qquad |\Delta\lambda| = \frac{85.029 \times 10^{-5}}{3 \times 10^8} = 28.34 \times 10^{-13} \, \text{m} = 0.0283 \, \text{Å}$$

4.20.2 Anomalous Zeeman Effect

The complex splitting of spectral lines is called *anomalous Zeeman effect*. In such splitting, a single spectral line is splitted into number of spectral lines. The splitted lines have more than one π-component and σ-component. Such splitting is found for the transition in doublet and triplet states.

Example: D_1 and D_2 lines of sodium are caused by transitions $3^2P_{1/2} \rightarrow 3^2S_{1/2}$ and $3^2P_{3/2} \rightarrow 3^2S_{1/2}$ respectively, but in presence of magnetic field, these lines split into four and six components respectively as shown in Figure 4.42.

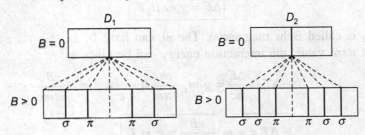

Figure 4.42 Anomalous Zeeman splitting of D_1 and D_2 lines of sodium.

Explanation

For the doublet and triplet states, the spin of electron is not zero, but it has finite value. The electron having orbital and spin motion, possesses two types of angular momenta (orbital and spin angular momentum). These two angular momenta rotates about the common axis whose resultant gives the total angular momentum of atom. Thus the total angular momentum of atom is vector combination of these two momenta. When such atom is placed in external magnetic field, then the total angular momentum of atom interacts with the magnetic field. Due to interaction, the total angular momentum vector **J** precesses around the direction of magnetic field. If **J** makes an angle θ with the direction of magnetic field, and J_Z is component of **J** along the magnetic field direction, then

$$J_Z = |\mathbf{J}| \cos\theta = J \cos\theta$$

$$\cos\theta = \frac{J_Z}{J} = \frac{m_j \hbar}{J} \tag{4.132}$$

where m_j is called *total magnetic quantum number*. Let ΔE is the interaction energy of atom in the magnetic field, then

$$\Delta E = \mu_j B \cos\theta \tag{4.133}$$

The total magnetic moment μ_j of an atom is given by the following expression:

$$\mu_j = g \frac{e}{2m} J \tag{4.134}$$

The quantity g is called as Lande g-factor, and it can be found by the following expression (see Section 4.16).

$$g = g_j = 1 + \frac{j(j+1) + s(s+1) - \ell(\ell+1)}{2j(j+1)}$$

Putting values in Eq. (4.133), we have

$$\Delta E = g_j \frac{e}{2m} JB \frac{m_j \hbar}{J}$$

$$\Delta E = g_j m_j \frac{e\hbar}{2m} B$$

$$\Delta E = g_j m_j \frac{eh}{4\pi m} B$$

$$\boxed{\Delta E = g_j m_j \mu_B B} \tag{4.135}$$

Here μ_B is called Bohr magnetron. The m_j can have 0, ±1, ±2, ..., ±j values. In the terms of term value, the interaction energy can be given as:

$$-\Delta T = \frac{\Delta E}{hc} = g_j m_j \frac{eh}{4\pi m} B \frac{1}{h\,c} = g_j m_j \frac{eB}{4\pi mc}$$

$$\boxed{-\Delta T = g_j m_j \frac{eB}{4\pi mc} = g_j m_j L'} \tag{4.136}$$

Here L' is called Lorentz unit. So the shift of energy can also be determined in the unit of L'. Since g_j is different for different j-states, Zeeman splitting will be different for different j-states. If E_0 is energy of atomic state at $B = 0$, and E is energy of atomic state in magnetic field, then

$$E = E_0 + \Delta E = E_0 + g_j m_j \mu_B B \tag{4.137}$$

Since m_j have $(2j + 1)$ values for a total quantum number j, energy state splits into equispaced $(2j + 1)$ Zeeman states. Suppose a transition occurs between the initial state E^i and final state E^f. Using Eq. (4.137), we can write

$$E^f - E^i = (E_0^f + g_j^f m_j^f \mu_B B) - (E_0^i + g_j^i m_j^i \mu_B B)$$

$$E^f - E^i = (E_0^f - E_0^i) + (g_j^f m_j^f - g_j^i m_j^i)\mu_B B$$

$$h\nu = h\nu_0 + (g_j^f m_j^f - g_j^i m_j^i)\,\mu_B B$$

$$\boxed{\nu = \nu_0 + (g_j^f m_j^f - g_j^i m_j^i)\,\frac{eB}{4\pi m}} \tag{4.138}$$

Here ν_0 is the frequency of spectral line with $B = 0$, while ν is the frequency of spectral lines with $B > 0$. The selection rule for transition among the Zeeman states is $\Delta m_j = 0, \pm 1$ but $m_j: 0 \to 0$ is not allowed if $\Delta J = 0$. The transitions under $\Delta m_j = 0$ provide π-components, while transitions under $\Delta m_j = \pm 1$ give σ-components. Since the quantity $(g_j^f m_j^f - g_j^i m_j^i)$ can have number of values, a single spectral line is splitted into number of components in the magnetic field. The anomalous Zeeman splitting of D_1 and D_2 lines of sodium is shown in Figure 4.43. Similarly Zeeman splitting for transition $2^3S_1 \to 2^3P_2$ is also shown in Figure 4.44.

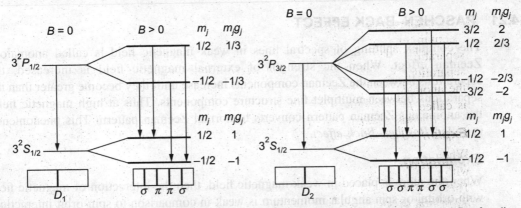

Figure 4.43 Electronic transitions for anomalous Zeeman splitting of spectral line D_1 and D_2 of sodium.

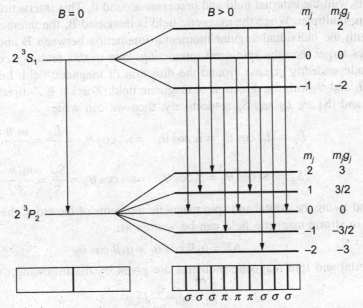

Figure 4.44 Anomalous Zeeman splitting of spectral line corresponding to transition $2\,^3S_1 \rightarrow 2\,^3P_2$.

EXAMPLE 4.9 Calculate the Lande g-factor for state $^3D_2$. Find the magnetic moment of this state and number of splitted substates in weak magnetic field.

Solution For state $^3D_2$, $\ell = 2$; $2s + 1 = 3 \Rightarrow s = 1$, $j = 2$

$$g = 1 + \frac{j(j+1) + s(s+1) - \ell(\ell+1)}{2j(j+1)}$$

$$g = 1 + \frac{2(2+1) + (1+1) - 2(2+1)}{2 \times 2(2+1)} = 1 + \frac{1}{6} = \frac{7}{6}$$

$$\mu = \sqrt{j(j+1)}\, g\mu_B = \sqrt{2(2+1)}\,\frac{7}{6}\mu_B = \frac{7}{\sqrt{6}}\mu_B$$

Number of splitted states = $2j + 1 = 2 \times 2 + 1 = 5$

4.21   PASCHEN–BACK EFFECT

The complex splitting of spectral lines in weak magnetic field is called anomalous Zeeman effect. When the strength of external magnetic field is increased, the separations between the Zeeman components increase until they become greater than the separations between multiplet fine structure components. Thus at high magnetic field, the anomalous Zeeman pattern converts to normal Zeeman pattern. This phenomenon is called *Paschen–Back effect*.

Explanation

When the atom is placed in weak magnetic field, then the interaction of magnetic field with orbital or spin angular momentum is weak in comparison to spin-orbit interaction. Therefore, the resultant of orbit and spin angular momenta (total angular momentum) interacts with the external field and precesses around it. This interaction causes the fine structure multiplet. When the magnetic field is increased **B**, the interaction of magnetic field with the individual angular momenta (interaction between **B** and **L** or **B** and **S**) becomes larger than the spin-orbit interaction. Due to this reason, vectors **L** and **S** of atom independently precess around the direction of magnetic field. Let **L** and **S** make angle θ_L and θ_S with the direction of magnetic field (Z-axis). If Z-direction components of $|\mathbf{L}|$ and $|\mathbf{S}|$ are L_Z and S_Z respectively, then we can write

$$L_Z = |\mathbf{L}| \cos \theta_L = L \cos \theta_L \quad \Rightarrow \quad \cos \theta_L = \frac{L_Z}{L} = \frac{m_\ell \hbar}{L} \tag{4.139}$$

and,

$$S_Z = |\mathbf{S}| \cos \theta_S = S \cos \theta_S \quad \Rightarrow \quad \cos \theta_S = \frac{S_Z}{S} = \frac{m_s \hbar}{S} \tag{4.140}$$

If μ_L and μ_S are the orbital and spin magnetic moments of the atom, then the interaction energy in strong magnetic field can be written as:

$$\Delta E = \mu_L B \cos \theta_L + \mu_S B \cos \theta_S \tag{4.141}$$

The orbital and spin magnetic moments are given by the following expressions:

$$\mu_L = g_\ell \frac{e}{2m} L; \, g_\ell = 1 \tag{4.142}$$

$$\mu_S = g_S \frac{e}{2m} S; \, g_S = 2 \tag{4.143}$$

Putting values in Eq. (4.141), we have

$$\Delta E = g_\ell \frac{e}{2m} LB \frac{m_\ell \hbar}{L} + g_S \frac{e}{2m} SB \frac{m_s \hbar}{S}$$

$$\Delta E = g_\ell m_\ell \frac{e\hbar}{2m} B + g_S m_s \frac{e\hbar}{2m} B$$

$$\Delta E = (g_\ell m_\ell + g_S m_s) \frac{e\hbar}{2m} B$$

$$\boxed{\Delta E = (m_\ell + 2m_s) \mu_B B} \tag{4.144}$$

Equation (4.144) is expression of the strong field interaction energy. In the terms of term value, the interaction energy can be written as:

$$-\Delta T = \frac{\Delta E}{hc} = (m_\ell + 2m_S)\frac{\mu_B B}{hc}$$

Since $\qquad \dfrac{\mu_B B}{hc} = \dfrac{e\hbar}{2m}\dfrac{B}{hc} = \dfrac{eh}{4\pi m}\dfrac{B}{hc} = \dfrac{eB}{4\pi mc} = L'\,(\text{Lorentz unit})$

Thus $\qquad\boxed{-\Delta T = (m_\ell + 2m_S)L'} \qquad\qquad\qquad$ (4.145)

Since **L** and **S** have $(2\ell + 1)$ and $(2s + 1)$ orientations respectively, a single level splits into $(2\ell + 1)(2s + 1)$ levels in strong magnetic field. Since $s = 1/2$, so the total number of levels in strong magnetic field becomes $4\ell + 2$. In weak magnetic field, the number of splitted levels corresponding to total quantum number j is $(2j + 1)$.

For a single valence electron system, j can have values $\ell + \dfrac{1}{2}$ and $\ell - \dfrac{1}{2}$. Thus the number of splitted level in weak magnetic field becomes equal to

$$\left\{2\left(\ell + \frac{1}{2}\right) + 1\right\} + \left\{2\left(\ell - \frac{1}{2}\right) + 1\right\} = 4\ell + 2.$$ Thus in strong magnetic field, the number

of splitted levels becomes equal to normal Zeeman levels. If E_0 is energy of atomic state at $B = 0$, and E is energy of atomic state in magnetic field, then

$$E = E_0 + \Delta E = E_0 + (m_\ell + 2m_S)\mu_B B; \ m_S = \pm 1/2$$
$$E = E_0 + \Delta E = E_0 + m_\ell \mu_B B \pm \mu_B B \qquad\qquad (4.146)$$

Equation (4.146) is same as for the normal Zeeman Effect, except an additional term $\pm\mu_B B$. The selection rule for transition in splitted states are $\Delta m_\ell = 0, \pm 1$ and $\Delta m_S = 0$. When the transition occurs under the selection rule, the additional term is cancelled. Thus the fine structure multiplet converts to normal Zeeman effect.

▌4.22 STARK EFFECT

The splitting of spectral line in electric field is called *stark effect*. This effect is analogous to Zeeman effect. It is found that every line of Balmer series of hydrogen splits into number of components in presence of strong electric field. When viewed perpendicular to the field direction, some of components are found to be plane polarised with the electric vector parallel to the field and other polarised with the electric vector normal to the field. The interaction energy of hydrogen like atom in an electric field is given by

$$\Delta T = AF + BF^2 + CF^3 + \cdots \qquad\qquad (4.147)$$

where ΔT represents the change in the term value of the atom in wave number or it is a shift in energy level from the field free states to the states in the electric field. The quantity F is the strength of the electric field. The coefficients A, B and C can be determined from the classical and quantum mechanical considerations. If $\Delta T \propto F$, then it is called *first order stark effect*. Similarly, the effect is termed as *second order stark effect* for $\Delta T \propto F^2$. The term shift for first order stark effect is given by the following expression:

$$\Delta T = \frac{3a_0 e}{2Zhc} Fn(n_2 - n_1) \qquad (4.148)$$

Here a_0 is the Bohr radius. The n is usual total quantum number. The quantities n_1 and n_2 are called *parabolic quantum number*. The quantum numbers n, n_1 and n_2 are considered in such a way that they result in electric quantum number m_L, i.e., ($m_L = n - n_1 - n_2 - 1$). The allowed values of all these quantum numbers are given as follows.

$$n = 1, 2, 3, \ldots \infty$$
$$m_L = 0, \pm 1, \pm 2, \pm 3, \ldots, \pm(n-1)$$
$$n_1 = 1, 2, 3, \ldots, (n-1)$$
$$n_2 = 1, 2, 3, \ldots, (n-1)$$

If field is in volt/cm, then putting the values of Bohr radius, electric charge, Planck's constant and light velocity in Eq. (4.148), we have

$$\Delta T = \frac{6.4 \times 10^{-5}}{Z} Fn(n_2 - n_1)\,\text{cm}^{-1} \qquad (4.149)$$

Thus the levels under stark effect for the hydrogen like atom are shifted by integral multiple of a fundamental amount from their original level.

4.23 BREADTH IN SPECTRAL LINE

The experimental measurement of spectrum indicates that the spectral lines are not sharp, but it has finite and measurable width which can be precisely defined over a range of wavelength. The sharp and diffuse series of each alkali metal are the good examples of this. The breadth in spectral line is defined as separation in cm^{-1} between the two points of line along its width whose intensity are half of its central maximum intensity. The breadth in spectral line is caused by following reasons:

Natural breadth

The spectrum arises due to transition of electron between the atomic states. From the Heisenberg uncertainty principle, the energy of atomic states cannot be defined exactly, but the product of uncertainty in energy ΔE and the mean of atom Δt is in order of Planck's constant, i.e.,

$$\Delta E \cdot \Delta t \approx h$$
$$h \Delta \nu \cdot \tau \approx h$$
$$\Delta \nu \approx \frac{1}{\tau} \qquad (4.150)$$

Since for ground state meta-stable states the mean life time is large, these states will be sharp, while the other excited states will have finite breadth in energy or frequency. Quantum mechanically the half intensity breadth of a spectral line due to transition from excited state to ground state in term of frequency and wavelength range is given by the following expressions:

$$\Delta v = \frac{1}{2\pi\tau} \tag{4.151}$$

$$\Delta\lambda = \frac{\lambda^2}{c}\Delta v \tag{4.152}$$

This is the expression for the determination of natural breadth. The natural breadth is much small for the measurement.

EXAMPLE 4.10 Calculate the natural breadth of spectral line in terms of wavelength if mean life time for the excited stated of an atom is 10^{-8} s and atom emits radiation of wavelength 6000 Å.

Solution The natural breadth of a spectral line in terms of frequency and wavelength range is given by the following expressions:

$$\Delta v = \frac{1}{2\pi\tau} \text{ and } \Delta\lambda = \frac{\lambda^2}{c}\Delta v$$

$$\Delta v = \frac{1}{2\times3.14\times10^{-8}} = 1.59\times10^7 \text{ s}^{-1}$$

$$\Delta\lambda = \frac{(6\times10^{-7})^2}{3\times10^8}1.59\times10^7 = 7.547\times10^{-15} = 0.75\times10^{-4} \text{ Å}$$

Doppler effect

The change in frequency of spectral line due to relative motion of source of light and observer is called Doppler effect. The observed frequency increases when the source moves towards observer, while it decreases when source moves away from observer. This effect plays an important role in broadening of spectral line. Let a source of light of frequency v_0 moves with velocity v relative to observer. If v is the observed frequency, then change in frequency can be written as:

$$\Delta v = v - v_0 = \pm v_0\frac{v}{c} \tag{4.153}$$

Here positive sign is for the motion towards observer, while the negative sign is for motion away from observer. In the source like a discharge lamp, the atoms emitting the light have velocities due to thermal agitation. If we consider that distribution of atom velocity follows the Maxwell distribution, then the relative intensity of source can be written as:

$$I(v) = I_0 \exp\left(-\frac{\mu v^2}{2RT}\right) \tag{4.154}$$

Here I_0 is maximum intensity. The quantities μ and R are the atomic weight and universal gas constant. Taking positive sign in Eq. (4.153), we have $v = \dfrac{v - v_0}{v_0} c$. Then Eq. (4.154) takes the following form:

$$I(v) = I_0 \exp\left(-\frac{\mu c^2}{2RT}\left(\frac{v - v_0}{v_0}\right)^2\right) \qquad (4.155)$$

The variation of $I(v)$ with frequency is shown in Figure 4.45. For the determination of breadth of spectral, let the intensity at frequency v decays to half of the maximum intensity, then

$$\frac{I_0}{2} = I_0 \exp\left(-\frac{\mu c^2}{2RT}\left(\frac{v - v_0}{v_0}\right)^2\right)$$

$$\frac{1}{2} = \exp\left(-\frac{\mu c^2}{2RT}\left(\frac{v - v_0}{v_0}\right)^2\right)$$

$$\ln 2 = \frac{\mu c^2}{2RT}\left(\frac{v - v_0}{v_0}\right)^2$$

$$v - v_0 = \frac{v_0}{c}\sqrt{\frac{2RT}{\mu}\ln 2}$$

$$2(v - v_0) = \frac{2v_0}{c}\sqrt{\frac{2RT}{\mu}\ln 2}$$

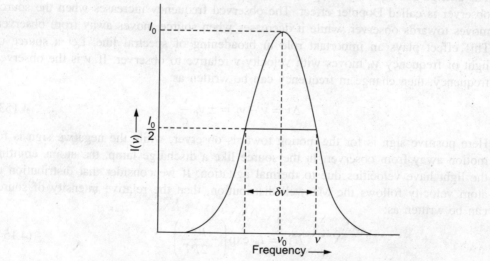

Figure 4.45   Variation in intensity of moving source with frequency.

$$\delta v = \frac{2v_0}{c}\sqrt{\frac{2RT}{\mu}}\ln 2$$

$$\delta v = 1.67\frac{v_0}{c}\sqrt{\frac{2RT}{\mu}} \qquad (4.156)$$

In terms of wavelength, Eq. (4.156) can be written as:

$$\delta\lambda = 1.67\frac{\lambda_0}{c}\sqrt{\frac{2RT}{\mu}} \qquad (4.157)$$

Equation (4.156) indicates that the line breadth due to Doppler broadening increases with temperature and decreases with increasing the atomic weight μ.

EXAMPLE 4.11 Calculate the Doppler half intensity breadth of the sodium spectral of wavelength 5893 Å corresponding to temperature 500 K. $R = 8.31$ erg/mole-K, atomic weight of sodium = 22.99 g/mole.

Solution The expression of Doppler half intensity breadth is given by

$$\delta\lambda = 1.67\frac{\lambda_0}{c}\sqrt{\frac{2RT}{\mu}}$$

$$\delta\lambda = 1.67\frac{(5893\times10^{-8}\ \text{cm})}{(3\times10^{10}\ \text{cm/s})}\sqrt{\frac{2\times(8.31\times10^7\ \text{erg/mole-K})\times500\ \text{K}}{22.99\ \text{gm/mole}}}$$

$$\delta\lambda = 1.97\times10^{-10}\ \text{cm}$$

$$\delta\lambda = 0.0197\ \text{Å}$$

EXAMPLE 4.12 Calculate the change in wavelength of spectral line (6000 Å) by source if it is moving with speed 6×10^4 m/s away from the observer.

Solution We know that, if source is moving towards the observer, then apparent frequency of spectral line can be given by the following expression:

$$v = v_0\left(1-\frac{v}{c}\right) \Rightarrow \frac{c}{\lambda} = \frac{c}{\lambda_0}\left(1-\frac{v}{c}\right) \Rightarrow \lambda = \lambda_0\left(1+\frac{v}{c}\right) \Rightarrow \lambda-\lambda_0 = \lambda_0\frac{v}{c} \Rightarrow \Delta\lambda = \lambda_0\frac{v}{c}$$

$$\Delta\lambda = \lambda_0\frac{v}{c} = 6000\frac{6\times10^4}{3\times10^8} = 6000\times2\times10^{-4} = 1.2\ \text{Å}$$

External effects

The atomic collision caused by high pressure and voltage causes the change in frequency of radiation. Thus it also contributes in broadening of spectral line.

| SOLVED NUMERICAL PROBLEMS |

PROBLEM 4.1   If the wavelength of first line of Lyman series of hydrogen is 1215 Å, then find the wavelength of second line of Lyman series.

Solution   The wavelength of Lyman series line of hydrogen is given by the following expression:

$$\frac{1}{\lambda} = R_H \left(\frac{1}{1^2} - \frac{1}{n_i^2} \right); n_i = 2, 3, 4, 5, 6, 7, \ldots$$

For the first and second line of Lyman series the expression will be

$$\frac{1}{\lambda_1} = R_H \left(\frac{1}{1^2} - \frac{1}{2^2} \right) = \frac{3R_H}{4}$$

$$\frac{1}{\lambda_2} = R_H \left(\frac{1}{1^2} - \frac{1}{3^2} \right) = \frac{8R_H}{9}$$

$$\Rightarrow \qquad \frac{\lambda_2}{\lambda_1} = \frac{3R_H}{4} \times \frac{9}{8R_H} = \frac{27}{32}$$

$$\Rightarrow \qquad \lambda_2 = \frac{27}{32} \lambda_1 = \frac{27}{32} \times 1215 = 1025.16 \approx 1025 \text{ Å}$$

PROBLEM 4.2   Find the series limit wavelength of Balmer series of hydrogen. ($R_H = 1.097 \times 10^5 \text{cm}^{-1}$).

Solution   The wavelength of Balmer series line of hydrogen is given by the following expression:

$$\frac{1}{\lambda} = R_H \left(\frac{1}{2^2} - \frac{1}{n_i^2} \right); n_i = 3, 4, 5, 6, 7, \ldots \infty$$

For the series limit wavelength of Balmer series of hydrogen, $n_i = \infty$

$$\frac{1}{\lambda} = R_H \left(\frac{1}{2^2} - \frac{1}{\infty} \right) = \frac{R_H}{4}$$

$$\lambda = \frac{4}{R_H} = \frac{4}{1.097 \times 10^7} = 3.646 \times 10^{-7} \text{m} = 3646 \times 10^{-10} \text{m} = 3646 \text{ Å}$$

PROBLEM 4.3   Calculate the third excitation potential and ionization potential for the hydrogen.

Solution   The energy of stationary states for the hydrogen is given by the following expression:

$$E_n = -\frac{13.6}{n^2} \text{eV}$$

So,

$$E_1 = -\frac{13.6}{1^2} = -13.6 \text{ eV}$$

$$E_3 = -\frac{13.6}{3^2} = -1.5 \text{ eV}$$

$$E_\infty = -\frac{13.6}{\infty^2} = 0 \text{ eV}$$

Third excitation potential $= E_3 - E_1 = - = -1.5 + 13.6 = 12.1$ eV
Ionization potential $= E_\infty - E_1 = - = 0 + 13.6 = 13.6$ eV

PROBLEM 4.4 If ionization potential for hydrogen is 13.6 eV, then find ionization potential for He$^+$.

Solution $\because \quad E_n = -\frac{13.6\,Z^2}{n^2}$ eV

$\therefore \quad$ Ionization potential $= IP = E_\infty - E_1 \propto Z^2$

$$\frac{IP_{He^+}}{IP_H} = \frac{Z^2_{He^+}}{Z^2_H} = \frac{2^2}{1^2} = 4$$

$$IP_{He^+} = 4 \times IP_H = 4 \times 13.6 = 54.4 \text{ eV}$$

PROBLEM 4.5 Calculate the shift of H_α line of hydrogen if hydrogen contains its mixture of two isotopes protium and titrium. (mass of electron $= 9.1 \times 10^{-28}$ g, mass of proton $= 1.67 \times 10^{-24}$ g, $R_\infty = 1.097 \times 10^5$ cm^{-1}).

Solution The wavelength of H_α line of hydrogen is given by the following expression:

$$\frac{1}{\lambda_H} = R_H \left(\frac{1}{2^2} - \frac{1}{3^2} \right) = \frac{5}{36} R_H$$

λ_P and λ_T are the wavelength of H_α line of hydrogen isotopes protium and titrium, then

$$\frac{1}{\lambda_P} = \frac{5}{36} R_P \quad \Rightarrow \quad \lambda_P = \frac{36}{5R_P}$$

$$\frac{1}{\lambda_T} = \frac{5}{36} R_T \quad \Rightarrow \quad \lambda_T = \frac{36}{5R_T}$$

Hence, shift of H_α line $= \lambda_P - \lambda_T = \frac{5}{36} \left(\frac{1}{R_P} - \frac{1}{R_T} \right)$

$\because \qquad R_M = \frac{R_\infty}{1 + m/M}$

$\therefore \qquad R_P = \frac{R_\infty}{1 + m/M_P} \text{ and } R_T = \frac{R_\infty}{1 + m/M_T}$

Shift of H_α line $= \lambda_P - \lambda_T = \frac{36}{5} \left(\frac{1 + m/M_P}{R_\infty} - \frac{1 + m/M_T}{R_\infty} \right)$

$$\lambda_P - \lambda_T = \frac{36}{5R_\infty}\left(1 + \frac{m}{M_P} - 1 - \frac{m}{M_T}\right)$$

$$\lambda_P - \lambda_T = \frac{36}{5R_\infty}\left(\frac{m}{M_P} - \frac{m}{3M_P}\right); \quad M_T = 3M_P$$

$$\lambda_P - \lambda_T = \frac{36}{5R_\infty}\frac{2m}{3M_P}$$

$$\lambda_P - \lambda_T = \frac{36}{5 \times 1.097 \times 10^5}\frac{2 \times 9.1 \times 10^{-28}}{3 \times 1.67 \times 10^{-24}}$$

$$\lambda_P - \lambda_T = 2.38 \times 10^{-8} \text{ cm} = 2.38 \text{ Å}$$

PROBLEM 4.6 Find out the total/inner quantum number and total magnetic quantum number corresponding to p-electron.

Solution For the p-electron, $s = 1/2$, $\ell = 1$
Since, $j = |l - s|$ to $|l + s|$
Thus, $j = 1/2, 3/2$
For $j = 1/2$, total magnetic quantum number $(m_j) = 1/2, -1/2$
For $j = 3/2$, total magnetic quantum number $(m_j) = 3/2, 1/2, -1/2, -3/2$

PROBLEM 4.7 Evaluate the orientations of total angular momentum vector for $j = 3/2$ in the magnetic field.

Solution Let θ be the orientation of total angular momentum vector with the magnetic field, then

$$\cos\theta = \frac{m_j}{\sqrt{j(j+1)}}$$

For $j = 3/2$, $m_j = 3/2, 1/2, -1/2, -3/2$

$$\cos\theta = \frac{m_j}{\sqrt{(3/2)\{(3/2)+1\}}} = \frac{2m_j}{\sqrt{15}}$$

$$\cos\theta = \frac{3}{\sqrt{15}}, \frac{1}{\sqrt{15}}, -\frac{1}{\sqrt{15}}, -\frac{3}{\sqrt{15}}$$

$$\cos\theta = 0.775, 0.258, -0.258, -0.775$$

$$\theta = 39.2°, 75.0°, 105°, 140.8°$$

PROBLEM 4.8 The quantum numbers (n, ℓ, s) for the two-valence electron are $(4, 0, 1/2)$ and $(5, 1, 1/2)$. Evaluate the possible terms under L-S coupling scheme.

Solution The evaluation of term for electron system under L-S coupling scheme can be done as follows:

n_1, n_2	s_1, s_2	s ($|s_1 - s_2|$ to $|s_1 + s_2|$)	$2s + 1$	ℓ_1, ℓ_2	l ($|\ell_1 - \ell_2|$ to $|\ell_1 + \ell_2|$)	j ($|l - s|$ to $|l + s|$)	Terms $^{2S+1}X_j$
4, 5	1/2, 1/2	0	1	0, 1	1	For $s = 0, l = 1: j = 1$	$^1P_1$
		1	3			For $s = 1, l = 1: j = 0, 1, 2$	$^3P_{0,1,2}$

PROBLEM 4.9 Explain the clear picture of Zeeman splitting for transition $^2D_{3/2} \to {}^2P_{1/2}$. Also show the transitions in splitted levels.

Solution The shift of splitted levels under the Zeeman splitting in terms of Lorentz unit is determined by the following equation.

$$-\Delta T = g_j m_j L'$$

where,

$$g_j = 1 + \frac{j(j+1) + s(s+1) - \ell(\ell+1)}{2j(j+1)}$$

So, for the term $^2P_{1/2}$: $2s + 1 = 2 \Rightarrow s = 1/2$; $l = 1$; $j = 1/2$, $2j + 1 = 2$ and $m_j = \pm 1/2$

$$g_j = 1 + \frac{(1/2)\{(1/2)+1\} + (1/2)\{(1/2)+1\} - (1+1)}{2(1/2)\{(1/2)+1\}} = \frac{2}{3}$$

$$-\Delta T = \pm \frac{1}{3} L'$$

Thus the term $^2P_{1/2}$ splits into two levels under Zeeman splitting which are at separation $\pm 1/3$ in Lorentz unit from the original state.

Similarly, for the term $^2D_{3/2}$: $2s + 1 = 2 \Rightarrow s = 1/2$; $l = 2$; $j = 3/2$, $2j + 1 = 4$ and $m_j = \pm 3/2$, $\pm 1/2$

$$g_j = 1 + \frac{(3/2)\{(3/2)+1\} + (1/2)\{(1/2)+1\} - 2(2+1)}{2(3/2)\{(3/2)+1\}} = \frac{4}{5}$$

$$-\Delta T = \pm \frac{6}{5} L', \pm \frac{2}{5} L'$$

Hence the term $^2D_{3/2}$ splits into four levels under Zeeman splitting which are at separation $\pm 6/5$ and $\pm 2/5$ in Lorentz unit from the original state. The transitions in the splitted states are shown in Figure 4.46.

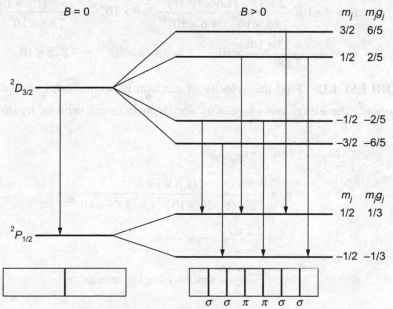

Figure 4.46 Transitions in the splitted states.

PROBLEM 4.10   In the Rutherford scattering experiment, an alpha particle of initial energy 10 MeV is scattered at 90°. Find the distance of closest approach and impact parameter. ($Z = 79$)

Solution   The following equations are the expressions of the distance of closest approach and impact parameter:

$$D = \frac{1}{4\pi\varepsilon_0} \frac{2Ze^2}{K}$$

$$b = \frac{1}{4\pi\varepsilon_0} \frac{Ze^2}{K} \cot\frac{\theta}{2} = \frac{D}{2} \cot\frac{\theta}{2}$$

$$\therefore \quad D = 9 \times 10^9 \frac{2 \times 79 \times (1.6 \times 10^{-19})^2}{10 \times 10^6 \times 1.6 \times 10^{-19}} = 9 \times 10^9 \frac{2 \times 79 \times (1.6 \times 10^{-19})}{10^7}$$

$$D = 9 \times 2 \times 79 \times 1.6 \times 10^{9-19-7} = 2275.2 \times 10^{-17} = 2.275 \times 10^{-14} \text{ m}$$

$$b = \frac{2.275 \times 10^{-14}}{2} \cot 45° = 1.137 \times 10^{-14} \text{ m}$$

PROBLEM 4.11   Find the distance of closest approach of an alpha particle of energy 7.68 MeV incident on aluminium foil. ($Z = 13$)

Solution   The following equations are the expressions of the distance of closest approach and impact parameter:

$$\therefore \quad D = \frac{1}{4\pi\varepsilon_0} \frac{2Ze^2}{K}$$

$$\therefore \quad D = 9 \times 10^9 \frac{2 \times 13 \times (1.6 \times 10^{-19})^2}{7.68 \times 10^6 \times 1.6 \times 10^{-19}} = 9 \times 10^9 \frac{2 \times 13 \times (1.6 \times 10^{-19})}{7.68 \times 10^6}$$

$$D = \frac{9 \times 2 \times 13 \times 1.6}{7.68} \times 10^{9-19-6} = 48.75 \times 10^{-16} = 4.875 \times 10^{-15} \text{ m}$$

PROBLEM 4.12   Find the velocity of electron in second orbit of hydrogen.

Solution   The expression of velocity for electron in nth orbit of hydrogen is:

$$v_n = \left(\frac{e^2}{2h\varepsilon_0}\right)\frac{1}{n}$$

$$v_n = \left(\frac{(1.6 \times 10^{-19})^2}{2 \times (6.626 \times 10^{-34}) \times (8.85 \times 10^{-12})}\right)\frac{1}{n}$$

$$v_n = \frac{2.2 \times 10^6}{n} \text{ m/s}$$

$$v_n = \frac{2.2 \times 10^6}{2} \text{ m/s} = 1.1 \times 10^6 \text{ m/s}$$

THEORETICAL QUESTIONS

4.1 Write down the Bohr postulates for atomic model. Deduce the expression of energy for the hydrogen on the basis of these postulates.

4.2 Show that $r_n \propto \dfrac{n^2}{Z}$, $v_n \propto \dfrac{Z}{n}$ and $E_n \propto -\dfrac{Z^2}{n^2}$; where r_n, v_n and E_n are the radius, velocity and energy of nth orbit electron of an atom of atomic number Z.

4.3 Explain the emission spectrum of hydrogen on the basis of Bohr model.

4.4 What are the reasons for the failure of Bohr model?

5.5 If $\bar{v}_\alpha$ and $\bar{v}_\beta$ are the wave numbers corresponding to H_α and H_β of hydrogen, then proves that $\bar{v}_\beta - \bar{v}_\alpha$ is equal to wave number of first line of Paschen series.

4.6 The spectral lines of the Pickering series of ionized helium have slightly higher wave numbers than the corresponding lines of the Balmer series of hydrogen. Why?

4.7 Write short note on critical potential.

4.8 Write down the details of Sommerfeld atomic model. Why the relativistic correction is required in this model? Is Sommerfeld's relativistic correction model explain exactly the fine structure of H_α line?

4.9 Obtain the expression of orbital and spin magnetic moment of electron in term of Bohr magnetron.

4.10 What is Larmor precession? Find the expression of Larmor frequency.

4.11 Explain (a) space quantisation, (b) spin quantisation and (c) vector atom model.

4.12 Write down the expression of Lande g-factor and show that the value of g for pure orbital and pure spin state is 1 and 2 respectively.

4.13 Discuss the Pauli exclusion principles.

4.14 Explain in detail (a) L-S coupling and (b) j-j coupling.

4.15 Deduce the terms of $3p\,4d$ configuration system under the L-S coupling scheme.

4.16 What is the difference between sp and p^2 electron system? Deduce the terms for both the systems.

4.17 Write a short note on the fine structure of H_α line.

4.18 How many series are found in the emission spectra of alkali metal? Draw the energy level of Na and show the transition for the spectrum of Na.

4.19 The fine structure of principal and sharp series lines of alkali spectrum are doublet. Why?

4.20 Give the reason behind the normal Zeeman splitting of level.

4.21 What is anomalous Zeeman Effect? Discuss the Zeeman pattern of D_1 and D_2 lines of Na.

4.22 What is Paschen–Back effect? Show that the number of levels in Paschen–Back effect is equivalent to normal Zeeman levels.

4.23 Calculate the Zeeman pattern arising from the transitions (a) $^3D_3 \rightarrow {}^3P_2$ and (b) $^3P_2 \rightarrow {}^3S_1$.

4.24 What is Stark effect.

4.25 Give the reasons for the breadth in the spectral line. Discuss any one of them in detail.

NUMERICAL PROBLEMS

P4.1 The wavelength of second spectral line of Balmer series of hydrogen is 4861 Å. Calculate the wavelength of second spectral line of Balmer series and ionizational potential of hydrogen.

(**Ans.** 6562.3 Å, 13.6 eV)

P4.2 Calculate the ionization potential for the second orbit electron of hydrogen. (**Ans.** 3.4 eV)

P4.3 Calculate the energy required to remove the electron from singly ionized helium.

(**Ans.** 54.4 eV)

P4.4 A mixture of gas contains the two isotopes of hydrogen protium and deuterium. How far apart in wavelength will the H_α lines of the two kinds of hydrogen. (**Ans.** 1.8 Å)

P4.5 An alpha particle incident on aluminum foil ($Z = 13$) with energy 7.68 MeV. If it is scattered at angle 90°, then find the impact parameter. (**Ans.** 2.437×10^{-15} m)

P4.6 Find the ratio of velocity in first orbit of hydrogen and speed of light. (**Ans.** 1/137)

P4.7 The distance of closet approach of an alpha particle when scattered with gold foil is 2.8×10^{-14} m. Find the kinetic energy of alpha particle. (**Ans.** 8.12MeV)

P4.8 Calculate orbital magnetic moment of d-orbital electron in term of Bohr magnetron.

(**Ans.** $\sqrt{6}\,\mu_B$)

P4.9 Calculate the possible orientation of orbital angular momentum vector of p-orbital electron.

(**Ans.** 45°, 90°, 135°)

P4.10 Evaluate the Lande g-factor for state $^2D_{3/2}$. (**Ans.** 4/5)

P4.11 Calculate the change in wavelength of spectral line (5000 Å) emitted by a star if it is moving with speed 6×10^4 m/s away from the observer. (**Ans.** 1.0 Å)

P4.12 Compute the total angular momentum of an one electron atom in state $^2D_{5/2}$.

(**Ans.** $\sqrt{35}\,h/4\pi$)

P4.13 Find the term for p^6 electron configuration. (**Ans.** $^1S_0$)

P4.14 A light source of single spectral line is placed in magnetic field of 0.3 W/m^2. Under normal Zeeman splitting, it breaks into three lines. If the change in frequency between π and σ-components is 2.83×10^{-12}, then find wavelength light emitted by source. (**Ans.** 4500 Å)

P4.15 If $\ell = 2$, $s = 1$ and $j = 3$ then find the term. For this state/term, calculate the number of splitted states in a weak magnetic field. (**Ans.** $^3D_3$, 7)

P4.16 Find the magnetic moment for the state $^2P_{3/2}$ in terms of Bohr magnetron.

(**Ans.** $2\sqrt{15}\,\mu_B/3$)

P4.17 Compute the Zeeman pattern components for $^2D_{3/2} \rightarrow {}^2P_{3/2}$. (**Ans.** 10 components)

MULTIPLE CHOICE QUESTIONS

MCQ 4.1 The emission spectra of hydrogen have five series of lines named Lyman, Balmer, Paschen, Brackett and Pfund. The series limit wavelengths are in ratio

    (a) 1 : 4 : 9 : 16 : 25   (b) 25 : 16 : 9 : 4 : 1

    (c) 1 : 9 : 4 : 16 : 25   (d) 1 : 4: 9 : 25 : 16

MCQ 4.2 The energy of nth stationary states of hydrogen is given by expression $E_n = -13.6/n^2$ eV. The ionization potential for an electron of third stationary state is:
  (a) 13.6 eV
  (b) 3.4 eV
  (c) 1.5 eV
  (d) 10.2 eV

MCQ 4.3 If m and M represent the mass of electron and proton, the value of Rydberg constant for the titrium is:
  (a) $\dfrac{3MR_\infty}{m + 3M}$
  (b) $\dfrac{3mR_\infty}{M + 3m}$
  (c) $\dfrac{MR_\infty}{3m + M}$
  (d) $\dfrac{mR_\infty}{3M + m}$

MCQ 4.4 The Lande g-factor for state $^2P_{3/2}$ is
  (a) 3/4
  (b) 2/3
  (c) 3/2
  (d) 4/3

MCQ 4.5 The spin magnetic moment of electron is:
  (a) $\sqrt{6}\mu_B$
  (b) $\sqrt{3}\mu_B$
  (c) $\sqrt{3}\mu_B/2$
  (d) $3\mu_B/2$

MCQ 4.6 The number of Zeeman line corresponding to transition $^1D_2 \to {}^1P_1$ (when it is viewed along the direction of magnetic field) is:
  (a) 3
  (b) 1
  (c) 2
  (d) None of these

MCQ 4.7 Due to L-S coupling, the p-state splits into:
  (a) 2 states
  (b) 1 states
  (c) 3 states
  (d) None of these

MCQ 4.8 The number of Zeeman states for $^3D_2$ is:
  (a) 7
  (b) 5
  (c) 6
  (d) 4

MCQ 4.9 If v_1 is velocity of electron in first orbit of hydrogen, then its velocity in second orbit will be:
  (a) $2 v_1$
  (b) $4 v_1$
  (c) $v_1/2$
  (d) $v_1/4$

MCQ 4.10 If v_1 is velocity of electron in first orbit of hydrogen, then value of v_1/c is equal to:
  (a) 137
  (b) 1/137
  (c) 13.7
  (d) 1.37

MCQ 4.11 The fine structure of H_α line has lines:
  (a) 2
  (b) 3
  (c) 4
  (d) 5

MCQ 4.12 In Rutherford experiment, if D is distance of closet approach and θ is the scattering angle, then impact parameter is equal to:
  (a) $\dfrac{D}{2}\cot\dfrac{\theta}{2}$
  (b) $D\cot\dfrac{\theta}{2}$
  (c) $2D\cot\dfrac{\theta}{2}$
  (d) $4D\cot\dfrac{\theta}{2}$

MCQ 4.13 The expression of Bohr magnetron in MKS unit is:

(a) $\dfrac{eh}{4\pi mc}$

(b) $\dfrac{eh}{2\pi m}$

(c) $\dfrac{e}{2m}$

(d) $\dfrac{eh}{4\pi m}$

MCQ 4.14 The expression of change in frequency (MKS unit) in normal Zeeman effect in magnetic field B is:

(a) $\dfrac{eB}{4\pi m}$

(b) $\dfrac{eB}{2\pi m}$

(c) $\dfrac{eB}{2m}$

(d) $\dfrac{eB}{4\pi mc}$

MCQ 4.15 The number of orientation of spin angular momentum in magnetic field is:

(a) 3

(b) 2

(c) 4

(d) None of these

Answers

4.1 (a)	**4.2** (c)	**4.3** (a)	**4.4** (d)	**4.5** (b)	**4.6** (c)	**4.7** (a)
4.8 (b)	**4.9** (c)	**4.10** (b)	**4.11** (d)	**4.12** (a)	**4.13** (d)	**4.14** (a)
4.15 (b)						

CHAPTER **5**

Molecular Physics

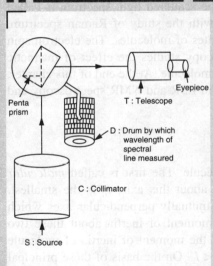

5.1 INTRODUCTION

The spectra of light emitted by a gas discharge tube are generally found to consist of number of discrete spectral lines. The origin of such spectral line can be explained on the basis of Bohr–Sommerfeld theory and by quantum mechanics. Such line spectra correspond to atomic spectra. In some cases, spectrum is found in the form of band. Such spectrum is obtained in the case of molecule. When spectral band seen by high resolution spectroscope, it is found to make up of closely spaced spectral lines. These lines are found to be arranged in a definite pattern. The spectral line within the band is found to be spread out from a definite point, called *band origin*, while the lines are densely closed packed towards the band head. A band spectrum can posses a number of regularly spaced bands. The band spectra of molecule can be explained on the basis of molecular energy levels. The energy states of a molecule can be understood by Max Born and J.R. Oppenheimer concept. Actually the energy of a molecule depends not only on the motion of orbital electrons, but also on the motion of atoms as a whole inside the molecule. As an example, the rotational motion of atoms about a common axis and a vibrational motion (harmonic motion) along their line joining provides the rotational and vibrational energy in addition to electronic energy. The present chapter deals with the energy and spectra of molecules. The scattered light spectrum is called Raman spectrum. The present chapter also deals with the study of Raman spectrum. The magnetic field causes degeneracy in energy states of molecules. The electron spin resonance or nuclear magnetic resonance spectroscopy studies the effect of magnetic field on the spin of electron or nuclei of the sample molecule. At the end of this chapter, we will discuss about the basic principle involved in ESR and NMR spectroscopy and their applications.

5.2 CLASSIFICATION OF MOLECULES

There are three types of principal axes for a molecule. The first is called *molecular axis* (a-axis). The moment of inertia of molecule about this axis, I_a is the smallest. The second and third molecular axes are the two mutually perpendicular axes which pass through centre of gravity of molecule. The moment of inertia about these two axes (I_b and I_c) are greater than that of I_a. Overall the moment of inertia of molecule about the three principal axes are such that $I_a < I_b < I_c$. On the basis of these principal moments of inertia, molecules can be classified into four categories:

Linear Molecule

The molecule which has one zero and two non-zero moment of inertia (i.e., $I_a = 0$, $I_b = I_c \neq 0$) called *linear molecule*, e.g. HCN (Figure 5.1), HCl, CO_2, etc.

Symmetric Top Molecule

The molecule having non-zero moment of inertia about the three principal axes of rotation in which two of them are equal (i.e., $I_a < I_b = I_c$ or $I_a = I_b < I_c$) is called *symmetric top molecule*. The symmetric top molecule for which $I_a < I_b = I_c$ is called

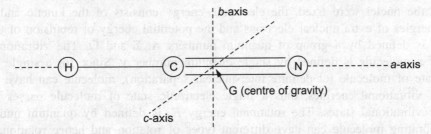

Figure 5.1   Three principal axes of HCN linear molecule.

prolate symmetric top molecule, e.g. CH_3Cl, CH_3F, NH_3, etc. While the molecule for which $I_a = I_b < I_c$ is called *oblate symmetric top molecule*, e.g. BF_3, BCl_3, etc.

Spherical Top Molecule

If moment of inertia about the principal axes of molecule are same ($I_a = I_b = I_c$) for the molecule, then it is called *spherical top molecule*, e.g. CH_4, CCl_4, SF_6, etc.

Asymmetric Top Molecule

If all the three principal moment of inertia are different ($I_a \neq I_b \neq I_c$), then the molecule is named as *asymmetric top molecule*. Most of the molecule belongs to this category, e.g. H_2O, CH_2CHCl, CH_3OH, etc.

The analysis of spectra and study of energy states for the symmetric top, spherical top and asymmetric top molecules are hard and complex. The study of energy and spectra of diatomic linear or symmetric top are found to be easier than other. Thus the present chapter especially deals with the study of energy and spectra of diatomic molecules.

5.3   MOLECULAR ENERGY STATES

According to the Born–Oppenheimer picture of a molecule, the electrons move rapidly in the field of the massive nuclei, and the nuclei move rather slowly under their mutual electrostatic repulsion and the electronic energy of attraction. When the two atoms are brought nearer to form a stable molecule, then the electronic energy decreases rapidly, while the repulsive energy increases. For a certain inter-nuclear separation, the total potential energy becomes minimum. At this position, the two nuclei vibrate about their respective equilibrium positions along the inter-nuclear axis and rotate about the centre of mass. So the molecule possesses the three types of energies:

(i) Electronic energy E_e: This energy is caused by motion of electron in molecule

(ii) Vibrational energy E_V: This energy is due to vibrational motion of atoms in a molecule about their equilibrium position

(iii) Rotational energy E_r: This energy is due to rotational motion of atoms of molecule as whole about its centre of mass

Hence the total energy of molecule can be written as:

$$E = E_e + E_v + E_r \tag{5.1}$$

If the nuclei were fixed, the electronic energy consists of the kinetic and potential energies of extra nuclear electrons and the potential energy of repulsion of the nuclei. It is defined by a group of quantum numbers Λ, Σ and Ω. The vibrational energy of a molecule is defined by single quantum number v. Since for a single electronic state of molecule (at definite inter-nuclear separation), molecule can have a number of vibrational energies, thus a single electronic state of molecule passes a number of vibrational states. The rotational energy E_r is defined by quantum number J. A vibrating molecule can have different types of rotation and hence rotational energy. Thus a vibrational state is composed of a number of rotational states. Figures 5.2(a) and 5.2(b) represent a schematic representation of energy states of a molecule.

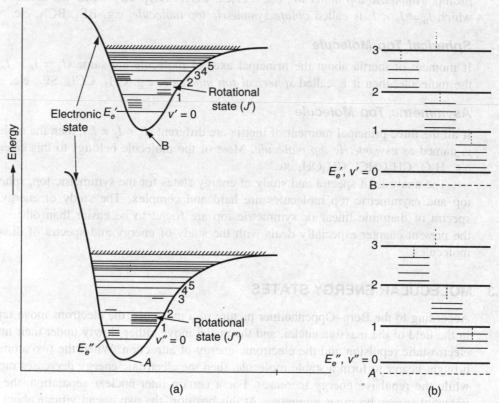

Figure 5.2   Representation of energy states of a molecule.

In terms of wave number ε, the total energy of molecule can be defined in following manner:

$$\therefore \qquad E = h\nu = \frac{hc}{\lambda}$$

$$\therefore \qquad \bar{\nu} = \varepsilon = \frac{1}{\lambda} = \frac{E}{hc} \qquad (5.2)$$

From Eqs. (5.1) and (5.2), we have

$$\frac{E}{hc} = \frac{E_e}{hc} + \frac{E_v}{hc} + \frac{E_r}{hc}$$

$$\varepsilon = \varepsilon_e + G(v) + F(v, J) \qquad (5.3)$$

where ε_e, $G(v)$ and $F(v, J)$ are the electronic term, vibrational term and rotational term corresponding to different electronic, vibrational and rotational states. The energy of a state can be defined in terms of eV or cm^{-1}. The relation in them is:

$$1\ cm^{-1} = 1.239 \times 10^{-4}\ eV$$

and $$1\ eV = 8066\ cm^{-1}$$

By the above relation, the energy of state can be converted in terms of wave number (cm^{-1}). The energy differences between two electronic, vibrational and rotational states of a molecule are found to ranging between $10^4\ cm^{-1} - 10^6\ cm^{-1}$, $10^2\ cm^{-1} - 10^4\ cm^{-1}$ and $10^0\ cm^{-1} - 10^2\ cm^{-1}$ respectively. Hence the wavelength corresponding to electronic, vibrational and rotational energy difference will be $10^0\mu - 10^{-2}\mu$, $10^2\mu - 10^0\mu$ and $10^4\mu - 10^2\mu$ respectively. These wavelengths are found to be ranging in UV-visible, near-infrared and far infrared (microwave) regions. Thus the transition between two electronic states involves the radiation falling in visible or ultraviolet region. The transition between two vibrational states associated with a given electronic state involves radiation falling in near infrared region. Similarly, the transitions in rotational states of a given electronic and vibrational state involve radiation falling in the far infrared region.

Since an electronic state is consist of number of vibrational states, and a vibrational state is composed of several rotational states, thus a transition in molecular electronic state corresponds to number of transition in vibrational and rotational states. Hence a molecular transition corresponds to band spectra. Each band has number of lines. Let a transition occurs from higher energy state E' to E'' for a molecule. If $\bar{v}$ is wave number corresponding to this transition, then from Eqs. (5.2) and (5.1), we can write

$$\bar{v} = \frac{E' - E''}{hc} = \frac{(E'_e + E'_v + E'_r) - (E''_e + E''_v + E''_r)}{hc}$$

$$\bar{v} = \frac{E'_e - E''_e}{hc} + \frac{E'_v - E''_v}{hc} + \frac{E'_r - E''_r}{hc}$$

$$\bar{v} = \bar{v}_e + \bar{v}_v + \bar{v}_r \qquad (5.4)$$

Since $E'_e - E''_e > E'_v - E''_v > E'_r - E''_r$, so $\bar{v}_e > \bar{v}_v > \bar{v}_r$. For each band, $\bar{v}_e$ and $\bar{v}_v$ remain constant, while $\bar{v}_r$ varies. The position at which $\bar{v}_r = 0$ is called *band origin*. For different bands, $\bar{v}_v$ varies, while $\bar{v}_e$ remains unchanged. The position at which $\bar{v}_v = 0$ and $\bar{v}_r = 0$ is called *system origin*.

5.4 SPECTRA

An arrangement of radiation according to their wavelength or frequency is called *spectrum*. A spectrum of radiation (developed by atomic molecular transition) can be obtained by three ways: emission spectroscopy, absorption spectroscopy and Raman spectroscopy.

Emission Spectroscopy

When an atom or molecule is subjected to heat or electronic discharge, it becomes excited. On returning to their lower energy by transition from the higher energy to lower energy state, atom or molecule emits radiation. The radiation contains suitable frequency of photon according to energy difference of states. The recording of such emitted radiation is done in emission spectroscopy and the spectrum is called *emission spectrum*.

Absorption Spectroscopy

In this spectroscopy, the sample under study is placed between the source of light and the spectrometer, which records the percentage of light absorbed against the range of frequencies to give the absorption spectrum. This spectrum is obtained only when the source of light is able to excite the atom or molecule or causes a transition from lower state to higher states.

Raman Spectroscopy

It is a technique which explores energy level of molecule by the scattering of light. In this, the sample under study is illuminated by a monochromatic radiation and the spectrum of scattered light is recorded.

5.4.1 Types of Molecular Spectra

On the basis of transition in electronic, vibrational and rotational states of molecule, the molecular spectra can be classified into three categories:

Electronic Spectra

The spectrum of molecule, which is obtained by the transition in electronic state of molecule, is called *electronic spectra*. This spectrum is observed both in emission and absorption spectra in the UV-visible region (λ: $10^0\mu - 10^{-2}\mu$) of electromagnetic spectrum. Since an electronic state is composed of a set of vibrational states or each vibrational state is composed of set of rotational states. Thus an electronic transition corresponds a large number of vibrational and rotational transitions. Hence the electronic spectrum of molecule consists of quite large number of bands. Each band is composed of series of lines. The separation of lines increases with the distance from band-head (where the intensity falls suddenly zero). While the intensity of lines decreases gradually from band-head towards the other end. Both the hetero-nuclear and homo-nuclear molecule provide the electronic spectra.

Vibrational–rotational Spectra

The molecules, which have permanent dipole moment like HCl, HF, HCN and CO give *vibrational–rotational spectra*. This type of spectrum is caused by the transition in vibrational states of molecule. The spectrum consists of an intense band (fundamental band)

and few weak bands (overtones). Since each vibrational state corresponds to a set of rotational states, thus a vibrational transition is composed of number of rotational transitions. Hence the fine structure of each band is consist of series of lines, named as P and R–branch lines. The homo-nuclear molecules (like H_2, O_2, N_2, etc) do not produce such spectra because they do not have the permanent dipole moment. The spectra of such molecules are studied in Raman spectroscopy. In the absorption spectroscopy of molecule, the vibrational rotational spectra is observed in near infrared region (1μ – 100μ).

Pure Rotational Spectra

The pure rotational spectra of a molecule is caused by transition in rotational states for a given electronic and vibrational states. This spectrum is found in far-infrared (microwave) region ($10^2\mu$ – $10^4\mu$). The spectrum consists of nearly equidistant lines. This spectrum is also observed for the hetero-nuclear molecules.

5.5 PURE ROTATIONAL SPECTRA (FOR IR SPECTRA): Molecule as Rigid Rotator

The molecule having permanent electronic dipole moments can absorb or emit electromagnetic radiation. The hetero-nuclear diatomic molecules like, HCl, HBr CO and possesses such property. These molecules produce pure rotational spectra which arise from transitions between rotational energy states. This spectrum is observed in the far infra-red ($\approx 10^2\mu$ – $10^3\mu$) or microwave ($\approx 10^3\mu$ – $10^4\mu$) region of the electromagnetic spectrum. In practice, such spectra are observed in absorption. The rotational spectra consist of a simple series of absorption maxima which are very nearly equidistant on wave number scale. Such spectra are found to be useful in determination of moment of inertia and inter-nuclear distance of the molecule with the knowledge of frequency or wavelength corresponding to maxima.

5.5.1 Rotational Energy Levels

According to classical electrodynamics, a rotating molecule can lead to emission of radiation, when a changing dipole moment is associated with it. The diatomic molecule with unlike atom, for which the centres of positive and negative charges do not coincide, possesses a dipole moment. The hetero-nuclear diatomic molecules come in this category. During rotation of molecule, the component of dipole moment in a fixed direction changes periodically with frequency of rotation of molecule. Thus a radiation of same frequency is emitted.

Similarly, the IR radiation can be absorbed by a rotating molecule. The electromagnetic wave is nothing, but an oscillating electromagnetic field. When hetero-nuclear diatomic molecule rotates in such field, then the molecule interacts with oscillating electric field due to dipole moment. As a result, absorption of rotation energy takes place and produces the absorption spectrum.

Mathematically, the pure rotational spectra of hetero-nuclear diatomic molecule can be explained on the basis of rigid rotator. In this model, it is considered that both the atoms of diatomic molecule are connected with a rigid massless rod. The atoms of molecule are rotating around a common axis. Being a rigid rotator, the electronic and vibrational energies remain unchanged. Let two atoms of masses m_1 and m_2 are held together by a rigid rod of length r. Both atoms are rotating about an axis XCY, as shown in Figure 5.3.

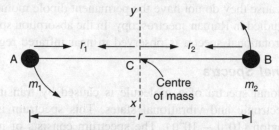

Figure 5.3 Diatomic molecule as rigid rotator.

If r_1 and r_2 are the distances of m_1 and m_2 atoms from the axis of rotation, then the moment of inertia can be written as:

$$I = m_1 r_1^2 + m_2 r_2^2$$

By the lever rule, we can write

$$m_1 r_1 = m_2 r_2$$

$$r_1 = \frac{m_2}{m_1} r_2$$

$$r_1 = \frac{m_2}{m_1}(r - r_1) \qquad (\text{as, } r = r_1 + r_2)$$

$$r_1 \left(1 + \frac{m_2}{m_1}\right) = \frac{m_2}{m_1} r$$

$$r_1 \frac{(m_1 + m_2)}{m_1} = \frac{m_2}{m_1} r$$

$$r_1 = \frac{m_2}{(m_1 + m_2)} r$$

Similarly, we have
$$r_2 = \frac{m_1}{(m_1 + m_2)} r$$

Putting values of r_1 and r_2 in expression of moment of inertia, we have

$$I = m_1 \left(\frac{m_2}{m_1 + m_2} r\right)^2 + m_2 \left(\frac{m_1}{m_1 + m_2} r\right)^2$$

$$I = \frac{m_1 m_2}{(m_1 + m_2)^2} r^2 (m_2 + m_1)$$

$$I = \frac{m_1 m_2}{m_1 + m_2} r^2$$

$$I = \mu r^2 \tag{5.5}$$

Here $\mu\{= m_1 m_2/(m_1 + m_2)\}$ is called *reduced mass* of the molecule. Equation (5.5) indicates that the diatomic molecule is equivalent to a single point mass μ at a fixed distance r from the axis of rotation. Such a system is called a *simple rigid rotator*. The possible energies of such rotating system can be obtained with the help of Schrödinger equation.

i.e.
$$\nabla^2 \psi + \frac{8\pi^2 \mu}{h^2}(E - V)\psi = 0 \qquad (5.6)$$

Here $V = 0$ because r is fixed. Thus Eq. (5.6) becomes

$$\nabla^2 \psi + \frac{8\pi^2 \mu}{h^2} E\psi = 0 \qquad (5.7)$$

The solution of Eq. (5.7) for energy provides that

$$E = \frac{h^2}{8\pi^2 \mu \, r^2} J(J + 1) = \frac{h^2}{8\pi^2 I} J(J + 1) \qquad (5.8)$$

Here J is a positive integer and is called *rotational quantum number*. J can have 0, 1, 2, ..., values. Hence the rotator can have a discrete set of energy levels. It is convenient to analyse the energy levels in terms of wave number.

$$\because \qquad E = h\nu = \frac{hc}{\lambda}$$

$$\therefore \qquad \bar{\nu} = \frac{1}{\lambda} = \frac{E}{hc}$$

Putting value of E from Eq. (5.8), we have

$$\bar{\nu} = F(J) = \frac{h}{8\pi^2 Ic} J(J + 1)$$

$$F(J) = BJ(J + 1) \qquad (5.9)$$

Here $B\{=h/(8\pi^2 Ic)\}$ is called *rotational constant*. For $J = 0, 1, 2, 3, 4, ...,$ the term $F(J)$ have values 0, $2B$, $6B$, $12B$, $20B$, ..., . This implies that there exist a set of discrete rotational energy levels whose spacing increases with increase in rotational quantum number.

Angular momentum and rotational frequency: In classical mechanics, the rotational energy E and angular momentum L is given by expressions:

$$E = \frac{1}{2} I\omega^2 \text{ and } L = I\omega$$

$$\Rightarrow \qquad L = \sqrt{2EI} \qquad (5.10)$$

Solving the Eqs. (5.8) and (5.10), we have

$$L = \frac{h}{2\pi} \sqrt{J(J + 1)} = \hbar \bar{J} \qquad (5.11)$$

This result shows that a discrete set of angular momentum is possible for rigid rotator, i.e., angular momentum is quantised. Due to discreteness in energy and angular momentum, the rotational frequency also have discrete values, i.e., certain rotational frequencies are possible. If v_{rot} is the frequency of rotation, then

$$\therefore \qquad \omega = 2\pi v_{rot} = \frac{L}{I}$$

$$\therefore \qquad \boxed{v_{rot} = \frac{h}{4\pi^2 I}\sqrt{J(J+1)}} \qquad (5.12)$$

This is the expression of rotational frequency for the rigid rotator.

5.5.2 Rotational Transition and Spectrum

The transitions between the rotational levels cause the emission or absorption of radiation. The rotational transition can only occur such that J changes by unit, i.e., $\Delta J = \pm 1$. This is called *selection rule*. For absorption, ΔJ has value $+1$. Let a transition takes place form a lower level J to an upper level $J+1$.

$$\therefore \qquad F(J) = BJ(J+1)$$

$$\therefore \qquad F(J+1) = B(J+1)(J+2)$$

If $\overline{v}$ is the frequency (in terms of wave number) of expected spectral line, then

$$\overline{v} = F(J+1) - F(J)$$

$$\overline{v} = B\,(J+1)(J+2) - BJ(J+1)$$

$$\overline{v} = B\,(J+1)(J+2-J)$$

$$\boxed{\overline{v} = 2B(J+1)} \qquad (5.13)$$

For $J = 0, 1, 2, 3, \ldots$, $\overline{v} = 2B, 4B, 6B, 8B, \ldots$. *Thus the absorption spectrum of a rigid rotator is expected to be consist of a series of equidistant lines with constant separation of 2B* (Figure 5.4).

As the spectrum is recorded over an assemblage of molecule that is distributed over different energy states having identical intrinsic transition probability, the intensity of these spectral lines is directly proportional to the initial number of molecules in each level. The thermal distribution of molecule follows the Maxwell–Boltzmann distribution, i.e.,

$$\frac{N_J}{N_0} \propto \exp\left(-\frac{E_J}{KT}\right)$$

$$\frac{N_J}{N_0} \propto \exp\left(-\frac{hcF(J)}{T}\right)$$

$$\frac{N_J}{N_0} \propto \exp\left(-\frac{hcBJ(J+1)}{T}\right) \qquad (5.14)$$

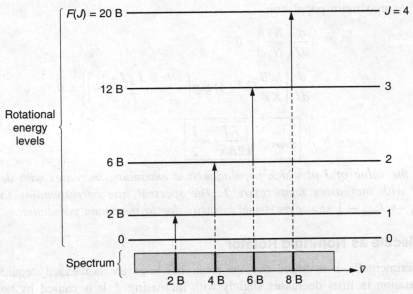

Figure 5.4 Transition in rotational energy levels and absorption spectrum.

Here N_0 and N_J are the number of molecules at the lowest and Jth level. Since the rotational state with quantum number J is $(2J + 1)$ degeneration system, Eq. (5.14) becomes

$$\frac{N_J}{N_0} \propto (2J+1)\exp\left(-\frac{hcB\ J(J+1)}{T}\right)$$

$$\frac{N_J}{N_0} = \frac{hcB}{KT}(2J+1)\exp\left(-\frac{hcB.J(J+1)}{T}\right) \qquad (5.15)$$

The plot of graph, N_J/N_0 against J is shown in Figure 5.5. It is clear from the plot that as J increases, the population first increases and attains the maximum and then decreases. So, intensity of rotational spectral line follows the same trend. For highest populated state, the intensity of spectral line is maximum.

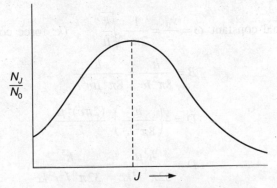

Figure 5.5 Plot of N_J/N_0 against J.

At maximum population,

$$\frac{d}{dJ}\left(\frac{N_J}{N_0}\right) = 0$$

$$\frac{d}{dJ}\left\{\frac{hcB}{KT}(2J+1)\exp\left(-\frac{hcB\;J\;(J+1)}{T}\right)\right\} = 0$$

$$\boxed{J_{max} = \sqrt{\frac{KT}{2Bhc} - \frac{1}{2}}} \qquad (5.16)$$

So, *the value of J at which population is a maximum, increases with decreasing B and with increasing temperature T. The spectral line corresponding to transition $J_{max} \to J_{max} + 1$ shows maximum intensity due to maximum population.*

5.5.3 Molecule as Non-rigid Rotator

Experimentally it is found that the rotational lines are not exactly equidistant. The separation in lines decreases slightly with increasing J. It is caused by non-rigidness in rotator or molecule. Thus a new model comes into effect for the proper explanation of pure rotational spectra, called *non-rigid rotator*. When the rotator rotates, the atomic separation or bond length of molecule slightly increases due to centrifugal force. Due to this, the energy levels are slightly changed. This change in energy is termed as *centrifugal distortion*. Taking this into account, the energy of rotational state and term value become:

$$E_J = hc\{BJ(J+1) - DJ^2(J+1)^2\} \qquad (5.17)$$

and

$$F_J = F(J) = \frac{E_J}{hc}$$

$$F_J = BJ(J+1) - DJ^2(J+1)^2 \qquad (5.18)$$

where D is the centrifugal distortion constant and is given by the following expression:

$$D = \frac{4B^3}{\omega^2}$$

Since, vibrational constant $\omega = \dfrac{v_0}{c} = \dfrac{1}{2\pi c}\sqrt{\dfrac{k}{\mu}}$ (k: force constant)

and

$$B = \frac{h}{8\pi^2 Ic} = \frac{h}{8\pi^2 \mu r^2 c}$$

So

$$D = 4\left(\frac{h}{8\pi^2 Ic}\right)^3 \frac{(2\pi c)^2 \mu}{k}$$

$$D = \frac{h^3 \mu}{32\pi^4 I^3 kc} = \frac{h^3}{32\pi^4 I^2 r^2 kc} \qquad (5.19)$$

The value of D is much less than B. Thus, energy difference between rotational levels of rigid and non-rigid rotator increases with J, but the speed of change is small, i.e., for low J value, the rotational levels have approximately same energy, while for high J value, there exist slight difference in energy of rotational levels of rotators (Figure 5.6).

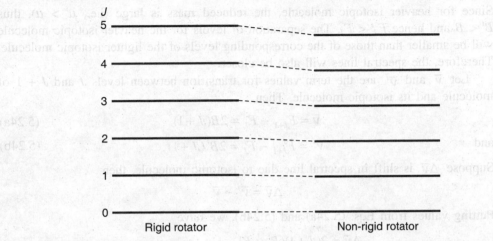

Figure 5.6 Rotational energy states of a non-rigid rotator in comparison with rigid rotator.

Let $\bar{v}$ is the wave number for rotational transition $J \to J +1$ of non-rigid rotator, then from Eq. (5.18), one can find that

$$\bar{v} = F(J+1) - F(J)$$

$$\boxed{\bar{v} = 2B(J+1) - 4D(J+1)^3} \qquad (5.20)$$

This expression indicates that the spectral lines are no longer exactly equidistant. Also for small J, the correction term [second term in Eq. (5.20)] is almost negligible, while for high J, it is appreciable. This shows that the separation between lines decreases slightly with increasing J because $D \ll B$. Thus non-rigid rotator explains the pure rotational spectra of diatomic molecules.

5.5.4 Isotope Effect on Rotational Levels

Suppose a diatomic molecule has reduced mass μ and bond length r, the rotational constant B for molecule can be written as:

$$B = \frac{h}{8\pi^2 I c} = \frac{h}{8\pi^2 \mu r^2 c} \qquad (5.21)$$

When molecule is replaced by its isotopic molecule, then the inter bond length remains same due to same electron charge distribution, but its reduced mass changes. Let μ^i, I^i and B^i are the reduced mass, moment of inertia and rotational constant respectively for isotopic molecule. Then,

$$B^i = \frac{h}{8\pi^2 I^i c} = \frac{h}{8\pi^2 \mu^i r^2 c} \qquad (5.22)$$

If F_J and F_J^i are the energy (in cm^{-1}) of rotational levels of molecule and its isotopic molecule, then

$$F_J = BJ(J + 1) \tag{5.23a}$$

$$F_J^i = B^i J(J + 1) \tag{5.23b}$$

Since for heavier isotopic molecule, the reduced mass is large (i.e., $\mu^i > \mu$), thus $B^i < B$ and hence $F_J^i < F_J$. The separation of levels for the heavier isotopic molecule will be smaller than those of the corresponding levels of the lighter isotopic molecule. Therefore, the spectral lines will also be closer.

Let $\bar{\nu}$ and $\bar{\nu}^i$ are the term values for transition between levels J and $J + 1$ of molecule and its isotopic molecule. Then

$$\bar{\nu} = F_{J+1} - F_J = 2B(J + 1) \tag{5.24a}$$

and

$$\bar{\nu}^i = F_{J+1}^i - F_J^i = 2B^i(J + 1) \tag{5.24b}$$

Suppose $\Delta\bar{\nu}$ is shift in spectral line due to isotopic molecule, then

$$\Delta\bar{\nu} = \bar{\nu}^i - \bar{\nu}$$

Putting values from Eqs. (5.24a) and (5.24b), we have

$$\Delta\bar{\nu} = 2(J + 1)(B^i - B)$$

$$\Delta\bar{\nu} = 2B(J + 1)\left(\frac{B^i}{B} - 1\right)$$

$$\Delta\bar{\nu} = 2B(J + 1)\left(\frac{\mu}{\mu^i} - 1\right); \qquad \left(\because \frac{B^i}{B} = \frac{\mu}{\mu^i} \text{ and let } \frac{\mu}{\mu^i} = \rho^2\right)$$

$$\boxed{\Delta\bar{\nu} = 2B(J + 1)(\rho^2 - 1)} \tag{5.25}$$

When $\mu^i > \mu$ or $\rho < 1$, then $\Delta\bar{\nu}$ becomes negative and increases with J. This implies that the spectral line for heavier isotopic molecule will be at lower wave number, and shift increases with rotational quantum number as shown in Figure 5.7.

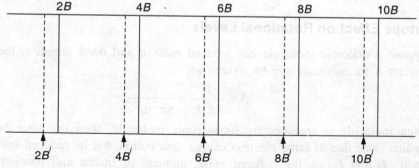

Figure 5.7 Rotational spectrum of a diatomic molecule (continuous lines) and its heavier isotopic molecule (dotted lines).

EXAMPLE 5.1 The average spacing between adjacent rotational lines of CO molecule is 3.8626 cm^{-1}. Calculate the bond length of CO.

Solution Given that

$$2B = 3.8626 \text{ cm}^{-1}$$
$$B = 1.9313 \text{ cm}^{-1} = 193.13 \text{ m}^{-1}$$

$\because$
$$B = \frac{h}{8\pi^2 \mu r^2 c}$$

$\therefore$
$$r^2 = \frac{h}{8\pi^2 \mu B c}$$

where
$$\mu = \frac{m_1 m_2}{m_1 + m_2}$$

Here
$$m_1 = \text{Mass of C atom} = 0.012 \text{ kg/mole} = 0.012/N_A \text{ kg}$$

$$m_2 = \text{Mass of O atom} = 0.016 \text{ kg/mole} = 0.016/N_A \text{ kg}$$

$$N_A = 6.023 \times 10^{23}$$

So
$$\mu = \frac{(0.012 \times 0.016)/(6.023 \times 10^{23})^2}{(0.012 + 0.016)/(6.023 \times 10^{23})}$$

$$\mu = \frac{0.012 \times 0.016}{0.028 \times 6.023 \times 10^{23}}$$

$$\mu = 1.139 \times 10^{-26} \text{ kg}$$

Thus
$$r^2 = \frac{6.624 \times 10^{-34}}{8 \times (3.14)^2 \times 1.139 \times 10^{-26} \times 193.13 \times 3 \times 10^8}$$

$$r^2 = 1.272384 \times 10^{-20}$$

$$r = 1.28 \times 10^{-10} \text{ m} = 1.128 \text{ Å}$$

So the bond length of CO molecule is 1.128 Å.

EXAMPLE 5.2 The $J : 0 \to 1$ transition in HCl occurs at 20.68 cm^{-1}. Calculate the wavelength of the transition $J : 14 \to 15$ assuming the molecule to be a rigid rotator.

Solution The wave number corresponding to transition $J \to J + 1$ is given by

$$\bar{v}_{J \to J+1} = 2B(J + 1)$$

So
$$\bar{v}_{0 \to 1} = 2B(0 + 1) = 2B$$

and
$$\bar{v}_{14 \to 15} = 2B(14 + 1) = 30B$$

$\therefore$
$$\bar{v}_{14 \to 15} = 15\bar{v}_{0 \to 1}$$

$$\overline{v}_{14\rightarrow15} = 15 \times 20.68$$

$$\overline{v}_{14\rightarrow15} = 310.2 \text{ cm}^{-1}$$

Thus

$$\lambda = \frac{1}{\overline{v}_{14\rightarrow15}} = \frac{1}{310.2} \text{ cm} = 32 \times 10^{-4} \text{ cm} = 32\mu$$

EXAMPLE 5.3   The transition $J : 3 \rightarrow 4$ in HCl is associated with radiation of 83.03 cm^{-1}. Use the rigid-rotator approximation to calculate the moment of inertia and inter nuclear distance of HCl. ($\mu_{HCl} = 1.61 \times 10^{-24}$ g)

Solution   Since,

$$\overline{v}_{J\rightarrow J+1} = 2B(J+1)$$

$$\therefore \qquad\qquad \overline{v}_{3\rightarrow4} = 2B(3+1) = 8B$$

$$83.03 = 8B$$

$$\Rightarrow \qquad\qquad B = 10.38 \text{ cm}^{-1}$$

We know that $B = \dfrac{h}{8\pi^2 Ic}$

So, $I = \dfrac{h}{8\pi^2 Bc} = \dfrac{6.624 \times 10^{-27} \text{ erg.s}}{8 \times (3.14)^2 \times (10.38 \text{ cm}^{-1}) \times (3 \times 10^{10} \text{ cm.s}^{-1})} = 2.7 \times 10^{-40} \text{ g.cm}^2$

$$\because \qquad I = \mu r^2 \qquad \Rightarrow \qquad r = \sqrt{\frac{I}{\mu}}$$

$$\therefore \qquad r = \sqrt{\frac{2.7 \times 10^{-40} \text{ g.cm}^2}{1.61 \times 10^{-24} \text{ g}}} = 1.29 \times 10^{-8} \text{ cm} = 1.29 \text{ Å}$$

EXAMPLE 5.4   The OH-radical has a moment of inertia of 1.48×10^{-40} g cm^2. Calculate its inter nuclear distance. Also calculate, for $J = 5$, its angular momentum and angular velocity. Determine the energy absorbed in the $J : 5 \rightarrow 6$ transition in cm^{-1} and erg. ($h = 6.62 \times 10^{-27}$ erg. s and $c = 3 \times 10^{10}$ cm/s)

Solution   The reduced mass of a diatomic molecule is given by

$$\mu = \frac{m_1 m_2}{m_1 + m_2}$$

Here $m_1 =$ Mass of O atom $= 16$ g/mole $= 16/N_A$ g

$m_2 =$ Mass of H atom $= 1$ g/mole $= 1/N_A$ g

$N_A = 6.023 \times 10^{23}$

So $\mu = \dfrac{(16 \times 1)/(6.023 \times 10^{23})^2}{(16+1)/(6.023 \times 10^{23})} = \dfrac{16}{17 \times 6.023 \times 10^{23}} = 1.54 \times 10^{-24}$ g

Let r is the inter nuclear distance of OH radical, then

$$I = \mu r^2 \quad \Rightarrow \quad r = \sqrt{\frac{I}{\mu}}$$

$$\therefore \quad r = \sqrt{\frac{1.48 \times 10^{-40} \text{ g.cm}^2}{1.54 \times 10^{-24} \text{ g}}} = 0.97 \times 10^{-8} \text{ cm} = 0.97 \text{ Å}$$

The angular momentum of a rigid rotator is written as:

$$L = \sqrt{J(J+1)}\,\frac{h}{2\pi}$$

So for $J = 5$, $\qquad L = \sqrt{5(5+1)}\,\dfrac{6.62 \times 10^{-27} \text{ erg.s}}{2 \times 3.14} = 5.77 \times 10^{-27} \text{ erg.s}$

If ω is the corresponding angular frequency, then

$$\omega = \frac{L}{I} = \frac{5.77 \times 10^{-27} \text{ erg.s}}{1.48 \times 10^{-40} \text{ g.cm}^2} = 3.90 \times 10^{-13} \text{ s}^{-1}$$

For rigid rotator,

$$\therefore \quad \bar{v}_{J \to J+1} = 2B(J+1)$$

$$\therefore \quad \bar{v}_{5 \to 6} = 2B(5+1) = 12B$$

$$\bar{v}_{5 \to 6} = 12\,\frac{h}{8\pi^2 Ic}$$

$$\bar{v}_{5 \to 6} = 12 \times \frac{6.62 \times 10^{-27} \text{ erg.s}}{8 \times (3.14)^2 \times (1.48 \times 10^{-40} \text{ g.cm}^2) \times (3 \times 10^{10} \text{ cm.s}^{-1})} = 227 \text{ cm}^{-1}$$

$$\therefore \quad E = \frac{hc}{\lambda} = \bar{v}hc$$

$$\therefore \quad E = (227 \text{ cm}^{-1}) \times (6.62 \times 10^{-27} \text{ erg.s}) \times (3 \times 10^{10} \text{ cm.s}^{-1}) = 4.5 \times 10^{-14} \text{ erg}$$

Thus, $\qquad r = 0.97$ Å, $L = 5.77 \times 10^{-27}$ erg.s, $\omega = 3.9 \times 10^{13}$ s^{-1}, $\bar{v} = 227$ cm^{-1}

and $\qquad E = 4.5 \times 10^{-14}$ erg

5.6 VIBRATIONAL–ROTATIONAL SPECTRA (NEAR IR SPECTRUM)

The hetero-nuclear diatomic molecule, which has permanent dipole moment, produces the spectrum near infra-red region ($1\mu - 10^2\mu$). A hetero-nuclear diatomic vibrating molecule is equivalent to an oscillating dipole moment. Such a molecule leads to emission of radiation on the basis of classical electrodynamics. On the basis of quantum theory, the vibrational–rotational spectra are produced by the transition between vibrational energy states associated with the same electronic state of the molecule. The spectrum consists of an inter..e band, known as fundamental band, accompanied by a number of weak bands (c vertones). The intensity of band decreases sharply from fundamental to overtone. It is also observed that each band is composed of a large number of lines. These lines arise due to transition in rotational levels of a particular vibrational states. Thus the near IR-spectra of a diatomic (unlike atom) molecule are explained on the basis of vibrating rotator, i.e., molecule is considered to have not only vibrating but also rotational motion.

Note: Homo-nuclear diatomic molecule does not produce vibrational spectra because they have zero dipole moment. Such molecule gives the Raman spectra, which is explained on the basis of polarisability of molecule.

5.6.1   Vibrational Energy Levels and Vibrational Transition (Harmonic Vibrator)

Suppose both the atoms of molecule are connected with a spring which are oscillating simple harmonically along its axis of joining (Figure 5.8). The differential equation for such motion is given by

$$\frac{d^2x}{dt^2} + \frac{k}{\mu}x = 0 \tag{5.26}$$

where k is spring constant and μ is reduced mass of molecule. If ω_0 is the natural angular frequency of oscillation, then

$$\omega_0 = 2\pi\nu_0 = \sqrt{\frac{k}{\mu}}$$

$$\nu_0 = \frac{1}{2\pi}\sqrt{\frac{k}{\mu}} \tag{5.27}$$

The potential energy and displacement curve for simple harmonic vibrator is a parabola (Figure 5.8). The relation in potential energy V and displacement x of such system is given by

$$V = \frac{1}{2}kx^2 \tag{5.28}$$

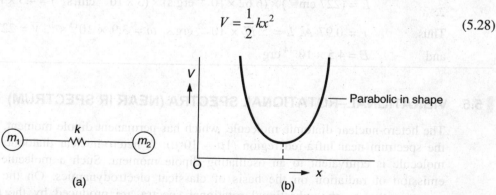

Figure 5.8   (a) Mass-spring arrangement of a diatomic molecule (b) potential energy curve of diatomic molecule when atoms vibrate harmonically.

The wave equation describing such motion under this potential can be written as:

$$\frac{d^2\psi}{dx^2} + \frac{2m}{\hbar^2}(E - V)\psi = 0$$

$$\frac{d^2\psi}{dx^2} + \frac{2m}{\hbar^2}\left(E - \frac{1}{2}kx^2\right)\psi = 0 \qquad (5.29)$$

The solution of Eq. (5.29) for energy provides that

$$E = \frac{h}{2\pi}\sqrt{\frac{k}{\mu}}\left(v + \frac{1}{2}\right)$$

$$E = h\nu_0\left(v + \frac{1}{2}\right) \qquad (5.30)$$

Here v is vibrational quantum number and can have values 0, 1, 2, 3, … . Let $G(v)$ be the term value (wave number) corresponding to energy E. Then

$$G(v) = \frac{E}{hc} = \frac{h\nu_0\left(v + \dfrac{1}{2}\right)}{hc} = \frac{\nu_0}{c}\left(v + \frac{1}{2}\right)$$

$$G(v) = \omega\left(v + \frac{1}{2}\right) \qquad (5.31)$$

Here $\omega\ (= \nu_0/c)$ is known as *vibrational constant* or *classical frequency* in cm^{-1}. For $v = 0, 1, 2, 3, 4, …$, $G(v)$ has values $\frac{1}{2}\omega, \frac{3}{2}\omega, \frac{5}{2}\omega, \frac{7}{2}\omega, …$. Thus, there exist a series of discrete vibrational levels for a vibrating molecule and levels have a common separation of ω. The transition in these levels will be allowed only if the vibrational quantum number changes by ±1 (i.e. $\Delta v = \pm 1$). Since vibrational spectra are usually observed by absorption spectroscopy, thus $\Delta v = +1$. If $\bar{\nu}$ (in cm^{-1}) is the frequency of expected band corresponding to transition $v \rightarrow v + 1$, then we may write

$$\bar{\nu} = G(v + 1) - G(v)$$

$$\bar{\nu} = \omega\left\{(v + 1) + \frac{1}{2}\right\} - \omega\left(v + \frac{1}{2}\right)$$

$$\bar{\nu} = \omega$$

Thus all transitions correspond to the same frequency of ω cm^{-1} (Figure 5.9). Hence, a harmonically oscillating molecule explains only the fundamental band of near IR spectra. It is unable to explain the weak occurrence of overtone bands which are obtained at nearly two or three times of this frequency. This leads deviations in oscillation to be harmonic.

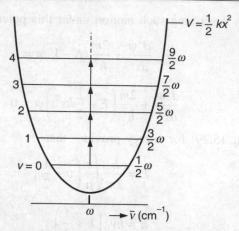

Figure 5.9 Vibrational energy states of diatomic molecule when it is considered as harmonic oscillator.

5.6.2 Correction to Vibrational Energy due to Anharmonicity (Anharmonic Oscillator)

The practical/experimental infra-red spectrum is consist of bands at frequency ω cm^{-1} (fundamental band) and approximately at 2ω, (overtone). The harmonic oscillator concept of diatomic molecule only explains only the fundamental band. In this approach, it is considered that potential is symmetrical due to being parabolic, i.e., the dipole moment of molecule is strictly linear with respect to the inter nuclear displacement. Thus it should not be strictly linear. This deviation from harmonicity is expressed in terms of *electrical anharmonicity*. It comes into effect when higher order terms are also involved in potential. The existence of overtones also implies that the selection rule $\Delta v = \pm 1$ is not strictly obeyed and the transitions corresponding to $\Delta v > 1$ have not been considered. Under the first approximation, the potential energy of a real diatomic molecule can be written as the sum of harmonic and anharmonic terms, i.e.,

$$V = k_1 x^2 - k_2 x^3$$

where k_1 and k_2 are the constant such that $k_1 << k_2$. The first term is harmonic term, while second is anharmonic term. The Schrödinger equation of wave under this potential provides the following eigenvalue of energy for the anharmonic oscillator:

$$E = h v_0 \left(v + \frac{1}{2} \right) - h v_0 x_e \left(v + \frac{1}{2} \right)^2$$

$$E = hc\omega_e \left(v + \frac{1}{2} \right) - hc\omega_e x_e \left(v + \frac{1}{2} \right)^2 \tag{5.32}$$

The term values corresponding to energy is given by

$$G(v) = \frac{E}{hc}$$

$$G(v) = \omega_e \left(v + \frac{1}{2} \right) - \omega_e x_e \left(v + \frac{1}{2} \right)^2 \tag{5.33}$$

The quantity ω_e is the spacing of energy levels that would come when the potential curve is a parabola. The coefficient $\omega_e x_e$ of squared quantum number term is known as the anharmonicity constant. This is always positive and less than ω_e. Equation (5.33) implies that the energy levels of anharmonic oscillator are not equidistant, but the separation decreases slowly with increasing vibrational quantum number (Figure 5.10). After certain vibrational quantum number, the energy becomes continuous. This shows that the molecule is dissociated into atoms and the extra energy appears as kinetic energy of these atoms.

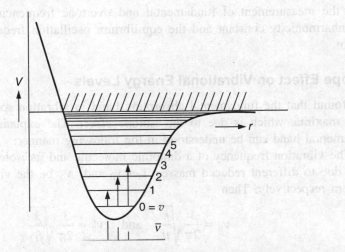

Figure 5.10 Vibrational energy states for an anharmonic oscillator and transitions among them for fundamental band and overtones.

The selection rule for the transitions in case of anharmonic oscillator is $\Delta v = \pm 1$, $\pm 2, \pm 3, \ldots$. The band corresponding to transition $\Delta v = \pm 1$ is called fundamental band. While, bands corresponding to transitions $\Delta v = \pm 2$ and $\Delta v = \pm 3$ are termed as the *first and second overtones*. The expected frequency of fundamental and its overtone bands can be obtained with the following process:

For fundamental band frequency ($v = 0 \to v = 1$),

$$\bar{v}_{0\to1} = G(1) - G(0)$$

$$\bar{v}_{0\to1} = \left\{ \omega_e \left(1 + \frac{1}{2} \right) - \omega_e x_e \left(1 + \frac{1}{2} \right)^2 \right\} - \left\{ \omega_e \left(0 + \frac{1}{2} \right) - \omega_e x_e \left(0 + \frac{1}{2} \right)^2 \right\}$$

$$\bar{v}_{0\to1} = \omega_e (1 - 2x_e) \ \text{cm}^{-1} \tag{5.34}$$

For first overtone band frequency ($v = 0 \to v = 2$),

$$\bar{v}_{0\to2} = G(2) - G(0)$$

$$\bar{v}_{0\to2} = \left\{ \omega_e \left(2 + \frac{1}{2} \right) - \omega_e x_e \left(2 + \frac{1}{2} \right)^2 \right\} - \left\{ \omega_e \left(0 + \frac{1}{2} \right) - \omega_e x_e \left(0 + \frac{1}{2} \right)^2 \right\}$$

$$\bar{v}_{0\to2} = 2\omega_e (1 - 3x_e) \ \text{cm}^{-1} \tag{5.35}$$

and for second overtone band frequency ($v = 0 \rightarrow v = 3$),

$$\bar{v}_{0\rightarrow3} = G(3) - G(0)$$

$$\bar{v}_{0\rightarrow3} = \left\{ \omega_e\left(3 + \frac{1}{2}\right) - \omega_e x_e\left(2 + \frac{1}{2}\right)^2 \right\} - \left\{ \omega_e\left(0 + \frac{1}{2}\right) - \omega_e x_e\left(0 + \frac{1}{2}\right)^2 \right\}$$

$$\bar{v}_{0\rightarrow3} = 3\omega_e(1 - 4x_e) \text{ cm}^{-1} \tag{5.36}$$

With the measurement of fundamental and overtone frequencies, one can evaluate the anharmonicity constant and the equilibrium oscillation frequency of anharmonic system.

5.6.3   Isotope Effect on Vibrational Energy Levels

It is found that the fundamental absorption band in vibration spectra consists of two close maxima, which is due to the isotope effect. The explanation of this shift in fundamental band can be understood in the following manner:

The vibration frequency of a diatomic molecule and its isotope differs from each other due to different reduced masses. Let v_0 and v_0^i be the vibrational frequencies of them respectively. Then

$$v_0 = \frac{1}{2\pi}\sqrt{\frac{k}{\mu}} \quad \text{and} \quad v_0^i = \frac{1}{2\pi}\sqrt{\frac{k}{\mu^i}}$$

So,

$$\frac{v_0^i}{v_0} = \sqrt{\frac{\mu}{\mu^i}} = \rho \quad \text{(suppose)} \tag{5.37}$$

where μ and μ^i are the reduced masses of diatomic molecule and its isotope. For the heavier isotope, $\rho < 1$. If ω and ω^i are the vibration constant for the diatomic molecule and its isotope, and

$$\omega = \frac{v_0}{c} \quad \text{and} \quad \omega^i = \frac{v_0^i}{c}$$

Then

$$\frac{\omega^i}{\omega} = \frac{v_0^i}{v_0} \tag{5.38}$$

From Eqs. (5.37) and (5.39), we can write

$$\frac{\omega^i}{\omega} = \rho$$

$\Rightarrow$

$$\omega^i = \rho\omega \tag{5.39}$$

Thus the heavier isotopic molecule has lower vibration constant and frequency. Assuming the harmonic vibration, the energy (cm^{-1}) of vibrational level corresponding to vibrational quantum number v for the diatomic molecule and its isotope can be expressed as:

$$G(v) = \omega \left(v + \frac{1}{2} \right) \tag{5.40a}$$

$$G^i(v) = \omega^i \left(v + \frac{1}{2} \right)$$

$$G^i(v) = \rho\omega \left(v + \frac{1}{2} \right) \qquad \text{[using Eq. (5.39)]} \tag{5.40b}$$

Therefore, the isotopic shift in vibrational level will be

$$G^i(v) - G(v) = \rho\omega \left(v + \frac{1}{2} \right) - \omega \left(v + \frac{1}{2} \right)$$

$$G^i(v) - G(v) = (\rho - 1)\omega \left(v + \frac{1}{2} \right) \tag{5.41}$$

Since for heavier isotope, $\rho < 1$ thus $G^i(v) - G(v) = $ negative or $G^i(v) < G(v)$. Therefore, a heavier isotope has lower vibrational level than that of lighter isotope. Suppose $\bar{v}$ and $\bar{v}^i$ are the frequencies/wave number of fundamental band. Then, the shift in the expected fundamental band $\Delta\bar{v}$ can be written as:

$$\Delta\bar{v} = \bar{v}^i_{0 \to 1} - \bar{v}_{0 \to 1}$$

$$\Delta\bar{v} = \{G^i(1) - G^i(0)\} - \{G(1) - G(0)\}$$

$$\Delta\bar{v} = \left\{ \omega^i \left(1 + \frac{1}{2} \right) - \omega^i \left(0 + \frac{1}{2} \right) \right\} - \left\{ \omega \left(1 + \frac{1}{2} \right) - \omega \left(0 + \frac{1}{2} \right) \right\} \qquad \text{[using Eq. (5.40)]}$$

$$\Delta\bar{v} = \omega^i - \omega$$

$$\Delta\bar{v} = \rho\omega - \omega \qquad \text{[using Eq. (5.39)]}$$

$$\Delta\bar{v} = (\rho - 1)\omega \tag{5.42}$$

Thus the isotopic fundamental band occurs at lower wave number for heavier isotope. Equation (5.42) supports the shift in maxima of fundamental band due to isotopic effect.

Note: If we assume anharmonic vibration of molecule, then shift in vibrational level due to isotopic effect will be $G^i(v) - G(v) = (\rho - 1) \left(v + \frac{1}{2} \right) \left\{ \omega_e - 2\omega_e x_e \left(v + \frac{1}{2} \right) \right\}$.

For $\rho < 1$, $G^i(v) < G(v)$. Thus anharmonic oscillator supports the same fact.

5.6.4 Molecule as Vibrating Rotator

The near infra-red spectrum consists of number of bands that are composed of number of lines. The complete explanation of such spectra can be done on the basis of vibrating rotator. It is assumed in the model that molecule also rotates during the vibration. Thus total energy of molecule will be the sum of its vibrational and rotational energy. If $\bar{\varepsilon}(v, J)$ is the energy of (v, J) state of molecules in cm^{-1}, then

$$\bar{\varepsilon}(v, J) = G(v) + F(J) \tag{5.43}$$

If we suppose that the molecule possesses harmonic vibration and rotational motion, then

$$G(v) = \omega\left(v + \frac{1}{2}\right) \quad \text{and} \quad F_J = BJ(J + 1)$$

Thus

$$\bar{\varepsilon}(v, J) = \omega\left(v + \frac{1}{2}\right) + BJ(J + 1) \tag{5.44}$$

Here, v and J are the vibrational and rotational quantum numbers. The quantities ω and B are called vibrational and rotational constants

$$\left(\omega = \frac{v_0}{c} = \frac{1}{2\pi c}\sqrt{\frac{K}{\mu}} \quad \text{and} \quad B = \frac{h}{8\pi^2 Ic} = \frac{h}{8\pi^2 \mu r^2 c}\right)$$

Due to being unsymmetrical curve of potential energy for a molecule, the equilibrium inter-nuclear distance r of molecule increases with the vibrational quantum number v. Since B is proportional to r^{-2}, it decreases. Hence we may say that the rotational constant is associated with vibrational constant. Thus B can be replaced by Bv, showing that it is for vibrational state v. This is for vibration–rotation interaction. The two improved models, i.e., anharmonic oscillator and non-rigid rotator could be introduced and would lead to term $(v + 1/2)^2$ and $[J(J + 1)]^2$. These correction terms are, however, less important than the rotation–vibration couple. Hence they are not taken into account in Eq. (5.44). Taking the effect of coupling, the expression of $\bar{\varepsilon}(v, J)$ becomes

$$\bar{\varepsilon}(v, J) = \omega\left(v + \frac{1}{2}\right) + B_v J(J + 1) \tag{5.45}$$

Expression (5.45) shows that there exist a set of rotational levels associated with each vibrational level. A transition between two vibrational levels would, therefore, be accompanied by a number of transitions between the two corresponding sets of rotational levels. This would result in a number of lines in a band. By the experimental observation, it is found that the lines are in the form of two branches and are not equidistant. The spacing of lines in one branch decreases slowly, while in other, it increases slowly. This leads to the vibration–rotation interaction.

The selection rule for transition between levels of vibrating rotator is given by the following conditions:

$$\Delta v = \pm 1, \pm 2, \pm 3, \dots, \quad \text{and} \quad \Delta J = \pm 1$$

For the fundamental absorption band, $\Delta v = \pm 1$ and $\Delta J = \pm 1$. Here the ± 1 value of ΔJ is applicable because the rotational levels involved in transition are of different vibration states. The set of transition lines are obtained for $\Delta J = +1$, called *R-branch*, while other set of lines are for $\Delta J = -1$, known as *P-branch*. The determination of wave number corresponding to each line of each branch can be understood as follows:

(a) *R-branch:* For this branch, the transition is such that

$$\Delta v = +1 \implies v: 0 \to 1 \quad \text{and} \quad \Delta J = +1 \implies J: J \to J+1$$

Let B_0 and B_1 are the rotational constants for $v = 0, 1$ vibrational states. If $\bar{v}_R$ be the wave number of R-branch lines in fundamental band, then

$$\bar{v}_R = \bar{\varepsilon}(1, J+1) - \bar{\varepsilon}(0, J)$$

$$\bar{v}_R = \left\{ \omega\left(1 + \frac{1}{2}\right) + B_1 (J+1)(J+2) \right\} - \left\{ \omega\left(0 + \frac{1}{2}\right) + B_0 J(J+1) \right\}$$

$$\bar{v}_R = \omega\{B_1 (J+1)(J+2) - B_0 J(J+1)\}$$

$$\bar{v}_R = \omega + 2B_1 + (3B_1 - B_0)J + (B_1 - B_0)J^2 \tag{5.46}$$

(b) *P-branch:* The transition conditions for this branch of lines are:

$$\Delta v = +1 \implies v: 0 \to 1 \quad \text{and} \quad \Delta J = +1 \implies J: J \to J-1$$

If $\bar{v}_P$ be the wave number of P-branch lines in fundamental band, then

$$\bar{v}_P = \bar{\varepsilon}(1, J-1) - \bar{\varepsilon}(0, J)$$

$$\bar{v}_P = \left\{ \omega\left(1 + \frac{1}{2}\right) + B_1(J-1)(J-1+1) \right\} - \left\{ \omega\left(0 + \frac{1}{2}\right) + B_0 J(J+1) \right\}$$

$$\bar{v}_P = \omega\left\{ B_1 J(J-1) - B_0 J(J+1) \right\}$$

$$\bar{v}_P = \omega - (B_1 + B_0)J + (B_1 - B_0)J^2 \tag{5.47}$$

The expression of $\bar{v}_R$ indicates that R-branch consists of series of lines named as $R(0)$, $R(1)$, $R(2)$, ... corresponding to $J = 0, 1, 2, ...$. The spacing between lines decreases slowly because the third term and fourth term of Eq. (5.46) are positive and negative respectively as $B_1 < B_0$. Expression (5.47) implies that P-branch is also consist of number of lines named as $P(1)$, $P(2)$, $P(3)$, ... corresponding to $J = 1, 2, 3, ...$. For this branch, J cannot be zero because $J = -1$ state for $v = 1$ for vibrational state does not exist. Since, both the second and third terms of $\bar{v}_P$ are negative as $B_1 < B_0$, spacing of lines increases with J. The transition corresponding to $\Delta J = 0$ is not allowed. This gives the missing line in the band. The missing line has wave number ω and is called *band origin*. The R-branch and P-branch come at higher and lower wave number side of missing line respectively as they have high and low wave number in comparison to wave number of missing line. If we neglect the vibration–rotation interaction, then B_1 and B_0 become equal, say B. Hence the expression of $\bar{v}_R$ and $\bar{v}_P$ becomes

$$\bar{v}_R = \omega + 2B + 2BJ$$

$$\bar{v}_P = \omega - 2BJ$$

Equations (5.46) and (5.47) can be represented by a single equation, i.e.,

$$\bar{v}_{P,R} = \omega + (B_1 + B_0)m + (B_1 - B_0)m^2 \tag{5.48}$$

For R-branch, m can have value $J + 1$ ($J = 0, 1, 2, \ldots$) and for P-branch, it can have value $-J$ ($J = 1, 2, 3, \ldots$). The zero value of m provides the wave number of null or missing line. The transitions corresponding to both branch of lines are shown in Figure 5.11.

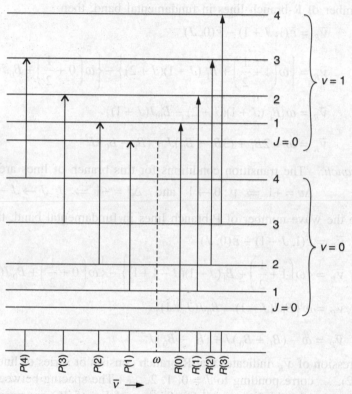

Figure 5.11   Transitions and spectral lines of fundamental band.

The intensity of these spectral lines increases on the both sides of missing line and after getting maxima, it decreases (Figure 5.12). The maxima occur at following rotational quantum number (J_{max}).

$$J_{max} = \sqrt{\frac{kT}{2Bhc}} - \frac{1}{2} \tag{5.49}$$

Here k is the Boltzmann constant. As the temperature of molecular system increases, the value of J_{max} also increases. It means, maxima position shifts away from missing line, i.e., band extends on both sides and more line appears in each branch. Suppose $\bar{v}_{R,max}$ and $\bar{v}_{P,max}$ are the wave numbers of that spectral line in both branches whose intensities are maximum. Then

$$\bar{v}_{R,max} - \bar{v}_{P,max} = \sqrt{\frac{8BkT}{hc}} = 2.3583\sqrt{BT} \tag{5.50}$$

From Eq. (5.50), we can determine the rotational constant if temperature and difference of wave numbers are known.

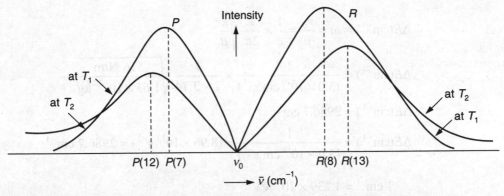

Figure 5.12 Intensity of P-and R-branch of spectral lines at two temperature T_1 and T_2 ($T_1 < T_2$).

EXAMPLE 5.5 Calculate the frequency of vibration of CO molecule which has a stiff bond of 18.7×10^5 dyne/cm.

Solution Given that, $k = 18.7 \times 10^5$ dyne/cm, $\nu_0 = ?$

$$\nu_0 = \frac{1}{2\pi}\sqrt{\frac{k}{\mu}}$$

where

$$\mu = \frac{m_1 m_2}{m_1 + m_2}$$

Here m_1 = Mass of C atom = 12 g/mole = $0.012/N_A$ g

m_2 = Mass of O atom = 16 g/mole = $0.016/N_A$ g

$N_A = 6.023 \times 10^{23}$

So, $\mu = \dfrac{(12 \times 16)/(6.023 \times 10^{23})^2}{(12+16)/(6.023 \times 10^{23})} = \dfrac{12 \times 16}{28 \times 6.023 \times 10^{23}} = 11.39 \times 10^{-24}\,\text{g} = 11.4 \times 10^{-24}\,\text{g}$

$$\mu = 11.39 \times 10^{-24}\,\text{g} = 11.4 \times 10^{-24}\,\text{g}$$

$$\nu_0 = \frac{1}{2 \times 3.14}\sqrt{\frac{18.7 \times 10^5}{11.4 \times 10^{-24}}} = 6.45 \times 10^{13}\,\text{s}^{-1}$$

EXAMPLE 5.6 The force constant of HCl molecule is 516 N/m. Find out the energy difference between $v = 0$ and $v = 1$ vibrational states in cm^{-1} and eV. ($\mu_{\text{HCl}} = 1.63 \times 10^{-27}$ kg)

Solution The energy of vibration state in terms of wave number is given by

$$G(v) = \omega\left(v + \frac{1}{2}\right)$$

So, $$\Delta E(\text{cm}^{-1}) = G(1) - G(0) = \omega\left(1 + \frac{1}{2}\right) - \omega\left(0 + \frac{1}{2}\right)$$

$$\Delta E(\text{cm}^{-1}) = \omega = \frac{v_0}{c} = \frac{1}{c} \times \frac{1}{2\pi} \sqrt{\frac{k}{\mu}}$$

$$\Delta E(\text{cm}^{-1}) = \frac{1}{(3.0 \times 10^{10}\,\text{cm.s}^{-1})} \times \frac{1}{2 \times 3.14} \sqrt{\frac{516\,\text{N/m}}{1.63 \times 10^{-27}\,\text{kg}}}$$

$$\Delta E(\text{cm}^{-1}) = 2986.7\,\text{cm}^{-1}$$

$$\Delta E(\text{cm}^{-1}) = \frac{1}{(3.0 \times 10^{10}\,\text{cm.s}^{-1})} \times (8.96 \times 10^{13}\,\text{s}^{-1}) = 2986.7\,\text{cm}^{-1}$$

$$\because \qquad 1\,\text{cm}^{-1} = 1.239 \times 10^{-4}\,\text{eV}$$

$$\Delta E(\text{eV}) = 2986.7 \times 1.239 \times 10^{-4} = 3700.5 \times 10^{-4}\,\text{eV}$$

$$\Delta E(\text{eV}) = 0.37\,\text{eV}$$

EXAMPLE 5.7 The fundamental mode of HCl occurs at 2886 cm^{-1}. Find the frequency of the same mode of DCl.

Solution We know that

$$\frac{\omega^i}{\omega} = \frac{v_0^i}{v_0} = \sqrt{\frac{\mu}{\mu^i}}$$

$$\frac{\omega_{\text{DCl}}}{\omega_{\text{HCl}}} = \sqrt{\frac{\mu_{\text{HCl}}}{\mu_{\text{DCl}}}} = \sqrt{\frac{(1 \times 35)/(1 + 35)}{(2 \times 35)/(2 + 35)}} = \sqrt{\frac{35}{36} \times \frac{37}{70}} = 0.7168$$

$$\omega_{\text{DCl}} = 0.7168\,\omega_{\text{HCl}} = 0.7168 \times 2886 = 2069\,\text{cm}^{-1}$$

EXAMPLE 5.8 The reduced mass of a hetero-nuclear diatomic molecule is 0.84 × 10^{-2} kg and its bond strength is 573 N/m. Find its vibrational quantum state corresponding to energy 4.5 eV ($h = 6.6 \times 10^{-34}$ Js).

Solution If E_v is energy of vibrational state v, then

$$E_v = h v_0 \left(v + \frac{1}{2} \right)$$

$$\because \qquad v_0 = \frac{1}{2\pi} \sqrt{\frac{k}{\mu}}$$

$$\therefore \qquad v_0 = \frac{1}{2 \times 3.14} \sqrt{\frac{573}{0.84 \times 10^{-2}}} = 1.315\,\text{s}^{-1}$$

So $\qquad E_v = (6.6 \times 10^{-34}\,\text{Joule.s}) \times (1.315\,\text{s}^{-1}) \left(v + \frac{1}{2} \right)$

$$E_v = 8.68 \times 10^{-20} \left(v + \frac{1}{2} \right) J = \frac{8.68 \times 10^{-20}}{1.6 \times 10^{-19}} \left(v + \frac{1}{2} \right) eV = 0.54 \left(v + \frac{1}{2} \right) eV$$

$$4.5 \text{ eV} = 0.54 \left(v + \frac{1}{2} \right) eV$$

$$\Rightarrow \qquad v = 8.3 \approx 8$$

EXAMPLE 5.9 The fundamental band and first overtone of a hetero-nuclear diatomic molecule are found at 2143.3 cm^{-1} and 4259.7 cm^{-1}. Calculate ω_e, $\omega_e x_e$, vibrational frequency and band strength of the molecule. ($N_A = 6.0251 \times 10^{23}$, $c = 3.0 \times 10^{10}$ cm/s and $\mu = 6.85841$ amu)

Solution The vibrational state of anharmonic oscillator is given as:

$$G(v) = \omega_e \left(v + \frac{1}{2} \right) - \omega_e x_e \left(v + \frac{1}{2} \right)^2$$

Thus

$$G(0) = \frac{1}{2} \omega_e - \frac{1}{4} \omega_e x_e$$

$$G(1) = \frac{3}{2} \omega_e - \frac{9}{4} \omega_e x_e$$

$$G(2) = \frac{5}{2} \omega_e - \frac{25}{4} \omega_e x_e$$

So

$$\bar{v}_{0 \to 1} = G(1) - G(0) = \omega_e - 2\omega_e x_e$$

and

$$\bar{v}_{0 \to 2} = G(2) - G(0) = 2\omega_e - 6\omega_e x_e$$

Given that $\bar{v}_{0 \to 1} = 2143.3$ and $\bar{v}_{0 \to 2} = 4259.7$

Hence

$$\omega_e - 2\omega_e x_e = 2143.3 \qquad \text{(i)}$$

and

$$2\omega_e - 6\omega_e x_e = 4259.7 \qquad \text{(ii)}$$

Solving Eqs. (i) and (ii), we have

$$\boxed{\omega_e = 2170.2 \text{ cm}^{-1}}$$

$$\boxed{\omega_e x_e = 13.45 \text{ cm}^{-1}}$$

$$\because \qquad v_0 = c\omega_e$$

$$\therefore \qquad v_0 = (3 \times 10^{10} \text{ cm.s}^{-1}) \times (2170.2 \text{ cm}^{-1})$$

$$\boxed{v_0 = 6.51 \times 10^{13} \text{ s}^{-1}}$$

$$\because \qquad v_0 = \frac{1}{2\pi}\sqrt{\frac{k}{\mu}}$$

$$\therefore \qquad k = 4\pi^2 \mu v_0^2 = 4 \times (3.14)^2 \times \frac{6.85841}{6.0251 \times 10^{23}} \times (6.51 \times 10^{13})^2$$

$$\boxed{k = 1.9 \times 10^6 \text{ dyne/cm}}$$

EXAMPLE 5.10 The wave number for lines of fundamental band of HCl is given by the following expression:

$$\bar{v} = 2885.9 + 20.577 \text{ m} - 0.3034 \text{ m}^2$$

Find the value of rotational constant for $v = 0$ and $v = 1$ states and vibrational constant.

Solution Since we know that

$$\bar{v} = \omega + (B_1 + B_0)\,\text{m} + (B_1 - B_0)\,\text{m}^2 \qquad \text{(i)}$$

Here B_0 and B_1 are the rotational constants for $v = 0$ and $v = 1$. The quantity ω is the vibrational constant. It is given that

$$\bar{v} = 2885.9 + 20.577 \text{ m} - 0.3034 \text{ m}^2 \qquad \text{(ii)}$$

Comparing Eqs (i) and (ii), we have

$$\omega = 2885.9 \text{ cm}^{-1}$$

$$B_1 + B_0 = 20.577 \qquad \text{(iii)}$$

and
$$B_1 - B_0 = -0.3034 \qquad \text{(iv)}$$

Solving Eqs. (iii) and (iv), we have

$$B_1 = 10.1368 \text{ cm}^{-1} \quad \text{and} \quad B_0 = 10.4402 \text{ cm}^{-1}$$

EXAMPLE 5.11 If the separation between two successive vibrational states is 1904 cm^{-1}, then find the zero point energy.

Solution The energy of vibrational state in terms of wave number is given by the following expression:

$$G(v) = \omega\left(v + \frac{1}{2}\right)$$

$$G(0) = \omega\left(0 + \frac{1}{2}\right) = \frac{\omega}{2}$$

$$\text{Zero point energy} = G(0) = \frac{\omega}{2}$$

$$\text{Zero point energy} = \frac{1904}{2} = 952 \text{ cm}^{-1}$$

5.7   SPECTRA OF DIATOMIC SYMMETRIC TOP MOLECULE

The molecule having non-zero moment of inertia around three perpendicular axes such that two of them are equal (i.e., $I_a < I_b = I_c$ or $I_a = I_b < I_c$), is called symmetric top molecule. The symmetric top molecule, for which $I_a < I_b = I_c$ is called *prolate symmetric top molecule* (e.g. CH_3Cl, CH_3F, NH_3, etc). While the molecule, for which $I_a = I_b < I_c$ is called *oblate symmetric top molecule* (e.g. BF_3, BCl_3, etc). A diatomic molecule having symmetric top configuration is called *symmetric top diatomic molecule*.

In these molecules, there exist certain electronic states, in which the electron cloud is not cylindrically symmetrical and has a component of angular momentum along the inter-nuclear axis. In these states, the total angular momentum of the molecule is the resultant of component of electronic angular momentum along inter nuclear axis and the nuclear rotation angular momentum perpendicular to the inter nuclear axis (Figure 5.13)

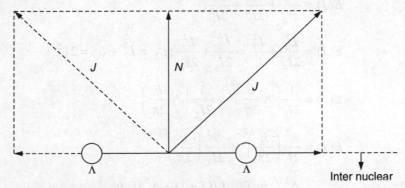

Figure 5.13   Total angular momentum and its components for symmetric top molecule.

Let the moment of inertia of diatomic molecule is such that $I_a < I_b = I_c$ ($I_a = I_{\parallel}$ and $I_b = I_{\perp}$). If Λ and N are the components of electronic angular momentum along inter nuclear axis and rotational angular momentum perpendicular to inter nuclear axis, i.e.,

$$\mathbf{L}_{\parallel} = \mathbf{L_a} = \Lambda \text{ and } \mathbf{L}_{\perp} = \mathbf{L_b} + \mathbf{L_c} = \mathbf{N} \qquad (5.51)$$

The magnitude of $\mathbf{L}_{\parallel}$ and $\mathbf{L}_{\perp}$ are given by

$$L_{\parallel} = \Lambda \frac{h}{2\pi} \text{ and } L_{\perp} = N \frac{h}{2\pi}; L_{\perp}^2 = L_b^2 + L_c^2 \qquad (5.52)$$

Here Λ and N are quantum numbers of angular momentum of electron about inter nuclear axis and perpendicular to it. Λ can have integral value. If $\mathbf{J}$ is resultant angular momentum of $\mathbf{L}_{\parallel}$ and $\mathbf{L}_{\perp}$, then

$$\mathbf{J} = \mathbf{L}_{\parallel} + L_{\perp} = \Lambda + \mathbf{N} \text{ and } |\mathbf{J}| = \sqrt{J(J+1)} \frac{h}{2\pi} \qquad (5.53)$$

The quantity J is resultant rotational quantum number that can have Λ, $\Lambda + 1$, $\Lambda + 2$, ... values. This indicates that states corresponding to $J < \Lambda$ are missing for the symmetric top molecules.

If the molecule is assumed to be a vibrating rotator, then the term value for a state can be written as:

$$\bar{\varepsilon} = G(v) + F_v(J) \tag{5.54}$$

where $G(v)$ and $F_v(J)$ are the vibrational and rotational term values. The total rotational energy $E(J)$ is the sum of rotational energies about the entire three axes.

$$E(J) = \frac{1}{2}I_a\omega_a^2 + \frac{1}{2}I_b\omega_b^2 + \frac{1}{2}I_c\omega_c^2$$

$$E(J) = \frac{1}{2}I_a\omega_a^2 + \frac{1}{2}I_b\omega_b^2 + \frac{1}{2}I_b\omega_b^2; \ I_b = I_c \ \text{and} \ \omega_b = \omega_c$$

$$E(J) = \frac{1}{2}\frac{(I_a\omega_a)^2}{I_a} + \frac{1}{2}\frac{(I_b\omega_b)^2}{I_b} + \frac{1}{2}\frac{(I_b\omega_b)^2}{I_b}$$

$$E(J) = \frac{L_a^2}{2I_a} + \frac{L_b^2}{2I_b} + \frac{L_b^2}{2I_b}$$

$$E(J) = \frac{L_a^2}{2I_a} + \frac{L_b^2}{I_b} = \frac{L_\parallel^2}{2I_\parallel} + \frac{L_\perp^2}{2I_\perp}; L_\perp^2 = L_b^2 + L_c^2 = 2L_b^2$$

$$E(J) = \frac{1}{2I_\parallel}\left(\Lambda\frac{h}{2\pi}\right)^2 + \frac{1}{2I_\perp}\left(N\frac{h}{2\pi}\right)^2$$

$$E(J) = \frac{\Lambda^2}{2I_\parallel}\left(\frac{h}{2\pi}\right)^2 + \frac{N^2}{2I_\perp}\left(\frac{h}{2\pi}\right)^2$$

$$E(J) = \frac{\Lambda^2}{2I_\parallel}\left(\frac{h}{2\pi}\right)^2 + \frac{(J(J+1) - \Lambda^2)}{2I_\perp}\left(\frac{h}{2\pi}\right)^2$$

$$E(J) = \frac{h^2}{8\pi^2}\left[\frac{J(J+1)}{I_\perp} + \left(\frac{1}{I_\parallel} - \frac{1}{I_\perp}\right)\Lambda^2\right] \tag{5.55}$$

Since $F_v(J) = \frac{E(J)}{hc}$,

So $\qquad F_v(J) = \frac{h}{8\pi^2 c}\left[\frac{J(J+1)}{I_\perp} + \left(\frac{1}{I_\parallel} - \frac{1}{I_\perp}\right)\Lambda^2\right]$

$$F_v(J) = \frac{h}{8\pi^2 I_\perp c}J(J+1) + \left(\frac{h}{8\pi^2 I_\parallel c} - \frac{h}{8\pi^2 I_\perp c}\right)\Lambda^2$$

Let $B_v = \frac{h}{8\pi^2 I_\perp c}$ and $A = \frac{h}{8\pi^2 I_\parallel c}$. Here $I_\parallel$ and $I_\perp$ are the moment of inertia about parallel and perpendicular to inter nuclear axis. Since $I_\parallel < I_\perp$, $A > B_v$.

So
$$F_v(J) = B_v J(J+1) + (A - B_v)\Lambda^2 \tag{5.56}$$

and
$$G(v) = \frac{E_v}{hc} = \omega\left(v + \frac{1}{2}\right) \tag{5.57}$$

Putting values of $G(v)$ and $F_v(J)$ in Eq. (5.54), we have

$$\bar{\varepsilon} = \omega\left(v + \frac{1}{2}\right) + B_v J(J+1) + (A - B_v)\Lambda^2 \tag{5.58}$$

Equation (5.58), has an extra term $(A - B_v)\Lambda^2$ in comparison to the expression of energy for a linear diatomic molecule (vibrating rotator). This indicates that the rotational levels of a symmetric top diatomic molecule have extra shift of amount $(A - B_v)\Lambda^2$ for a given Λ in comparison to linear diatomic molecule. For $\Lambda = 0$, it becomes as the case of simple vibrating rotator.

The selection rule for the transition between the states of symmetric top molecule is:

For $\Lambda \neq 0$, $\Delta v = \pm 1$ and $\Delta J = 0, \pm 1$

Thus, the fundamental absorption band ($\Delta v = +1$, $\Lambda = 0 \rightarrow \Lambda = 1$) of such molecule will have three branch spectra. The set of transition lines corresponding to $\Delta J = +1$ -1 are called *R-branch* and *P-branch lines*. While the lines corresponding to $\Delta J = 0$ are named as *Q-branch*. The formation of R-branch and P-branch are same as linear diatomic vibrating rotator. The symmetric top molecule possesses an extra Q-branch in addition to these branches.

If $\bar{\varepsilon}_f$ and $\bar{\varepsilon}_i$ are the energies (cm^{-1}) of final and initial states, then wave number $\bar{v}$ corresponding to transition $i \rightarrow f$ can be obtained as:

$$\bar{v} = \bar{\varepsilon}_f - \bar{\varepsilon}_i \tag{5.59}$$

Let $\bar{v}_{P,R}$ and $\bar{v}_Q$ are the wave numbers corresponding to lines of P-R-branch and Q-branch, then we may write from Eqs. (5.58) and (5.59), (see Eq. (5.48) in Section 5.6.4),

$$\bar{v}_{P,R} = \omega + (B_1 + B_0)m + (B_1 - B_0)m^2 \tag{5.60}$$

For R-branch, m have value $J + 1$ (where $J = 0, 1, 2, ...$) and for P-branch, m have $-J$ (where $J = 2, 3, 4, ...$) value. In case of P-branch, J cannot have value 1 because this branch comes by the transition $J \rightarrow J - 1$, and there should exist $J = 0$ level for $\Lambda = 1$, that is missing state.

$$\bar{v}_Q = \omega + (B_1 - B_0)m + (B_1 - B_0)m^2 \tag{5.61}$$

Here $m = +J(J : 1, 2, 3, 4, ...)$. Equations (5.60) and (5.61) contain an extra term $(B_1 + B_0)\Lambda^2$, but it has been ignored due to $J > \Lambda$ and $J^2 \gg \Lambda^2$. Since B_1 and B_0 are nearly equal, so all the lines of R-branch fall near to origin ω. The P, Q and R branch spectra is also called *three-branch spectra*. Figure 5.14 represents the schematic diagram of three-branch spectra.

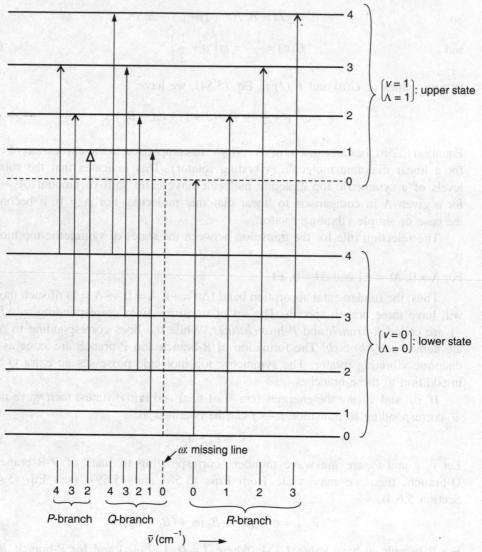

Figure 5.14 Transition among the energy states and spectrum for diatomic symmetric top molecule.

5.8 ELECTRONIC SPECTRA

The atoms held together by chemical bonds are called molecule. The total energy of a molecule is the sum of electronic energy, vibration and rotational energy due to motion of electron, vibration and rotation of atoms. The various possible arrangements of electron for a given molecule form a pattern of allowed electronic states for the molecule. For each electronic state of the molecule, there exists a potential energy curve that has minima at equilibrium nuclear separation. Hence for energy electronic state, there is a set of vibrational states. Since vibration is correlated to rotation of atoms, a set of rotational levels also exists for each and every vibrational states (Figure 5.14).

Let $\overline{\varepsilon}$ is the energy of a state (in cm^{-1}) of a molecule, then

$$\overline{\varepsilon} = \frac{E}{hc} = \frac{E_e}{hc} + \frac{E_v}{hc} + \frac{E_r}{hc} = \varepsilon_e + G(v) + F(v, J) \tag{5.62}$$

where ε_e, $G(v)$ and $F(v, J)$ are the term values for the electronic, vibrational and rotational states. The transition in the electronic states of molecule provides the electronic spectra. Since an electronic state is composed of number of vibrational and rotational states, electronic spectrum is more complex than the other. Such spectrum is obtained in visible and UV regions. All the molecules exhibit electronic spectra. The homo-nuclear molecule also provides the electronic spectra yet they do not have rotational or vibrational–rotational spectra. The reason behind it is that the instantaneous dipole moment change during the redistribution of electronic charge accomplishes the electronic transition. The electronic spectra are in the form of band that is much complicated with respect to vibrational or rotational bands. Each band is composed of number of lines.

Suppose $\overline{v}$ is the frequency of a line in electronic spectra (Figure 5.15) corresponding to transition between upper state (e', v', J') to lower state (e'', v'', J''), then from Eq. (5.62), we have

$$\overline{v} = \{\varepsilon_{e'} - \varepsilon_{e''}\} + \{G(v') - G(v'')\} + \{F(v', J') - F(v'', J'')\}$$

$$\overline{v} = \overline{v}_e + \{G(v') - G(v'')\} + \{B'J'(J' + 1) - B''J''(J'' + 1)\} \tag{5.63}$$

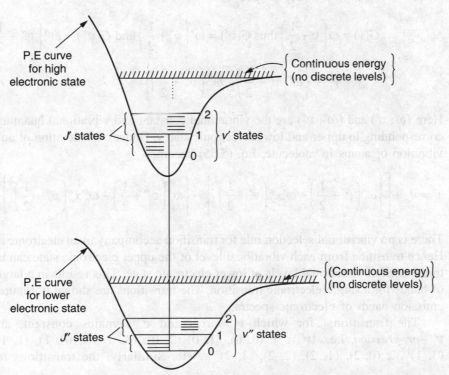

Figure 5.15 Emission spectra by electronic transition (Vibrational electronic spectra in emission).

Here B' and B'' are the rotational constants for the vibration states v' and v''. There is no restriction for the transition $v' \rightarrow v''$, so Δv is unrestricted. The selection rule for transition in rotational states is $\Delta J = 0, \pm 1$. But the transition from $J' = 0 \rightarrow J'' = 0$ is forbidden. The electronic transitions give rise to the number of bands that have closely packed lines converging to a head on one side of the bands. Such spectrum is obtained both in emission and absorption spectra.

When a molecule performs an electronic transition from its upper (excited) state to lower (ground) states, emission spectra is obtained. An electronic transition consists of transition in electronic, vibrational and rotational states (Figure 5.14). Suppose $\bar{v}$ is the frequency of a spectral line in the emission spectra. Then

$$\bar{v} = \{\varepsilon_{e'} - \varepsilon_{e''}\} + \{G(v') - G(v'')\} + \{F(v', J') - F(v'', J'')\}$$

$$\bar{v} = \bar{v}_e + \{G(v') - G(v'')\} + \{F(v', J') - F(v'', J'')\} \tag{5.64}$$

Here $\bar{v}_e$ is a constant for a given electronic transition. The quantities $G(v')$ and $G(v'')$ are the energies of vibrational states associated with different electronic states. Similarly, $F(v', J')$ and $F(v'', J'')$ are the energies of rotational states corresponding to different vibrational states of electronic states. Since $\varepsilon_e \approx 1000 G(v) \approx 1000000 \, F(v, J)$, we can ignore the effect of rotational transition for the vibrational coarse structure of the electronic spectra. So, Eq. (5.64) becomes

$$\bar{v} = \bar{v}_e + \{G(v') - G(v'')\}$$

$$\because \quad G(v) = \omega\left(v + \frac{1}{2}\right) \text{ thus } G(v') = \omega'\left(v' + \frac{1}{2}\right) \text{ and } G(v'') = \omega''\left(v'' + \frac{1}{2}\right)$$

$$\bar{v} = \bar{v}_e + \left\{\omega'\left(v' + \frac{1}{2}\right) - \omega''\left(v'' + \frac{1}{2}\right)\right\} \tag{5.65}$$

Here (ω', v') and (ω'', v'') are the vibrational constant and vibrational quantum number corresponding to upper and lower electronic states. Under consideration of anharmonic vibration of atoms in molecule, Eq. (5.65) becomes,

$$\bar{v} = \bar{v}_e + \left[\left\{\omega_e'\left(v' + \frac{1}{2}\right) - \omega_e' x_e'\left(v' + \frac{1}{2}\right)^2\right\} - \left\{\omega_e''\left(v'' + \frac{1}{2}\right) - \omega_e'' x_e''\left(v'' + \frac{1}{2}\right)^2\right\}\right] \tag{5.66}$$

There is no vibrational selection rule for transition accompany in an electronic transition. Hence transition from each vibrational level of the upper electronic state can take place to the each of vibrational levels of lower electronic state. This results in a large number of bands for a single electronic transition. The transitions are shown in Figure 5.15 for emission bands of electronic spectra.

The transitions, for which v' varies and v'' remains constant, are called v' –progression, i.e., (v', v''): (0, 0), (1, 0) (2, 0), (3, 0) ...; (0, 1), (1, 1), (2, 1), (3, 1) ...; (0, 2), (1, 2), (2, 2), (3, 2) ... etc. Similarly, the transitions for which

v'' varies and v' remains constant, are called v''-*progression*, i.e., (v', v''): (0, 0), (0, 1), (0, 2), (0, 3) ...; (1, 0), (1, 1), (1, 2), ... From Figure 5.15, it is clear that the total band system is composed of number of group and, each group consists of a few very close bands. These groups are called *sequence*. For each sequence, Δv is constant. Thus, the bands (0, 1), (1, 2), (2, 3), (3, 4) ...; (0, 0), (1, 1), (2, 2), (3, 3), (4, 4) ...; (1, 0), (2, 1), (3, 2), (4, 3) ...; (2, 0), (3, 1), (4, 2), ... are called *sequence*. The analysis of band can be easily done on the basis of Deslandre table (Figure 5.16). It is a row and column arrangement of v' and v'' as shown in the figure. The rows and columns of this table have constant separation (in cm^{-1}: on the basis of energy of levels).

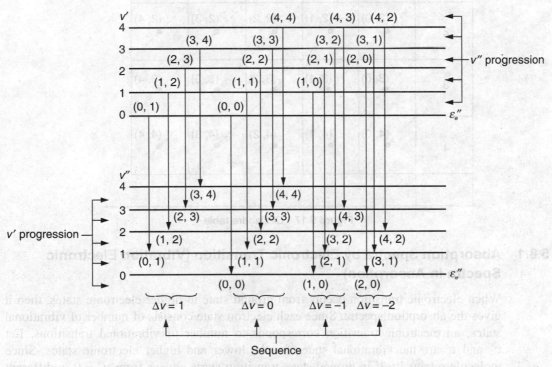

Figure 5.16   Electronic transition from vibrational levels of higher electronic state to vibrational levels of lower electronic state.

From Figure 5.17, we can say that progression bands lie on the horizontal rows or vertical columns of the Deslandre table. While, the bands of a sequence exist along the diagonal or along a line parallel to the diagonal of the table. The wave numbers of the bands in a sequence do not differ because levels v' and v'' are equally spaced for the lower values. While the wave numbers of band in a progression differs. The intensity of these progression bands can be explained on the basis of *Franck–Condon principle*.

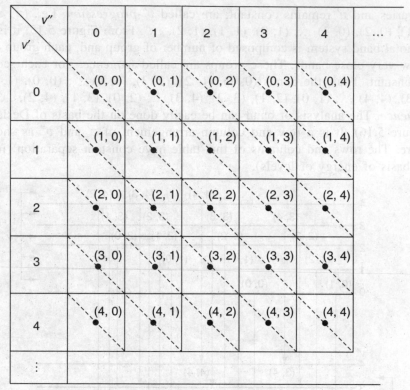

Figure 5.17 Deslandre table.

5.8.1 Absorption Spectra by Electronic Transition (Vibration Electronic Spectra in Absorption)

When electronic transition occurs from ground state to higher electronic states, then it gives the absorption spectra. Since each electron state consists of number of vibrational states, an electronic transition corresponds to number of vibrational transitions. Let v'' and v' are the vibrational states for the lower and higher electronic states. Since molecule retains itself in ground state, transition starts always from $v'' = 0$ to different vibrational levels v' of higher electronic states (as shown in Figure 5.18). This leads to a number of absorption bands in absorption electronic spectra. Here, transition is such that v'' remains constant, and v' varies. It provides v' progression bands, i.e., (v', v''): $(0, 0), (1, 0), (2, 0), (3, 0), (4, 0), (6, 0) \ldots$.

We know that the energy of vibrational states increases as the vibrational number increases (i.e., $v' : 0 \to \infty$) and also, the levels come closer with increase in vibrational number due to anharmonicity. After large value vibrational state, energy becomes continuous. Thus wave number corresponding to progression increases continuously and limits for $(\infty, 0)$. Furthermore, the separation in bands continuously decreases and limits at $v' = \infty$.

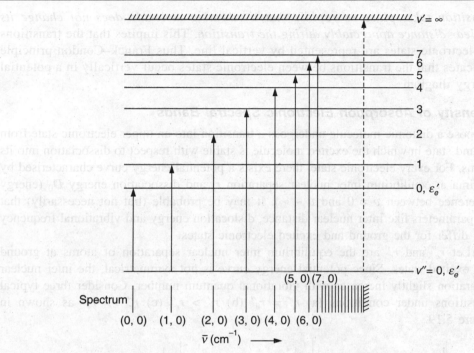

Figure 5.18 Presentation of transitions from lower to higher electronic state and electronic absorption spectra.

If $\bar{v}$ is wave number of the band origin for the progression of band, then

$$\bar{v} = \bar{v}_e + \omega'_e\left(v' + \frac{1}{2}\right) - \omega'_e x'_e\left(v' + \frac{1}{2}\right)^2 - \frac{\omega''_e}{2} + \frac{\omega''_e x''_e}{4} \tag{5.67}$$

where v' can have values 0, 1, 2, 3, … .

Such absorption electronic bands are experimentally obtained, when continuous visible or UV light passes through a long tube containing the sample of molecular gas and then directed towards a spectrograph. The intensity of absorption bands can be explained on the basis of *Franck–Condon principle*. The absorption continuum joins number of band corresponds to the dissociation of the molecule. Thus dissociation energy can be computed with the help of this spectrum. Hence the study of absorption spectra is found to be easier than the emission electronic spectra and gives more information with less computation regarding excited state of molecules.

5.9 FRANCK–CONDON PRINCIPLE

There are no restrictions for the vibration transition in electronic spectra. The spectral lines found by these transitions (vibrational lines in a progression) do not have same intensity. Intensity distribution of such spectra is readily explained in terms of the Franck–Condon principle. *According to this principle, an electronic transition takes place so rapidly that a vibrating molecule does not change its nuclear distance appreciably during the transition.* This implies that the transitions in electronic states are represented by vertical line. Thus Franck–Condon principle indicates that the transitions between electronic states occur vertically in a potential energy diagram.

Intensity of Absorption Electronic Spectral Bands

Suppose a diatomic molecule undergoes a transition into an upper electronic state from ground state in which the excited molecule is stable with respect to dissociation into its atoms. For every electronic state, there exists a potential energy curve characterised by minima at equilibrium inter nuclear separation r_e and dissociation energy D_0 (energy difference between $v = 0$ and $v \sim \infty$). It may be probable (but not necessarily) that the parameters like inter nuclear distance, dislocation energy and vibrational frequency will differ for the ground and excited electronic states.

Let r_e'' and r_e' are the equilibrium inter nuclear separation of atoms at ground and excited states. Since potential energy curve is not symmetrical, the inter nuclear separation slightly increases with vibrational quantum number. Consider three typical transitions under conditions (a) $r_e' = r_e''$ (b) $r_e' > r_e''$ (c) $r_e' \gg r_e''$ as shown in Figure 5.19.

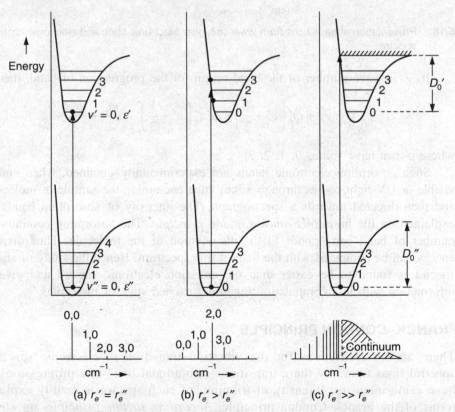

Figure 5.19 Transition from lower to higher energy state under three conditions.

Commonly, in absorption electronic spectra, the transition occurs from $v'' = 0$. When r_e'' and r_e' [Figure 5.18(a)] are equal then vertical line meets at $v' = 0$ under FC principle. The most probable transition will be $v'' = 0 \to v' = 0$. Thus $(0, 0)$ band has maximum intensity under this condition. Transitions to levels $v' = 1, 2, ...$ also occur, but with less probability because they involve a change in inter nuclear separation, i.e., deviates from FC principle. This leads to a weak absorption bands. Now, under condition $r_e' > r_e''$ [Figure 5.18(b)], the transition $v'' = 0$ to $v' = 0$ will be the most probable transition under FC principle. Thus band $(2, 0)$ will have maximum intensity. While transition to lower or higher v' are less likely, thus intensity of other bands diminishes rapidly on both sides of the band $(2, 0)$. In case of $r_e' \gg r_e''$ [Figure 5.18(c)], a vertical electronic transition from $v'' = 0 \to$ high v' occurs or terminates in the continuum of the upper electronic state, which will dissociate the molecule. Thus in this case, the spectrum is expected to be consist of a progression of weak bands joined by a continuum of maximum absorption intensity. Hence, variation in intensity of bands can be explained on the basis of FC principle, which is due to variation in inter nuclear distance for ground and excited electronic states.

Intensity of Emission Electronic Bands (Franck–Condon Parabola)

In emission spectra, the transition may be from any vibrational level of upper electronic state to any vibrational level of lower electronic state (i.e., $v' \to v''$). Due to no restriction on such transition, a large number of band-progression exists in emission spectra. To explain intensity of bands in a progression, let us suppose a molecule occupies a level PQ(v') of the upper electronic state (Figure 5.20). Since molecule spends more time at turning points P and Q during vibration, transitions from these points will be most probable. According to FC principle, the molecule immediately goes to either R-point of level RS (v'' level) or U point of level TU (other v'' level) after vertical transition. This indicates that the transition for which v' is constant have maximum probability. Thus there will be two intensity maxima in a v''-progression, one at smaller v'' and the other at larger v''.

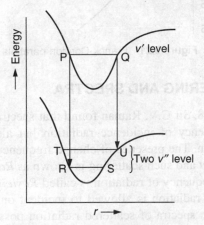

Figure 5.20 Molecule with upper electronic state.

On increasing v', point U moves more rapidly than point R due to being unsymmetrical potential energy curve. Thus the two intensity maxima would separate more and more from each other and also would go to higher v'' values.

The distribution of intensity in a band progression with $v' = 0$ in emission corresponds to that of a progression with $v'' = 0$ in absorption. When equilibrium inter nuclear separation for upper and lower states are equal, then transition from $v' = 0$ to $v'' = 0$ will be most probable, i.e., one maxima would be obtained for (0, 0) band. Except these maxima (spectral bands), other bands are not dominant due to unlike transition under FC principle. These results are found to be in good agreement with observation.

When we plot the intensities of bands in $v' - v''$ array similar to Deslandre table and join the most intense bands, we get a parabolic curve whose axis is the principal diagonal (Figure 5.21). This curve is called *Franck–Condon parabola*. This curve represents the same fact as discussed above, i.e., there are two intensity maxima in all the horizontal rows (v''-progression), except $v' = 0$, which has one maximum. As we move towards high v', the separation in maxima increases (limb of parabola goes away from diagonal).

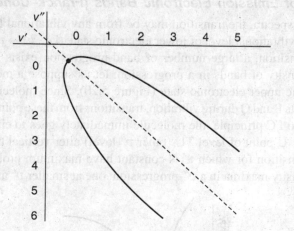

Figure 5.21   Franck-Condon parabola.

5.10   RAMAN SCATTERING AND SPECTRA

On February 28, 1928, Sir C.V. Raman found that spectra of scattered light or radiation have not only frequency of incidence radiation but also there exists some changed frequency of radiation. The presence of change frequency of radiation in scattered light is called *Raman effect* and such scattering is known as *Raman scattering*. The scattering without change in frequency of radiation is called *Rayleigh scattering*. When frequency v of monochromatic radiation is allowed to incident on transparent solid or dust free liquids or gases, then spectra of scattered radiation possess three types of frequencies ie. $v - v_m$, v, $v + v_m$ where v_m is the rotational or vibrational or rotational-vibrational frequency of molecule (Figure 5.22).

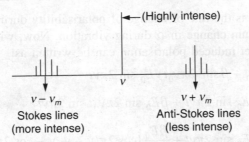

Figure 5.22   Raman spectrum.

The spectral line corresponding to frequency $v - v_m$ and $v + v_m$ are called Stokes lines and anti-Stokes lines respectively. In terms of wavelength, the Stokes lines have wavelength greater than the Rayleigh line wavelength, while anti-Stokes lines have a wavelength less than this value on the other side. The Stokes lines are found to be more intense in comparison to anti-Stokes line. The spectral lines with changed frequency are termed as *Raman lines*. The frequency shift of these lines with Rayleigh is called *Raman shift*.

5.10.1   Explanation of Raman Scattering with Classical Approach

A molecule is made up of an assemblage of positively charged nuclei embedded in a cloud of negative electricity. When a molecule is placed in an electric field, then the centre of gravity of positive and negative charges are slightly displaced, which results a dipole moment in the molecule. The development of induced dipole moment in the molecule by the electric field is called *polarisation*. The induced dipole moment per unit volume ($\mathbf{P}$: Polarisation) is directly proportional to electric field $\mathbf{E}$, i.e.,

$$\mathbf{P} \propto \mathbf{E}$$
$$\mathbf{P} = \alpha\mathbf{E}$$
$$|\mathbf{P}| = \alpha|\mathbf{E}|$$
$$P = \alpha E \tag{5.68}$$

where quantity α is called *polarisability of molecule*. When a beam of visible or ultra violet light falls upon a molecule, the rapidly oscillating electric field of the beam induces an oscillation electric dipole moment in the molecule, i.e., a forced oscillation sets up in the molecule. Suppose electric field associated with a light of frequency v is given as:

$$E = E_0 \sin 2\pi vt \tag{5.69}$$

From Eqs. (5.68) and (5.69), we can write

$$P = \alpha E_0 \sin 2\pi vt \tag{5.70}$$

Classically an oscillating molecule radiates electromagnetic wave. In scattered light spectrum, a high intense line exists at frequency (Rayleigh line). For a vibrating molecule, the polarisability varies about an average value α_0 with a frequency equal to vibrational frequency (v_0), i.e.,

$$\alpha = \alpha_0 + \beta \sin 2\pi v_0 t \tag{5.71}$$

where quantity β is the rate of change of polarisability during vibration, and it is also known as maximum change in α during vibration. Now, with the help of Eqs. (5.70) and (5.71), the net induced polarisation can be written as:

$$P = (\alpha_0 + \beta \sin 2\pi v_0 t) E_0 \sin 2\pi v t$$

$$P = \alpha_0 E_0 \sin 2\pi v t + \beta E_0 \sin 2\pi v_0 t \sin 2\pi v t$$

$$P = \alpha_0 E_0 \sin 2\pi v t + \frac{\beta E_0}{2}[\cos 2\pi(v - v_0)t - \cos 2\pi(v + v_0)t] \qquad (5.72)$$

Thus induced polarisation contains three distinct frequencies: v, $v - v_0$ and $v + v_0$. Hence the scattered light spectra of a vibrating molecule contain spectral lines at frequencies v (Rayleigh line) and $v \pm v_0$ (Raman lines).

Similarly, the Raman effect can also be explained by considering the molecule as rotator. Since for a rotating molecule, polarisability varies about an average α_0' with a frequency equal to twice of rotational frequency v_r, hence

$$\alpha = \alpha_0' + \beta' \sin \{2\pi(2v_r)t\} \qquad (5.73)$$

Thus net polarisation for rotating molecule becomes

$$P = \alpha_0' E_0 \sin 2\pi v t + \frac{\beta' E_0}{2}[\cos 2\pi(v - 2v_r)t - \cos 2\pi(v + 2v_r)t] \qquad (5.74)$$

where β' is the rate of change of polarisability during rotation. This expression also contains three frequencies v, $v - 2v_r$ and $v + 2v_r$ which indicates that spectrum of scattered light for a rotating molecule will also have spectral lines at frequencies v (Rayleigh line) and $v \pm 2v_r$ (Raman lines).

Overall the above analysis shows that the scattered light spectra of vibrating or rotating molecule have one line at same frequency v as of incident light and two other at changed frequencies ($v \pm v_m$), where v_m is rotational or vibrational frequency. Thus the classical approach clearly explains the existence of Raman lines, but it is unable to explain their relative intensity. The relative intensity of Raman lines can be explained on the basis of quantum approach.

5.10.2 Explanation of Raman Scattering under Quantum Approach

Suppose E_1 and E_2 are the molecular energy states of molecule in ground state position. Further more, E_1' and E_2' are the energy states in excited state of molecule. In ground state position, most of the molecule exists in state E_1 and very few in E_2 state because $E_1 < E_2$. When the molecules are excited by the photon (light energy quanta) of energy hv, then molecules of E_1 state go to E_1' state, while molecules of E_2 state go to E_2' state (Figure 5.22). According to quantum mechanics, a molecule or atom cannot reside for a large time in excited state; it goes to ground state by emission of photon. Thus, molecules of E_1' and E_2' excited states make transitions to E_1 or E_2 states to come in ground state. Now, there is possibility of four types of transitions as

$$E_1' \rightarrow E_2, E_1' \rightarrow E_1, E_2' \rightarrow E_2 \text{ and } E_2' \rightarrow E_1$$

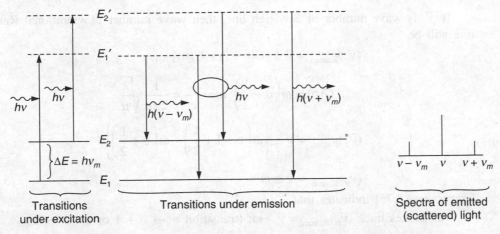

Figure 5.23 Transition among the molecular energy state for Raman spectra.

The energy of emitted photons by the transitions $E_1' \rightarrow E_1$ and $E_2' \rightarrow E_2$ has same energy as incident photon. Thus in spectrum, there exists a single line (Rayleigh line) at frequency v. This line will have very high intensity because it is formed by the transition of both excited state molecules. If the energy difference between states E_1 and E_2 is hv_m, then the energy of photon emitted by transitions $E_1' \rightarrow E_2$ and $E_2' \rightarrow E_1$ will be $h(v - v_m)$ and $h(v + v_m)$ respectively. Hence, in spectrum of emitted light, there will also exist lines at frequencies $(v - v_m)$ (Stoke line) and $(v + v_m)$ (anti-Stoke line). These complementary lines are termed as *Raman lines*.

The population of E_1' state is larger than the E_2' state due to large number of molecules in E_1 state before excitation. Thus large number of molecules corresponds to transition $E_1' \rightarrow E_2$ and very few transitions for $E_2' \rightarrow E_1$. Hence the intensity of Stokes line (spectral line at frequency $v - v_m$) is larger than that of anti-Stokes line (spectral line at frequency $v + v_m$). Therefore, the quantum approach explains not only the existence of Raman line in scattered light spectra, but also their relative intensity.

The energy state of molecule is the combination of vibrational and rotational energy states. The additional spectral lines in scattered light spectra are caused by transition in these states. Hence Raman spectra can be divided into two types: vibrational Raman spectra and rotational Raman spectra.

5.10.3 Vibrational Raman Spectra

When vibration of atoms in a molecule produces a change in polarisability with change of the inter-nuclear distance, then transition of molecule from one vibrational state to the other of same electronic state causes shifts in the spectral line in either side of the Rayleigh spectral line. These shifted lines on both sides of Rayleigh line are called vibrational Raman lines. The transition in vibrational states are such that $\Delta v = \pm 1$, known as selection rule for vibrational Raman spectra.

If $\bar{v}$ is wave number of Rayleigh line, then wave number of vibrational Raman line will be

$$(\bar{v}_v)_{\text{Raman}} = \bar{v} \pm \{G(v+1) - G(v)\}$$

$\because$
$$G(v) = \omega\left(v + \frac{1}{2}\right); \; \omega = \frac{v_0}{c} = \frac{1}{2\pi c}\sqrt{\frac{k}{\mu}}$$

$$(\bar{v}_v^{\circ})_{\text{Raman}} = \bar{v} \pm \left\{\omega\left(v + 1 + \frac{1}{2}\right) - \omega\left(v + \frac{1}{2}\right)\right\}$$

$$(\bar{v}_v)_{\text{Raman}} = \bar{v} \pm \omega \tag{5.75}$$

Expression (5.75) indicates that

For Stokes line, $(\bar{v}_v)_{\text{Raman}} = \bar{v} - \omega$; (transition $v \to v+1$ or $v : 0 \to 1$)

For anti-Stokes line, $(\bar{v}_v)_{\text{Raman}} = \bar{v} + \omega$; (transition $v + 1 \to v$ or $v : 1 \to 0$) and shift of Raman line from Rayleigh line

$$= |(\bar{v}_v)_{\text{Raman}} - \bar{v}| = \omega = \frac{v_0}{c} \tag{5.76}$$

Normally, the molecules are found in ground state ($v = 0$) and very few in excited state ($v = 1$). Thus intensity of Stokes line is larger than the anti-Stokes line. Here Raman shift is found equal to vibrational constant ω which is same as wave number of fundamental band in near infra-red spectrum of molecule. Hence, there is a good resemblance between vibration Raman shift and near infra-red absorption band.

Under realisation of anharmonicity, the selection rule becomes, $\Delta v = \pm 1, \pm 2, \pm 3 \ldots$.

This leads to the exception of lines corresponding to overtones, combination bands and hot bands in Raman spectra also. However, the scattering of ordinary light is of very low intensity, and the weaker effects like overtones, combination of bands and hot bands were ignored. It was only with the advent of the laser that such bands have come to prominence.

5.10.4 Rotational Raman Spectra

Homonuclear diatomic molecules do not give a vibrational or rotational spectrum because they do not have permanent dipole moment. However, such molecule gives pure rotational Raman spectra since the polarisation changes during rotation. Thus the molecular parameters of homonuclear molecule can be obtained by Raman spectroscopy.

The pure rotational Raman lines appearing on the both sides of Rayleigh line arise from transitions of molecule from one rotation level to other of the same vibrational state. The expression of rotation energy level for a diatomic, linear molecule can be written as

$$F(J) = BJ(J+1) \tag{5.77}$$

where J is rotational quantum number and can have values 0, 1, 2, $\ldots$. Raman effect involves polarisability and not the dipole moment as in rotational spectra, thus quantum mechanical treatment leads new selection rule:

$$\Delta J = 0, \pm 2$$

The zero value of ΔJ corresponds to Rayleigh line which defines the centre of rotational Raman spectra. The transitions of molecule under condition $\Delta J = \pm 2$ provides the Raman lines in spectra. The transitions $J \rightarrow J + 2$ and $J + 2 \rightarrow J$ gives Stokes and anti-Stokes Raman lines in respective spectra (Figure 5.24). If $\bar{v}$ is wave number of Rayleigh line, then the wave number for rotation Raman line will be

$$(\bar{v}_r)_{\text{Raman}} = \bar{v} \pm \{F(J+2) - F(J)\}$$

$$(\bar{v}_r)_{\text{Raman}} = \bar{v} \pm \{B(J+2)(J+2+1) - BJ(J+1)\}$$

$$(\bar{v}_r)_{\text{Raman}} = \bar{v} \pm \{4BJ + 6B\}$$

$$(\bar{v}_r)_{\text{Raman}} = \bar{v} \pm 4B\left\{J + \frac{3}{2}\right\} \tag{5.78}$$

where the plus sign refers to anti-Stokes lines and the minus sign to Stokes.

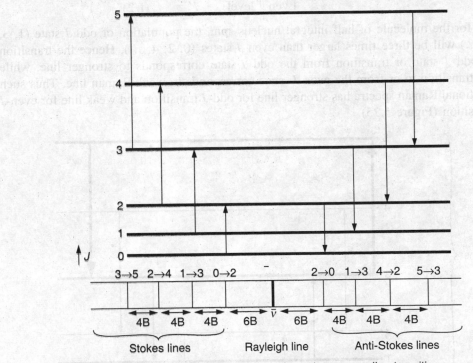

Figure 5.24 Rotational Raman spectra and corresponding transitions.

Since even at room temperature, molecules can have different J values. Thus there exist a good number of Stokes and anti-Stokes lines in the spectrum.

From Eq. (5.78), we have

$$\text{Rotational Raman shift} = \left|(\bar{v}_r)_{\text{Raman}} - \bar{v}\right| = 4B\left\{J + \frac{3}{2}\right\}$$

$$= 6B, 10B, 14B, \ldots; J = 0, 1, 2, \ldots$$

So it is clear that the Raman spectrum of diatomic linear molecule consists of a series of equidistant lines on the either side of the Rayleigh line starting at distance $6B$ and have separation of $4B$ between two successive lines (Figure 5.23). The separation of spectral line in infra-red spectrum is $2B$. So the Raman lines have twice separation between two successive lines in comparison to IR lines. This is in agreement with the experimental observation.

The rotational Raman spectra of homonuclear diatomic molecules show an intensity alteration, i.e., rotational Raman lines are alternatively weak or strong, and in some cases, the alternate lines are missing. This is due to nuclear spin of molecule. The nuclear spin of homonuclear diatomic molecule nuclei can be half integral or integral. The half integral nucleus spin molecule nuclei are called fermions (e.g. H_2). Such molecule has anti-symmetric wave function. Thus nuclear-spin statistical weight ratio (population ratio) for odd J level will be high.

$$\text{Population ratio} = \frac{\text{Odd } J \text{ level}}{\text{Even } J \text{ level}} = \frac{I' + 1}{I'} = \frac{(1/2) + 1}{(1/2)} = \frac{3}{1}$$

So, for the molecule of half integral nucleus spin, the population of odd J state (1, 3, 5, ...) will be three times larger than even J-states (0, 2, 4, ...). Hence the transition to odd J state or transition from the odd J-state corresponds to stronger line, while the transition to or from the even J- state corresponds to weak Raman line. Thus such rotational Raman spectra has stronger line for odd-J transition and weak line for even-J transition (Figure 5.25).

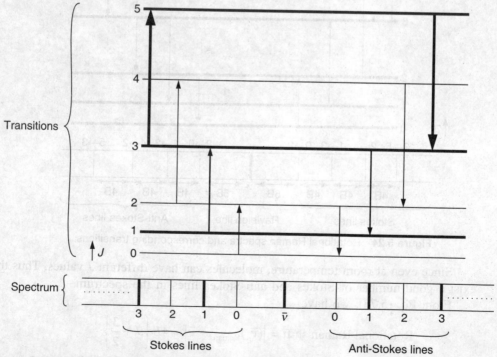

Figure 5.25 Rotational Raman spectra of homonuclear molecule having $I' = 1/2$ (H_2).

The integral nucleus spin molecule nuclei are called bosons (e.g. N_2, D_2 and O_2). For both N_2 and D_2, nucleus spin is 1, while for O_2, it is 0. Such molecule has symmetric (even) wave function. Therefore, the population of even J level of such molecule will be large in comparison to odd J levels. For N_2 or D_2 molecule; nucleus spin $(I') = 1$, so

$$\text{Population ratio} = \frac{\text{Even } J \text{ level}}{\text{Odd } J \text{ level}} = \frac{I'+1}{I'} = \frac{1+1}{1} = \frac{2}{1}$$

Thus population of even J level of N_2 or D_2 molecule will be twice to that of population of odd J level. Hence the transitions under condition $\Delta J = \pm 2$ lead strong spectral line for even J level and weak for odd J level (Figure 5.26).

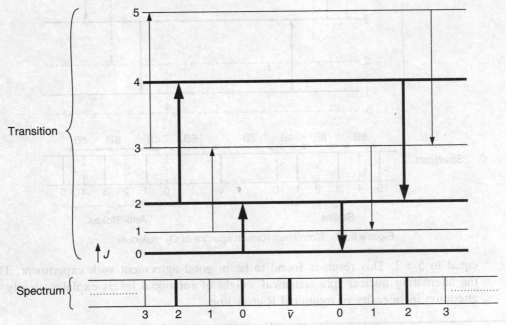

Figure 5.26 Rotational Raman spectra of N_2 or D_2 molecule.

Since the O_2 molecule has zero nucleus spin, this molecule has only one nuclear wave function which is symmetric with respect to both nuclei. Also the total wave function for the molecule must be symmetric. Furthermore for ground states, even levels belong to anti-symmetric wave function, while odd J level corresponds to symmetric wave function. For this molecule, the population of odd J level will be maximum and will be zero for even J levels. Hence the transition under condition $\Delta J = \pm 2$ leads the rotations of Raman spectrum in which line corresponding to even J level transitions are missing (Figure 5.27).

Figure 5.27 indicates that the separation between first Stokes lines and first anti-Stokes lines for O_2 molecule Raman spectra will be $20B$, and separation between two successive lines of any branch will be $8B$. The ratio of these two separations are

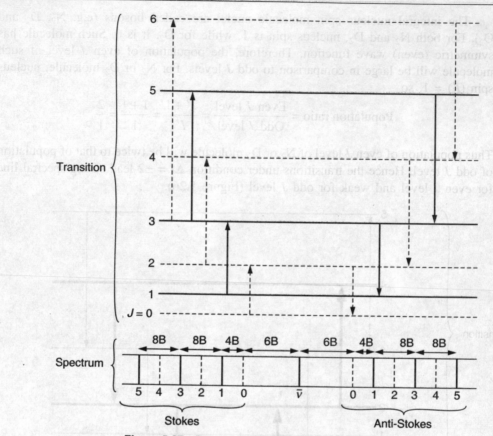

Figure 5.27   Rotational Raman spectra of O_2 molecule.

equal to 5 : 2. This result is found to be in good agreement with experiment. Thus the alternating nuclear spin statistical weight of rotational levels explains clearly the alteration in intensity of rotational Raman lines.

EXAMPLE 5.12   In the scattered light spectra, Stokes line and Rayleigh line are found at wavelengths 5400 Å and 5340 Å. Find the wave number corresponding to anti-Stokes line.

Solution   It λ is wavelength of Rayleigh line, then $\lambda + \Delta\lambda$ and $\lambda - \Delta\lambda$ will be the position of Stokes and anti-Stokes lines.

Here $\lambda + \Delta\lambda = 5400$ Å and $\lambda = 5340$ Å

Thus $\Delta\lambda = 5400 - 5340 = 60$ Å

∴   Wavelength of anti-Stoke line $= \lambda - \Delta\lambda$

$$= 5340 - 60 = 5280 \text{ Å}$$

∴   Wave number for anti-Stoke line $= \dfrac{1}{(5280 \times 10^{-8})} = 18939 \text{ cm}^{-1}$

EXAMPLE 5.13 A diatomic molecule is irradiated with a 2294 cm^{-1} radiation. If the Raman lines are due to transition in vibrational state and a line is found at 20973 cm^{-1}, then find wave number corresponding to another Raman line and vibrational frequency of molecule.

Solution Given that

$$\bar{v} = 22946 \text{ cm}^{-1}$$

$$\bar{v} - \bar{v}_m = 20973 \text{ cm}^{-1}$$

So

$$\bar{v}_m = 22946 - 20973 = 1973 \text{ cm}^{-1}$$

Hence wave number of another Raman line = 22946 + 1973 = 24919 cm^{-1}

Since, under vibrational transition, $\bar{v}_m = \omega = \dfrac{v_0}{c}$

$$v_0 = \bar{v}_m c = 1973 \times 3 \times 10^{10} = 5.92 \times 10^{13} \text{ Hz}$$

EXAMPLE 5.14 In the rotational Raman spectrum of a heteronuclear molecule, the shift of lines from exciting line are found to be given the expression $\Delta\bar{v} = \pm(48J + 72)$. Find the rotational constant for the molecule and also find the moment of inertia.

Solution The rotational Raman shift is given by

$$\Delta\bar{v} = \pm 4B\left(J + \frac{3}{2}\right)$$

Here

$$\Delta\bar{v} = \pm(48J + 72)$$

$\Rightarrow$

$$\Delta\bar{v} = \pm 48\left(J + \frac{3}{2}\right)$$

$\therefore$

$$4B = 48$$

$$B = 12 \text{ cm}^{-1}$$

$\because$

$$B = \frac{h}{8\pi^2 Ic} \quad \Rightarrow \quad I = \frac{h}{8\pi^2 Bc}$$

$$I = \frac{6.62 \times 10^{-27} \text{ erg s}}{8 \times (3.14)^2 \times (12 \text{ cm}^{-1}) \times (3 \times 10^{10} \text{ cm s}^{-1})}$$

$$I = 2.33 \times 10^{-40} \text{ g.cm}^2$$

EXAMPLE 5.15 The vibrational Raman lines are found at wave numbers 2.2×10^4 cm^{-1} and 2.6×10^4 cm^{-1}. Find the vibration constant and frequency of vibration.

Solution Vibrational Raman lines are found at wave numbers $\bar{v} \pm w$.

For stokes line, wave number = $\bar{v} - w$

and for anti-Stokes line, wave number = $\bar{v} + w$

Here

$$\bar{v} - w = 2.2 \times 10^4$$
$$\bar{v} + w = 2.6 \times 10^4$$

Thus

$$2w = 0.4 \times 10^4$$

or

$$w = 2000 \text{ cm}^{-1}$$

$$w = \frac{v_0}{c}$$

$$v_0 = wc = 2000 \times 3 \times 10^{10} = 6 \times 10^{13} \text{ s}^{-1}$$

5.11 SPECTROMETERS

When an atom or molecule changes its energy then it emit or absorbs electromagnetic radiation. An arrangement of radiation in its increasing or decreasing order of wavelength is called as spectrum. An instrument with help of which, we can see and read the wavelength of spectral lines is called as spectrometer. The device, which is used to take a photograph of spectrum, is called as spectrograph.

We have read in earlier topics, that a molecule possess three types of energy states viz. electronic, vibrational and rotational states. The rotational energy causes rotational spectra, which is found in far or microwave region. While vibrational spectra is obtained due to change in vibrational states, that is found in near IR region. The third one i.e. electronic spectra is found in UV-visible region which is due to transition in electronic states of molecule. Thus the base of spectrum region or difference of molecular energy states, the spectrometers are divided into three categories.

(i) Microwave spectrometer (rotational spectra)
(ii) Infrared spectrometer (vibrational spectra)
(iii) UV-visible spectrometer (electronic spectra)

The vibrational spectra or infra red spectra can also be recorded by means of Raman spectrometer. This technique or spectroscopy uses a visible or ultraviolet spectrometer. The study of spectrum obtains by these spectrometers provide the information about energy of different molecular energy states, their separation, bond strength moment of inertia, dissociation energy of molecule etc.

In addition to these spectroscopic technique, there exist two other spectroscopic technique as electron spin-resonance and magnetic-resonance spectroscopy. We will study about these two spectroscopic technique in later topics.

The following are the list of some spectrometers/spectrographs

(A) Spectrometer for infrared region (700 nm to 400 μm)
  1. Wordsworth Prism-Mirror spectrograph
  2. Grating spectrograph
  3. Double beam infrared spectrometer
  4. Raman spectrometer

(B) Spectrometers/spectrograph for UV region (10 nm to 300 nm)
  1. Quartz spectrograph
  2. Littrow spectrograph
  3. Concave grating spectrograph

(C) Spectrometers/spectrograph for visible region (300 nm to 700 nm)
1. Constant deviation spectrometer/spectrograph
2. Prism spectrometer

In the emission spectroscopy, the source of radiation is initially require. An electric arc of carbon, iron or other materials, mercury vapour lamps discharge of electricity through hydrogen contained quartz tube and vacuum spark discharge are used as UV-visible source of radiation in the laboratory. While Nernst glower and Glober are the most common source of infrared radiations. The spectrum of radiation emitted by source are produced either by prism or by grating through the process of dispersion or diffraction. The dispersed or diffracted radiations are then focused on detector by using lenses or mirrors. For the detection, photographic plate or thermopile or computer controlled photomultiplier tube is used. If photographic plate is used, we get photograph of spectrum. The wavelength of spectral line obtained in photograph of spectrum can be determined with help of comperator. In modern spectrometers, the photomultiplier tube (PMT) is used as detector. It consists of light sensitive surface which emits electrons when light falls on it.

The tiny electron current is amplified and applied to a pen-recorder, which is connected to computer. Now the amplified current or voltage is obtained for each spectral line by scanning whole spectrum. The obtained amplified current or voltage is plotted against frequency with help of computer. Such plot is termed as spectrum in frequency domain. By this plot the wavelength of spectral line can be determined.

In the absorption spectroscopy, light from a source, which emits radiations of all wavelengths is passed through sample under study. The transmitted light is spectrally studied to find the absorption at different wavelengths. The ray diagram of some spectrometers can be seen in following Figures 5.28–5.36.

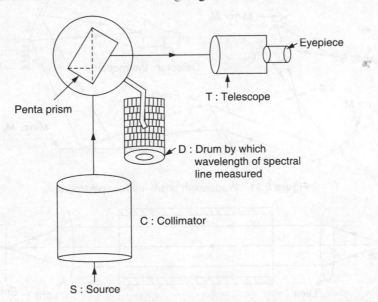

Figure 5.28   Constant deviation spectrometer (when eyepiece is replaced by photographic plate then it is called constant deviation spectrograph.)

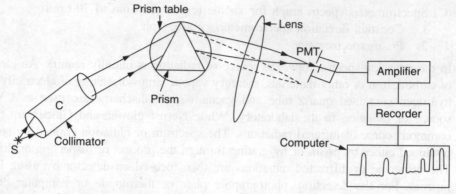

Figure 5.29 Modern spectrometer having PMT as detector.

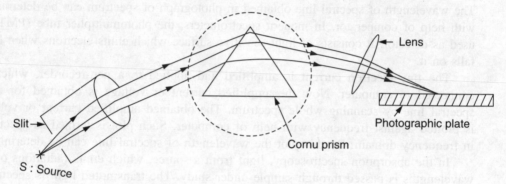

Figure 5.30 Quartz spectrograph.

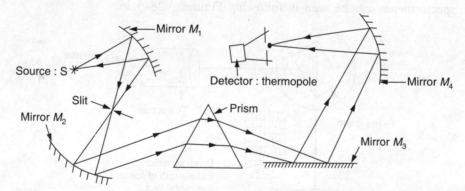

Figure 5.31 Wadsworth prism–mirror spectrograph.

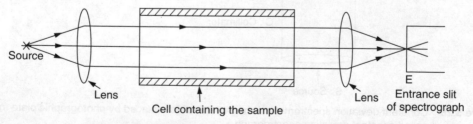

Figure 5.32 Basic ray diagram of absorption spectrometer.

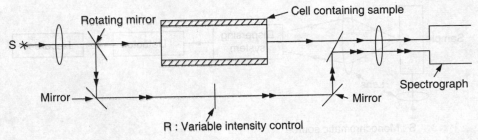

Figure 5.33 Basic Ray-diagram of double beam infrared spectrometer.

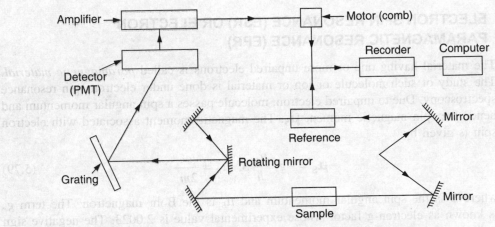

Figure 5.34 Schematic diagram of double beam infrared spectrometer.

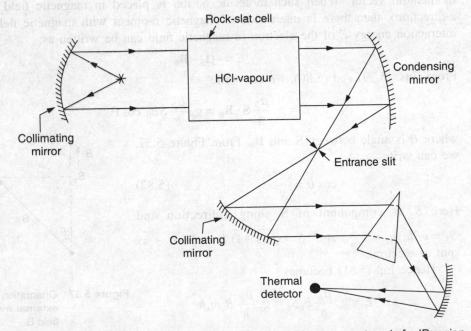

Figure 5.35 Prism-mirror spectrometer for recording absorption spectrum in far IR region.

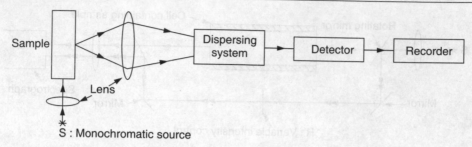

Figure 5.36   Block diagram of Raman spectrometer.

5.12   ELECTRON SPIN RESONANCE (ESR) OR ELECTRON PARAMAGNETIC RESONANCE (EPR)

The material having one or more unpaired electrons is called *paramagnetic material*. The study of such molecule or ion or material is done under electron spin resonance spectroscopy. Due to unpaired electron, molecule passes a spin angular momentum and hence the spin magnetic moment μ_S. The magnetic moment associated with electron spin is given by

$$\mathbf{\mu_S} = -g_e \frac{\mu_B}{\hbar} \mathbf{S}; \quad \mu_B = \frac{e\hbar}{2m_e} \tag{5.79}$$

where $\mathbf{S}$ is the spin angular momentum and μ_B is the Bohr magnetron. The term g_e is known as electron-g factor whose experimental value is 2.0023. The negative sign indicates that the direction of the magnetic moment vector is opposite to angular momentum vector. When such molecule or ion is placed in magnetic field $\mathbf{B_0}$ (in z-direction), then there is interaction of magnetic moment with magnetic field. The interaction energy E of the electron in magnetic field can be written as:

$$E = -\mathbf{\mu_S} \cdot \mathbf{B_0} \tag{5.80}$$

From Eqs. (5.79) and (5.80), we can write

$$E = g_e \frac{\mu_B}{\hbar} \mathbf{S} \cdot \mathbf{B_0} = g_e \frac{\mu_B}{\hbar} SB_0 \cos \theta \tag{5.81}$$

where θ is angle between $\mathbf{S}$ and $\mathbf{B_0}$. From Figure 5.37, we can write

$$\cos \theta = \frac{S_z}{S} \tag{5.82}$$

Here S_z is component of $\mathbf{S}$ along z-direction and $S_z = m_s \hbar; m_s = \pm \frac{1}{2}$ and $S = \sqrt{s(s+1)}\ \hbar$; where s is spin of electron.

Hence Eq. (5.81) becomes

$$E = g_e \frac{\mu_B}{\hbar} B_0 S_z = g_e \frac{\mu_B}{\hbar} B_0 m_s \hbar$$

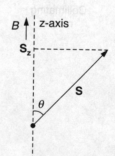

Figure 5.37   Orientation of $\mathbf{S}$ in external magnetic field $\mathbf{B}$.

$$E = g_e \mu_B B_0 m_s \qquad (5.83)$$

Since magnetic spin quantum number m_s have value either +1/2 or –1/2, so

When $\qquad m_s = \dfrac{1}{2},\quad E_+ = \dfrac{1}{2} g_e \mu_B B_0 \qquad$ (for α-spin) $\qquad$ (5.84a)

and when $\qquad m_s = -\dfrac{1}{2},\quad E_- = -\dfrac{1}{2} g_e \mu_B B_0 \qquad$ (for β-spin) $\qquad$ (5.84b)

Equation (5.84) indicates that the spin–magnetic interaction causes degeneracy in energy of unpaired electron. Now, electron can have energies E_+ and E_- ($E_+ > E_-$). This splitting of energy is called *electron spin resonance*. When a suitable frequency of radiation (microwave region) falls on it, a transition of spin occurs from E_- state to E_+ state. This causes absorption of radiation.

$$\Delta E = E_+ - E_-$$
$$\Delta E = g_e \mu_B B_0 \qquad (5.85)$$

If v is frequency of radiation, then $\Delta E = hv$, so Eq. (5.85) becomes

$$hv = g_e \mu_B B_0$$
$$B_0 = \frac{h}{g_e \mu_B} v \qquad (5.86)$$

The spectrum of such absorbed radiation is made in ESR spectroscopy. Equation (5.86) is called *resonance condition* for ESR absorption. The ESR spectrometer can have two different designs. In the first method, frequency could be varied at constant magnetic field B_0, while second method, the magnetic field could be varied at constant frequency v in microwave region. However, the latter method is convenient and thus preferred. The unpaired electron is induced to reverse its orientation ($\beta \to \alpha$) by maintaining a constant frequency v (in MHz) and sweeping the applied field over a range until the incident radiation is absorbed (Figure 5.38). At this field, the energy separation between α and β orientation can be evaluated with the radiation frequency.

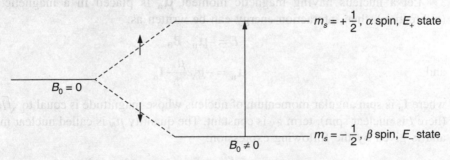

Figure 5.38 Splitting of a level under ESR.

The hyperfine structure of ESR spectra is caused due to interaction of the unpaired electron with the magnetic moment of nuclei within its orbital. This interaction is absent for the molecule, which has zero nuclear spin quantum number. If a nucleus has spin i, then ESR line splits into $(2i + 1)$ lines due to this interaction. For $i = 1/2$, hyperfine energy level is shown in Figure 5.39. The selection rule for ESR transition is, $\Delta m_s = \pm 1$ and $\Delta m_i = 0$. The ESR spectroscopy is mostly used in the structure study of free radicals and kinetic properties of bio-molecules. The symmetry analysis of ionic crystal can also be done by this spectroscopic technique.

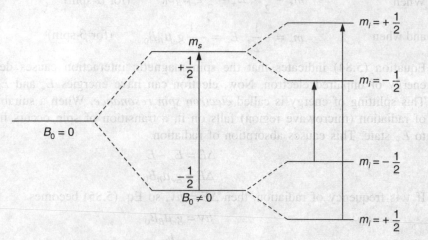

Figure 5.39 Hyperfine energy structure under ESR.

5.13 NUCLEAR MAGNETIC RESONANCE (NMR)

This phenomenon is similar to that of electron spin resonance. This involves the action of nuclear spin instead of electrons. A nucleus can have spin zero, half or integral value. So, it possess a spin angular momentum $\mathbf{I_n}$ and hence nuclear magnetic moment $\boldsymbol{\mu_n}$. *When a molecule/atom having nuclear magnetic moment is placed in a magnetic field, then interaction in nuclear magnetic moment and magnetic field results degeneracy in energy states. This is called nuclear magnetic resonance (NMR).*

Let a nucleus having magnetic moment $\boldsymbol{\mu_n}$ is placed in a magnetic field $\mathbf{B_0}$ (z-direction), then interaction energy can be written as:

$$E = -\boldsymbol{\mu_n} \cdot \mathbf{B_0} \tag{5.87}$$

and

$$\boldsymbol{\mu_n} = -g_N \frac{\mu_N}{\hbar} \mathbf{I_n}$$

where $\mathbf{I_n}$ is spin angular momentum of nucleus whose magnitude is equal to $\sqrt{I(I+1)}\,\hbar$ (here I is nuclear spin), term g_N is constant. The quantity μ_N is called nuclear magnetron and is given by the following expression:

$$(\mu_N)_{\text{proton}} = \frac{e\hbar}{2m_p} = 5.051 \times 10^{-27} \text{ N/m}^2$$

Putting values in Eq. (5.87), we have

$$E = -g_N \frac{\mu_N}{\hbar} \mathbf{I_n} \cdot \mathbf{B_0} = -\frac{g_N \mu_N}{\hbar} I_n B_0 \cos\theta \tag{5.88}$$

where θ is the angle between $\mathbf{I_n}$ and $\mathbf{B_0}$. If $\mathbf{I_z}$ is component of $\mathbf{I_n}$ along direction of magnetic field (Figure 5.40), we can write

$$\cos \theta = \frac{I_z}{I_n}$$

Equation (5.88) becomes

$$E = -\frac{g_N \mu_N}{\hbar} I_z B_0$$

$$E = -\frac{g_N \mu_N}{\hbar} (m_I \hbar) B_0; \text{ as } I_z = m_I \hbar; \ m_I = -I, ..., 0, ..., I$$

$$E = -g_N \mu_N B_0 m_I \tag{5.89}$$

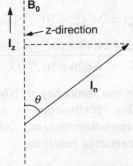

Figure 5.40 Orientation I_n in external magnetic field $\mathbf{B_0}$.

Since m_I have $(2I + 1)$ values, the nuclear state is splitted into $(2I + 1)$ states. This type of degeneracy or splitting of energy states due to magnetic field is termed as NMR. For $I = 1/2$, m_I has two values, i.e., $\pm 1/2$, hence energy state is splitted into two states (Figure 5.41).

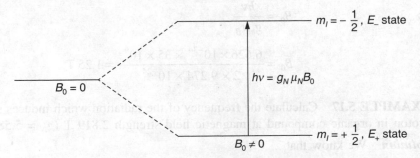

Figure 5.41 Splitting of energy level under ESR.

When $m_I = -\frac{1}{2}, E_- = \frac{1}{2} g_N \mu_N B_0$ (5.90) and when

$$m_I = +\frac{1}{2}, E_+ = -\frac{1}{2} g_N \mu_N B_0 \tag{5.91}$$

Thus the energy difference between E_- and E_+ states will be

$$\Delta E = E_- - E_+ = g_N \mu_N B_0 \tag{5.92}$$

If there are $2I + 1$ states, then for any two consecutive states, the energy difference will be $g_N \mu_N B_0$. The basis of NMR is to induce transition from one possible orientation to another, that lead absorption or emission of a discrete amount of energy. If v is frequency of radiation emitted or absorbed in the transition, then $\Delta E = hv$. So Eq. (5.92) becomes

$$hv = g_N \mu_N B_0$$

$$v = \frac{g_N \mu_N B_0}{h} = \frac{g_N |\mu_n| B_0}{h\sqrt{I(I+1)}} \tag{5.93}$$

For proton, $I = 1/2$ and; $|\mu_n| = 24.42 \times 10^{-27}$ N/m^2. If $B_0 = 1.4092$ T, then

$$v = \frac{2 \times 24.42 \times 10^{-27} \times 1.4092}{6.626 \times 10^{-34} \sqrt{3}} = 60 \text{ Hz}$$

Thus the absorption or emission frequency in NMR spectroscopy falls in short wave radio frequency region. This spectroscopy is used in determination of molecular structure, nuclear magnetic resonance imaging (NMRI), hydrogen bonding studies and in study of intermolecular exchange reactions.

EXAMPLE 5.16   An unpaired electron gives ESR at 35 GHz. Find the strength of magnetic field ($g = 2.00$ and $\mu_B = 9.274 \times 10^{-24}$ N/m^2).

Solution   We know that

$$v = \frac{g_e \mu_B B_0}{h}$$

$$\Rightarrow \qquad B_0 = \frac{hv}{g_e \mu_B}$$

$$\Rightarrow \qquad B_0 = \frac{6.626 \times 10^{-34} \times 35 \times 10^9}{2 \times 9.274 \times 10^{-24}} = 1.25 \text{ T}$$

EXAMPLE 5.17   Calculate the frequency of the radiation which induces spectrum of proton in organic compound at magnetic field strength 2.819 T ($g_N = 5.585$).

Solution   We know that

$$v = \frac{g_N \mu_N B_0}{h}$$

$$v = \frac{5.585 \times 5.051 \times 10^{-27} \times 2.819}{6.626 \times 10^{-34}} = 120 \text{ MHz}$$

SOLVED NUMERICAL PROBLEMS

PROBLEM 5.1 The average spacing between adjacent rotational lines of a heteronuclear diatomic molecule is 3.8626 cm^{-1}. Calculate its moment of inertia and bond length if reduced mass of it is 1.139×10^{-26} kg.

Solution

$$2B = 3.8626 \text{ cm}^{-1}$$

$$B = 193.13 \times 10^{-2} \text{ cm}^{-1} = 193.13 \text{ m}^{-1}$$

$\because$
$$B = \frac{h}{8\pi^2 Ic} \quad \Rightarrow \quad I = \frac{h}{8\pi^2 Bc}$$

$$I = \frac{6.626 \times 10^{-34} \text{ J.sec}}{8 \times (3.14)^2 \times (193.13 \text{ m}^{-1}) \times (3 \times 10^8)} = 1.45 \times 10^{-46} \text{ kg}$$

$\because$
$$r = \sqrt{\frac{I}{\mu}}$$

$\therefore$
$$r = \sqrt{\frac{1.45 \times 10^{-46} \text{ kg.m}^2}{1.139 \times 10^{-26} \text{ kg}}} = 1.128 \text{ Å}$$

PROBLEM 5.2 The fundamental bond in vibrational spectra of diatomic molecule is obtained at wave number 1876 cm^{-1}. Find out the force constant of bond of molecule (Reduced mass = 1.63×10^{-27} kg).

Solution $\omega = 1876$ cm^{-1}

$$v_{osc} = \omega c = 1876 \times 3 \times 10^{10} = 5.628 \times 10^{13} \text{ Hz}$$

$\because$
$$v_{osc} = \frac{1}{2\pi} \sqrt{\frac{k}{\mu}}$$

$\therefore$
$$k = 4\pi^2 \mu v_{osc}^2$$

$$k = 4 \times (3.14)^2 \times 1.63 \times 10^{-27} \times (5.628 \times 10^{13})^2$$

$$k = 203.62 \text{ N/m}$$

PROBLEM 5.3 The first rotational Raman line for a homonuclear diatomic molecule appear at 346 cm^{-1} from the exciting line calculate moment of inertia and bond length of molecule ($\mu = 8.365 \times 10^{-28}$ kg).

Solution Separation in first Raman line and exciting line = $6B$

$\therefore$
$$6B = 346 \text{ cm}^{-1}$$

$$B = 57.6667 \text{ cm}^{-1} = 5766.67 \text{ m}^{-1}$$

$$\therefore \qquad I = \frac{h}{8\pi^2 Bc}$$

$$\therefore \qquad I = \frac{6.626 \times 10^{-34}}{8 \times (3.14)^2 \times 5766.67 \times (3 \times 10^8)} = 4.8557 \times 10^{-48} \text{ kg.m}^2$$

$$\therefore \qquad r = \sqrt{\frac{I}{\mu}}$$

$$\therefore \qquad r = \sqrt{\frac{4.8557 \times 10^{-48} \text{ kg.m}^2}{8.365 \times 10^{-28} \text{ kg}}} = 0.762 \times 10^{-10} \text{ m Å}$$

$$r = 0.762 \text{ Å}$$

PROBLEM 5.4   If zero point energy of a vibrating molecule is 3000 cm^{-1}, find the wavelength of fundamental band for the molecule.

Solution   We know that

$$\text{Zero point energy} = \frac{\omega}{2} = 3000 \text{ cm}^{-1}$$

and

$$\omega = \frac{\nu_0}{c} = \frac{c}{\lambda_0 c} = \frac{1}{\lambda_0}$$

So

$$\frac{1}{\lambda_0} = 6000 \text{ cm}^{-1}$$

$$\lambda_0 = \frac{1}{6000 \text{ cm}^{-1}} = 16667 \times 10^{-8} \text{ cm} = 16667 \times 10^{-10} \text{ m}$$

$$\lambda_0 = 16667 \text{ Å}$$

PROBLEM 5.5   The 60-GHz frequency of radiation is required for ESR absorption in a paramagnetic material. Then find out the energy difference between α and β spin states.

Solution $E_\alpha - E_\beta = h\nu = 6.626 \times 10^{-34} \times 60 \times 10^9 = 3.976 \times 10^{-23}$ J

EXERCISES

THEORETICAL QUESTIONS

5.1 What are the different types of molecules?

5.2 Give a short note on energy of molecule and classification of molecular spectra.

5.3 Show that rotational lines are equidistant for a rigid rotator.

5.4 Why does a homonuclear molecule not have rotational and vibrational spectra?

5.5 The heavier isotope passes a closer rotational spectral line. Why?

5.6 Discuss the effect of non-rigidness on rotational states.

5.7 Show that fundamental band in vibrational spectra falls at vibration constant ω.

5.8 What is zero point energy?

5.9 Explain near infrared band spectrum considering a molecule as anharmonic oscillator.

5.10 Discuss the fine structure of infrared bands of diatomic molecules. What are P and R branches in the vibrational rotational spectra? Explain also their origin.

5.11 Give the theory of isotope effect in relation to vibrational bands of a diatomic molecule and discuss the importance of its study.

5.12 Give a short note on energy and spectra of symmetric top diatomic molecule.

5.13 What are electronic spectra? Give a short note on absorption electronic spectra.

5.14 Give the details of emission electronic spectra.

5.15 What is Franck–Condon principle?

5.16 Explain the intensity of spectral lines of absorption electronic spectra on the basis of Frank–Condon principle.

5.17 Write down the salient features of emission electronic spectra. Define sequence and progression. How can the intensity of emission electronic bands be explained on the basis of FC principle?

5.18 Find out shift in spectral line of vibrational–rotational spectra due to isotope effect.

5.19 How do the pure rotational spectra give the information about the bond length of heteronuclear molecule?

5.20 Write down the expression of energy for a diatomic molecule in terms of wave number if it is considered as
   (i) Rigid rotator   (ii) Non-rigid rotator
  (iii) Harmonic oscillator   (iv) Anharmonic oscillator
   (v) Vibrating rotator

5.21 What information is provided by the Franck–Condon parabola?

5.22 Write down a short note on three Branch (P, Q and R) spectra of diatomic molecule.

5.23 What is Raman effect?

5.24 Explain the existence of Raman lines on the basis of polarisability of the molecule.

5.25 Why does the Stoke lines stronger than the anti-Stokes lines?

5.26 Show that vibrational Raman shift is equal to vibrational constant.

5.27 Write a short note on rotational Raman spectrum.

5.28 Give the theory of a vibrational–rotational spectrum of a diatomic molecule.

5.29 What is the difference between Raman spectra and infra-red spectra?

5.30 Define electron spin resonance. Evaluate the expression for frequency of radiation which causes the transition of spin in paramagnetic material in presence of magnetic field.

5.31 Write down a short note on nuclear magnetic resonance and its applications.

5.32 The rotational Raman lines are found to be alternately weak or strong or missing for homonuclear molecules. Why?

5.33 What is hyperfine structure in ESR?

5.34 Prove that wave number difference between the consecutive pure rotational Raman lines is $4B$, while the wave number shift of first rotational Raman line with the origin is $6B$, where B is rotational constant.

NUMERICAL PROBLEMS

P5.1 The far infra-red spectrum of H^1Br^{79} consists of a series of lines space 17 cm^{-1} apart. Find the intern-uclear distance of H^1Br^{79}.

$$(h = 6.62 \times 10^{-27} \text{ erg-s}, c = 3.0 \times 10^{10} \text{ cm-s}^{-1}, \text{ and } N_A = 6.023 \times 10^{-23})$$

(Ans. 1.42 Å)

P5.2 The moment of inertia of HCl is 2.66×10^{-40} g-cm^2. Calculate the energy difference between the rotational levels $J = 0$ and $J = 1$ of HCl ($h = 6.62 \times 10^{-27}$ erg-s).

(Ans. 2.6×10^{-3} eV)

P5.3 The far IR spectrum of H^1F^{18} molecule consists of lines having a near constant separation of 40.5 cm^{-1}. Calculate the moment of inertia and inter-nuclear distance of molecule.

$$(h = 6.62 \times 10^{-27} \text{erg-s}, c = 3.0 \times 10^{10} \text{ cm-s}^{-1}, \text{ and } N_A = 6.023 \times 10^{-23}).$$

(Ans. 1.38×10^{-40} g-cm^2, 0.93)

P5.4 In the far IR-spectrum of H^1Cl^{35} molecule, the first line falls at 20.8 cm^{-1}. Calculate the moment of inertia and bond length of the molecule.

$$(h = 6.62 \times 10^{-27} \text{erg-s}, c = 3 \times 10^{10} \text{ cm-s}^{-1}, \text{ and } N_A = 6.023 \times 10^{-23}).$$

(Ans. 1.69×10^{-40} g-cm^2, 1.29 Å)

P5.5 A heteronuclear diatomic molecule has a vibrational absorption line at wavelength 3.465×10^{-6} m. Find force constant for its bond ($\mu = 1.6379 \times 10^{-27}$ kg).

(Ans. 484.2 N/m)

P5.6 The fundamental band of $C^{12}O^{16}$ is found at 2144 cm^{-1}. Calculate the fundamental vibration frequency, force constant and zero point energy of the molecule

$$(c = 3 \times 10^{10} \text{ cm-s}^{-1}, \text{ and } N_A = 6.023 \times 10^{-23}).$$

(Ans. 6.43×10^{13} s^{-1}, 1.86×10^6 dyne/cm, 2.13×10^{-13} erg-s)

P5.7 The value of ω_e and $\omega_e x_e$ are 1580.36 and 12.073 cm^{-1} respectively for a diatomic molecule. Calculate zero point energy (1 eV $= 8068$ cm^{-1}). **(Ans. 0.097 eV)**

P5.8 If vibrational frequency of H^1F^{19} molecule is 1.24×10^{14} s^{-1}. Then find force constant for H–F bond ($N_A = 6.023 \times 10^{23}$). **(Ans. 966 N/m)**

P5.9 The energy difference between the lowest and the first vibrational state of a heteronuclear diatomic molecule is 2904.48 cm^{-1}. Find the force constant for the molecule ($\mu = 1.6379 \times 10^{-27}$ kg). **(Ans. 470 N/m)**

P5.10 The rotational constants for vibrational states $v = 0$ and $v = 1$ are 10.440 cm^{-1} and 10.138 cm^{-1} respectively of a molecule (diatomic). If in fundamental band, the missing line is corresponding to wave number 2143.3 cm^{-1}, find the position of lines corresponding to R(1), R(0), P(1), and P(2) lines.

(Hint: $\bar{v} = \omega + (B_1 + B_0) m + (B_1 - B_0) m^2$**)**

(Ans. 2183.25 cm^{-1}, 2163.57 cm^{-1}, 2122.42 cm^{-1}, 2100.94 cm^{-1})

P5.11 Raman lines for HF molecule in vibrational Raman spectra are found at wavelengths 2670 Å and 3430 Å. Find the fundamental frequency of the molecule.   (**Ans.** 1.24×10^{14} Hz)

P5.12 Calculate the NMR frequency of absorption for a proton of an organic molecule at magnetic field strength 1.5 T ($g_N = 5.585$ and $\mu_N = 5.05 \times 10^{-27}$ JT^{-1}).   (**Ans.** 63.86 MHz)

P5.13 A free electron gives resonance at 9.3 GHz. What is the strength of field at resonance if g_e has value equal to 2.0023?   (**Ans.** 0.33 T)

P5.14 An unpaired electron is placed in a magnetic field of strength 1.2 T. Find the frequency of radiation, that causes spin transition of electron ($g_e = 2.003$).   (**Ans.** 33.6 GHz)

P5.15 The zero point energy of NO molecule is 1.8923×10^{-24} J. Find out the vibrational constant for the molecule.   (**Ans.** 1904 cm^{-1})

MULTIPLE CHOICE QUESTIONS

MCQ 5.1 Pure rotational spectra is found in
(a) Near infrared region   (b) Far infrared region
(c) UV-visible region   (d) γ-ray region

MCQ 5.2 The wave number separation between two successive pure rotational lines of diatomic molecule considered as rigid rotator is (*B: rotational constant*)
(a) $2B$ (b) $4B$
(c) $6B$ (d) $8B$

MCQ 5.3 Zero point energy in terms of vibrational constant is
(a) ω (b) 2ω
(c) $\dfrac{\omega}{2}$ (d) $\dfrac{3\omega}{2}$

MCQ 5.4 The energy of rotational and vibrational states in terms of wave number can be given by expressions
(a) $B(J+1), \omega\left(v+\dfrac{1}{2}\right)$ (b) $2B(J+1), \omega\left(v+\dfrac{1}{2}\right)$
(c) $BJ(J+1), \omega\left(2v+\dfrac{1}{2}\right)$ (d) $BJ(J+1), \omega\left(v+\dfrac{1}{2}\right)$

(*B:* rotational constant, ω: vibrational constant, *J* and *v*: rotational and vibrational quantum number)

MCQ 5.5 If vibrational constant for diatomic molecule is 2886 cm^{-1}, then energy of first vibrational state will be
(a) 0.536 eV   (b) 5.36 eV
(c) 53.6 eV   (d) 536 eV

MCQ 5.6 If v_{rot} is the rotational frequency of diatomic molecule, the shift of rotational Raman lines from fundamental lines will be
(a) v_{rot} (b) $2v_{rot}$
(c) $\dfrac{v_{rot}}{2}$ (d) $v_{rot} + \dfrac{3}{2}$

MCQ 5.7 The fundamental band in the vibrational spectra is obtained at wave number ω. The vibrational frequency of the molecule will be

(a) $\dfrac{\omega}{2}$ (b) ωc

(c) $2\omega c$ (d) $\dfrac{\omega c}{2}$

MCQ 5.8 The shift of first rotational Raman line from the exciting line is equal to
(a) $6B$ (b) $12B$
(c) $4B$ (d) $8B$

MCQ 5.9 The fundamental band in vibrational spectra for two isotopes A and B of a diatomic molecule is found at wave numbers ω_A and ω_B. The ratio of their reduced masses will be

(a) $\sqrt{\omega_A/\omega_B}$ (b) $\sqrt{\omega_B/\omega_A}$

(c) ω_A^2/ω_B^2 (d) ω_B^2/ω_A^2

MCQ 5.10 Bohr magnetron is equal to
(a) 9.274×10^{-26} JT^{-1} (b) 9.274×10^{-24} JT^{-1}
(c) 9.274×10^{-25} JT^{-1} (d) 9.274×10^{-27} JT^{-1}

MCQ 5.11 Nuclear magnetron is equal to
(a) 5.051×10^{-25} JT^{-1} (b) 5.051×10^{-26} JT^{-1}
(c) 5.051×10^{-27} JT^{-1} (d) 5.051×10^{-24} JT^{-1}

Answers

5.1 (b)   **5.2** (a)   **5.3** (c)   **5.4** (d)   **5.5** (a)   **5.6** (b)   **5.7** (b)

5.8. (a)   **5.9** (d)   **5.10** (b)   **5.11** (c)

CHAPTER 6

Nuclear Physics

CHAPTER OUTLINES

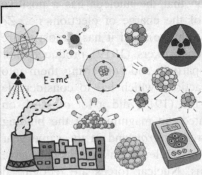

6.1   INTRODUCTION

Nuclear physics is an important and fascinating branch of Physics. Current researches in this area give new facts, some of which are against our older concepts. The beginning of Nuclear Physics should be traced back to the discovery of natural radioactivity by H. Becquerel in 1896. Madam Curie and her husband Pierre Curie later discovered radioactive properties of several others elements. Rutherford and his collaborators showed that rays emitted by radioactive substances are of three types: one having positive charge twice that of electron and mass close to helium atom, other having either positive or negative charge equal to that of electron and third having no charge on them. Later these rays were named as α (alpha), β (beta)- and γ (gamma)-rays. Rutherford and his co-workers, Geiger and Marsden in the period 1911–13 got a result in which α-particles were scattered by much larger angle than is expected from uniform charge distribution. While interpreting these results, they proposed that an atom contains a central part, which has huge concentration of positive charge. They also showed that this positive charge must have dimension less that 10^{-12} cm. They called this part as *nucleus*. Rutherford later showed that α-, β- and γ-rays originate from nucleus. The research work of Brock (1913), Bohr (1913), Moseley (1911–15), and Soddy and Fajan (1914) led to the conclusion that the nucleus has a positive charge whose magnitude is atomic number times of the charge of election, i.e., Ze (where Z is the atomic number and e the electronic charge), and it has almost entire mass of the atom within it. Rutherford and his co-workers (1919) produced first nuclear reaction by hitting nitrogen nucleus by α–particle and converting them into oxygen. This paved the ways for other nuclear reactions which led to considerable understanding about the nature of the nucleus. Aston (1919) did a lot of work on isotopes through his mass spectrograph and estimated the magnitude of the binding energy of the nucleus. In 1932, Chadwick discovered *neutron* and in the same year Heisenberg proposed the proton–neutron structure of nucleus. Yukawa (1932) gave mesonic theory of binding forces between nucleons. Nuclear forces are recognized as altogether different types of forces during this period. Weizsäcker (1935) obtained an expression for the binding energy of the nucleus using *liquid drop model*. The phenomenon of nuclear fission was discovered by Otto Haln in 1939, and the same year Bohr and Mottleson offered an explanation for it using liquid drop model. The stability of nucleus was thoroughly studied in the mean time by various workers. This led the concept of *magic number*. Later Mager (1949) and Haxel, Jonson and Suess (1950) proposed *shell model of the nuclei*. Collective model of nucleus was proposed by Bohr and Mottleson (1953) to explain both quadruple moment and magic number of nuclei. During the last forty years, tremendous improvement has been made regarding the structure of nuclei and nuclear forces. Quark model of nucleus proposed by Gell Mann (1962) is almost established. Nature of weak nuclear forces has been explained by Abdus Salam (1980) and other. It is not possible to list them all here but it may be remarked that study of nucleus, nuclear forces and fundamental particles has become most fascinating, important and modern branch of Physics. A lot, however, has to be still done to understand the mystery of tiny nucleus.

In fact, the development of Nuclear Physics owes much to study of nuclear reactions. These particles may be light nuclei, α-particle, deuteron, proton, neutron, electron, γ-ray photon or any of other fundamental particles. The products of nuclear reaction are new nuclei and some small particles. Thus to produce and observe nuclear reaction, we need at least two fundamental instruments: one which can produce energetic particles called *particle accelerator*, and the other which can detect the products of nuclear reaction are called *particle detector*. Improvement in the design of particle detectors and particle accelerators has been key to the progress of Nuclear Physics during the last sixty years.

6.2 STRUCTURE OF NUCLEUS

Just after the discovery of neutron by Chadwick in 1932, Heisenberg proposed *neutron–proton* structure of new nucleus. Later on, this became the common accepted structure of the nucleus. According to this hypothesis, a nucleus is a closed packed structure consisting of proton and neutron. Proton is a particle with positive charge e and mass 1.00758 a.m.u. [1 a.m.u. = 1.66038×10^{-27} kg], while neutron is a neutral particle with a rest mass 1.00897 a.m.u.. Both these particles are usually called by a common name of 'nucleon'. The total number of nucleons in any nucleus is equal to mass number A. Out of these, number of proton is Z (the atomic number) and that of neutron is $(A-Z)$. The basic questions, which were asked against this model, were:

1. How the protons with strong repelling force because of their positive charge are held within the nucleus?
2. How neutrons with no charge are retained in the nucleus?

The search for the answer of these questions led to the 'discovery of nuclear force'. These are described as the forces, which have altogether different nature than that of electromagnetic and gravitational forces.

Depending upon the relative numbers of protons and neutrons, atomic nuclei are classified into following categories:

Isotopes: Those nuclei, which have same number of proton (Z) are called isotopes. Examples of isotopic nuclei are: $^{13}_{7}$N, $^{14}_{7}$N, $^{15}_{7}$N and $^{15}_{8}$O, $^{16}_{8}$O.

Isobars: Those nuclei, which have same number of nucleons (or A) are called isobars or isobaric nuclei. Examples of isobaric nuclei are $^{16}_{6}$C^{3}, $^{13}_{7}$N and $^{15}_{8}$O, $^{15}_{7}$N.

Isotones: Those nuclei, which have same number of neutron (N) are called isotones. An example of isotonic nuclei is $^{14}_{6}$C$_8$, $^{15}_{7}$N$_8$, and $^{16}_{8}$O$_8$.

Isomers: Those nuclei, which have same number of neutrons and protons, but slightly differ in their internal structure resulting in different half life time in their β-decay are called isomers. Examples of isomers are two types of $^{80}_{35}$Br nuclei. Both emit β-particle, but half life time is 4.4 hours for first and for second it is just 18 minutes.

6.3 GENERAL PROPERTIES OF NUCLEUS

The important properties of the atomic nucleus are as follows:

Nuclear mass: A nucleus accounts for more than 99.99% of the mass of the atom. Nuclear mass is determined by the measurement of ratio of its charge to mass by mass spectrometer and with the help of nuclear reaction energies. It always comes out to be very close to an integer denoted by A, when expressed in terms of atomic mass unit (a.m.u.), where

$$1 \text{ a.m.u.} = \frac{10^{-3}}{N} = \frac{10^{-3}}{6.023 \times 10^{23}} = 1.6604 \times 10^{-27} \text{ kg} = 931.4 \text{ MeV}$$

where N is Avogadro's number and the integer A is called mass number.

The mass of a carbon atom was found to be equal to 12 a.m.u.. Practically the total nuclear mass is equal to the sum of mass of protons and neutrons constituting the nucleus, i.e., $M = Zm_p + Nm_n$, where m_p and m_n are masses of proton and neutron respectively.

Nuclear charge: A nucleus is positively charged due to protons contained in it. The number of unit of positive charge carried by a nucleus is equal to the number of its protons. Thus, the charge carried by a nucleus is Ze, where Z is its atomic number and $e = +1.6 \times 10^{-19}$ C for a proton. For example, charge on a hydrogen nucleus ($Z = 1$) is $+e$ and that on a $_6^{12}C$ nucleus ($Z = 6$) is $+6e$.

Nuclear size: Existing experimental evidence indicates that distribution of positive charge within the nucleus is uniform. Hence, in general, the atomic nuclei are considered to be spherical in shape. Since at the time of Rutherford, a large number of experiments were performed to find the radius of nucleus. These experiments showed that the volume of the nucleus is directly proportional to the number of nucleons (mass number) in a given nucleus. Thus if R be the radius of the nucleus with mass number A, then

$$\frac{4}{3}\pi R^3 \propto A$$

or

$$R^3 \propto A$$

or

$$R \propto A^{1/3}$$

or

$$R = R_0 A^{1/3} \tag{6.1}$$

where R_0 is a constant, known as *the nuclear radius parameter* that varies swiftly from one nucleus to another and is roughly constant for nuclei having $A > 72$. The value of R_0 is found to be near 1.3×10^{-15} m = 1.3 Fermi [$\because$ 1 Fermi = 1×10^{-15} m]. Therefore, radius R of the nucleus can be written as:

$$R = 1.3A^{1/3} \text{ Fermi} \tag{6.2}$$

Hence for example, the radius of the light nucleus like

$$_6^{12}C \text{ is } R = 1.3 \times (12)^{1/3} = 3 \text{ Fermi}$$

and for heavy nucleus like $_{92}^{235}U$, $R = 1.3 \times (235)^{1/3} = 7.2$ Fermi.

Nuclear quantum states: Nuclei are found to have quantised energy levels and corresponding set of quantum states like atoms. On the basis of study of α- and γ-ray spectra as well as artificial radioactivity, it is shown that the spacing of the levels in nuclei is many times larger than that in atoms. In heavy nuclei, the level spacing is usually several tens of keV, while in light nuclei, it can be up to 10 MeV or more. If a nucleus is raised to one of its excited states (e.g., when hit by energetic neutron), it returns to the ground state by emitting a photon. This is similar to the emission of photons in atomic transition. However, the energy of the photons is much larger than that of emitted photons in atomic transition. Photons emitted in nuclear transitions are called γ-rays. The fundamental difference lies in the nature of force acting in the nuclear region and extra-nuclear region in an atom. In the extra-nuclear region, Coulomb's force dominates, while inside the nucleus, the short range nuclear force reign supreme.

Angular momentum of nucleus: The discovery of angular momentum of the nucleus is the result of the indepth study of spectral line of elements. When individual components of multiple times were examined by spectrographs of high resolving power, it was found that each of these components is splitted into a number of times lying very close to each other. This further splitting of spectral times is called *hyperfine structure* and could be explained on the basis of extra-nuclear electrons. Pauli in 1924 suggested that hyperfine structure was related to the properties of atomic nucleus. These properties are the mass and angular momentum of the nuclei. Just like the extranuclear electrons, the nucleons in the nucleus are at constant motion in discrete quantised orbits within the nucleus. Due to the orbital motions, these particles possess angular momentum and an associated orbital magnetic moment. The electrically charged particles inside the nucleus also possess magnetic moment, which is over and above that due to orbital motion. These particles are also spinning about their axes like spinning electrons. Due to this spinning, these particles possess spin angular momentum and an associated magnetic moment is called spin magnetic moment.

Nuclear magnetic moment: The magnetic dipole moment μ_l of the nucleus is taken as the resultant of the moment due to orbital and spin motion of the individual nucleons (protons and neutrons).

$$\mu_l = \Sigma(\mu_{pl} + \mu_{ps}) + \Sigma\mu_{ns} \qquad (6.3)$$

The first term of Eq. (6.3) represents the sum of orbital magnetic moments of the proton, and second term represents the resultant of spin magnetic moments of all the neutrons. No contribution to the magnetic moment of the nucleus comes from the orbital motion of neutrons. It may be noted that though neutron is neutral, yet it has a spin magnetic moment. It may be explained that the neutron contains equal amount of positive and negative charges and spin magnetic moment could arise if these charges are not distributed uniformly.

This total magnetic dipole moment is called *nuclear magnetic moment*.

6.4 NUCLEAR FORCE

The forces between nucleons are mainly of two types. One of them is the Coulomb force of repulsion between protons. The other is a strong attractive force which all nucleons exert on each other. The second force, i.e., a strong attractive force that holds the nucleus together, is too strong to be gravitational. It cannot be electrostatic for them the neutrons would be free from it. It cannot be magnetic for them. It would be attractive only for certain orientations.

Nuclear forces have two important characteristics. First they exist only when the nucleons are very close together. The range of nuclear force is found to be of the order of 10^{-15} m. The nuclear force is the same between any two nucleons that is between two neutrons or two protons or between a proton and a neutron. Second, nuclear forces are saturated. This means that a nucleon can interact only with a finite number of other nucleons.

The saturation ability of nuclear forces is explained by assuming them to be exchange forces. The nucleons are assumed to have a substructure. They are supposed to exchange some particles back and forth. Such a particle is called a *meson*. It was postulated by the Japanese scientist, Yukawa in 1936. According to the meson theory, all nucleons consist of identical cores surrounded by a cloud of one or more mesons. The mesons may be electrically neutral or they carry a positive or a negative charge. The difference between a proton and a neutron is due to composition of the meson cloud surrounding the core.

The force between a neutron and another neutron or between a proton and another proton is due to the exchange of a neutral meson (designated π^0) between them. By emitting a negative meson, a neutron becomes a proton.

Thus
$$n \rightarrow p + \pi^-$$

By absorbing a negative meson, a proton becomes a neutron.

Thus
$$p + \pi^- \rightarrow n$$

Conversely by emitting a positive meson, a proton becomes a neutron. By absorbing a positive meson, a neutron becomes a proton. Thus
$$p \rightarrow n + \pi^+$$
$$n + \pi^+ \rightarrow p$$

By using advanced mathematical procedure, it can be shown that the exchange of mesons between nucleons leads to mutually attractive forces.

6.5 NUCLEAR BINDING ENERGY

In the nucleus, nucleons are held together by attractive forces, called *nuclear forces*. Therefore, if a nucleus is to be decomposed into its constituent nucleons, work has to be done against these forces, i.e., energy has to be supplied to the nucleus. This energy is called *nuclear binding energy*.

Hence nuclear binding energy is defined as the energy required for decomposing a nucleus into its constituent nucleons.

It is found that mass of a nucleus is always less than the total mass of all the nucleons, constituting it. It means that when a number of nucleons are brought together to form a nucleus, a certain amount of mass disappears.

This difference between the sum of the rest masses of the nucleons constituting it and the actual rest mass of the nucleus is called *disappearing mass*.

It is denoted by Δm. This disappeared mass is connected into energy which binds the nucleons together in the nucleus.

Hence nuclear binding energy may also be defined as the energy which is equivalent to mass disappeared in the formation of the nucleus.

The binding energy results from the nuclear forces that keep the nucleons together in the nucleus. The binding energy of the nucleus divided by the number of nucleons is called *binding energy per nucleon*. It is very important quantity and accounts for the stability of the nucleus. A nucleus is more stable, if its average binding energy per nucleon is large and *vice-versa*.

6.5.1 Expression for Binding Energy

Consider an element $_Z^A X$ with atomic number Z and mass number A. Then it contains Z protons and $(A - Z)$ neutrons. If M_N, m_p and m_n represent the masses of parent (nucleus) a proton and a neutron respectively, then mass disappeared Δm in the formation of the nucleus is given by

$$\Delta m = Zm_p + (A - Z)m_n - M_N(_Z^A X) \qquad (6.4)$$

For example in case of deuteron, which contains are proton and one neutron, the combined $[1.0073 + 1.0087]$ a.m.u. $= 2.0160$ a.m.u., whereas the actual mass of deuteron nucleus is 2.0136 a.m.u.

$\therefore$ Mass disappeared $\Delta m = (2.0160 - 2.0136)$ a.m.u. $= 0.0024$ a.m.u.

The relation between mass disappeared in the formation of a nucleus and the binding energy is given by Einstein's relation i.e., binding energy E_B of the nucleus is given by

$$E_B = \Delta mc^2 \qquad (6.5)$$

The binding energy of the nucleus with charge number Z and mass number A is given by

$$E_B = \Delta mc^2 = [Zm_p + Zm_e + (A - Z)m_n - M_N(_Z^A X)]c^2$$
$$= [Z(m_p + m_e) + (A - Z)m_n - \{M_N(_Z^A X)\}]c^2 \qquad (6.6)$$

where m_e is mass of electron.

Now, $m_p + m_e = m_H =$ Mass of hydrogen atom

and $Zm_H + (A - Z)m_n = M_n(_Z^A X) =$ Atomic mass

(= mass of atom having mass number A and atomic number Z)

Therefore, in terms of atomic masses, the binding energy is given by

$$E_B = [Zm_H + (A - Z)m_n - M_N(_Z^A X)]c^2 \qquad (6.7)$$

Thus nuclear binding energy can be calculated by using any of the above expressions [Eqs. (6.5)–(6.7)] (the difference is negligible). The binding energies for most of the isotopes (heavy) are very large e.g., 127.6 MeV for $_8^{15}O$, 500 MeV for $_{26}^{57}Fe$, 1786 MeV for $_{92}^{235}U$ and 1640 MeV for $_{15}^{209}Bi$.

The binding energy is usually referred to as the total nuclear binding energy. However, it is often useful experimentally to refer to the average binding energy per nucleon.

6.5.2 Binding Energy Per Nucleon

Binding energy per nucleon is the average energy required to extract one nucleon from the nucleon.

It is the ratio of binding energy of the nucleus to the number of nucleons.

Therefore, dividing both sides of Eqs. (6.6) and (6.7) by A, we get binding energy per nucleon:

$$\frac{E_B}{A} = \frac{1}{A}[Z(m_p + m_e) + (A - Z)m_n - M_N({}^A_Z X)]c^2 \tag{6.8}$$

and in terms of atomic masses,

$$\frac{E_B}{A} = \frac{1}{A}[Z(m_H + (A - Z)m_n - M_N({}^A_Z X)]c^2 \tag{6.9}$$

EXAMPLE 6.1 Calculate the binding energy of α-particle and express it in MeV and joules. Given $m_p = 1.00758$ a.m.u., $m_n = 1.000897$ a.m.u, and $m_{He} = 4.0028$ a.m.u.

Solution Binding energy of α-particle

$$= [Zm_p + (A - Z)m_n - M_N({}^A_Z X)]c^2$$

$$= (2 \times 1.00758 + 2 \times 1.00897 - 4.0028) \times 931.49 \text{ MeV}$$

$$[\because 1 \text{ a.m.u.} = 931.49 \text{ MeV}]$$

$$= 27.29 \text{ MeV} = 4.37 \times 10^{-12} \text{ J}.$$

6.5.3 Packing Fraction

Mass defect (disappeared) does not provide exact information about nuclear stability, although it is said that higher the mass defect, more tightly bound nucleons exist in the nucleus. Mass defect generally increases as number of nucleons in the nucleus increases. For example, mass defect for ${}^4_2$He is 0.002604 a.m.u. (= 2.4249 MeV), while that for ${}^{235}_{92}$U, it is 0.04396 a.m.u. (= 40.930 MeV). ${}^4_2$He is much more stable than ${}^{235}_{92}$U. The term packing fraction was introduced by Aston in 1926, which gives better information about the nuclear stability. Packing fraction is defined as:

$$\text{Packing fraction} = \frac{\text{Atomic mass} - \text{Mass number}}{\text{Mass number}}$$

$$f = \frac{\Delta m}{A} = \frac{\text{Mass defect}}{\text{Mass number}} \tag{6.10}$$

The smaller the value of packing fracttion, the more stable is the nucleus and vice-versa.

EXAMPLE 6.2 The mass of a deuteron nucleus is 2.014103 a.m.u.. If the masses of proton and neutron are respectively 1.007825 a.m.u. and 1.008663 a.m.u., find the mass defect and packing fraction.

Solution Mass defect $\Delta M = m_p + m_n - m_d$

$$= (1.008825 + 1.008663 - 2.014104) \text{ a.m.u.}$$

$$= 0.002385 \text{ a.m.u.}$$

$$\text{Packing fraction } f = \frac{\text{Mass defect}}{\text{Mass number}} = \frac{0.002385}{2} = 0.001195$$

6.6   RADIOACTIVE DECAY

Radioactive substances usually emit either α-particles or β-particles and sometimes γ-rays also. It is due to disintegrating nuclei of atoms. This spontaneous disintegration is called *radioactive decay*. When the nucleus of an atom emits an α-particle, its atomic number is decreased by 2 and mass number decreased by 4 and a new atom is formed. When the nucleus emits a β-particle, its atomic number is increased by 1 and the mass number remains unchanged. The mass number and atomic number both remain unchanged, when γ-ray is emitted from the nucleus.

6.6.1   Laws of Radioactive Decay

Rutherford, Soddy and others carried out extensive studies on the spontaneous disintegration of the radioactive substances and concluded certain facts, which are termed as *laws of radioactive decay*.

Rutherford Rule

This law is stated as follows:

1. The activity shown by radioactive substance is a nuclear phenomenon. The disintegration is accompanied by emission of α-, β-and γ-particles originating from the nuclei. By emitting α-or β-particles, the initial atom, called *parent atom*, changes into another atom called *daughter atom*.
2. The algebraic sum of charges before disintegration is equal to the total electric charge after disintegration.
3. The sum of the masses of the disintegrated particle is same as that of initial (parent) atom. However, this law of conservation of mass is valid if mass energy equivalence is taken into account.

Soddy–Fajan's Displacement Law

The parent atom emits only one particle at a time, α or β. However, there might be emission of energy in the form of γ-rays or some other form. According to this law,

1. In α-decay, i.e., after emitting α-particle, the parent atom changes into daughter atom, whose mass number is less by four and atomic number is less by two than the parent atom. Thus

$$\underset{\text{(Parent atom } X)}{{}^{A}_{Z}X} \xrightarrow{\ \alpha\text{-decay}\ } \underset{\text{(Daughter atom } Y)}{{}^{A-4}_{Z-2}Y}$$

$$\underset{\text{(Radium)}}{{}^{226}_{88}\text{Ra}} \xrightarrow{\ -\alpha\ } \underset{\text{(Radon)}}{{}^{222}_{86}\text{Rn}}$$

2. In β-decay (i.e., after emitting β-particle), the parent atom changes into daughter atom, whose mass number is same, but the atomic number is increased by one. Thus

$$\underset{\text{(Parent atom } X)}{{}^{A}_{Z}X} \xrightarrow{\ \beta\text{-decay}\ } \underset{\text{(Daughter atom } Y)}{{}^{A}_{Z+1}Y}$$

$$\underset{\text{Radium B (Isotope of lead)}}{{}^{214}_{88}\text{Ra}_{B}} \xrightarrow{\ -\beta\ } \underset{\text{Radium C (Isotope of bismuth)}}{{}^{214}_{83}\text{Ra}_{C}}$$

During both these emissions, certain amount of energy is always released. Hence decay reaction can be written as:

$$\underset{86}{{}^{226}}\text{Ra} \xrightarrow{\ -\alpha\ } {}^{222}_{86}\text{Rn} + {}^{4}_{2}\text{He} + Q + \gamma\text{-rays}$$

$$\underset{82}{{}^{214}}\text{Ra}_{B} \xrightarrow{\ -\beta\ } {}^{214}_{83}\text{Ra}_{C} + {}^{0}_{-1}e$$

Radioactive Decay Law

The rate of disintegration of a particular substance, i.e., number of atom disintegrating per second at any instant is proportional to the number of atoms present at that instant.

Let N be the number of atoms present in radioactive substance at any instant t.

Let dN be the number of atoms disintegrated in a short duration of time dt, then

$$-\frac{dN}{dt} \propto N \quad \text{or} \quad \frac{dN}{dt} = -\lambda N \tag{6.11}$$

where λ is a constant for a given substance and is known as 'decay constant' (or radioactive constant or disintegration constant).

$$\therefore \qquad \frac{dN}{dt} = -\lambda\, dt$$

Integrating it, we get

$$\log_e N = -\lambda t + C$$

where C is a constant of integration.

If N_0 be the number of radioactive atoms present in the beginning, i.e., at a time $t = 0$, then

$$\log_e N_0 = C$$

$$\therefore \qquad \log_e N = -\lambda t + \log_e N_0$$

$$\text{or} \qquad \log_e \frac{N}{N_0} = -\lambda t$$

$$\frac{N}{N_0} = e^{-\lambda t}$$

$$\text{or} \qquad N = N_0\, e^{-\lambda t} \tag{6.12}$$

Thus the number N of radioactive atoms left, decreased exponentially with time is shown in Figure 6.1.

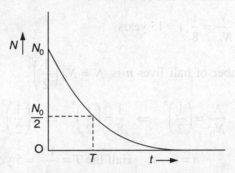

Figure 6.1 Radioactive decay.

6.6.2 The Half-Life Period T

The half-life period T of a radioactive element is defined as the time taken for half the atoms to disintegrate as shown in Figure. 6.1, i.e., in a time T, the radioactivity of the element diminishes to half of its value. Hence, from Eq. (6.12),

$$\frac{N_0}{2} = N_0 e^{-\lambda t}$$

$$\therefore \qquad T = \frac{1}{\lambda} \log_e 2 = \frac{0.693}{\lambda} \qquad (6.13)$$

The half-life period varies considerably in a particular radioactive series. In a uranium series, uranium-I has a half-life period of the order of 4500 million years, radium has one of about 1600 years, radium A about 138 days, radium B about 27 minutes and radium C about 10^{-4} second. Hence we can see that the half-life of radioactive nuclei vary from as short as 10^{-14} second to as long as 10^{14} years. Further, after the second half-life,

$$N = \frac{N_0/2}{2} = \frac{N_0}{4} = N_0 \left(\frac{1}{2} \right)^2$$

Also the third half $N = \frac{N_0/4}{2} = \frac{N_0}{8} = N_0 \left(\frac{1}{2} \right)^3$

and so on, for n half-lives

$$N = N_0 \left(\frac{1}{2} \right)^n = \frac{N_0}{2^n} \qquad (6.14)$$

The number of half-life n can also be put as:

$$n = \frac{t}{T}$$

Therefore, $$N = \frac{N_0}{2^{t/T}}$$

or $$N(t) = N_0 2^{-t/T} \qquad (6.15)$$

EXAMPLE 6.3 Calculate the half-life period of a radioactive substance if its activity drops (1/8)th of its initial value in 15 years.

Solution Given $\dfrac{N}{N_0} = \dfrac{1}{8}$, $t = 15$ years

We know the number of half lives n is $N = N_0 \left(\dfrac{1}{2}\right)^n$

or $\qquad \dfrac{N}{N_0} = \left(\dfrac{1}{2}\right)^n \;\Rightarrow\; \dfrac{1}{8} = \left(\dfrac{1}{2}\right)^n \;\Rightarrow\; \left(\dfrac{1}{2}\right)^3 = \left(\dfrac{1}{2}\right)^n$

$\therefore \qquad\qquad\qquad n = 3 \quad \therefore \quad$ Half life $T = \dfrac{15}{3} = 5$ years

6.6.3 The Average or the Mean Life Time (T_a)

Atom which disintegrates in the beginning have a very short life, and those which disintegrates at the end have the longest life.

The average life of a radioactive atom is equal to the sum of the life times of all the atoms divided by the total number of atoms.

Let N_0 is the total number of atoms at time $t = 0$, and N is the number remaining at the instant t. Let dN of these atoms disintegrate between time t and $(t + dt)$, and each of these dN atoms had a life time of t second (as dt is small). Hence the total life time of dN atoms is equal to $t\,dN$.

As the disintegration obeys statistical law, therefore, atom may have a life from 0 to ∞. Therefore, sum of life times of all the atoms is:

$$= \int_{t=0}^{t=\infty} -t\,dN$$

Hence the average or mean life time

$$T_a = \frac{\text{Sum of the life times of all atoms}}{\text{Total number of atoms}}$$

$$= \frac{\int_0^\infty -t\,dN}{N_0} \tag{6.16}$$

But from Eq. (6.12), we get

$$N = N_0\,e^{-\lambda t}$$

Differentiating it, we get

$$dN = -N_0\lambda e^{-\lambda t}dt$$

Substituting it in Eq. (6.16), we get

$$T_a = \frac{\int\limits_0^\infty t N_0 \lambda e^{-\lambda t} \, dt}{N_0} = \lambda \int\limits_0^\infty t e^{-\lambda t} \, dt$$

Integrating it by parts, we get

$$T_a = \lambda \left[\frac{t e^{-\lambda t}}{-\lambda} \right]_0^\infty - \lambda \int\limits_0^\infty \frac{e^{-\lambda t}}{-\lambda} \, dt$$

$$= [-t e^{-\lambda t}]_0^\infty + \int\limits_0^\infty e^{-\lambda t} \, dt$$

The first term becomes zero.

$$\therefore \qquad T_a = \int\limits_0^\infty e^{-\lambda t} \, dt = \left[\frac{e^{-\lambda t}}{-\lambda} \right]_0^\infty = \left[0 - \frac{1}{-\lambda} \right] \quad [\text{here } e^{-\infty} = 0]$$

$$T_a = \frac{1}{\lambda} \tag{6.17}$$

Thus the mean life of a radioactive atom is equal to the reciprocal of its disintegration constant.

Using Eq. (6.13), we have

$$T_a = \frac{T}{0.693} = 1.44T \tag{6.18}$$

Hence the mean life of a radioactive atom is equal to 1.44 times of its half-life period.

EXAMPLE 6.4 One gram of radium emits 3.7×10^{10} particles per second. Find the half-life and average life of radium. (Atomic weight of radium = 226).

Solution Given $R = \dfrac{dN}{dt} = \lambda N = 3.7 \times 10^{10}$ particles/second

Then $$N = \frac{6.02 \times 10^{23}}{226} = 2.66 \times 10^{21}$$

$$\therefore \qquad \lambda = \frac{R}{N} = \frac{3.7 \times 10^{10}}{2.66 \times 10^{21}} = 1.39 \times 10^{-11} \text{ s}^{-1}$$

Half-life of radium is:

$$T = \frac{0.693}{\lambda} = \frac{0.693}{1.39 \times 10^{-11}} = 4.98 \times 10^{10} \text{ s}$$

But $\qquad$ 1 years $= 365 \times 24 \times 60 \times 60 = 3.15 \times 10^7 \text{ s}$

$$\therefore \qquad T = \frac{4.98 \times 10^{10}}{3.15 \times 10^7} = 1581 \text{ years}$$

Average life of radium is:

(Method I)
$$T_a = \frac{1}{\lambda} = \frac{1}{1.39 \times 10^{-11}} = 7.19 \times 10^{10} \text{ s}$$

$$T = \frac{7.19 \times 10^{10}}{3.15 \times 10^7} = 2281 \text{ years}$$

(Method II)
$$T_a = 1.44 T$$
$$= 1.44 \times 1581$$
$$= 2281 \text{ years}$$

6.7 NUCLEAR RADIOACTIVE TRANSFORMATION SERIES

There are large numbers of radioactive elements present in nature. However, there exists four natural series of radioactive elements, which gives the sequence in the formation of different radioactive elements. These are:

(a) Uranium series
(b) Thorium series
(c) Actinium series
(d) Neptunium series

However, at the moment possibility of neptunium series is ruled out.

Uranium Series

In this series, the starting element is uranium I ($^{238}_{92}U$). It successively disintegrates by emitting α-, β-and γ-particles and converts itself into different radioactive atoms. The final product is an isotope of lead. The uranium series is shown in Figure 6.2.

Thorium Series

The parent atom of this series is thorium ($^{232}_{92}Ta$). It disintegrates successively, and finally it is changed into lead ($^{232}_{92}Pb$) which is stable and non-radioactive (Figure 6.3).

Actinium Series

The parent atom of this series is uranium ($^{235}_{92}U$) and final product is isotope of lead ($^{207}_{82}Pb$) which is stable (Figure 6.4).

Neptunium Series

The parent element in this series is plutonium with $Z = 94$ ($^{241}_{84}Pu$) and $A = 241$ (Figure 6.5). Because the element neptunium has highest life time in this series, the series has been named as the *neptunium series*. The final product of this series

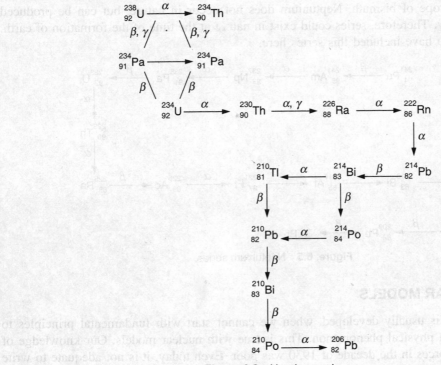

Figure 6.2 Uranium series.

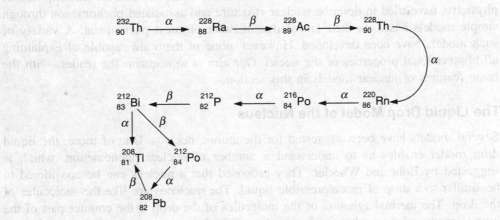

Figure 6.3 Thorium series.

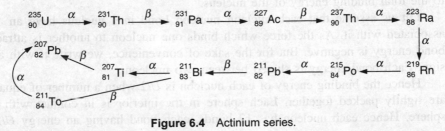

Figure 6.4 Actinium series.

is an isotope of bismuth. Neptunium does not occur in nature, but can be produced artificially. Therefore, series could exist in nature at the time of the formation of earth. Hence we have included this series here.

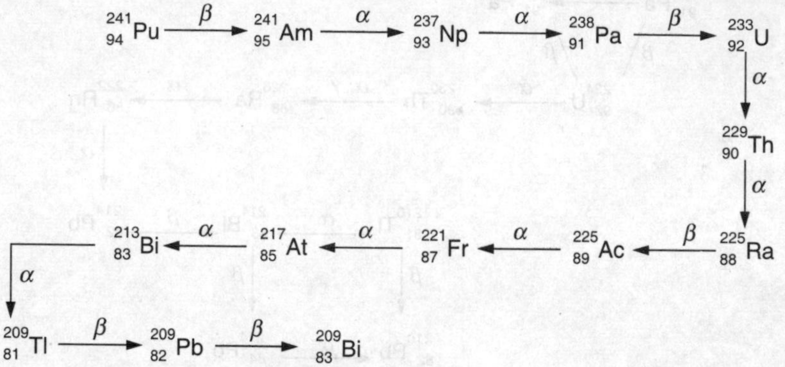

Figure. 6.5   Neptunium series.

6.8   NUCLEAR MODELS

A model is usually developed, when we cannot start with fundamental principles to describe a physical phenomenon. This is true with nuclear models. Our knowledge of nuclear forces in the decade of 1930 was poor. Even today, it is not adequate to write a mathematical formula that describes the nuclear forces. In such a situation, various physicists have tried to describe nuclear structure and associated phenomenon through simple models based on their knowledge of other physical phenomena. A variety of such models have been developed. However, none of them are capable of explaining all observational properties of the nuclei. Our aim is to acquaint the readers with the basic features of nuclear models in this section.

6.8.1   The Liquid Drop Model of the Nucleus

Several models have been suggested for the atomic nucleus. One of these, the liquid drop model enables us to understand a number of nuclear phenomenon, which is suggested by Bohr and Wheeler. They proposed that a nucleus can be considered to be similar to a drop of incompressible liquid. The nucleons are like the molecules of the drop. The thermal agitation of the molecules of the drop is the counter part of the kinetic energy of the nucleons. The energy required to evaporate the drop is analogous to the total binding energy of the nucleus.

Let us assume that each nucleon–nucleon bond in a nucleus have an energy U associated with it. As the force which binds one nucleon to another is attractive, the bond energy is negative. But for the sake of convenience, we write it with a positive sign. Each bond energy is shared between two nucleons.

Hence the binding energy of each nucleon is $U/2$ when a number of equal spheres are tightly packed together. Each sphere in the interior is in contact with 12 other sphere. Hence each nucleon has 12 bonds, each bond having an energy $U/2$. Hence the total bond energy of a nucleon in the interior of the nucleus is $12(U/2) = 6U$. This energy is called *volume energy*. If the mass number is A, there will be A nucleons in the nucleus. If each of them had 12 nucleon–nucleon bonds, the volume energy would be given by

$$\text{Volume energy } E_V = 6\,AU = \sigma_1 A$$

Thus the volume energy is proportional to A (mass number).

Some of the nucleons, however, are on the surface of the nucleus. These would have less than 12 neighbours.

A nucleon near the surface, therefore, interacts with fewer other nucleons than one in the interior (Figure 6.6). The number of such nucleons with less than 12 neighbours, depends on the surface area of the nucleus.

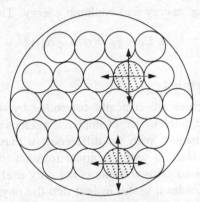

Figure 6.6   Illustration for liquid drop model.

If the radius of the nucleus is R, the surface area would be

$$4\pi R^2 = 4\pi R_0^2 A^{2/3}$$

Therefore, the number of nucleons with less than 12 bonds is proportional to $A^{2/3}$. The total binding energy is, therefore, reduced by

$$E_S = -\sigma_2 A^{2/3}$$

This energy is called *surface energy*. This surface energy is more significant in light nuclei because a larger fraction of the total number of nucleons will be at the surface than in heavier nuclei. Any natural system always tends towards that configuration in which its potential energy is a minimum. Therefore, nuclei tend towards the configuration in which the binding energy is a maximum, i.e., one in which the surface energy is a minimum. Thus a nucleus must exhibit surface tension like a liquid drop. In the absence of external forces, it should, therefore, take up a spherical shape which has the least surface area for a given volume.

The electrostatic repulsion between each pair of protons in a nucleus tends to disrupt the nucleus, i.e., to reduce the binding energy. If the atomic number is Z, there are $Z(Z-1)/2$ pairs of protons in the nucleus. The coulomb energy of a nucleus is the work that has to be done in bringing Z protons from infinity to a spherical aggregate whose size is that of the nucleus. This energy is proportional to

1. $\dfrac{Z(Z-1)}{2}$

2. Average value of $\dfrac{1}{r}$

Assuming the protons to be uniformly distributed inside the nucleus, we can take $(1/r)$ as being proportional to coulomb energy. Hence as r is inversely proportional to $A^{1/3}$, we can write

$$E_C = \text{coulomb energy} = -\sigma_3 \frac{Z(Z-1)}{A^{1/3}}$$

The total binding energy E_B of a nucleus can, therefore, be taken to be the sum of volume energy, surface energy and coulomb energy. Thus

$$E_B = E_V + E_S + E_C = \sigma_1 A - \sigma_2 A^{1/3} - \sigma_3 \frac{Z(Z-1)}{A^{1/3}}$$

Semi-empirical Formula

Several nuclear properties are critically controlled by the masses or equivalently the binding energy of the nuclei. For example, a nucleus can decay only into a set of particles with lower masses, with the difference in mass appearing as kinetic energy of the final particles. Here, a semi-empirical formula to determine the masses of the nuclei is briefly discussed. This formula is not only useful in discussion of stability of the nuclei but also provides a useful insight into the physical properties that determine their masses.

The total energy of a nucleus is largely controlled by the short range nature of the forces leading the saturation in binding the Pauli exclusive principle and the *coulombic* repulsion between protons with some finer effects due to pairing like nucleons.

$$M_1 = Zm_p + (A - Z)M_n \tag{6.19}$$

Since nuclear forces are of short range, the binding energy of the nucleus is proportional to the number of nucleons A, so that mass is reduced by

$$M_2 = -a_1 A \tag{6.20}$$

However, nucleons near the surface are less tightly bound which may be taken into account by a term:

$$M_3 = a_2 A^{2/3} \tag{6.21}$$

proportional to the surface area, the electrostate, repulsive interaction may be incorporated by noting that the electrostate energy of a sphere of the charge Ze is proportional to Z^2/R this brings in contribution

$$M_4 = a_3 Z^2 A^{-1/3} \tag{6.22}$$

The Pauli exclusion principle gives:

$$M_5 = a_4 \frac{\left(Z - \frac{1}{2}A\right)^2}{A} \tag{6.23}$$

because of the imbalance of protons and neutrons. Finally it is noted that the Pauli's principle allows pairs of protons and neutrons with spin 1/2 to occupy the same energy state, whereas odd proton and neutron are forced to go into a higher energy state. This effect is included by pairing term:

$$\delta = \begin{cases} \delta(A) & \text{for odd } Z \text{ and odd } (A - Z) \\ 0 & \text{for odd } A \\ -\delta(A) & \text{for even } Z \text{ and even } (A - Z) \end{cases} \qquad (6.24)$$

The final formula for the mass of a nucleus is:

$$M(Z, A) = Zm_p + (A - Z)m_n - a_1 A + a_2 A^{2/3} + a_3 Z^2 A^{-1/3} + a_4 \frac{(Z - A/2)^2}{A} + \delta(A) \quad (6.25)$$

Equation (6.25) is known as Weizsäcker mass formula.

The constants in Eq. (6.25) are determined empirically, from least square fit to the observed masses. The results for a_1, a_2, a_3, a_4 and $\delta(A)$ are obtained semiempirically by this method are

$$a_1 = 15.7 \text{ MeV}$$
$$a_2 = 17.8 \text{ MeV}$$
$$a_3 = 0.71 \text{ MeV}$$
$$a_4 = 94.8 \text{ MeV}$$
$$\delta(A) = 33.6A^{-3/4} \text{ MeV} \qquad (6.26)$$

6.8.2 Magic Numbers and Shell Model

There are certain nuclear phenomena, which show the existence of excited states of nuclei of sharply defined nuclear energy levels. This led to the shell model of the nucleus, the idea of which was borrowed from external electronic structure of the atom.

We know that the electrons in an atom are grouped in 'shells' and 'sub-shells; and atoms with 2, 10, 18, 36, 54 and 86 electrons have all their shells completely filled. Such atoms are usually stable and are chemically inert. A similar situation exists with nuclei also.

Nuclei having 2, 8, 20, 28, 50, 82 and 126 nucleons of the same kind (either protons and neutrons) are more stable than nuclei of neighbouring mass numbers. These numbers are called *magic numbers*.

This observation is based on the following lines:

1. Helium ($Z = 2$) and oxygen ($Z = 8$) are particularly stable.
2. Calcium ($Z = 20$) and tin ($Z = 50$) have number of stable isotopes than any other element.
3. The end products of the three naturally occurring radioactive series are lead isotopes ($Z = 82$).
4. The binding energy curve has small peaks (showing extra stability) for the following nuclei [Figure 6.7].

$$_{28}^{62}\text{Ni}, \ _{38}^{88}\text{Sr}, \ _{50}^{120}\text{Sn}, \ _{58}^{140}\text{Ce}, \ _{82}^{208}\text{Pb}$$

It is seen that for these nuclei either Z (number of protons) or $(A - Z)$ [number of neutrons] is a magic number.

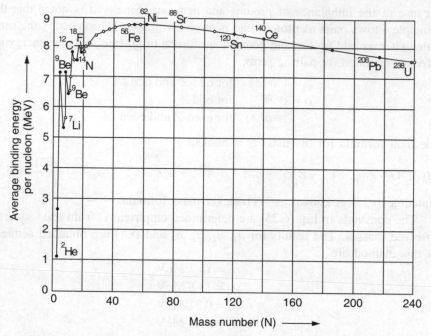

Figure 6.7   Mass number versus average binding energy per nucleon curve.

5. The electric quadruple moments of magic number nuclei are very low (nearly zero) compared with those of other nuclei. This means that those nuclei have almost spherical charge distribution, which is expected for the more stable nuclei.

The shell model is an attempt to account for the existence of magic numbers and certain other nuclear properties. According to this model, the nucleons [protons as well neutrons] in a nucleus live in a square well potential with rounded corners in which they occupy separate discrete sets of energy levels.

When a spin-orbit interaction is assumed to exist, the energy levels of either class of nucleon are arranged into different group (shells). In this group formation, the gap appears at just the right places, i.e., after the completed groups having 2, 8, 20, 28, 50, 82 and 126 protons or neutrons.

6.9   CROSS-SECTION OF NUCLEAR REACTION

In nuclear reactions usually a beam of more energetic particles is made to impinge on the target material of a thickness dx. If the number of particles per unit volume in the target be n, then according to the nuclear size, various nuclei present an effective cross-section in the path of incident projectiles for scattering/reaction. The effective area is normally not the same as the geometrical area of the nuclei and is termed as the *cross-section area* σ. This area presented by nuclear targets is expressed in the units of 10^{-24} cm^2, which is called a *barn*.

If at any instant, there are n particles in a beam with a velocity v, the flux of the beam is nv, and if N be the number of nuclei exposed to the beam at any instant of time, then the reaction/scattering is expressed as:

$$\frac{dN}{dt} \propto nv$$

and is also proportional to N.

or
$$\frac{dN}{dt} = -\sigma nvN \tag{6.27}$$

On integration, we have

$$N = N_0 e^{-\sigma nvt} \tag{6.28}$$

where N_0 is the number of target nuclei at $t = 0$, and N is the number after time t. The probability that a particular type of nuclear process (reaction) will take place due to an incident particle, will depend on the number of particles per unit volume of the target and the thickness of the target:

$$\Delta P = \sigma N \Delta x$$

If $N = 1$ particle/m^3 and $\Delta x = 1$ m, then $\Delta P = \sigma$ per square metre.

The cross-section could be defined in terms of the probability equation as:

The probability to produce a reaction of one particle is facing the projectile per unit area of the target in the path of an incident particle.

If n be the number of particles falls on the target in a time Δt, then the rate of reaction will be

$$\frac{n\Delta P}{\Delta t} = \frac{dN}{dt} = -\sigma Nn \frac{\Delta x}{dt} = -\sigma Nnv \tag{6.29}$$

Equation (6.29) is similar to Eq. (6.27).

The number of particles scattered in the thickness dx, can be evaluated by rearranging Eq. (6.29).

$$n\Delta P = -\sigma Nn\Delta x$$

$$\Rightarrow \qquad dn = -\sigma Nndx$$

$$\Rightarrow \qquad \frac{dn}{n} = -N\sigma dx$$

$$\Rightarrow \qquad n = n_0 e^{-\sigma Nx} \tag{6.30}$$

where n_0 is supposed to be the neutron intensity at $x = 0$, and n is the neutron intensity at the thickness x. This gives the attenuation of the incident beam.

If we express Eq. (6.30) as:

$$\Rightarrow \qquad n = n_0 e^{-\mu x} \tag{6.31}$$

where $\mu = \sigma N$ and is called the attenuation coefficient.

This expression is useful in deriving the mean free path of absorption of a particle in a target:

$$\text{Mean free path } \xi = \frac{1}{\mu} = \frac{\int\limits_0^a x e^{-\mu x}\,dx}{\int\limits_0^a e^{-\mu x}\,dx} \tag{6.32}$$

and is defined as the length in a target, where the intensity falls $1/e$ times of the incident intensity of particles. We can express the above equation as:

$$n = n_0 e^{-x/\xi} \tag{6.33}$$

For a target of density ρ, with atomic mass number A and Avogadro's number N_A, we get

$$N = \frac{\rho N_A}{A} \tag{6.34}$$

and ξ could be computed, if we know the cross-section of a process.

The yield of reaction can be measured by finding the activity of the residual nuclei and the products of the reaction can be found by detecting the product particles and the residual nucleus. It can be seen that the cross-section of a reaction can be measured as it will be given by

$$\sigma = \frac{\text{Number of events of a given type per unit time per nucleus}}{\text{Number of incident particles per unit area}}$$

With each type of nuclear reaction, a cross-section is associated. When an incident particle is simply scattered, we call it a scattering cross-section σ_{sc}. When it is absorbed and a reaction product is produced, which is different from the initial particles, then the cross-section is said to be reaction cross-section σ_r. Similarly, if the target is fissile and fusion reaction takes place, then the cross-section is known as fission cross-section σ_f. In nuclear reactions, these cross-sections have the magnitudes of nuclear cross-sections, i.e., πR^2, although it bears no simple relationship with nuclear area. These are measured in barns, where 1 barn = 10^{-24} cm^2. The total cross-section is sum of the individual cross-sections and is written as:

$$\sigma_T = \sigma_{sc} + \sigma_r + \sigma_f \tag{6.35}$$

Nuclei, in general, is not fissile, therefore, the total cross-section consists of two parts only, the scattering cross-section and reaction cross-section.

Hence,
$$\sigma_T = \sigma_{sc} + \sigma_r \tag{6.36}$$

The scattering is of two kinds: elastic scattering in which the incident particle is scattered without the change in the energy and inelastic scattering, in which the same particle is emitted with its energy altered. The inelastic scattering is also included in the reaction cross-section. The cross-section of different processes are dependent upon the incident energy and show marked increase at some energies. This phenomenon is called *resonance*.

6.10 NUCLEAR FISSION

The discovery of nuclear fission started from the attempt of E. Fermi and his collaborators in 1934, to produce transuranic elements by bombarding uranium with neutrons.

According to Fermi's expectation, when uranium ($^{238}_{92}$U) is bombarded by a neutron, the isotope of uranium ($^{239}_{92}$U) should be formed and this should be a β-emitter as:

$$^{238}_{92}U + ^{1}_{0}n \rightarrow ^{239}_{92}U \tag{6.37}$$

and
$$^{238}_{92}U \rightarrow ^{239}_{93}X + ^{0}_{-1}e \tag{6.38}$$

In reaction, given by Eq. (6.38), X having atomic number 93 should be transuranic element.

In actual experiment, Fermi and his co-workers found four β-particles with different half-life periods 10 s, 40 s, 10 s and 40 s.

From these β-rays activities, Fermi and his collaborators concluded that new process is taking place because uranium is natural radioactive substance, which disintegrates by emission of α-particles. As β-activities increase the atomic number by one, hence Fermi and his co-workers felt that 4 transuranic elements with atomic numbers 93, 94, 95 and 96 are formed.

At the same time, Fermi and his co-workers do not have any doubt that the new substance so formed may have atomic number less than uranium.

In 1939, two German scientists, Otto Haln and F. Strassman discovered that when uranium ($Z = 92$) was bombarded with neutrons, it splits up into two separate fragments, which were identified as barium ($Z = 56$) and krypton ($Z = 36$).

They also observed that

1. Fragment travels in opposite directions with tremendous speed.
2. In the splitting of uranium, enormous energy is released because the original nucleus has a greater mass than the sum of masses of two fragments, and this decrease of mass appears as the energy inaccordance with Einstein equation:

$$E = mc^2 \tag{6.39}$$

The reaction is represented as:

$$^{238}_{92}U + ^{1}_{0}n \rightarrow ^{148}_{56}Ba + ^{88}_{36}Kr + 3^{1}_{0}n + \text{energy}$$

The integration process, in which heavy nucleus after capturing a neutron, splits up into two higher nuclei of nearly equal masses is called fission of the nucleus or nuclear fission.

6.10.1 Energy Released in Nuclear Fission

When a heavy nucleus undergoes a fission process, a large amount of energy is released together with the emission of neutrons, which further produce fission. This energy is known as nuclear energy or atomic energy.

The amount of energy released in nuclear fission may be obtained by mass defect method, i.e., difference between mass of initial nucleus and masses of nuclei produced by fission. The difference of mass is converted into energy according to the Einstein mass–energy relation:

$$E = mc^2$$

where m = Mass converted into energy
and c = Velocity of light $= 3 \times 10^8$ m/s
For example, consider the fission of uranium ^{235}U ($Z = 92$) into barium ^{141}Ba ($Z = 56$) and ^{92}Kr ($Z = 36$) by slow reaction.

The reaction is given by

$$_{92}^{235}\text{U} + _{0}^{1}\text{n} \rightarrow _{92}^{236}\text{U} + _{56}^{141}\text{Ba} + _{36}^{92}\text{Kr} + 3_{0}^{1}\text{n} + Q$$

Let us estimate the actual masses before and after the fission reaction:

Actual mass before the fission reaction = Mass of $_{92}^{235}\text{U}$ + Mass of $_{0}^{1}\text{n}$

$$= 235.124 \text{ a.m.u.} + 1.009 \text{ a.m.u.}$$

$$= 236.133 \text{ a.m.u.}$$

After fission reaction = Mass of $_{56}^{141}\text{Ba}$ + Mass of $_{36}^{92}\text{Kr}$ + 3 × Mass of neutron

$$= 140.958 \text{ a.m.u.} + 91.926 \text{ a.m.u.} + 3 \times 1.009 \text{ a.m.u.}$$

$$= 235.910 \text{ a.m.u.}$$

Now, mass decrease during nuclear reaction = (236.133 − 235.910) a.m.u.

$$= 0.223 \text{ a.m.u.}$$

∴   Corresponding released energy = 0.223 × 931 MeV

$$= 200 \text{ MeV}$$

6.10.2   Bohr–Wheeler Theory of Nuclear Fission

We know that ^{238}U can be fissioned by fast neutrons only (having energy 1.0 MeV or more), while ^{235}U is fissionable by fast as well as by slow neutrons. An explanation of this observation was given by Bohr and Wheeler in terms of liquid drop model.

A nucleus is like a spherical drop of an incompressible liquid, which is held in equilibrium by a balance between the short range attraction among the nucleons and the repulsive electrostatic forces among the protons. The inter nucleon forces also give rise to surface tension forces, which maintain the spherical shape of the drop, [Figure 6.8(a)].

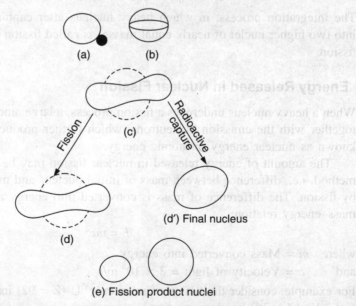

Figure 6.8   Nuclear fission by Bohr–Wheeler theory.

When the nucleus drop captures a neutron, it becomes a *compound nucleus* of high energy. The energy added to the nucleus is partly the kinetic energy of the incident neutron and partly the binding energy of the same neutron, which it releases on being captured. This excitation energy initiates rapid oscillations within the drop which at times become ellipsoidal in shape [Figure 6.8(b)].

The surface tension forces tend to make the drop return to its spherical shape, while the excitation energy tends to distort the shape still further as shown in Figure 6.8(c).

If the excitation energy is small, the nucleus oscillates until it eventually loses its excitation energy by γ-emission (radioactive capture) and returns to the spherical shape. If the excitation energy is sufficiently large, the nucleus may attain the shape of a dumb-bell as shown in Figure 6.8(d).

When this happens, the repulsive electrostatic forces between the two parts of the dumb-bell overcome the attractive forces between nucleons and splitting take place. Each split part becomes spherical in shape as shown in Figure 6.8(e).

Thus, there is a stage [Figure 6.8(d')], after which the compound nucleus is bound to split.

Note: The nuclei $^{238}_{92}U$ and $^{239}_{94}Pu$ are fissionable by slow neutrons, while $^{232}_{90}Th$ and $^{231}_{91}Pa$ can be fissioned by fast neutrons only.

6.10.3 Nuclear Fission and Other Nuclear Reactions

Following are the features of nuclear fission reactions, which make it different from other nuclear reactions:

1. Enormous amount of energy (200 MeV per nuclear fission) is liberated.
2. A heavy nucleus splits into other nuclei of comparable size.
3. Fission fragments emit one or more fast neutrons.
4. With a suitable size and shape of fissionable material, self propagating chain reaction is possible.

6.10.4 Nuclear Fission Take Place in Heavier Elements Only

Heavy nuclei of mass greater than 110 are unstable. A nucleus has a spherical size. With increasing charge and size of nucleus, it becomes flattened in order to be stable. The $Z \cong 100$, the deformation would be so far advanced so as to split the nucleus into two. As the nucleus of uranium (92), thorium (90) and protactium (91) lie close to limit of complete deformation, therefore, they are easily excited upon receiving a moderate energy of excitation to oscillate like those produced in a liquid drop by surface tension. It leads to split the nucleus into two lighter nuclei. This is the reason why the fission takes place only with a few of the heaviest element.

6.11 MODERATOR

A moderator is a substance, which reduces the kinetic energy of fast fission neutrons in the nuclear reactor to a level appropriate to cause the fission of ^{235}U effectively.

Thus, moderator prevents the chain reaction of fission from dying out. This is because the fission of ^{235}U takes place effectively only if the bombarding neutrons are slow and have more time to react with the nucleus and cause its fission.

6.12 NUCLEAR REACTORS

We know that a self sustaining controlled chain reaction is a source of nuclear energy. A nuclear reactor is a device or apparatus in which nuclear fission is produced under a self sustaining controlled nuclear chain reaction. It may be looked as a sort of nuclear furnace, which burns fuels like ^{235}U, ^{238}U or ^{239}Pu and produces, in turn, many useful products like neutrons, heat, radio isotopes, etc.

The essential components of nuclear reactors are shown in Figure 6.9.

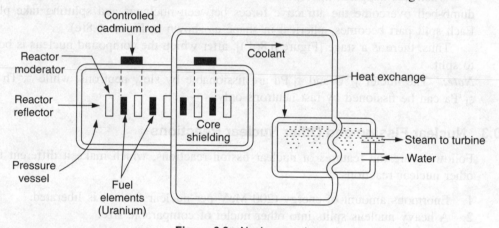

Figure 6.9 Nuclear reactor.

The brief descriptions of all essential components are discussed as under.

Reactor Core

This is the main part of reactor, which contains the fissionable material called the *reactor fuel*. The fuel bearing region is called *reactor core*. Here the nuclear reaction takes place and huge quantity of heat is generated. Reactor cores generally have a shape approximately a right circular cylinder with a diameter of few metres. The fuel elements are made of plates of rods of uranium metal. In general, the reactor core has fuel elements, moderator, control rods and cooling material housed in the pressure vessel.

Reactor Reflector

The region surrounding reactor core is known as *reflector*. Its function is to reflect back some of the neutrons that leak out from the surface of the core. The material of the reflector is same as that of moderator.

Reactor Moderator

The function of a reactor moderator is to slow down the fast neutrons. The moderator should have high boiling point, large scattering cross-section, small absorption cross-section and low atomic number.

The commonly used moderators are ordinary water, heavy water (best moderator), graphite and beryllium oxide.

Reactor Coolant

The material used to remove heat produced by fission, as it is liberated, is known as *reactor coolant*. The coolant is generally pumped through the reactor in the form of liquid or gas. It is circulated throughout the reactor to maintain a uniform temperature.

Properties of coolant

1. High boiling point
2. High specific heat
3. Easy to pump
4. Cheap
5. Chemical stable.

Materials used as coolant

1. Ordinary and heavy water
2. Organic liquids (hydrocarbons)
3. Liquid metals
4. Gases

When the energy generated in a reactor is to be converted into electric power, it is transferred from coolant to a working liquid to produce a hot gas. The resulting vapour or the gas can be used to generate power by means of turbine.

Reactor Control Materials

In a reactor, it is very essential to control the fission process, otherwise the chain reaction may become explosive, and consequently the reactor will be damaged. To consider the reactor control, we define a quantity K_c known as *effective multiplication factor*. This governs the conditions of the reaction, i.e., whether the chain reaction will be continue at a steady rate, increases or decreases. The effective multiplication factor K_c is defined as the ratio of production of neutrons P to combined rate of absorption A and rate of leakage L of neutrons, i.e.,

$$K_c = \frac{P}{A + L}$$

Since

$$P = NF$$

Then

$$K_c = \frac{NF}{A + L} = \frac{N(F/A)}{1 + L/A} \tag{6.40}$$

where F = Rate of fusion occur
N = Average number of neutron/fission

Note:

1. If $K_c = 1$: Chain reaction is critical.
2. If $K_c > 1$: Chain reaction is building up, i.e., supercritical.
3. If $K_c < 1$: Chain reaction is dying, i.e., subcritical.

Reactor Shielding

In a nuclear reactor, various types of rays are emitted. These rays may harm the persons working near the reactor. Hence thick walls of cement and concrete are constructed around the reactor, which are known as *shields*.

6.12.1 Working of Nuclear Reactor

There always remain some neutrons in the reactor. To operate the reactor, all the controlling rods except one are taken out. The remaining controlling rod is now slowly taken out until the intensity of neutrons begin to increase. The reactor now works at a constant level by adjusting the single control.

6.12.2 Applications

The nuclear reactors are widely used for the following purposes:

1. To produce plutonium (^{239}Pu).
2. To produce a neutron beam of high intensity for nuclear bombardment studies.
3. To produce artificially radioactive isotopes for medical, industrial and biological uses.
4. To generate electricity.

6.13 NUCLEAR FUSION AND THERMONUCLEAR REACTION

If two light nuclei are brought close enough, they may fuse into a single nucleus. This process is called *nuclear fusion*. The mass of the single nucleus is less than the sum of masses of the combining nuclei. The difference in mass is released in the form of energy. The nuclei carry positive charge. Hence the powerful electrostatic forces of repulsion between them have to overcome before they can be fused together. The combining nuclei have, therefore, to be provided with very large initial energies. One way of doing this would be to raise the temperature of the fusion material. The temperature required is of the order of ten to one hundred million to a very high temperature and then fused together is called a *thermonuclear reaction*. Spectroscopic examination of the radiation from the sun and the stars shows that they contain large quantities of hydrogen. This led Be the to suggest in 1939 that fusion of hydrogen into helium may be the primary source of stellar energy. There is no longer any doubt that nuclear reactions are the source of stellar energy. These reactions are envisaged as follows:

A star system may start as a diffuse mass of hydrogen with probably other elements in smaller proportions. This mass starts contracting under the gravitational forces. The gravitational energy is converted into kinetic energy of the hydrogen atoms. As a result, the temperature rises. When the temperature reaches about 10^7 K, the hydrogen nuclei will be shed of outer electrons and will be in the form of plasma. At such a temperature, the kinetic energy of the protons will be sufficiently large to overcome the coulomb barrier effects. Then the so called (p, p) reaction gives 0.42 MeV of energy per reaction. Thus

$$^1_1\text{H} + ^1_1\text{H} \rightarrow ^2_1\text{H} + ^0_1\text{e} + 0.42 \text{ MeV} \qquad (6.41)$$

Soon sufficiently deuterium will be formed to set in what is called the (p, d) reaction. Thus

$$\,_1^1\text{H} + \,_1^2\text{H} \rightarrow \,_2^4\text{He} + 5.5 \text{ MeV} \qquad (6.42)$$

The reaction leads on to

$$\,_2^3\text{He} + \,_2^3\text{He} \rightarrow \,_2^4\text{He} + 2\,_1^1\text{H} + 12.8 \text{ MeV} \qquad (6.43)$$

The total energy released in one cycle of reaction is:

$$2[0.42 + 5.5] + 12.8 = 24.64 \text{ MeV}$$

The so called proton–proton cycle of reactions is believed to be the main source of stellar energy. When this cycle of reactions is going on the star, it is said to be *burning hydrogen*.

Figure 6.10 illustrates the proton–proton cycle. With the continuation of the hydrogen burning, the heavier helium collects at the core and the temperature of this core starts rising. When it reaches about 10^8 K, helium burning commences. The reactions of helium burning are as follows:

$$\,_2^4\text{He} + \,_2^4\text{He} + 95 \text{ keV} \rightarrow \,_4^8\text{Be}$$
$$\,_4^8\text{Be} + \,_2^4\text{He} \rightarrow \,_6^{12}\text{C} + 7.2 \text{ MeV} \qquad (6.44)$$

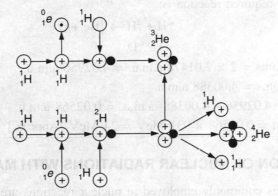

Figure 6.10 Proton–proton cycle.

As $\,_6^{12}\text{C}$ becomes available, it takes the role of a catalyst for the hydrogen burning that takes place in carbon–nitrogen cycle [Figure 6.11]. The cycle suggested by Bethe is as follows:

$$\,_6^{12}\text{C} + \,_1^1\text{H} \rightarrow \,_7^{13}\text{N} + 2.0 \text{ MeV}$$
$$\,_7^{13}\text{N} \rightarrow \,_6^{12}\text{C} + 2.0 \text{ MeV}$$
$$\,_6^{12}\text{C} + \,_1^1\text{H} \rightarrow \,_7^{14}\text{N} + 7.6 \text{ MeV}$$
$$\,_7^{13}\text{N} + \,_1^1\text{H} \rightarrow \,_8^{15}\text{O} + 7.3 \text{ MeV}$$
$$\,_8^{15}\text{O} \rightarrow \,_7^{15}\text{N} + \,_1^0\text{e} + 1.7 \text{ MeV}$$
$$\,_7^{13}\text{N} + \,_1^1\text{H} \rightarrow \,_6^{12}\text{C} + \,_2^4\text{He} + 4.9 \text{ MeV} \qquad (6.45)$$

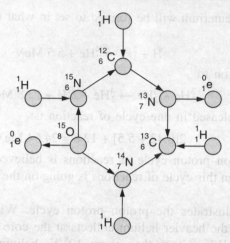

Figure 6.11 Carbon–nitrogen cycle.

EXAMPLE 6.5 Calculate the energy liberated, when a single helium nuclei is formed by the fusion of two deuterium nuclei.

Solution The required reaction is:

$$_1^2H + _1^2H \rightarrow _2^4He + Q$$
$$D \quad\ \ D$$

Mass of 2D atoms = 2 × 2.01478 a.m.u. = 4.02956 a.m.u.

Mass of He atoms = 4.00388 a.m.u.

Mass defect = (4.02956 – 4.00388) a.m.u. = 0.02568 a.m.u.

Energy released = 0.02568 × 931 MeV = 24 MeV [Approx.]

6.14 INTERACTION OF NUCLEAR RADIATIONS WITH MATTER

Nuclear particles commonly employed in nuclear reactions are:

1. Charge group of particles like α-particle, β-particle, proton, deuteron etc.
2. Electrically neutral particles like γ-rays and neutrons.

Besides these, there are a lot of other particles, which are produced in high energy reactions and occasionally one employs $\mu^{\pm 0}$ mesons, $k^{\pm 0}$ mesons and several types of hyperons and nuclear fragments in nuclear reactions.

The main group of charged particles with which one is mainly concerned in low energy nuclear interactions are α- and β-particles, protons and deuterons. These particles in their interaction with atoms of the matter, produce ionization of atoms. When these particles pass close the atomic nuclei, they get involved in nuclear interactions. Charged particles are, therefore, called ionizing radiations as they produce ionization of the medium through which they pass all along their path.

γ–rays, which are electromagnetic radiations, are electrically neutral and they loose energy to electron through processes like

1. Pair production
2. Compton electron scattering
3. Electron–positron pair production

Electrons produced of knocked out in the above processes in turn produce ionization of the medium and give the information of the passage of γ-rays through medium. Neutrons are also neutral particles and do not produce ionization in the medium directly. These particles however, produce very efficient nuclear reaction with some nuclei or scatter protons from hydrogen nuclei in which energy of neutron is transferred to charged particles. It is through these charged particles that the presence of neutrons is detected. It is evident that the understanding of various phenomena, on the passage of a radiation through a medium, is essential for the study of these particles.

6.14.1 Absorption of Heavy Charged Particles

Heavy charged particles like protons, deuterons, α-particles or fission fragments loose energy in matter mainly through ionization because due to their heavy rest mass, their energies, at which radiation losses become perceptible, are quite large, i.e., 1 BeV for proton, 2 BeV for deuteron, and similarly other higher values for heavy ions and fragments. When these heavy charged particles produce ionization, they transfer small amounts of their energy to electrons. They are also not much deviated from their original directions unlike electrons. Only at the end of their paths, when they move slowly, they are deviated by large angles by the nuclei of the atoms of the medium. In case of heavy charged particles colliding with a free electron (a crude assumption), the loss of energy of the heavy ion will be equal to the gain of energy of electron in the coulomb interaction. If heavy particle has a charge $+Ze$ and is moving along the x-axis, as shown in Figure 6.12, and the electron lies inside a shell at an impact distance from the x-axis.

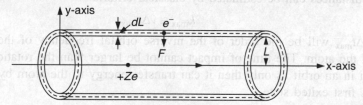

Figure 6.12   Schematic representation of the collision parameters between a heavy charged particle and electron.

The coulomb force F can be splitted in two components F_x and F_y, i.e., parallel to the motion of the heavy ion and perpendicular to the motion. As the heavy ion passes by the side of the electron, it transfers momentum to it in the respective directions as:

$$\int F_x \, dt = 0$$

$$\int F_y \, dt = p_y \tag{6.46}$$

The electrostatic field over the cylindrical surface surrounding a charge is expressed by Gauss's theorem as:

$$\int \mathbf{E} \cdot d\mathbf{S} = 4q \tag{6.47}$$

where E = Electric field at the surface

dS = Element of surface.

Using this relation, we get the momentum p_y as:

$$\int \frac{E_y}{e} 2\pi L dx = 4\pi Z e \tag{6.48}$$

But $dx = vdt$, so that Eq. (6.48) gives, after rearranging,

$$\int E_y \, dt = 2 \frac{Ze^2}{Lv} = p_y \tag{6.49}$$

The kinetic energy gained by the electron is given by

$$E = \frac{p^2}{2m} = \frac{2Z^2 e^4}{mL^2 v^2} \tag{6.50}$$

If the absorbing material has n atoms per unit volume and is of atomic number Z so that larger numbers of electrons per atom are encountered, there will be $nz. \ 2\pi L dL \cdot dx$ electrons within the shell in the distance dx. To each of them, the heavy particle imparts energy. The energy loss per unit path length in this manner can be given by

$$-\frac{dE}{dx} = \int_{L_{min}}^{L_{max}} nZ \ 2\pi L \ dL \left(\frac{2Z^2 e^4}{mL^2 v^2} \right) = \frac{4\pi e^4 Z^2 nZ}{mv^2} \log_e \frac{L_{max}}{L_{min}} \tag{6.51}$$

This expression is a fair estimate of the energy loss due to a heavy charged particle except in the case of head-on collisions, which are rare. The maximum and minimum impact distances can be estimated by classical criteria:

$$L_{max} \cong v\Delta t_{max}$$

where Δt_{max} will be the order of the inverse orbital frequency of the electron in an orbit of the atom. The time of impact cannot be larger than the rotation period of the electron in an orbit as only then it can transfer energy to the atom by exiting it to at least its first exited state.

$$L_{max} \cong \frac{v}{v} \tag{6.52}$$

The minimum time is associated with the uncertainly principle. It can be detected within a distance of the order of de Broglie wavelength of electron.

$$L_{max} \cong \lambda = \frac{h}{mv} \tag{6.53}$$

Then we can write the expression (6.51) as:

$$-\frac{dE}{dx} = \frac{4\pi e^4 Z^2 nZ}{mv^2} \log_e \frac{2mv^2}{I_{av}} \tag{6.54}$$

Here I_{av} replaces hv and represent the mean ionization and resonance potential energy of the absorbs, which effectively come into play. The energy loss by electron through ionization can also be expressed approximately by this expression. The only difficulty in case of electrons comes because of its relatively lighter mass, and it will suffer multiple scattering in its way through absorbs at low energies. For heavy charged particles, the concept of mean range R_m can be useful. This is defined by the average distance that it travels before it losses its entire kinetic energy in the absorber.

$$R_m = \int_0^{R_m} dx = \int_{E_0}^0 \frac{dx}{dE} dE = \int_0^{E_0} \left(\frac{-dE}{dx} \right)^{-1} dE = \int_0^{E_0} \left(\frac{4\pi e^4 Z^2 nZ}{mv^2} \log_e \frac{2mv^2}{2I_{av}} \right)^{-1} dE \quad (6.55)$$

If we consider the term v^2 under logarithmic term to be contributing negligibly as compared to v^2 in the denominator expression, we can approximately say that the mean range R_m will be proportional to

$$R_m \propto \int_0^{E_0} EdE - E_0^2 \quad (6.56)$$

where E_0 is the energy of α-particles.

In actual case, the range integral is not simple because of the logarithmic dependence of v^2 in the numerator. The exponent 2 may be smaller in Eq. (6.56).

α-particles are helium nuclei that were extensively studied in the early stage of the development of Nuclear Physics by Rutherford and his co-workers. The range of α-particle shows a distribution as shown in Figure 6.13.

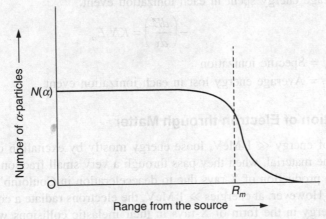

Figure 6.13 Range of the α-particles.

α-particles are slowed down due to a large number of ionizing collisions with the atoms of the medium, and since they do not loose exactly the same amount of energy in each collision, the range of particles shows a distribution towards its end. This small variation in the range of particles is called *straggling*. The property of straggling is common with all the heavy particles. The range is taken to be the mean value of the range of a large number of particles as shown in Figure 6.13. This is the value of range at which the number of α-particles falls to half of the original number. The range R_m of α-particles is proportional to the energy of these particles and can be expressed by an empirical relation:

$$R_m = 0.318E^{3/2} \qquad (6.57)$$

where

R_m = Matter (air) at NTP in cm

E = Energy of range R_m of α-particle in MeV

If the range of α-particle is shown in air, its range in another medium with atomic mass A, atomic number Z can be expressed by the following empirical relation:

$$\frac{R}{R_{\text{air}}} = 1.5 \times 10^{-4}\frac{A}{A_0}\left[\frac{Z+0}{2}\right]^{1/2} \qquad (6.58)$$

Similarly the range of another particle with mass M and charge Z can be measured in comparison to the range of α-particle as:

$$\frac{R}{R_\alpha} = \frac{M/Z^2}{M_\alpha/Z_\alpha^2} \qquad (6.59)$$

Specific Ionization and Stopping Power of α-particle

Energy loss of a particle in passing through a unit track length of medium is known to be the *stopping power* of the medium. This is proportional to the number of ion pairs produced by the particle per unit length of the medium and is also proportional to the average energy spent in each ionization event.

Hence

$$-\left(\frac{dE}{dx}\right) = KN_iE_a \qquad (6.60)$$

where N_i = Specific ionization

E_a = Average energy lost in each ionization event

6.14.2   Absorption of Electron through Matter

Electrons of energy $\ll 1$ MeV, loose energy mostly by excitation or ionization of the atoms of the material, when they pass through a very small fraction of electron energy goes to the production of X-rays due to deacceleration in Coulomb's field of electrons and nuclei. However, at energies $\gg 1$ MeV, the electrons radiate a considerable fraction of their energy in the form of X-rays in their inelastic collisions with the Coulomb's field of the atomic nuclei and also with the electrons. Such radiations are called *bremsstrahlung* radiations. Energy loss per unit thickness of the material is called *stopping power,* and in these processes, it is approximately expressed as:

$$-\left[\frac{dE}{dx}\right]_{\text{ion}} = \frac{4\pi e^2 nZ}{mv^2}\ln\frac{2mv^2}{I_\alpha} \qquad (6.61)$$

and

$$-\left[\frac{dE}{dx}\right]_{\text{rad}} \propto \frac{4Z^2e^2}{hc}\left(\frac{e^2}{mc^2}\right)^2\frac{N}{A}E\left\{\log 183\, Z^{-1/3} + \frac{1}{18}\right\} \qquad (6.62)$$

where N/A = Number of atoms per unit mass of the medium

   Z = Nuclear charge

   m = Mass

   E = Energy of electrons.

A comparison of the stopping powers of the materials by ionization and radiation shows that the ionization is considerable even at low energies, the radiation losses are zero for low energy electrons. Moreover, the ionization loss is proportional to nZ, which is nearly constant for low mass nuclei. It depends on $(1/mv^2)$ for low energies. Ionization is quite high at low energies, and decreases as energy increases but as the energy exceeds the rest energy m_0c^2 for electron, the particle becomes relativistic and the logarithmic term in the Eq. (6.62) takes over, and the ionization again starts increasing slowly. The minimum ionization region is useful for detection and identification of the rest mass of a particle as $E_{min} \approx$ rest energy (m_0c^2).

The energy loss through the radiation depends upon Z^2 of the material contrary to Z dependence in the case of ionization. Radiation losses are, therefore, more prominent in heavy materials. This increases with energy. The radioactive losses are negligible for heavy charged particles like protons, deuterons and α-particle as it is inversely proportional to the square of mass of the incident particle.

The relative energy loss contribution due to ionization and radiation has been determined for lead is shown in Figure 6.14.

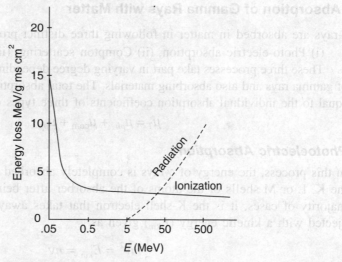

Figure 6.14   Energy loss in lead through ionization and radiation for electron.

Figure 6.14 depicts that at certain energy ionization loss becomes equal to radiation loss. This energy is known as *critical energy*. Beyond this energy, the energy losses due to the particle in a vacuum are mainly radioactive below 1 MeV. The energy loss is mainly due to ionization.

6.14.3   Intersection of γ-rays with Matter

γ-rays are electromagnetic radiations and suffer absorption in matter just like light photons. The attenuation in their intensity follows a similar exponential low:

$$dI \propto -I dx$$

where dI is the change in intensity in passing through a thickness dx.

$$\frac{dI}{dx} = -\mu I$$

where μ is called absorption coefficient of material and usually expressed in g/cc.

This leads to the relation:

$$I = I_0 e^{-\sigma x} \qquad (6.63)$$

where σ is called the microscopic cross-section and is related to the absorption coefficient of material as $\mu = N_0 \sigma$ where N_0 is the number of atoms per unit mass.

The thickness of material needed to attenuate the original intensity of a beam to half the value is called half value thickness:

$$x_0 = \frac{\log_e 2}{\mu} = \frac{0.693}{\mu} \qquad (6.64)$$

In the case of lead absorber, this thickness for 3 MeV γ-rays is 0.5 inch.

6.14.4   Absorption of Gamma Rays with Matter

γ-rays are absorbed in matter in following three distinct processes:

(i) Photo-electric absorption, (ii) Compton scattering, (iii) Pair production.

These three processes take part in varying degree depending upon the incident energy of gamma rays and also absorbing materials. The total absorption coefficient is, therefore, equal to the individual absorption coefficients of three types of processes involved, i.e.,

$$\mu_T = \mu_{ph} + \mu_{com} + \mu_{pp} \qquad (6.65)$$

Photoelectric Absorption

In this process, the energy of γ-rays is completely absorbed by one of the electrons in the K, L or M shells of the atoms of the absorber after being hit by the photon. In a majority of cases, it is the K-shell electron that takes away the γ-rays energy and is ejected with a kinetic energy (E_{kin}) given as:

$$\frac{1}{2} mv^2 = E_{kin} = h\nu \qquad (6.66)$$

Removal of an electron from a K-shell gives rise to the emission of characteristic γ-rays of two absorber atom as the electron from L, M and other highest shells try to fill the gap in the lower state of the atom.

The absorption cross-section of this process below energy 0.5 MeV approximately is given by

$$\mu_{ph} = N_0 \sigma_{ph} = \frac{N_0 A Z^4}{E_\gamma^3} \qquad (6.67)$$

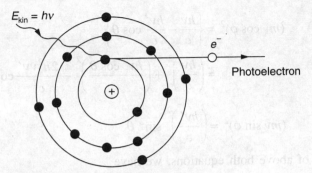

Figure 6.15   Photoelectric absorption.

where A is $\cong 1.25 \times 10^{-9}$ and E_γ is expressed in MeV. This mechanism of absorption is most prominent at low energies, i.e., $E_r \ll 0.5$ MeV and is negligible above energy 1 MeV.

Compton Scattering

In this process, the γ-ray is scattered in an inelastic process with a free electron. A portion of energy of γ-rays is transferred to the electron, and remaining energy is carried by the scattered photon. This is shown schematically in Figure 6.16.

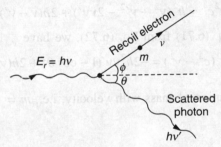

Figure 6.16   Compton absorption.

The energy of the incident photon is $h\nu$ and its momentum is $h\nu/c$. After scattering with the stationary electron, the electron gets a recoil velocity v and the γ-rays is reduced in its energy and momentum. Its energy is $h\nu'$ and momentum is $h\nu'/c$. If we apply the conservation of energy and moment before and after collision, we get the following equation:

$$h\nu + m_0 c^2 = h\nu' + mc^2 \tag{6.68}$$

$$\frac{h\nu}{c} + 0 = \frac{h\nu'}{c}\cos\theta + mv\cos\phi \tag{6.69}$$

Linear momentum before collision = Linear momentum after collision

If m_0 is rest mass of electron and m be the mass with velocity v, then

$$0 = \frac{h\nu'}{c}\sin\theta - mv\sin\phi \tag{6.70}$$

$$(mv \cos \phi)^2 = \left\{ \frac{h\nu}{c} - \frac{h\nu'}{c} \cos \theta \right\}^2$$

$$= \left(\frac{h\nu}{c} \right)^2 + \left(\frac{h\nu' \cos \theta}{c} \right)^2 - \frac{2h^2 \nu\nu'}{c^2} \cos \theta$$

and
$$(mv \sin \phi)^2 = \left(\frac{h\nu'}{c} \right)^2 \sin^2 \theta$$

On addition of above both equations, we have

$$m^2 v^2 \cos^2 \phi + m^2 v^2 \sin^2 \phi = \frac{h^2 \nu^2}{c^2} + \frac{h^2 \nu'^2}{c^2} \cos^2 \phi + \frac{h^2 \nu'^2}{c^2} \sin^2 \phi - \frac{2h^2 \nu\nu'}{c^2} \cos \theta$$

or
$$m^2 v^2 c^2 = (h\nu)^2 + (h\nu')^2 - 2h^2 \nu\nu' \cos \theta$$

$$= h^2(\nu^2 + \nu'^2 - 2\nu\nu' \cos \theta) \qquad (6.71)$$

From Eq. (6.68), we get

$$mc^2 = h(\nu - \nu') + m_0 c^2$$

or
$$m^2 c^4 = h^2(\nu^2 + \nu'^2 - 2\nu\nu') + 2h(\nu - \nu')m_0 c^2 + m_0^2 c^4 \qquad (6.72)$$

Now, subtracting Eq. (6.71) from Eq. (6.72), we have

$$m^2 c^2 (c^2 - v^2) = -2h^2 \nu\nu'(1 - \cos \theta) + 2h(\nu - \nu')m_0 c^2 + m_0^2 c^4$$

Since we know variation of mass with velocity, i.e., $m = \dfrac{m_0}{\sqrt{\left(1 - \dfrac{v^2}{c^2}\right)}}$, then Eq. (6.72)

becomes

$$\frac{m_0^2 c^2}{\left(1 - \dfrac{v^2}{c^2}\right)} (c^2 - v^2) = -2h^2 \nu\nu'(1 - \cos \theta) + 2h(\nu - \nu')m_0 c^2 - m_0^2 c^4$$

or
$$m_0^2 c^4 = -2h^2 \nu'(1 - \cos \theta) + 2h(\nu - \nu')m_0 c^2 + m_0^2 c^4$$

or
$$h^2 \nu\nu'(1 - \cos \theta) = h(\nu - \nu') \, m_0 c^2 \quad \text{or} \quad \frac{\nu - \nu'}{\nu\nu'} = \frac{h}{m_0 c^2}(1 - \cos \theta)$$

Hence
$$\frac{1}{\nu'} - \frac{1}{\nu} = \frac{h}{m_0 c^2}(1 - \cos \theta) \quad \text{or} \quad \frac{1}{\nu'} = \frac{1}{\nu} + \frac{h}{m_0 c^2}(1 - \cos \theta)$$

or
$$\nu' = \frac{\nu}{1 + \dfrac{h\nu}{m_0 c^2}(1 - \cos \theta)} = \frac{\nu}{1 + 2\alpha \sin^2 \theta/2} \qquad (6.73)$$

When $\alpha = \dfrac{h\nu}{m_0 c^2}$, the energy of electron is given by

$$W = h\nu - h\nu' = h\nu - h\nu \left[\frac{1}{1 + 2\alpha \sin^2 \theta/2} \right] = h\nu \left[\frac{2\alpha \sin^2 \theta/2}{1 + 2\alpha \sin^2 \theta/2} \right] \quad (6.74)$$

The energy of the scattered electron is directly proportional to the energy of incident photon and is also proportional to the angle of scattering θ in a complicated manner. The coefficient of the Compton process depends upon the number of electrons per atom, i.e., Z, available for scattering

$$\mu_{\text{com}} = N_0 Z \sigma_{\text{com}}$$

Klein Nishima has given a quantum mechanical relation for absorption coefficient:

$$\sigma_{\text{com}} = \frac{1 + (\alpha - 1)^2}{4\alpha^3} \log(1 + 2\alpha) + \frac{\alpha^2 (1 + \alpha) + 2(1 + \alpha)^2}{2\alpha^2 (1 + 2\alpha)^2}$$

where $\qquad \alpha = \dfrac{E_\gamma}{m_0 c^2}$

The Compton absorption coefficient depends upon (Z/A) as:

$$\mu_{\text{com}} = N_{\text{Avogadro}} \left(\frac{Z}{A} \right) f \left(\frac{E_r}{m_0 c^2} \right) \quad (6.75)$$

For heavy atom, (Z/A) changes slightly, but for light atoms, this factor does not change much and is nearly half. Compton absorption coefficient remains constant for the light atoms. This, however, decreases with increasing energy.

Pair Production

This process starts at threshold γ-photon of energy 1.23 MeV, which corresponds to the rest mass energy of two electrons. The γ-photon in the pair production process is lost in the Coulomb's field of the nucleus, and a pair of electrons with opposite charges on them is created.

$$E_r = m_{e^-}^0 c^2 + m_{e^+}^0 + E_{\text{kin}}^+ + E_{\text{kin}}^-$$

$$= 0.51 \text{ MeV} + 0.51 \text{ MeV} + E_{\text{kin}}^+ + E_{\text{kin}}^-$$

$$= 1.02 \text{ MeV} + E_{\text{kin}}^+ + E_{\text{kin}}^+$$

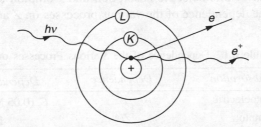

Figure 6.17 Pair production.

A very small energy is transferred to the recoiling nucleus, which is negligible. This process becomes increasingly important for the energies of γ-rays $E_\gamma \gg 1.02$ MeV and depends upon Z^2.

$$\mu_{pp} \propto N_0 Z^2 \log E_\gamma \tag{6.76}$$

As γ-ray produces an electron–positron pair in the process [Figure 6.17],

$$\gamma \to e^+ + e^-$$

and materialise, the positron, which is highly unstable, annihilates with an electron in the inverse process. Before annihilation, they form a transient atomic configuration called positronium atom. Two kinds of these atoms are possible. If spin is parallel, it decays to two γ-quanta as:

$$e^+ + e^- \to \gamma_1 + \gamma_2$$

which go in opposite directions in order to conserve the linear momentum.

It is evident that γ-rays of different energies are absorbed through one or more of these three competing processes, which essentially depend upon the absorber material (Z-number) and the energy of γ-rays. The contribution of different absorption coefficient with energy of γ-rays for lead is shown in Figure 6.18.

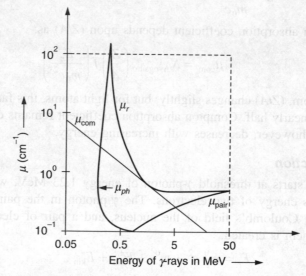

Figure 6.18 Total absorption coefficient vs. energy of γ-rays for lead.

The absorption coefficient of pair-production process is found to increase with energy, while that of photoelectric absorption and Compton scattering decreases with γ-rays energy. The dependence of the various processes on Z and E_γ can be summarised in Table 6.1.

Table 6.1 Dependence of the Various Processes on Z and E_γ

Absorption	Dependence	Dependence of E_γ
Photoelectric	Z^2	$E_r (0.05 \text{ MeV}) = 1/E_\gamma$
Compton	Z/A	$1/E_\gamma$
Pair production	Z^2	$\log E_\gamma$

6.15 NUCLEAR RADIATION DETECTORS

The design and working of any nuclear detector depends upon interaction of nuclear radiations with sensitive material of the radiation detector. It is, therefore, important to understand different types of interactions between radiation and matter. The matter may be solid, liquid or gas and nuclear radiations may be charged particles (α-particles, electrons, protons and other negative and positive ions) or neutral particles (neutrons, photons, γ-rays, etc.). The charged particle may be further categorised as: light charged particles, electron and positron, and heavy charged particles, α-particle, proton and other negative and positive ions.

When charged particles pass through the matter, they cause ionization in the matter and hence there occurs formation of negative and positive ions of the matter. The interaction of γ-rays with matter is quite different from that of charged particles. The γ-rays have large penetrating power and a well collimated beam of these shows exponential absorption in matter unlike charged particles. Energetic γ-rays (or photons) interact with matter by three well known processes namely *photoelectric effect, Compton effect* and *pair production*. Hence when energetic γ-rays enter the penetrating medium, these interact via any of the above three processes and as a result of it, energetic electrons are liberated. The liberated electrons are detected by their resulting ionization or excitation of atoms in the medium. In each process, a part of γ-rays energy is converted into kinetic energy of liberated electrons, which is a measure of its ionizing power.

In fact, the development of Nuclear Physics owes much to study of nuclear reactions. These particles may be light nuclei, deuteron, proton, neutron, electron, γ-rays photon or any of the other fundamental particles. The products of nuclear reactions are new nuclei and some small particles. Thus to produce and to observe nuclear reaction, we need two fundamental instruments:

1. Particle accelerator which can produce energetic particles.
2. Particle detector which can detect the products of nuclear reaction.

The classification of nuclear radiation detector is shown in Figure 6.19.

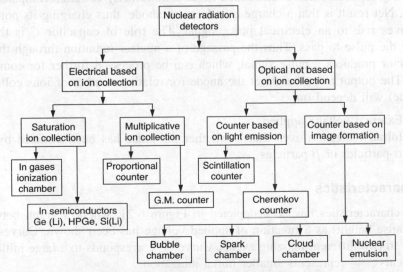

Figure 6.19 Flow diagram of nuclear radiation detectors.

▌6.16 GAS FILLED DETECTORS

Gas filled detectors are some of the oldest and most widely used detectors for nuclear radiations.

6.16.1 Principle, Construction and Working

Gas filled detectors are based on the principle of ionization and excitation.

The schematic diagram of a gas filled detector is shown in Figure 6.20. The system shown in Figure 6.20 consists of a gas filled chamber with a central electrode well illustrated from the chamber walls.

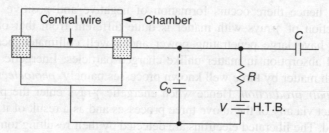

Figure 6.20 Gas filled detector.

An external voltage V is applied between the walls of the chamber (cathode) and the central wire (anode) through a resistance R. In this way, an electric field is set up in the volume of the gas. The charged particles and electromagnetic radiation traversing this gas undergo inelastic collisions with atoms or molecules, ionizing them and thereby forming positive ions and electrons. In the absence of electric field, the ion pairs thus created will just recombine to form neutral atoms. However, in the presence of the applied electric field, the positive and negative ions will move along the radial electric lines of force towards the cathode and anode respectively. Normally, the negative ions (usually electrons) move with much faster drift velocity 10^6 cm/s compared to positive ions. Net result is that a charge q collects on anode, thus charging its potential across R gives rise to an electrical pulse signal. The role of capacitor C is that it allows only the pulse to pass. Thus the passage of a nuclear radiation through the gas of the detector produces a pulse signal, which can be processed further for counting.

The output pulse height at the anode (or relative number of ions collected by the anode) will depend on

1. External voltage applied
2. Initial ionization even that is whether the pulse has been initiated by passage of α-particles or β-particles

6.16.2 Characteristics

The characteristics has been depicted in Figure 6.21, where a graph between output (or pulse height) as a function of applied voltage has been shown. Curves 'a' and 'b' refer to two different ionizing events. Curve 'a' corresponds to a large initial ionization and curve 'b' represents smaller initial ionization.

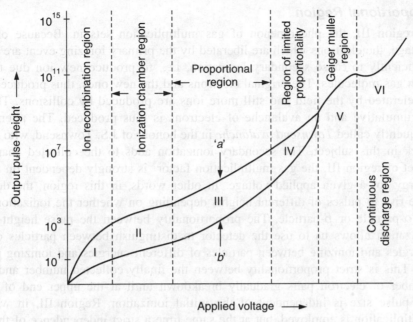

Figure 6.21 Output pulse height vs the applied voltage.

The examination of these results that the variation of the number of ions collected with the applied voltage can be described by dividing curves 'a' and 'b' of Figure 6.21 into following six regions:

Ion Recombination Region

In region I, initially the applied voltage is small, the electric field is not so effective in removing the ions for collection at the electrodes and, therefore, there is competition between two processes:

1. Loss of ion pairs through recombination
2. Removal of charges to the electrodes by electric field.

As the electric field is increased, the ions move faster having less time for possible recombination, and thus relatively large numbers of ions reach the electrodes. So region I is known as *ion recombination region*.

Ionization Chamber Region/Saturation Region

As the voltage is increased further, we get region II. In this region, the voltage is sufficiently high so the loss of ions through the recombination is negligible, and ion pair moves to electrodes rapidly, and virtually every ion pair reaches the electrodes. Thus the pulse size remains independent of the potential as long as voltage is increased upto the point at which secondary electrons are produced by collision in the gas. This region is called the *ionization chamber region* or *saturation region*.

Proportional Region

In region III, the phenomenon of gas multiplication sets in. Because of increased voltage, the electrons which are liberated by the primary ionizing event are accelerated sufficiently to cause secondary ionization, i.e., to produce new ion due to collision with gas molecules. The original electrons and the new ones, thus produced are again accelerated by the field and still more ions are produced by collisions. This process is cumulative and an avalanche of electrons is thus produced. The phenomenon is frequently called *Townsend avalanche* in the honour of J.S. Townsend, who did pioneer work in this subject. The secondary ionization adds to the collected charge. At the onset of region III, the gas multiplication factor[*] is strongly dependent on the particle energy for a given applied voltage. In other words, in this region, the detector will give rise to pulses of different heights depending on whether the ionization is caused by α-particle or β-particle. The proportionality between the pulse height and initial ionization allows us to use the detector to distinguish between particles of different energies and ionizing between particles of different energies and ionizing power.

This is strict proportionality between the finally collected number and the initial number of electron pairs gradually breakdown until at the upper end of region IV, the pulse size is independent of the initial ionization. Region III, in which a gas multiplication is employed, but at the same time a strict independence of the collected charge on the initial ionization remains, is called *proportional region*. In this region, multiplication factor remains constant

Limited Proportionality Region

Region IV, in which proportionality is progressively decreased, is known as the *limited proportionality region.*

Geiger–Müller Region

In region V, the pulse size is large and is also substantially independent of the ionization. All the particles produce pulses of the same height irrespective of their energy and primary ionization. The region is known as Geiger–Müller region. This is a sensitive region, in which even if an ion pair is formed in this sensitive volume, it is sufficient to register a pulse. If the voltage is increased still further, it causes a continuous electric discharge in the sensitive volume.

Continuous Discharge Region

Region VI is continuous discharge region. Gas filled nuclear detectors have different names depending upon the regions (as discussed earlier) in which they are opted.

6.17   IONIZATION CHAMBER

The ionization chamber is the simplest of all gas-filled radiation detectors, and is extensively used for the detection and measurement of certain kinds of ionizing

---

*Total number of ions produced by collisions, as each electron travels towards the central wire is called multiplication factor.

radiation; X-rays, γ- and β-particles. Hess provided first evidence for cosmic rays during a balloon flight by an ionization chamber in 1910. Rutherford and Geiger developed first cylindrical electrical counter for α-particles in 1908. They improved this chamber on introduction a spherical counter in 1912. Further progress was achieved in 1913 when a detector for β-particle was developed. Traditionally, the term 'ionization chamber' is used completely to explain those detectors which accumulate all the charges created by direct ionization within the gas through the application of an electric field. It only uses the discrete charges created by each interaction between the incident radiation and the gas, and does not involve the gas multiplication mechanisms.

6.17.1 Principle

The ionization chamber type particle detectors are suitable sensitive electrical instruments for studying the nature and energy of particles. The basic principle of the technique is that the fast moving particles produce ionization of gas varying with nature and velocity. Each gas atom produces, after ionization, an ion and electron usually referred to as pairs.

6.17.2 Construction

The schematic diagram of ionization chamber is shown in Figure 6.22. The apparatus consists of a hollow metallic cylinder, having a small window at one end for the entry of the particle.

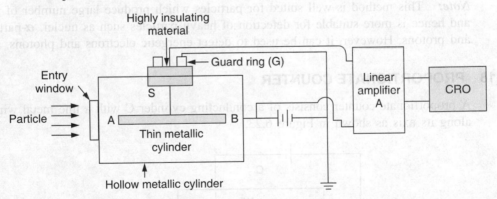

Figure 6.22 Ionization chamber.

In the middle of cylinder and along its axis hangs a thin metallic cylinder AB, suitably supported by highly insulating material and is connected to feeble current measuring instrument like quadrant electrometer or input of a linear amplifier. The cylinder is kept at a constant potential of few hundred volts. The possible current leakage between cylinder and rod is prevented by the use of earthed guard ring G. The cylinder is filled with suitable gas usually at atmospheric pressure and more than atmospheric pressure. The size of cylinder, nature and pressure of gas to be filled depends upon the nature of particle to be detected. Thus for accurate estimation, use is made of different cylinders for detecting different particles. Small size cylinder (3 cm to 5 cm long) filled with air at NTP is suitable for α-particle. For detecting γ-ray photon, one needs a large chamber filled with methyl iodide at a pressure of 30–40 atmospheres. In general, hydrogen, carbon monoxide, air, carbon dioxide, carbon disulphide and methyl iodide may be used. The output of chamber is fed to electrometer or linear amplifier (LA) followed by cathode ray oscilloscope (CRO).

6.17.3 Working

When the particles enter the cylinder, they ionize the gas inside. The charges produced are caught by metallic rod AB and cylinder S and produce a current pulse. As particles travel the cylinder, they produce more ion as height of current pulse linearly increases with time. The magnitude of this current pulse can be measured with time by sensitive electrometer. The sensitivity of the method can be improved by the use of linear amplifier. The feeble ionization current can be amplified by a factor 10^6 to 10^7 times without producing distortion by the use of amplifier. Special protection and shielding is, however, required to protect them from extra noise signal. These amplified current pulses can be registered in the CRO. Under the condition of linear amplification, the CRO record indicates the passage of each particle by a linear jump. The length of this linear jump is proportional to number of ions produced by the particle in crossing the ionization chamber. For this one gets information about the nature and energy of particle.

Note: This method is well suited for particles which produce large number of ions, and hence is more suitable for detection of heavy particles such as nuclei, α-particles and protons. However, it can be used to detect energetic electrons and photons.

6.18 PROPORTIONATE COUNTER

A proportionate counter consists of a conducting cylinder C with a fine metal wire W along its axis as shown in Figure 6.23.

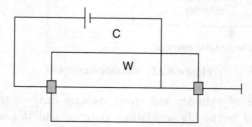

Figure 6.23 Proportionate counter.

The cylinder contains air or argon at a pressure of 5 cm to 12 cm of Hg. A potential difference of several thousand volts is applied between wire and cylinder with the wire at positive terminal.

Under pressure conditions each primary electron produced by the direct interaction of the incident radiation produces N more electrons by ionization. Let the amount of charge produced by the incident radiation by direct interaction in the gas in the chamber be q. Then the total charge collected by the control positive electrode is:

$$Q = Nq \qquad (6.76)$$

where N is the multiplication factor.

Thus

Ionization current recorded $\propto$ Charge produced by the incident radiation

Hence, the device is called *proportionate counter* or *proportional counter*.

6.19 GEIGER–MÜLLER COUNTER

Hans Geiger and Ernest Rutherford made Geiger Counter in 1908. This device was only capable to detect α-particle. In 1928, Hans Geiger and his research scholar, Walther Müller improved the Geiger counter so that it could detect more ionizing radiations and the device is named as Geiger–Müller counter after their names.

6.19.1 Principle

This apparatus uses the principle of ionization of gas. Whenever a fast moving charge particle passes through gas, they ionize it. The electron and ion thus produced constitute a small current known as *ionization current*. The measure of this ionization current gives direct information about the number of particles entering the chamber.

6.19.2 Construction

The Geiger–Müller counter has basically two parts:
(a) Geiger–Müller tube and (b) the counting electronic circuit

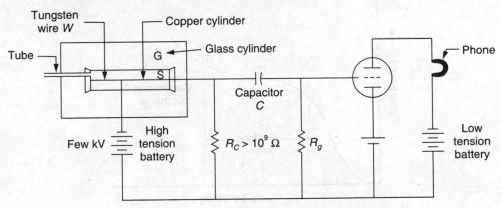

Figure 6.24   Conventional Geiger–Müller counter.

The Geiger–Müller tube consists of an open ended copper cylinder S, 1 cm to 5 cm in diameter and 25 cm to 90 cm in length, fitted inside a thin walled glass cylinder

G with fine tungsten wire W stretched along the middle. The wire is well insulated. The tube is evacuated through tube T and filled with suitable gas at pressure of 2 cm to 10 cm of Hg. The common gases, which are normally used, are hydrogen, argon and their mixture. The central wire is kept at high potential of the order of few kV. This applied voltage is critical and has to be fixed by studying the ionization current versus applied voltage characteristics of the tube.

The counting circuits are considerably improved today, and it is common now to get the number of particles entering directly on digital meters. The conventional circuit is a simple one-stage amplifier circuit with a phone in the output. The output of the G.M. tube is fed to input of this amplifier through a very high resistance R_C ($>10^9$ Ω) and a capacitor C.

6.19.3 Working

The central wire is kept at a suitable high potential depending upon the particle to be detected. As soon as a particle enters the G.M. tube, it ionizes. The gas and a pulse current is generated, which increases with the passage of particle through the tube. This momentary current is amplified by the amplifier and produces an audible '*click*' in the phone. Thus passage of each particle through the counter can be counted by number of clicks in phone connected with the output. The resolving power of such circuit is always small because of its large time constant. This sets upper limit of counting rate.

6.19.4 Plateau Characteristics

The variations of pulse height as a function of applied voltage is shown as plateau curve of G.M. counter in Figure 6.25. If the height of pulse is less than that corresponding to the threshold point A, the pulses are bypassed by electronic discriminator circuit of the G.M. counter.

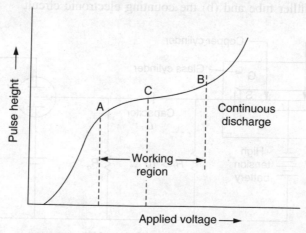

Figure 6.25 Plateau curve of a G.M. counter.

The discrimination level of the G.M. counting circuit is adjusted in such a way that it counts only those pulse whose height is above the threshold limit A of the plateau curve AB. The G.M. counter is considered superior if its plateau with smaller slope as compared to halogen quenched tube. These G.M. tubes are operated at a voltage corresponding to the middle point C of the plateau.

6.20   SCINTILLATION COUNTER

The scintillation method is oldest for the detection of particle and has been used in many of the classical researches by Rutherford and others. This method has been highly improved by using photomultiplier tube and counting circuits. It is one of the best methods for detecting the energetic particle.

6.20.1   Principle

When a charged energetic particle hits the screen coated with certain materials (popularly known as phosphors), it produces a light pulse of extremely short ($\cong 10^{-9}$ s) duration. When large number of particles are hitting the phosphor one by one in rapid succession, they produce light pulses with same time gap. These light pulses, when allowed to hit the cathode of a photomultiplier tube, are converted into electric pulses of sufficient height with similar time gap. The electric pulses can be suitable amplified and counted by electronic count circuit.

6.20.2   Construction and Working

A typical scintillation counter has three parts:

1.   Properly mounted fluorescent material
2.   Photomultiplier tube
3.   Electronic amplifier and counting circuits

The fluorescent material used depends on the type of particle to be detected. The main requirements of typical phosphor are (a) transparency to photon it produces, (b) the number of photons or pulse height produced and (c) linearity (as far as possible) of pulse intensity with the energy of incident particles These requirements are met in number of fluorescent materials. The common solid state materials are zinc sulfide, thallium activated sodium iodide, anthracene and plastics. Some liquids and gases (xenon) are also used for this purpose. The fluorescent material is put in a box with inner reflecting surface and transparent window. The photomultiplier tube has quick secondary emitter *dynodes* which are given suitable voltage with a stabilised power supply. The pulses generated are fed to counting circuit through a capacitor C. The electronic counting circuits have quick response with digital indicator [Figure 6.26].

Incident particle beam hits the aluminium box having inner reflecting surface and contains phosphor P. Sequential light pulses are generated and are allowed to fall on cathode of photomultiplier tube after passing through a window W. Photoelectrons are emitted and are drawn towards dynode D. Through secondary emission, they are multiplied and finally collected by collector plate. The signal generated is fed to counting circuit which often has a preamplifier through capacitor C. The number of pulses counted per second gives information about the number of particles.

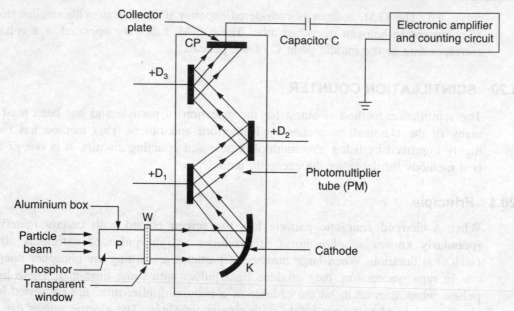

Figure 6.26 Schematic diagram of scintillation counter.

SOLVED NUMERICAL PROBLEMS

PROBLEM 6.1 Assuming the nuclear radius R for a atomic nucleus of mass number A, to be given by $R = R_0 A^{1/3}$, where $R_0 = 1.2 \times 10^{-15}$ m, calculate the density of nuclear matter in kg m^{-3}. Also determine the number of nucleons per unit volume, mass of one nucleon = 1.66×10^{-27} kg.

Solution The density ρ of nuclear matter is:

$$\rho = \frac{\text{Mass of nucleus}}{\text{Nuclear volume}} = \frac{A \times 1.66 \times 10^{-27}}{\frac{4}{3}\pi(1.2 \times 10^{-15} A^{1/3})^3}$$

$$= \frac{A \times 1.66 \times 10^{-27}}{\frac{4}{3} \times \frac{22}{7} \times (1.2)^3 \times 10^{-45} \times A} = 2.292 \times 10^{17} \text{ kg m}^{-3}$$

$$\text{Number of nucleons/m}^3 = \frac{\text{Density}}{\text{Mass of nucleon}} = \frac{2.292 \times 10^{17}}{1.66 \times 10^{-27}}$$

$$= 1.38 \times 10^{44} \text{ nucleons/m}^3$$

PROBLEM 6.2 The half-life period of radium is 1590 years. In how many years will one gram of pure element, (a) lose one centigram and (b) reduce to one centigram.

Solution Given $T = 1590$ years

We know half-life time of particle is given as:

$$T = \frac{0.693}{\lambda}$$

and

$$\lambda = \frac{1}{t} \log_e \frac{N_0}{N}$$

(a) When 1 g substance loses 1 cg, we have

$$N_0 = 1 \text{ g and } N = 1 - 0.01 = 0.99 \text{ g}$$

∴

$$\lambda = \frac{1}{t} \log_e \frac{1}{0.99}$$

where

$$\lambda = \frac{0.693}{t} = \frac{0.693}{1590} \text{ year}^{-1}$$

Then $t = \dfrac{1590}{0.693} \times 2.303 \times \log_{10} \dfrac{1}{0.99} = 23.25$ years.

(b) When 1 g of substance is reduced to one cg, we have

$$N_0 = 1 \text{ g and } N = 0.01 \text{ g}$$

Then $t = \dfrac{1590}{0.693} \times 2.303 \times \log_{10} \dfrac{1}{0.01} = 10564.5$ years.

PROBLEM 6.3 The half life of a radioactive substance is 15 hr. Calculate the period in which 12.5% of the initial quantity of the substance will be left over.

Solution Let t be the time period in which 87.5% of the initial quantity of the radioactive substance is decayed or 12.5% is left over. Now,

Half life $T = 15$ hr

Disintegration constant $\lambda = 0.693/15 = 0.0462$ per hr. If N_0 be the number of atoms at $t = 0$, then the number N at time t is given by radioactive law:

$$N = N_0 e^{-\lambda t}$$

Here

$$\frac{N}{N_0} = 12.5\% = \frac{12.5}{100} = \frac{1}{8}$$

∴

$$\frac{1}{8} = e^{-\lambda t}$$

or

$$e^{\lambda t} = 8$$

or

$$\lambda t = \log_e 8 = 2.303 \log_{10} 8$$

or

$$t = \frac{2.303 \log_{10} 8}{\lambda} = \frac{2.303 \times 0.9031}{0.0462} = 45 \text{ hr}$$

PROBLEM 6.4 The cross-section of ^{113}Cd for capturing thermal neutrons is 2×10^4 barn. The mean atomic mass of natural Cd is 112 a.m.u. and its density is 8.64×10^3 kg/m^3.

(a) What fraction of an incident beam of thermal neutrons is absorbed by a Cd sheet of 0.1 mm thick?

(b) What thickness of Cd is needed to absorb 99% of an incident beam of thermal neutrons?

Solution   Given $A = 112$ a.m.u., $\rho = 8.64 \times 10^3$ kg/m^3

$$\sigma = 2 \times 10^4 \text{ barn} = 2 \times 10^4 \times 10^{-28} = 2 \times 10^{-24} \text{ m}^2$$

As ^{113}Cd contains 12% of natural Cd, so the number of ^{113}Cd atoms/m^3 is:

$$N = \frac{0.12 \times 8.64 \times 10^3}{112 \times 1.67 \times 10^{-27}} = 5.58 \times 10^{27} \text{ atom/m}^3$$

$\therefore$

$$N\sigma = 5.58 \times 10^{27} \text{ atoms/m}^3 \times 2 \times 10^{-24} \text{ m}^2$$

$$= 1.12 \times 10^4 \text{ m}^{-1}$$

So the fraction on incident neutrons that is absorbed

$$= \frac{n_0 - n}{n_0} = \frac{n_0 - n_0 e^{-N\sigma x}}{n_0} = (1 - e^{-N\sigma x})$$

(a) Now $x = 0.1$ mm $= 10^{-4}$ m

$\therefore$

$$= \frac{n_0 - n}{n_0} = 1 - e^{(-1.12 \times 10^4 \times 10^{-4})} = 0.67 = \frac{2}{3}$$

which means that 67% of the incident neutrons are absorbed.

(b) In this 1% of incident pass through

$$\frac{n}{n_0} = \frac{1}{100}$$

But

$$\frac{n}{n_0} = e^{-N\sigma x}$$

$\therefore$

$$\frac{1}{100} = e^{-N\sigma x}$$

or

$$x = \frac{\log_e 0.01}{1.12 \times 10^4} = 4.1 \times 10^{-4} \text{ m} = 0.41 \text{ mm}$$

PROBLEM 6.5   Nuclear reaction is given by $^{10}_5\text{B} + ^4_2\text{He} \rightarrow ^{13}_6\text{C} + ^1_1\text{H}$. Given $^{10}_5\text{B} = 10.016125$ a.m.u., $^4_2\text{He} = 4.003874$ a.m.u., $^{13}_6\text{C} = 13.007440$ a.m.u. and $^1_1\text{H} = 1.008146$ a.m.u. Compute the energy released.

Solution $Q = [(10.016125 + 4.003874) - (13.007440 + 1.008146)]$
$$\times 931 \text{ MeV} = 4.11 \text{ MeV}$$

PROBLEM 6.6   A neutron breaks into a proton and an electron. Calculate the energy produced in this reaction in MeV. Given mass of an electron $= 9.0 \times 10^{-31}$ kg, mass of proton $= 1.6747 \times 10^{-27}$ kg, mass of neutron $= 1.675 \times 10^{-27}$ kg and speed of light $= 3 \times 10^8$ m/s.

Solution Mass defect Δm = mass of neutron – mass of (proton + electron)

$$= 1.675 \times 10^{-27}\,\text{kg} - (1.6747 \times 10^{-27} + 9.1 \times 10^{-31})\,\text{kg}$$

$$= 0.0013 \times 10^{-27}\,\text{kg}$$

Energy produced $E = \Delta m c^2$

$$= 0.0013 \times 10^{-27} \times (3 \times 10^8)^2\,\text{J} = \frac{0.0013 \times 10^{-27} \times 9 \times 10^{16}}{1.6 \times 10^{-13}}\,\text{MeV}$$

$$= 0.73\,\text{MeV}$$

PROBLEM 6.7 A reactor is developing nuclear energy at a rate of 32,000 kW. How many atoms of ^{235}U undergo fission per second? How many kg of ^{235}U would be used up in 1000 hours of operation? Assume an average energy of 200 MeV released per fusion. Take Avogadro's number as 6.023×10^{23} and 1 MeV = 1.6×10^{-13} J.

Solution The power developed by the reactor is 32000 kW, i.e., 3.2×10^7 W. Therefore, the energy released by the reactor = 3.2×10^7 J   [∵ 1 W = 1 J/S]

$$= \frac{3.2 \times 10^7}{1.6 \times 10^{-13}} = 2 \times 10^{20}\,\text{MeV}\qquad [\because 1.6 \times 10^{-13}\,\text{J} = 1\,\text{MeV}]$$

The energy released per fission is 200 MeV. Therefore, the number of fission occurring

in the reactor per second $= \dfrac{2 \times 10^{20}}{200} = 10^{18}$

The number of atoms (or nuclei) of ^{235}U consumed in 1000 hours

$$= 1018 \times 1000 \times 3600 = 36 \times 10^{23}$$

Now, 1 g atom (i.e., 235 g) of ^{235}U has 6.03×10^{23} atoms. Therefore, the mass of ^{235}U consumed in 1000 hrs

$$= \frac{36 \times 10^{23} \times 235}{6.03 \times 10^{23}}\,\text{g} = 1410\,\text{g} = 1.41\,\text{kg}$$

PROBLEM 6.8 When a hydrogen bomb explodes, a thermonuclear for the formation of $^4_2$He occurs from deuterium ($^2_1$H) and tritium ($^3_1$H). (a) Write the equation of the nuclear reaction. (b) Find the energy per reaction liberated.

Solution Given: $^2_1$H = 2.01474 a.m.u., $^3_1$H = 3.01700 a.m.u., and $^1_0$n = 1.008986 a.m.u., $^4_2$He = 4.003880 a.m.u. and 1 a.m.u. = 931 MeV

(a) The thermonuclear reaction in hydrogen bomb explosion is:

$$^2_1\text{H} + {}^3_1\text{H} \rightarrow {}^4_2\text{He} + {}^1_0\text{n}$$

(b) Q = Atomic mass ($^2_1$H + $^3_1$H) $\approx$ Atomic mass ($^4_2$He + $^1_0$n)

$$= (2.01474 + 3.01700)\,\text{a.m.u.} - (4.003880 + 1.008986)\,\text{a.m.u.}$$

$$= (0.018874\,\text{a.m.u.}) \times (931\,\text{MeV}) = 17.575\,\text{MeV}$$

PROBLEM 6.9 One g of ^{226}Ra has an activity of 1 curie. Determine the half life of radium. Avogadro's number is 6.02×10^{23}.

Solution 1 g atom of radium has a mass of 226 g, and contains 6.02×10^{23} atoms. Therefore, the number of atoms in 1 g of radium is:

$$N = \frac{6.02 \times 10^{23}}{226} = 2.66 \times 10^{21}$$

Now, the activity R of a radioactive sample is the rate of disintegration of its atoms:

$$R = -\frac{dN}{dt} = \lambda N$$

where λ is the disintegration constant.

Here $R = 1$ curie $= 3.7 \times 10^{10}$ disintegrations/second and $N = 2.66 \times 10^{21}$.

$$\lambda = \frac{R}{N} = \frac{3.7 \times 10^{10}}{2.26 \times 10^{21}} = 1.637 \times 10^{-11} \text{ per second}$$

and   Half life $T = \dfrac{0.693}{\lambda} = \dfrac{0.693 \times 2.26 \times 10^{21}}{3.7 \times 10^{10}}$

$$= 1.8729 \times 10^{10} \text{ s}$$

PROBLEM 6.10 A certain radioactive element has a half life of 20 days. How long will take for 3/4 of atoms originally present to disintegrate? What are disintegration constant and the average life of this element?

Solution Since the half life $T = 20$ days

$\therefore$ Decay constant or disintegration constant $\lambda = \dfrac{0.693}{t} = \dfrac{0.693}{20} \text{ day}^{-1}$

Now, $$\log_e \frac{N}{N_0} = \lambda t$$

$$T = 20 \text{ days}$$

$$\frac{N}{N_0} = \frac{3}{4}$$

$\therefore$ $$2.303 \log_{10} \frac{N}{N_0} = -\lambda t$$

$$2.303 \log_{10} \frac{3}{4} = -\frac{0.693}{20} t$$

$\therefore$ $$t = \frac{20 \times 2.303}{0.693} \log_{10} \frac{3}{4}$$

Now, $\qquad$ Average time $T_a = \dfrac{1}{\lambda} = \dfrac{T}{0.693} = 1.44T$

$$= 1.44 \times 20 = 28.8 \text{ days}$$

/// EXERCISES ///

THEORETICAL QUESTIONS

6.1 What is nucleus? Discuss its important properties.

6.2 Describe important characteristics of nuclear force.

6.3 What is radioactivity? Obtain radioactive decay law and hence decay constant, half life and mean life for radioactive nuclei.

6.4 Giving a schematic diagram, explain the working of a nuclear reactor.

6.5 What are the distinctive features between nuclear fission and fusion? Briefly describe the working of nuclear reactor and give its applications.

6.6 Explain the construction and working of a G.M. counter. Give its important applications. How quenching is achieved in this counter.

6.7 What are the advantages of proportionate counter over G.M. counter.

6.8 Distinguish between fission and fusion.

6.9 Explain how γ-rays interact with matter before detection.

6.10 Explain neutron cross-section and give its unit.

6.11 What is natural radioactivity? Establish the law of radioactive decay. Explain the terms 'decay constant' and 'half life' of a radioactive substance.

6.12 What is natural radioactivity? Establish relation $N = N_0 e^{-\lambda t}$, where the symbols have their usual meanings.

6.13 State the law of radioactive disintegration. Show how the average life of the radioactive substance changes with time.

6.14 Derive an expression for the mean life of a radioactive element in terms of decay constant.

6.15 Write short notes on:

6.16 (i) Isotopes and isobars, (ii) stable and unstable nuclei, (iii) nuclear constituents, (iv) mass defect, (v) binding energy of nucleus, (vi) liquid drop model and (vii) shell model.

NUMERICAL PROBLEMS

P6.1 Find the half-life time and mean life time of a radioactive substance of which the decay constant is 4.28×10^{-4} per year. $\qquad$ (**Ans.** $T = 1619$ years, $T_a = 2336$ years)

P6.2 10 g of a radioactive substance is reduced by 2.5 mg in 6 years through α-decay. Evaluate half-life time and mean life time of the substance.

$\qquad\qquad\qquad$ (**Ans.** $T = 1.8 \times 10^4$ years, $T_a = 2.6 \times 10^4$ years)

P6.3 Complete the following nuclear reactions:

(i) $^{135}_{17}C + ? \rightarrow ^{32}_{16}S + ^{4}_{2}He$

(ii) $^{10}_{5}B + ? \rightarrow ^{7}_{3}Li + ^{4}_{2}He$

(iii) $^{27}_{13}Al + ^{1}_{0}n \rightarrow ^{27}_{12}Mg + ?$

(iv) $^{6}_{3}Li + ? \rightarrow ^{7}_{4}Be + ^{1}_{0}n$

(v) $^{7}_{3}Li + proton \rightarrow \alpha - particle + ?$

(Ans. (i) $^{1}_{1}H$ (ii) $^{1}_{0}n$ (iii) $^{1}_{1}H$ (iv) $^{2}_{1}H$ (v) $^{3}_{1}H$ (tritium))

P6.4 Calculate the minimum energy of a photon that will break the nucleide $^{7}_{3}Li$ into an α-particle and a tritron. The atomic masses are: $^{7}_{3}Li = 7.016004$ a.m.u., $^{4}_{2}He = 4.002603$ a.m.u., $^{7}_{1}H = 0.016050$ a.m.u. and 1 a.m.u. = 931 MeV. **(Ans.** 2.47 MeV)

P6.5 A nucleus with $A = 235$ splits into two nuclei whose mass numbers are in the ratio 2 : 1. Find the radii of the nuclei. (Take $R_0 = 1.4$ fm).

(Ans. For $A = 157$, $R = 7.5$ fm and for $A = 78$, $R = 6$ fm)

P6.6 Calculate the minimum energy of gamma rays necessary to disintegrate a deuteron into a proton and a neutron. **(Ans.** 2.23 MeV)

P6.7 Energy released in the fusion of a single uranium nucleus is 200 MeV. Calculate the number of fissions per second to produce milliwatt power. **(Ans.** 3.25×10^7)

P6.8 Show that the radius of $^{238}_{92}U$ is only 6 times that of $^{1}_{1}H$.

P6.9 How much energy in kWH is released by the total fission of 1 g of ^{235}U?

(Ans. 2.275×10^7 kWH)

P6.10 A radioactive source contains 1 mg of ^{239}Pu. This source is estimated to emit 2200 α-particles per second in all directions. Estimate half life of plutonium.

(Ans. 1.503×10^6 years)

MULTIPLE CHOICE QUESTIONS

MCQ 6.1 Isobars are nuclides which have
(a) Same A but different Z
(b) Same A and Z
(c) Same Z but different A
(d) Same Z and same N

MCQ 6.2 End A of a metallic wire is irradiated with α-rays and end B is irradiated with β-rays, then
(a) Current will flow from end B to end A
(b) A current will flow from end A to end B
(c) There will be no current in the wire
(d) None of the above statement is correct

MCQ 6.3 ^{210}Bi has half life of 5 days. The time taken for 1/8th of a sample to decay is
(a) 20 days
(b) 15 days
(c) 10 days
(d) 5 days

MCQ 6.4 Neutron–proton capture is inverse process of
(a) Pair production
(b) Photoelectric effect
(c) Photo-disintegration
(d) None of the above

MCQ 6.5 Shell model of the nucleus is based upon
(a) Spherically symmetric potential
(b) Ellipsoidal symmetric potential
(c) Bohr's correspondence principle
(d) None of the above

MCQ 6.6 1 barn is equal to
    (a) 10^{-14} m^2 (b) 10^{-19} m^2
    (c) 10^{-28} m^2 (d) 10^{-34} m^2

MCQ 6.7 The control rod in nuclear reactor is made of
    (a) Cadmium   (b) Uranium
    (c) Graphite   (d) Plutonium

MCQ 6.8 Which of the following is a fusion reaction?
    (a) $_0^1\text{n} + _7^{14}\text{N} \rightarrow _6^{14}\text{C} + _1^1\text{H}$ (b) $_1^2\text{H} + _1^2\text{H} \rightarrow _2^4\text{He}$
    (c) $_0^1\text{n} + _{92}^{238}\text{U} \rightarrow _{93}^{239}\text{N} + _{-1}\beta + r$ (d) $_1^3\text{H} \rightarrow _2^4\text{He} + _{-1}\beta + \gamma$

MCQ 6.9 In nuclear fission, the percentage of mass energy is about
    (a) 1%   (b) 0.01%
    (c) 0.1%   (d) 1%

MCQ 6.10 Neutron is a particle which is
    (a) With charge and spin both   (b) With charge but without spin
    (c) Without charge and without spin   (d) Without charge and with spin

Answers

6.1 (a)	6.2 (b)	6.3 (b)	6.4 (c)	6.5 (a)	6.6 (c)	6.7 (a)
6.8 (b)	6.9 (c)	6.10 (d)				

Solid State Physics

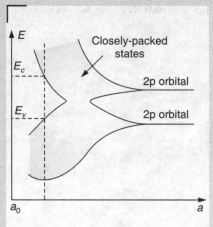

7.1 INTRODUCTION

In recent years, solid state physics has seen many vivid improvements and has developed into one of the major autonomous branches of physics. It has simultaneously stretched out into a lot of new areas, playing a crucial position in fields that were once the sphere of the materials science and engineering. An outcome of this volatile improvement is that no single university lecturer can today be expected to have a thorough facts of all aspects of this huge subject matter likewise, it is impracticable to visualise that could offer students a comprehensive understanding of the entire discipline and its numerous applications. In this chapter of solid state physics, we will study the foundation topics of this division of physics like free electron theory of metals, band theory of solids, semiconductors, photoconductivity and photovoltaics.

7.2 CLASSICAL FREE ELECTRON THEORY (DRUDE–LORENTZ THEORY)

The free electron model explains a number of important physical properties of metals. All metals are good conductors of heat and electricity and obey Ohm's law because of valence electrons. These valence electrons are actually responsible for physical properties of metal such as electrical and thermal conductivities, thermoelectricity, thermoionic and photoelectric effects.

Consider a conductor as a regular array of atoms and a collection of free electrons. The free (conduction) electrons, although bound to their respective atoms, when the atoms are not part of a solid, gain mobility when the free atoms condense into a solid. That is, when a large number of atoms combine to form a metal, the boundaries of neighbouring atoms slightly overlap each other. Due to this overlapping, the valence electrons of all the atoms get detached from their respective atoms. Thus the valence electrons will be free to move within the metal lattice. These electrons are called *free electrons*. Free electrons move in random directions through the conductor with average speed of the order of 10^6 m/s. The situation is similar to the gas molecules confined in a vessel. In fact, according to some physicists, free electrons are responsible for electrical and thermal conductions in a solid. These free electrons are also called *conduction electrons*.

The loss of electrons makes the atom to loose their electrical neutrality, and they became positive ion fixed about their mean positions. Thus, according to Drude–Lorentz theory, all metals contain free electrons which are free to move through the positive ion core of the metals. The metal is held together by the electrostatic forces of attraction between positively charged ions and negatively charged electron gas. The mutual repulsion between negative ions and negatively charged electrons is ignored in this theory. Thus the electron can move from place to place in the metal, without any change in their energy and collide occasionally with the atoms. Thus free electrons are responsible for all the properties of metals.

7.2.1 Main Assumptions of Free Electron Theory

The main assumptions of classical free electron theory are listed as follows:

1. All metals contain a large number of free electrons which move freely through the positive ion core of the metal.

2. The free electrons are treated as electron gas whose particles have three degree of freedom. Hence the law of classical kinetic theory of gases can be applied to a free electron also. Thus the electrons can be assigned a mean free path λ, a mean collision time τ and an average speed $\bar{c}$. In absence of an electric field, the kinetic energy associated with an electron at a temperature T is given by $\frac{3}{2}k_B T$, where k_B is the Boltzmann constant.

$$\therefore \quad \frac{1}{2}mv_{th}^2 = \frac{3}{2}k_B T \tag{7.1}$$

where v_{th} is the thermal velocity. It is same as root mean square velocity.

3. The electric field due to positive ion cores is considered to be constant.
4. The repulsion between free electrons is considered to be negligible.
5. The electric current flow in a metal due to an externally applied electric field is a consequence of the drift velocity of the electrons in a direction opposite to the direction of the field.

7.2.2 Important Parameters

Drift Velocity

The average velocity with which free electrons move in a steady state opposite to the direction of the electric field in a metal is called *drift velocity*.

Under thermal equilibrium condition, the valence electrons in a solid are in state of random motion. The kinetic energy of an electron at a temperature T is given by

$$\frac{1}{2}mv_{th}^2 = \frac{3}{2}k_B T$$

where v_{th} is the mean thermal velocity and k_B is the Boltzmann constant. The random velocity of the electrons is called *thermal velocity*. In the absence of electric field, the mean thermal velocity of electron at 27 °C is of the order of 10^6 m/s. The electrons move rapidly in all directions in a chaotic manner, as illustrated in Figure 7.1(a).

In the absence of applied field, the velocity of electron, i.e., net current moving across any given plane at any instant is zero, owing to random nature of motion as shown in Figure 7.1(b). If a constant electric field is applied, the electron will experience a force eE and get accelerated. An electric field E modifies the random motion and causes the electrons to drift in a direction opposite to that of E as shown in Figure 7.1(c).

The electrons get accelerated and gain kinetic energy. They collide with atoms (ions) of the metal. During collision electrons give their energy to the atoms. Hence their velocities decrease, but again they are accelerated. There is collision once again and the increased velocity is reduced to zero. The electrons acquire a constant average velocity v_d. The drift motion is directional and causes current flow, called *drift current*.

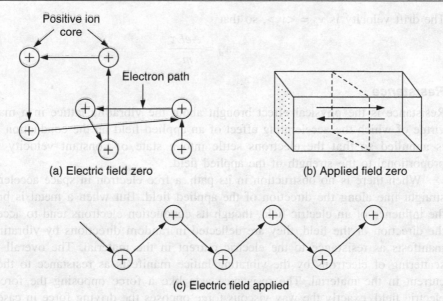

Figure 7.1 Electron random motion.

Expression for drift velocity: If a constant electric field E is applied to the crystal, the electron of mass m_e and charge e will experience a force $F = -eE$. These results drift velocity v_d. Since this is the only force of a free electron between collisions, Newton's second law, $\Sigma F = m_e a$, gives the acceleration between collisions as:

$$a = \frac{-eE}{m_e}$$

If we align x-axis along E, then the x-component of a free electron's velocity at a time t after collision is:

$$v_x = v_{x0} + a_x t = v_{x0} - \frac{eE}{m_e}t \tag{7.2}$$

where v_{x0} is the x-component of electron's velocity, immediately after the collision. On an average, we have

$$<v_x> = <v_{x0}> - \frac{eE}{m_e}\tau \tag{7.3}$$

where τ is the average time internal between successive collisions and is called the *mean free time* or *relaxation time*.

For a rather sizeable current, the drift velocity v_d is only about 10^{-4} m/s, whereas the average speed $<v>$ of the free electron is about 10^6 m/s. Since v is a factor of about 10^{10} larger than v_d, the contribution to the motion of free electron is negligible. Hence $<v_{x0}> = 0$ Thus

$$<v_x> = \frac{-eE\tau}{m_e}$$

The drift velocity is $v_d = \langle v_x \rangle$, so that

$$v_d = \frac{-eE\tau}{m_e} \qquad (7.4)$$

Resistance

Resistance is the physical effect brought about the vibrating lattice in a material by virtue of which the accelerating effect of an applied field on the conduction electrons is annulled so that the electrons settle into a state of constant velocity which is proportional to the strength of the applied field.

When there is no obstruction in its path, a free electron in space accelerates in a straight line along the direction of the applied field. But when a metal is brought in the influence of an electric field, though its conduction electrons tend to accelerate in the direction of the field, they are deflected in random directions by vibrating lattice manifests as resistance to the electric current in the material. The overall effect of scattering of electrons by the vibrating lattice manifests as resistance to the electric current in the material. The resistance acts like a force opposing the force due to electric field, exactly the way viscous force opposes the driving force in case of fluid and cancels the acceleration effect. In effect the conduction electrons settle into a state of constant velocity which is the same as drift velocity.

Relaxation Time or Mean Free Life Time (τ_r)

In a metal, due to randomness in the direction of motion of the conduction electrons, the probability of finding one among them moving in any given direction is equal to finding another one moving in exactly the opposite direction, in absence of an electric field. As a result, the average velocity of the electrons in any given direction becomes zero, i.e.,

$$v_{av} = 0 \qquad \text{[in absence of the field]}$$

However, when the metal is subjected to an external electric field, there will be a net positive value v'_{av} for average velocity of the conduction electrons in the direction of applied field, due to the drift velocity,

i.e., $\qquad\qquad\qquad\qquad\qquad v_{av} = v'_{av} \qquad \text{[In presence of the field]}$

If the field is turned off suddenly, the average velocity reduces exponentially to zero from a value v'_{av}, which is the value of v_{av} at the instant the field is just turned off as illustrated in Figure 7.2.

The decay process is represented by the equation:

$$v_{av} = v'_{av}\, e^{-t/\tau_r} \qquad (7.5)$$

where t is the time counted at the instant the field is turned off and τ_r, a constant is called the *relaxation time*.

In Eq. (7.5), if $t = \tau_r$, then

$$v_{av} = v'_{av}\, e^{-1}$$

or $\qquad\qquad\qquad\qquad\qquad v_{av} = \frac{1}{e}\, v'_{av} \qquad (7.6)$

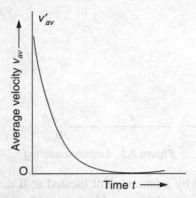

Figure 7.2 Decay of average velocity.

Due to the sudden disappearance of an electric field across a metal, the average velocity of its conduction electrons reduces exponentially to zero, and the time required in this process for the average velocity to reduce $1/e$ times of its value at the time, when the field is just turn off, is known as *relaxation time*.

Mean Free Path (λ)

Mean free path, according to kinetic theory of gases, is the distance travelled by a gas molecule in between two successive collisions. In the classical theory of free electron model, it is taken as average distance travelled by the conduction electrons between successive collisions with the lattice ions.

Mean Collision Time (τc)

The average time that elapses between the consecutive collisions of an electron with the lattice point (the averaging is over a large number of collisions) is called *collision time*.

If v is the total velocity of the electrons, which is the velocity due to combined effect of thermal and drift velocities, then the mean collision time is given by

$$\tau_c = \frac{\lambda}{v} \tag{7.7}$$

If v_d is the drift velocity, then $v_d \ll v_{th}$.

$$\therefore \quad v \approx v_{th}$$

$$\therefore \quad \tau_c = \frac{\lambda}{v_{th}} \tag{7.8}$$

In case of metals, it can be shown that the relaxation time τ always refers to a single total velocity called *Fermi velocity*.

Relation between τr and τc

Consider a conduction electron travelling inside a metal in direction AB with a velocity as shown in Figure 7.3.

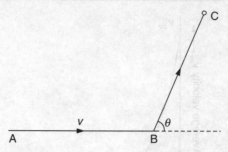

Figure 7.3 Lattice scattering.

Let it be scattered by a lattice point located at B due to which the electron travels along BC with the same velocity making an angle θ with the original direction. θ is called *scattering angle*.

The value of θ is likely to be different for each case of collision between an electron and the lattice. If the average value of cos θ is taken over a very large number of such collisions and the value is represented by $<\cos \theta>$, then the relaxation time τ_r and the mean collision time τ_c are related through the equation:

$$\tau_r = \frac{\tau_c}{1 - <\cos \theta>} \tag{7.9}$$

If the probability of scattering is same in all direction, then the scattering is called *isotropic*, for each case $<\cos \theta> = 0$ and $\tau_r = \tau_c$ i.e., the mean collision time can be treated as relaxation time itself.

Current Density (J)

It is the current per unit area of cross-section of an imaginary plane held normal to the direction of current. In a current carrying conductor, if I is the current, and A is area of cross-section, then the current density **J** is given by

$$J = \frac{I}{A} = \frac{ne^2 \tau_r}{m_e} E \tag{7.10}$$

Electric Field (E)

The potential drop per unit length of the conductor gives the electric field that exists across a homogeneous conductor,

If L is the length of a conductor of uniform cross-section and of uniform material composition and V is the potential difference between its two ends, then the electric field E, at any point inside it is given by

$$E = \frac{V}{L} \tag{7.11}$$

Electrical Conductivity (σ)

It is a physical property that characterises the conducting ability of a material.

If R is the electrical resistance of uniform material of length L and area of cross-section A, then the electrical conductivity is given by

$$\sigma = \frac{1}{R}\left(\frac{L}{A}\right) \qquad (7.12)$$

$$\sigma = \frac{ne^2\tau_r}{m_e} \qquad (7.13)$$

Now, if we consider the product σE, then from Eqs. (7.11) and (7.12), we have

$$\sigma E = \frac{V}{L} \times \frac{1}{R}\left(\frac{L}{A}\right) = \frac{V}{RA} \qquad \left[\because I = \frac{V}{R}\right]$$

or

$$\sigma E = \frac{I}{A} \qquad (7.14)$$

Comparing Eqs. (7.10) and (7.14), we have

$$J = \sigma E \qquad (7.15)$$

Expression (7.15) represents *Ohm's Law*.

Electrical Resistivity (ρ)

Resistivity signifies the resistance property of the material and is given by inverse of conductivity:

$$\rho = \frac{1}{\sigma} \qquad (7.16)$$

or

$$\rho = \frac{m_e}{ne^2\tau_r} \qquad (7.17)$$

7.2.3 Drude's Theory of Conduction

Consider a metallic rod AB with its end A at a higher temperature than its end B. As end A is at higher temperature than end B, the electrons near end A possess higher energy than the electron near end B. This is a state of unstable equilibrium. The electrons near end A lose their kinetic energy, while the electrons near end B gain energy. As a result, we observe the heat (a form of energy) flows from end A to end B.

The amount of heat Q passing through any given cross-section of the rod per unit area per unit time is given by

$$Q = \frac{1}{3}nv_{th}\lambda\frac{dE}{dl} \qquad (7.18)$$

where $\dfrac{dE}{dl}$ is the rate of change of electron energy along the length of the bar.

From the kinetic theory of gases, the energy of an electron at temperature T, is given by

$$E = \frac{3}{2}k_B T \qquad (7.19)$$

$$\Rightarrow \qquad \frac{dE}{dl} = \frac{3}{2} k_B T \frac{dT}{dl}$$

So that Eq. (7.18) gives

$$Q = \frac{1}{3} n v_{th} \frac{3}{2} k_B T \frac{dT}{dl}$$

$$\Rightarrow \qquad Q = \frac{1}{2} n k_B T v_{th} \lambda \frac{dT}{dl} \qquad (7.20)$$

Also, we know that

$$Q = K \frac{dT}{dl} \qquad (7.21)$$

A comparison of Eqs. (7.20) and (7.21) gives

$$K = \frac{1}{2} n k_B T n v_{th} \lambda \qquad (7.22)$$

The value of thermal conductivity K is given by Eq. (7.22) is found to be in close agreement with the value that is obtained experimentally. The free electron theory thus successfully explains the thermal conductivity of metals.

7.2.4 Wiedemann–Franz Law

It states that the ratio of thermal conductivity to electrical conductivity of any metal is directly proportional to the absolute temperature of the metal, i.e.,

$$\frac{\text{Thermal conductivity } K}{\text{Electrical conductivity } \sigma} \propto T$$

or $\qquad \dfrac{K}{\sigma} = LT \qquad$ or $\qquad \dfrac{K}{\sigma T} = L = \text{Constant}$

where L is Lorentz's number.

Proof: We know the thermal conductivity K is [as given in Eq. (7.22)]:

$$K = \frac{1}{2} n k_B T v_{th} \lambda \qquad (7.23)$$

The electrical resistivity is:

$$\sigma = \frac{ne^2}{m_e} \tau_r$$

The relaxation time τ_r is given by

$$\tau_r = \frac{\lambda}{v_{th}} \qquad (7.24)$$

$$\therefore \qquad \sigma = \frac{ne^2}{m_e} \left(\frac{\lambda}{v_{th}} \right) = \frac{ne^2 \lambda}{m_e v_{th}} \qquad (7.25)$$

Also
$$\frac{1}{2}mv_{th}^2 = \frac{3}{2}k_B T$$

or
$$mv_{th} = \frac{3}{2}\frac{k_B T}{v_{th}}$$

Substituting this value in Eq. (7.25), we have

$$\sigma = \frac{ne^2 \lambda v_{th}}{3k_B T} \qquad (7.26)$$

From Eqs. (7.23) and (7.26), we get

$$\frac{K}{\sigma} = \frac{\frac{1}{2} nk_B \lambda v_{th} k_B T}{ne^2 \lambda v_{th}}$$

or
$$\frac{K}{\sigma} = \frac{3}{2}\left(\frac{k_B}{e}\right)^2 T$$

So
$$\frac{K}{\sigma} = LT$$

or
$$\frac{K}{\sigma} \propto T \qquad (7.27)$$

Equation (7.27) is an expression for Wiedemann–Franz law.

Now, Lorentz number is given by

$$L = \frac{3}{2}\left(\frac{k_B}{e}\right)^2 = 1.2 \times 10^{-8} \text{ W}\Omega\text{K}^{-2}$$

EXAMPLE 7.1 Electrical resistivities of Cu and Ni at room temperature are 1.65×10^{-8} Ωm and 14×10^{-8} Ωm respectively. If the Weidemann–Franz law applies to these materials, find the electronic contribution to thermal conductivities of these materials.

Solution Given that $\rho_{Cu} = 1.65 \times 10^{-8}$ Ωm, $\rho_{Ni} = 14 \times 10^{-8}$ Ωm. Here, we take $L = 2.45 \times 10^{-8}$ WΩK^{-2}. Since, we know that Weidemann–Franz law is:

$$\frac{K}{\sigma} = LT$$

Thermal conductivity of copper,

$$K_{Cu} = \sigma_{Cu} LT = \frac{LT}{\rho_{Cu}} = \frac{2.45 \times 10^{-8} \times 300}{1.65 \times 10^{-8}} = 455 \text{ Wm}^{-1}\text{K}^{-1}$$

and thermal conductivity of nickel,

$$K_{Ni} = \frac{LT}{\rho_{Ni}} = \frac{2.45 \times 10^{-8} \times 300}{14 \times 10^{-8}} = 52.5 \text{ Wm}^{-1}\text{K}^{-1}$$

7.2.5   Failures of Classical Free Electron Theory

Though the classical free electron theory has been successful in a accounting for certain experimental facts such as electrical and thermal conductivities in metals, it failed to account for many other experimental facts among which the notable ones are, specific heat, mean free path, temperature dependence of electrical conductivity and the dependence of electrical conductivity on electron concentrates.

Specific Heat

According to classical free electron theory, when energy is supplied to a metal, all the valence electrons absorb the energy. The energy associated with each electron is $\frac{3}{2}k_BT$. A kilomole of a solid contains Avogadro's number N_A of atoms. Assuming that each atom contributes a valence electron, the total internal energy U_T at temperature T is given by

$$U_T = \frac{3}{2}N_A k_B T = \frac{3}{2}RT \tag{7.28}$$

Also the theory predicts that the specific heat does not depend on temperature, whereas it was found experimentally that the specific heat is proportional to temperature.

Mean Free Path (λ)

According to classical free electron theory, the resistivity of a metal is given by

$$\rho = \frac{m_e}{ne^2 T} \tag{7.29}$$

$$\therefore \qquad \text{Mean collision time } \tau = \frac{m_e}{ne^2 \rho} \tag{7.30}$$

Substituting the values of m_e, n, e and ρ for a typical metal, it was found that

$$\tau = 2.47 \times 10^{-14} \text{ s}$$

If λ be the mean free path between the collisions of free electron, then

$$\lambda = \bar{c}\tau \tag{7.31}$$

where $\bar{c}$ is average velocity of electron and is equal to 1.15×10^3 m/s. In this case,

$$\lambda = 1.15 \times 10^3 \times 2.47 \times 10^{-14} = 2.85 \text{ nm}$$

The experimental value of λ (= 0.285 nm) was found to be ten times less than this value. This discrepancy also could not be explained by the free electron theory.

Temperature Dependence of Electrical Conductivity

It was experimentally observed that for metals, the electrical conductivity σ is inversely proportional to the temperature T, i.e.,

$$\sigma \propto \frac{1}{T} \tag{7.32}$$

But according to classical free electron theory,

$$\frac{3}{2}k_BT = \frac{1}{2}mv_{th}^2$$

$$v_{th} = \sqrt{\frac{3k_BT}{m_e}}$$

$$v_{th} \propto \sqrt{T} \qquad (7.33)$$

Since the mean collision time τ is inversely proportional to the thermal velocity, we can write

$$\tau \propto \frac{1}{v_{th}}$$

$$\tau \propto \frac{1}{\sqrt{T}} \qquad (7.34)$$

But σ is given by

$$\sigma = \frac{ne^2\tau}{m_e}$$

The proportionality between σ and T can be represented as:

$$\sigma \propto \tau$$

or $$\sigma \propto \frac{1}{\sqrt{T}} \qquad \text{[from Eq. (7.34)]} \qquad (7.35)$$

Now, from Eqs. (7.32) and (7.35), it is clear that the prediction of classical free electron theory is not agreeing with the experimental observations.

Dependence of Electrical Conductivity On Electron Concentration

As per the classical free electron theory, the electrical conductivity σ is given by

$$\sigma = \frac{ne^2\tau_r}{m_e}$$

where n is the electron concentration.

$$\therefore \qquad \sigma \propto n \qquad (7.36)$$

If we consider the specific cases of zinc and cadmium, which are divalent metals, the electrical conductivities are respectively 1.69/ Ωm, which are much less than that for copper and silver, the values for which are 5.88/ Ωm and 4.50 / Ωm respectively. On the other hand, the electron concentrations for zinc and cadmium are $13.10 \times 10^{28}/m^3$ and $9.28 \times 10^{28}/m^3$, which are much higher than that for copper and silver, the values for which are $8.45 \times 10^{28}/m^3$ and $5.85 \times 10^{28}/m^3$ respectively. For a trivalent metal aluminium and gallium, the electron concentration values are $18.06 \times 10^{28}/m^3$ and $15.30 \times 10^{28}/m^3$; which are far higher than that for copper and silver, but the electrical conductivity values for them are 3.65/ Ωm and 0.67/ Ωm, which are far lower than that for copper and silver.

These examples imply that the predictions of the classical free electron theory that $\sigma \propto n$ does not hold good.

7.3 SOMMERFELD'S FREE ELECTRON THEORY OF METALS

In 1928, by applying quantum mechanical principle, Arnold Sommerfeld succeeded in overcoming many of drawbacks of the classical free electron theory, while retaining all the essential features of the classical free electron theory. This approach is based upon quantization of electrical energy levels. He realised the role played by Pauli exclusion principle in restricting the energy values of electron. The theory proposed is known as quantum free electron theory.

The main assumptions of quantum free electron theory are as follows:

1. The energy levels of the conduction electrons are quantized.
2. The distribution of electrons in various allowed energy levels occur as per Pauli exclusion principle.

   However, the following assumptions of classical free electron theory are also applicable to quantum free electron theory.
3. The electrons travel in a constant potential inside the metal but stay confined within the boundaries.
4. Both the attraction between the electrons and the lattice ions, and the repulsion between the electrons themselves are ignored.

7.3.1 Fermi Energy

Due to quantization rules, materials in solid state possess a set of allowed energy levels. For a metal containing N free electrons, there will be N such allowed energy levels, which are separated by energy differences that are characteristic of the material. The allowed energy values for conduction electron in the metal are also quantized, and are related to the energy levels of the metal. As per Pauli exclusion principle, each energy level can accommodate a maximum of two electrons, one with spin up and the other with spin down (Figure 7.4).

The allowed energy values of the electron are realised in terms of occupation of the allowed energy levels, under certain rules. When the filling up of the energy levels is undertaken, the universal rules that—any system, which is free, tends to go state of lowest energy—comes into picture.

Thus a pair of electrons, one with spin up, and other with spin down, occupies the lowest level. The next pair of electrons occupies the next higher level, and so on, till all the electrons in the metal are accommodated.

However, there will be many allowed energy levels available for occupation by the electrons. The energy of the highest occupied level at zero degree absolute temperature is Fermi energy, and the energy level is referred to as the Fermi level. The Fermi energy is denoted by E_F.

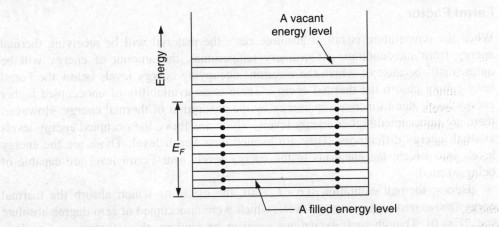

Figure 7.4 Fermi energy.

When there is no external energy supply for electrons, such as thermal energy or electrical energy, the electrons are free, and thus settle in the lowest allowed energy state available. Thus at a temperature of absolute zero, and when the metal is not in the influence of any external field, all the energy levels lying above the Fermi level are empty, and those lying below are completely filled. Since there are two electrons in each energy level, out of the N allowed energy levels, $N/2$ of them will be occupied.

7.3.2 Fermi Level

The Fermi level is an energy level between top surface of valence band and bottom surface of conduction band as shown in Figure 7.5.

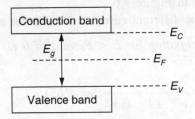

Figure 7.5 Fermi level.

If E_V is the energy corresponding to top surface of valence band, E_C is the energy corresponding to bottom surface of conduction band and E_F is the Fermi energy corresponding to Fermi level, then

$$E_F = \frac{E_C + E_V}{2} = E_C + \frac{E_V - E_C}{2}$$

$$= E_C + \frac{E_g}{2} \implies E_F - E_C = \frac{E_g}{2} \tag{7.37}$$

7.3.3 Fermi Factor

When the temperature is not at absolute zero, the material will be receiving thermal energy from surroundings. At ordinary temperature, the amount of energy will be quite small[*] because of which the electrons occupying energy levels below the Fermi level cannot absorb the thermal energy. There is no availability of unoccupied higher energy levels due to increase in energy by the absorption of thermal energy. However, there are unoccupied higher energy levels, which are above the occupied energy levels at small energy differences. They are located near Fermi level. Those are the energy levels into which, the electrons in the energy levels near Fermi level are capable of being excited.[†]

During thermal excitation (i.e., $T > 0$), the electrons which absorb the thermal energy move into higher energy levels, which were unoccupied at zero degree absolute (i.e., $T = 0$). Though such excitations seem to be random, the occupation of various energy levels obeys a statistical distribution[‡] is called Fermi–Dirac distribution, once the system is in thermal equilibrium (i.e., at a steady temperature state).

The probability $f(E)$ that a given energy state with energy E is occupied at a steady temperature T, is given by

$$f(E) = \frac{1}{e^{(E-E_F)/k_BT} - 1} = \frac{1}{\exp \cdot \dfrac{E - E_F}{k_BT} - 1} \tag{7.38}$$

$f(E)$ is called the *Fermi factor*. It is defined as follows:

Fermi factor is the distribution function, which gives the probability of occupation of a given energy state for a material in thermal equilibrium in terms of the Fermi energy, Boltzmann constant and temperature.

The dependence of Fermi factor on temperature and the effect on occupancy of energy level is shown in Figure 7.6.

Let us consider the different cases of distribution as follows:

1. *Probability of occupation for $E < E_F$ at $T = 0$ K*
   When $T = 0$ and $E > E_F$

$$f(E) = \frac{1}{e^{-\infty} + 1} = \frac{1}{0 + 1} \Rightarrow f(E) = 1 \qquad (\text{for } E < E_F)$$

At $T = 0$ K, all the energy levels below Fermi level are occupied.

---

[*]The quantity of thermal energy is given by k_BT, where k_B is Boltzmann constant. At room temperature, $k_BT = 0.025$ eV. Compare this with Fermi energy which is of the order of 5 eV.

[†]Since such electrons constitute only a small fraction of the total number of electrons in the metal as a whole, the specific heat of the material becomes low. Thus the reason for failure of classical free electron theory to account for low specific heat of metals is now understood.

[‡]The statistical distribution deals with the question as to what is the probability occupation of a particular energy level.

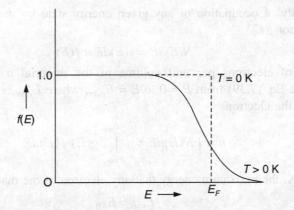

Figure 7.6 Variation of $f(E)$ with E.

2. *Probability of occupation for $E > E_F$ at $T = 0$ K*

When $T = 0$ and $E > E_F$,

$$f(E) = \frac{1}{e^{\infty} + 1} = \frac{1}{\infty} \Rightarrow f(E) = 0 \qquad \text{(for } E > E_F\text{)}$$

At $T = 0$ K, all the energy levels above Fermi level are unoccupied.

3. *Probability of occupation at ordinary temperature*

At ordinary temperatures, the value of probability starts reducing from 1 for values of E close to but lesser than E_F.

The value of $f(E)$ becomes 1/2 at $E = E_F$.

Since for $E = E_F$, $e^{(E-E_F)/k_B T} = e^0 = 1$

$$f(E) = \frac{1}{e^{(E-E_F)/k_B T} + 1} = \frac{1}{1+1} = \frac{1}{2}$$

Further, for $E > E_F$, the probability value falls off to zero rapidly (Figure 7.6).

The discussion on the variation of $f(E)$ with E, brings forth the point that the Fermi energy is the most probable or the average energy of the electron across which the energy transitions occur at temperature above absolute zero degree, which result in the conductivity of the metal. This may be considered as the physical basis for the concept of Fermi energy.

Calculation of Fermi Energy at $T = 0$ K

Let the value of E_F at 0 K be denoted by E_{F0}. Also, if we denote the number of electrons per unit volume, which possesses energy only in the range E and $(E + dE)$ by $N(E)dE$, then $N(E)dE$ is given by

Density of available states in the energy range E and $(E + dE)$ × Probability of occupation of those energy levels by the electrons

But, probability of occupation of any given energy state by the electron is given by the Fermi factor $f(E)$.

$$\therefore \qquad N(E)dE = g(E)dE \times f(E) \qquad (7.39)$$

The number of electrons per unit volume of the material n can be evaluated by integrating the Eq. (7.39) from $E = 0$ to $E = E_{max}$, where E_{max} is the maximum energy possessed by the electrons

$$n = \int N(E)dE = \int_{E=0}^{E=E_{max}} g(E)\, f(E)dE \qquad (7.40)$$

But at $T = 0$ K, the maximum energy that any electron of the material can have is E_{F0}.

Hence $\qquad\qquad\qquad E_{max} = E_{F0}$

Also $f(E) = 1$ at $T = 0$ K,

$$\therefore \qquad n = \int_{E=0}^{E=E_{max}} g(E)dE \times 1$$

$g(E)dE$ is given by

$$g(E)dE = \frac{8\sqrt{2}\,\pi m^{3/2}}{h^3} E^{1/2} dE \qquad (7.41)$$

where m is the mass of electron and h is the Planck's constant.

$$n = \frac{8\sqrt{2}\,\pi m^{3/2}}{h^3} \int_{E=0}^{E=E_{max}} E^{1/2} dE$$

$$\Rightarrow \qquad n = \frac{8\sqrt{2}\,\pi m^{3/2}}{h^3} \times \frac{2}{3}(E_{F_0})^{3/2}$$

$$\Rightarrow \qquad n = \frac{16\sqrt{2}\,m^{3/2}}{h^3} \frac{\pi}{3} E_{F_0}^{3/2} \qquad (7.42)$$

$$\therefore \qquad (E_{F_0})^{3/2} = \frac{h^3}{(8m)^{3/2}} \left(\frac{3n}{\pi}\right)$$

$$\text{or} \qquad E_{F_0} = \left(\frac{h^2}{8m}\right)\left(\frac{3n}{\pi}\right)^{2/3} \qquad (7.43)$$

$$\text{or} \qquad E_{F_0} = Bn^{2/3} \qquad (7.44)$$

where $B = \left(\frac{h^2}{8m}\right)\left(\frac{3}{\pi}\right)^{2/3}$ is a constant $= 5.85 \times 10^{-38}$ J $= 3.65 \times 10^{-19}$ eV

Fermi Energy at T > 0 K

The Fermi energy E_F at any temperature T, in general, can be expressed in terms of E_{F0} through the relation:

$$E_F \cong E_{F_0} = \left[1 - \frac{\pi^2}{12}\left(\frac{k_B T}{E_{F_0}}\right)^2\right] \tag{7.45}$$

Except at extremely high temperature, the second term within the bracket is very small compared to unity.

$$\therefore \qquad\qquad E_F \cong E_{F0} \tag{7.46}$$

Hence the value of E_{F0} can be taken to be essentially equal to E_F itself.

7.3.4 Fermi Temperature (T_F)

Fermi temperature T_F is the temperature at which the average thermal energy of the free electron in a solid becomes equal to the Fermi energy at 0 K. But the thermal energy possessed by electrons is given by the product $k_B T$.

When $T = T_F$, the equation, $E_{F0} = k_B T_F$ is satisfied. But for practical purposes, $E_{F0} \cong E_F$

$$\therefore \qquad\qquad k_B T_F = E_F \quad \therefore \quad T_F = \frac{E_F}{k_B} \tag{7.47}$$

The Fermi temperature is only a theoretical concept. Since, at ordinary temperature, it is not possible for the electrons to receive thermal energy in a magnitude of E_F. For example, let us consider the case of metals. We know that E_F will be of the order of few eV.

Let us consider the case in which $E_F = 3$ eV.

$$= 3 \times 1.602 \times 10^{-19} \text{ J} = 4.8 \times 10^{-19} \text{ J} \quad \therefore \quad T_F = \frac{4.8 \times 10^{-19}}{1.381 \times 10^{-23}} = 34800 \text{ K}$$

This is quite an overstated temperature to be realised in practice.

7.3.5 Fermi Velocity (v_F)

The energy of the electrons, which is at the Fermi level, is E_F. The velocity of the electrons, which occupy the Fermi level, is called the Fermi velocity V_F.

$$\therefore \qquad\qquad E_F = \frac{1}{2} m v_F^2$$

or

$$v_F = \sqrt{\frac{2E_F}{m}} \tag{7.48}$$

7.3.6 Mean Energy of Electron Gas at Absolute Zero

The total energy of a free electron in a metal at 0 K is:

$$E_{F_0} = \int_0^{E_F} E\, N(E)\, dE$$

At $T = 0$ K, $$N(E)dE = \frac{3N}{2} E_F^{-3/2} E^{1/2} dE \implies E_{F_0} = \int_0^{E_F} \frac{3N}{2} E_F^{-3/2} E^{1/2} dE$$

$$= \frac{3N}{2} E_F^{-3/2} \int E^{3/2} dE = \frac{3N}{2} E_F^{-3/2} \times \frac{2}{5} E_F^{5/2}$$

$$E_{F_0} = \frac{3}{5} N E_F \tag{7.49}$$

The average energy of free energy in a metal at 0 K is

$$E_{av} = \frac{3}{5} E_F \tag{7.50}$$

EXAMPLE 7.2 Calculate the Fermi energy and Fermi velocity for free electron gas in silver. Given the number of free electrons per unit volume in silver is 5.8×10^{28} per metre cube.

Solution Given $n = 5.8 \times 10^{28}/m^3$, $h = 6.63 \times 10^{-34}$ Js and $m = 9.1 \times 10^{-31}$ kg
The Fermi energy is:

$$E_{F_0} = \left(\frac{h^2}{8m}\right)\left(\frac{3n}{\pi}\right)^{2/3} = Bn^{2/3}$$

where $$B = \left(\frac{h^2}{8m}\right)\left(\frac{3n}{\pi}\right)^{2/3} = 5.85 \times 10^{-38} \text{ J}$$

Then $$E_{F_0} = 5.85 \times 10^{-38} \times (5.8 \times 10^{-38})^{2/3}$$
$$= 8.7472 \times 10^{-19} \text{ J} = 5.467 \text{ eV}$$

Fermi energy is related to Fermi velocity as:

$$E_{F_0} = \frac{1}{2} m v_{F_0}^2$$

or $$v_{F_0} = \sqrt{\frac{2E_{F_0}}{m}} = \sqrt{\frac{2 \times 8.7472 \times 10^{-19}}{9.1 \times 10^{-31}}} = 1.39 \times 10^6 \text{ m/s}$$

EXAMPLE 7.3 Use the Fermi distribution function to obtain the value of $f(E)$ for $E - E_F = 0.01$ eV at 200 K.

Solution Given $E - E_F = 0.01$ eV, $k_B = 1.381 \times 10^{-23}$ J/K, $T = 200$ K. We know Fermi–Dirac distribution function is given as:

$$f(E) = \frac{1}{\exp \cdot \dfrac{E - E_F}{k_B T} + 1} = \frac{1}{\exp \cdot \dfrac{0.01 \times 1.6 \times 10^{-19}}{1.381 \times 10^{-23} \times 200} + 1}$$

$$= \frac{1}{1.785 + 1} = \frac{1}{2.785} = 0.359$$

7.4 COMPARISON BETWEEN CLASSICAL FREE ELECTRON THEORY AND QUANTUM FREE ELECTRON THEORY

Similarities between the Two Theories

The following assumptions apply to both the theories:

1. The valence electrons are treated as though they constitute an ideal gas.
2. The valence electrons can move freely throughout the body of the solid.
3. The mutual repulsion between the electrons and the force of attraction between the electrons and ions are considered insignificant.

Difference between the Two Theories

We can differentiate both theories as tabulated in Table 7.1.

Table 7.1 Difference between Classical and Quantum Free Electron Theories

S. No.	Classical free electron theory	Quantum free electron theory
1.	The free electrons, which constitute electron gas, can have continuous energy values.	The energy values of the free electrons are discontinuous because of which the energy levels are discrete.
2.	It is possible that many electrons may possess same energy.	The free electrons obey Pauli exclusion principle. Hence no two electrons can possess same energy.
3.	The pattern of distribution of energy among the free electrons obeys Maxwell–Boltzmann statistics.	The distribution of energy among the free electrons is according to Fermi–Dirac statistics, which imposes a severe restriction on the possible way in which the electrons absorb energy from an external source.

7.5 PERIODIC POTENTIAL IN CRYSTALLINE SOLID

In solid, one deals with a large number of interacting particles and hence the problem of determining the electron wave functions and the energy levels is extremely complicated. A simple quantum mechanical picture of an electron in crystal can, however, be obtained by assuming that the atomic nuclei are at rest in the crystalline state and that the electron is in a periodic potential, which has periodicity of lattice. The periodic potential may be considered to be caused by a fixed nuclei and some average potential due to all other electrons. The solution of Schrödinger equation for this potential gives a set of state or levels, which may be occupied by the electrons according to Pauli's exclusion principle.

The potential energy of an electron at a distance r from an atomic nucleus of charge Ze is given by

$$V = -\frac{Ze^2}{4\pi\varepsilon_0 r}$$

To form a crystal, a number of such nuclei are brought close to each other. Therefore, the potential energy of an electron is the sum of potential energies due to individual nuclei. Hence the variation of potential energy with distance for a one-dimensional crystal is shown in Figure 7.7.

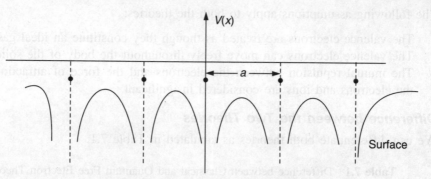

Figure 7.7 Representation of potential experienced by an electron in a perfectly periodic crystal lattice with lattice parameter a.

7.6 BLOCH THEOREM

In free electron theory, an electron moves in a constant potential V_0 and the Schrödinger equation for a one-dimensional case is:

$$\frac{d^2\psi}{dx^2} + \frac{2m}{\hbar^2}(E - V_0)\psi = 0 \tag{7.51}$$

The solution of this equation in plane waves of the type

$$\psi(x) = e^{\pm ikx}$$

where k represents the momentum of electron divided by $\hbar$. From Eq. (7.51), we get the kinetic energy of the electron E_{kin}:

$$E_{\text{kin}} = E - V_0 = \frac{\hbar^2 k^2}{2m} = \frac{p^2}{2m} \tag{7.52}$$

The complete solution for the wave function containing the time is obtained by multiplying $\psi(x)$ by $e^{-i\omega t}$, where $\omega = \dfrac{E}{\hbar}$.

Now, consider the Schrödinger equation for an electron moving in one-dimension periodic potential. In this case, the potential energy of an electron satisfies the equation:

$$V(x) = V(x + a) \tag{7.53}$$

where a is the period. In this case, the Schrödinger equation is:

$$\frac{d^2\psi}{dx^2} + \frac{2m}{\hbar^2}[E - V(x)]\psi = 0 \tag{7.54}$$

With reference to the solutions of this equation, there is an important theorem, which states that there exists solutions of the form:

$$\psi(x) = e^{\pm ikx}U_k(x)$$

where
$$U_k(x) = U_k(x+a) \tag{7.55}$$

Thus the solution are plane wave modulated by the function $U_k(x)$, which has the same periodicity as the lattice. This theorem is known as the *Bloch theorem* or *Floquet's theorem*. The function defined by equation in known as *Bloch function*.

7.7 KRONIG–PENNEY MODEL (QUALITATIVE)

The free electron model implies that potential inside the solid is uniform in metals. A more accurate model allows for variation in potential energy due to the fixed positively charged lattice ions. A one-dimensional lattice ions separated by lattice parameter a, is presented in Figure 7.8.

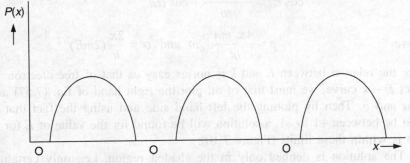

Figure 7.8 One-dimensional lattice.

The highest potential is half way between the atom and goes to $-\infty$ as position of the ion is approached. This potential distribution is quite complicated and for mathematical solution of Schrödinger's equation, a simpler model known as Kronig–Penney Model is used. The Kronig–Penney model still displays the following essential features, where the function:

- has the same period as lattice.
- potential is higher between ions and lower near the lattice ion.

The ions are located at $x = 0, a, 2a, 3a, \ldots$ etc. The potential wells are separated from each other by potential barrier of higher V_0 and with wall w as shown in Figure 7.9. The Schrödinger's equation should be solved for potential distribution separately for $V = +\dfrac{V_0}{2}$ and $V = -\dfrac{V_0}{2}$ and compare the solutions at the boundary conditions.

The assumed wave function has the form:

$$\psi(x) = U_k(x)e^{ikx} \tag{7.56}$$

where $U_k(x)$ is a periodic function with the same period as the lattice.

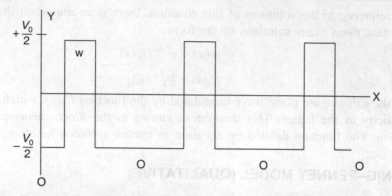

Figure 7.9 Geometrical arrangement for Kronig–Penney model.

The solution for Eq. (7.56) exists if k is related to energy E by the following equation and assuming that as $w \to 0$, $V \to \infty$, w, V = constant.

$$\cos ka = \frac{p \sin \alpha a}{\alpha a} + \cos \alpha a \qquad (7.57)$$

where

$$p = \frac{4\pi^2 ma}{h^2} V_0 \omega \quad \text{and} \quad \alpha = \frac{2\pi}{h}(2mE)^{1/2}$$

Here the relation between E and k is not as easy as that of free electron. To find the exact $E - k$ curve, we must first of all plot the right hand of Eq. (7.57) as a function of α and a. Then by plotting the left hand side and using the fact that this always must be between +1 to −1, a solution will be found for the value of E for which right side is within these limits (Figure 7.10).

The solution is defined only in the shaded region, i.e., only certain values are allowed. This implies that there are only certain allowed energy bands in crystal solids.

For the solution of Kronig–Penney model mentioned earlier, if the value of P is large, i.e., $V_0 w$ is large, the function is given by right hand side of Eq. (7.57) crosses at a steep slopes. This means that the allowed bands are narrower and forbidden bands are wider. Let us observe some specific conditions.

1. As $p \to \infty$, the band reduces to a single energy level, i.e., discrete energy spectrum for isolated atoms, which is the actual case. So P is proportional to the potential barrier of the given system.

   $$\frac{\sin \alpha a}{\alpha a} = 0 \Rightarrow \sin \alpha a = 0$$

   As $p \to \infty$, $\qquad\qquad \alpha a = n\pi$

   $$\frac{2\pi a}{h}\sqrt{2mE} = n\pi \quad \text{or} \quad E = \frac{n^2 h^2}{8ma^2} \qquad (7.58)$$

   which is exactly the same as the equation for an electron in an infinite potential well of width a. So all the electrons are independent of each other and confined to infinite barrier.

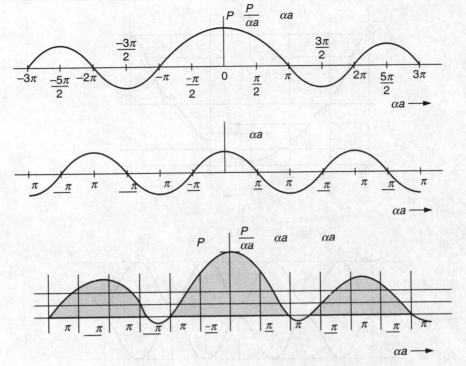

Figure 7.10   Solution of Kronig–Penney model.

2. $\cos \alpha = \cos ka$

$$\frac{2\pi a}{h}\sqrt{2mE} = ka \text{ or } E = \frac{h^2 k^2}{8\pi^2 m} = \frac{h^2 k^2}{2m} \tag{7.59}$$

Equation (7.59) is exactly same as that of equation for a free electron. So the variation in the value of p from zero to infinity accounts for the whole range of conditions in the crystalline solid.

3.   At the boundary conditions of the allowed bands, i.e.,

$$\cos ka = \pm 1, \cos ka = \cos n\pi \qquad \text{where } n = 1, 2, 3, ..., \text{ etc.}$$

$$ka = n\pi \text{ or } k = \frac{n\pi}{a} \tag{7.60}$$

E vs k^2 curve is plotted for one-dimensional lattice structure from Kronig–Penney model [Figure 7.11(a)]. At the boundaries we can see discontinuities in energy levels as the value of n changes. By transferring the zero curves by an appropriate multiple of π/a to axis E, we can see the separation between energy levels clearly. Each of these energy levels is termed as *Brillouin zone*.

From Figure 7.11(b), we can say that in the middle of Brillouin zones the $E - k$ curve is identical to the free electron. But the boundaries of the zones, i.e., at $k = n\pi$, the behaviour is totally different.

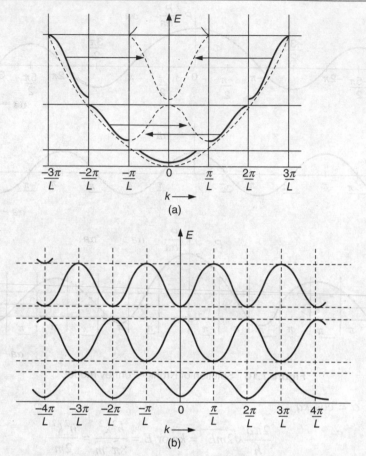

Figure 7.11 (a) *E–k* curve for one-dimensional lattice structure from Kronig–Penney model for the normal and reduced zone cases. Also shown the *E–k* curve for free electron (b) Plot of extended zone curves for the same lattice structure.

Electrons that occupy Brillouin zone can move freely inside a crystal. All the information for Brillouin zones are contained in first Brillouin zone if we use the reduced zone plot. So the Brillouin zones are the allowed energy regions in k.

7.8 BAND THEORY OF SOLIDS

7.8.1 Energy Bands in Solids

When a large number of atoms are brought together to form a solid, a single energy level of an isolated atom is splitted into an energy band consisting of closely spaced levels of slightly differing energy. It is known from quantum theory of energy of bound electron in an atom is quantised, i.e., the electron can take on only discrete values of energy and we know that the electron is not localised at a given radius. Rather the electron location is obtained from a probability density function, which is the probability of finding the electron at a particular distance from the nucleus. The application of quantum mechanics and Schrödinger's wave equation is needed to find out the electron energy bands in a solid for assumed potential functions.

The probability density function for lowest energy state of a single isolated atom is shown in Figure 7.12(a). When two atoms are brought close together, as shown in Figure 7.12(b), the wave functions of the electrons in the two atoms overlap and in the overlapping region the discrete single energy level splits into two levels [Figure 7.12(c)] due to interacting potential so that each electron can occupy a distinct quantum state according to Pauli exclusion principle. In the assembly of N atoms, each

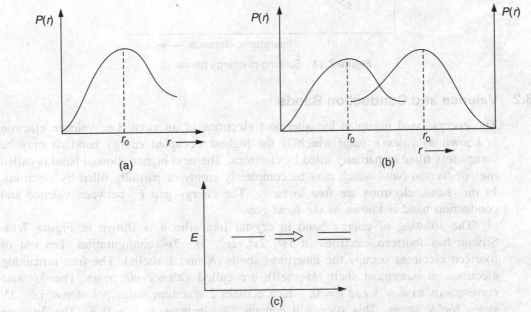

Figure 7.12 (a) Probability density function for lowest electron energy state of a single isolated atom (b) Probability density function of two atoms brought in close proximity to each other (c) Splitting of energy level.

of the energy level splits into N energy levels and forms an energy band of very closely spaced levels differing slightly in energy. For example, a solid containing 10^{19} atoms will have 2×10^{19} possible states, since only two electrons with opposite that can occupy the same state inaccordance with Pauli exclusion principle. If the width of the energy band be 2 eV, then the energy levels are separated by $2/2 \times 10^{-19}$, i.e., 10^{-19} eV. Since this spacing of energy level is extremely small, a quasi quantum level is formed.

Figure 7.13 shows the splitting of $n = 1$, $n = 2$, $n = 3$, i.e., K, L, M energy levels as the distance between different atoms is reduced. At $n = 3$ (M-shell), energy level is affected, and as the separation is reduced inner shells, i.e., $n = 2$ and $n = 1$ are gradually affected. Thus bands of allowed energies that the electrons may occupy are separated by bands of forbidden energy when the interatomic distance reaches the equilibrium value r_0. This is the essence of energy band theory of single crystal material.

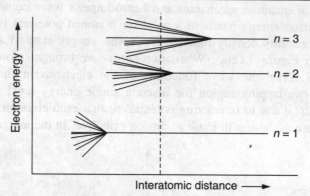

Interatomic distance ⟶

Figure 7.13 Splitting of energy bands.

7.8.2 Valence and Conduction Bands

The energy band occupied by outermost electrons of an atom, i.e., valence electron is known as *valence band* which is the highest occupied energy band. It may be completely filled or partially filled by electrons. The next higher allowed band is called the *conduction band* which may be completely empty or partially filled by electrons. In this band, electrons are free to move. The energy gap E_g between valence and conduction band is known as the *band gap*.

The splitting of energy band in crystal like silicon is shown in Figure 7.14. Silicon has fourteen electrons in $1s^2$, $2s^2$, $2p^6$, $3s^2$, $3p^6$ configuration. Ten out of fourteen electrons occupy the innermost shells (K and L shells). The four remaining electrons in outermost shell (M shell), are called *valence electrons*. The $3s$ state corresponds to $n = 3$ and $l = 0$, which contain 2 quantum states per atoms, i.e., $2N$ states for N atoms. This state will contain $2N$ electrons at $T = 0$ K. The $3p$ state corresponds to $n = 3$ and $l = 1$ which contains 6 quantum states per atom, i.e., $6N$ states, and $2N$ remaining valence electrons will occupy these states. As the atoms are brought closer, the $3s$ and $3p$ states interact and overlap which causes the splitting of

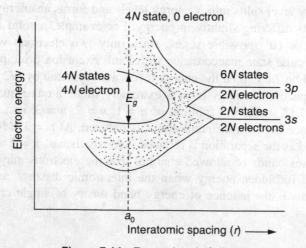

Figure 7.14 Energy band of silicon.

energy levels into bands. At equilibrium inter atomic distance, the single band splits into two bands separated by the energy gap E_g. The lower band contains $4N$ states. At 0 K, all the states in the lower band (valence band) will be occupied by valence electrons, but those in the upper band will be empty (conduction band). The energy band gap E_g, between the top of the valence band and the bottom of the conduction band is the width of the forbidden gap which contains no available states for the electrons to occupy. The excitation of electrons from the valence band through the forbidden gap into the conduction band is mainly responsible for electrical conduction.

7.8.3 Insulator, Semiconductor and Conductor

The distinction among insulators, semiconductors and conductors can be understood for their large variation in band structure and wide range of electrical conductivity.

A material having either completely full or empty band is an insulator. The conductivity of an insulator is very small as no free electron is available in the upper empty band for flow of current in this material. Between the full and empty band is a forbidden region whose width is so large that at physical realisable temperature, electrons from the top of the highest fill band (valence band) cannot be excited across the wide forbidden region to the bottom of lowest empty band (conduction band).

The width of the forbidden region is called the *band gap energy E_g* whose value is of the order of 2.5 V to 6 V or large for insulators. This situation is depicted in Figure 7.15 (a).

In semiconductor, the band gap energy E_g between full and empty band is small of the order of 1 eV. It is shown in Figure 7.15(b). Electrons from states near the top of the lower full band will have an appreciable probability to be thermally excited across the small gap to states near the bottom of upper empty band. Thus a limited number of electrons will be available near the bottom of the upper empty band for conduction of electric current when an electric field is applied. The vacant state left behind by the thermally excited electrons will create positively charged empty states near the top of the lower full band (valence band).

At $T > 0$ K, valence electrons gains thermal energy to top into empty states alternatively filling one empty state and creating a new empty state. The movement of these positive charges are called *holes* in the valence band will also give rise to an electric current in a semiconductor. The electrical conductivity is much smaller than that of a metal owing to a few numbers of free electrons and holes available in the energy bands. The conductivity is strongly temperature dependent and also a function of E_g. The semiconductor possesses a negative temperature of resistance, since the resistance decreases sharply with increase in temperature. Further the resistivity can be controlled over a wide range by means of doping, i.e., introducing proper impurity in semiconductor.

The resistivity of semiconductor ranges from 10^{-2} Ω cm to 10^6 Ω cm, that of metal of the order of 10^{-6} Ω cm, while insulators have resistivity greater than 10^{10} Ω cm.

The distinction between semiconductor and insulator is of degree only as all semiconductors are insulators at $T = 0$ K. Similarly, all insulators are semiconductor at sufficient high temperature, which is not realisable experimentally.

Figure 7.15(c) shows the energy band diagram of metals in which a part of the upper region of valence band overlaps into a part of lower region of conduction band. Thus a large number of electrons and holes will be available for conduction. So these materials exhibit high thermal conductivity.

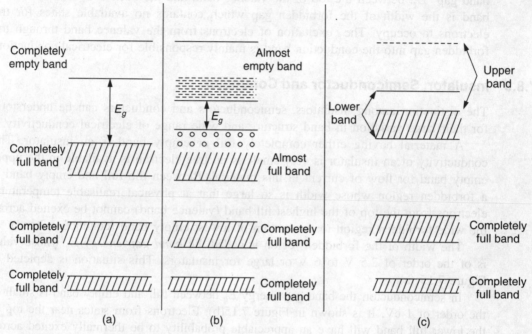

Figure 7.15   Energy band diagrams (a) Insulator, (b) Semiconductor and (c) Metal (conductor).

7.9   PHOTOCONDUCTIVITY

When light radiation falls on the crystal, electrical conductivity of the insulating crystal increases. This phenomenon is called *photoconductivity*.

When the energy of incident photon is greater than the energy gap E_g, free electron-hole pairs are produced in the crystal by the absorption of the incident photons and these electrons and holes serve as the carriers of electrical current.

Impurities and imperfections in the crystal also contribute towards photoconductivity. If these are present in the crystal, then even the photons having energy below the threshold for the production of electron-hole pairs may be able to produce mobile electrons/holes. Imperfections and impurities introduce discrete energy levels in the forbidden energy gap called *traps*. Two classical examples of photoconductive materials are the polymer polyvinylcarbazole, which is used extensively in photocopying (xerography); and the lead sulphide which is used in infrared detection applications, such as *U.S. Sidewinder* and *Russian Atoll* heat seeking missiles and selenium as employed in early television and xerography experiments.

7.9.1 Variation of Photoconductivity with Illumination (Simple Model of Photoconductor)

To study the variation of photoconductivity with illumination, consider a simple model of photoconductor (Figure 7.16).

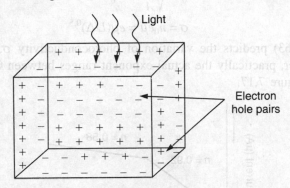

Figure 7.16   A photoconductor illuminated with light.

Assumptions

1. When the light radiations fall on the crystal, the electron-hole pairs are produced uniformly throughout the crystal volume.
2. In the recombination process, electrons directly recombine with holes.
3. Mobility of holes is smaller than the mobility of electrons and, therefore, mobility of holes can be neglected.

Using these assumptions, rate of change of electron concentration can be written as

$$\frac{dn}{dt} = L - Anp$$

$$\Rightarrow \qquad \frac{dn}{dt} = L - An^2 \qquad \text{[using } n = p] \qquad (7.61)$$

where L is the number of photons absorbed per unit volume per unit time and Anp is the recombination rate (we know that the recombination rate is proportional to the product np and A is the proportionality constant).

We know that drift velocity of all the carriers, v is proportional to applied electric field E, i.e.

$$v \propto E \Rightarrow v = \mu E$$

where μ is the constant of proportionality and is known as mobility of the medium. In steady state,

$$\frac{dn}{dt} = 0$$

and $$n = n_0$$

Hence using Eq. (7.61),

$$L = An_0^2$$

$$\Rightarrow \qquad n_0 = \sqrt{\frac{L}{A}} \Rightarrow n_0 \propto L^{0.5} \tag{7.62}$$

and

$$\sigma = n_0 e\mu = e\mu(L/\Delta)^{0.5} \tag{7.63}$$

Equation (7.63) predicts the variation of photoconductivity σ, with light level L as $L^{0.5}$. However, practically the actual exponent ranges between 0.5 and 1.0 or more as shown in Figure 7.17.

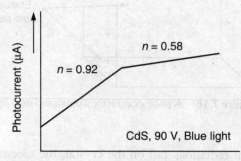

Figure 7.17 Variation of photocurrent of cadmium sulphide with different intensity at 70 V bias voltage with blue light.

Decay

The decay of photoelectrons after switching off the light suddenly is

$$\frac{dn}{dt} = 0 - An^2 \tag{7.64}$$

$$\Rightarrow \qquad \int_{n_0}^{n} \frac{dn}{n^2} = -A\int_0^t dt$$

Solution of above equation is

$$\frac{1}{n} = At + \frac{1}{n_0} \tag{7.65}$$

where n_0 is the carrier concentration at $t = 0$ when high is just switched off.

Response Time (t_0)

This is the time during which carrier concentration reduces to half of starting carrier concentration.

Hence at $t = t_0$, $n = \dfrac{n_0}{2}$

Using Eq. (7.65),

$$\frac{1}{n_0/2} = At_0 + \frac{1}{n_0}$$

$$\Rightarrow \qquad \frac{1}{n_0} = At_0$$

$$\Rightarrow \qquad t_0 = \frac{1}{An_0} = \frac{1}{A\sqrt{L/A}}$$

$$\Rightarrow \qquad t_0 = \sqrt{\frac{1}{LA}}$$

$$t_0 = \frac{1}{L}\frac{L}{\sqrt{LA}} = \frac{1}{L}\sqrt{\frac{L}{A}} = \frac{n_0}{L}$$

or $$\qquad t_0 = \frac{n_0}{L} \qquad\qquad\qquad (7.66)$$

So $$\qquad t_0 = \frac{1}{L}\left(\frac{\sigma}{e\mu}\right) = \frac{\sigma}{e\mu L} \qquad\qquad (7.67)$$

$$\therefore \qquad t_0 \propto \sigma$$

Response time is directly proportional to σ at a given light intensity level.

Good photoconductors have large t_0. They are rarely found in practice.

Sensitivity of Gain Factor (G)

This can be calculated as the ratio of numbers of carriers crossing the unit area of specimen to the number of photons absorbed by that area in the specimen. It can be represented by G. So for a specimen of thickness d and the cross-section area unity, we have

$$G = \frac{\text{Particle flux}}{L \times (d \times 1)} \qquad\qquad (7.68)$$

We know flux is the rate of flow of carriers per unit area per unit time is given by F_n.

Also, the current density

$$J_n = \sigma E = n_0\, e\mu(V/d)$$

A potential V produces the particle flux F_n and is given by

$$F_n = \frac{\mu V}{d}\, n_0$$

$$\Rightarrow \qquad F_n = \frac{\mu V}{d}(Lt_0) \qquad \text{[From Eq. (7.66)]}$$

$$\Rightarrow \qquad F_n = \frac{\mu V}{d} \times L \times \frac{1}{\sqrt{LA}}$$

Hence $$\qquad F_n = \frac{\mu V}{d} \times L \times \frac{1}{\sqrt{LA}} \qquad\qquad (7.69)$$

So from Eq. (7.68),

$$G = \frac{F_n}{L\,d}$$

$$\Rightarrow \qquad G = \frac{1}{Ld} \times \frac{\mu V}{d} \times L \times \frac{1}{\sqrt{LA}}$$

$$\Rightarrow \qquad G = \frac{\mu V}{d^2 (LA)^{1/2}} \qquad\qquad (7.70)$$

Now, let T_e be the life time of the electron before illumination, it is nothing but the response time, so

$$T_e = t_0 = \frac{1}{\sqrt{LA}}$$

Let T_d be the transit time of an electron between the electrodes and is given by

$$T_d = \frac{d}{V_d} = \frac{d}{\mu E}$$

$$\Rightarrow \qquad T_d = \frac{d}{\mu V/d} = \frac{d^2}{\mu V}$$

So from Eq. (7.70),

$$G = \frac{\mu V}{d^2 (LA)^{1/2}} = \frac{T_e}{T_d} \qquad\qquad (7.71)$$

The expression of G as given in Eq. (7.71) quite general, however, this theoretical value of t_0 do not agree and in some instances the discrepancy is $\approx 10^8$ times. The result then indicates that the model is not successful.

We should look out for some missing phenomenon not considered in this expression.

7.9.2 Effect of Traps on Photoconductivity

A trap is an impurity atom or imperfection in the crystal capable of capturing an electron or hole. In simple words, it is an energy level that can capture either electrons or holes.

There are two types of traps:

1. One type of traps helps electrons and holes to recombine and thereby assists in restoration of thermal equilibrium. These types of traps are called recombination centres. In the presence of traps, n_0 recombination proceeds at much higher rate.
2. Second type of traps does not contribute directly in recombination process, but not contribute directly in recombination process, but restricts the freedom of motion of holes or electrons. This is shown in Figure 7.18.

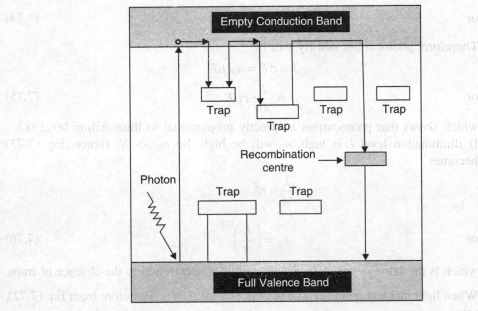

Figure 7.18 Influence of traps in electron hole recombination.

To study the effect of second type of traps on photoconductivity, let us consider a crystal with N electron trap levels per unit volume. The rate of change of electron concentration n is given by,

$$\frac{dn}{dt} = L - An(n + N) + Bn_t \qquad (7.72)$$

where recombination coefficient A is same for both electron-hole recombination and electron trap capture. The term Bn_t represents the rate of thermal evaporation of trapped carrier back into conduction band and is neglected.

In steady state,

$$\frac{dn}{dt} = 0$$

Therefore, from Eq. (7.72), we get

$$L - An_0(n_0 + N) = 0$$

or

$$\frac{L}{A} = n_0(n_0 + N) \qquad (7.73)$$

Limiting Cases

1. If illumination level L is low, n_0 will be small. So $n_0 \ll N$ and, hence, n_0 can be neglected in comparison to N. Therefore, Eq. (7.73) becomes

$$\frac{L}{A} = n_0 N$$

or
$$n_0 = \frac{L}{AN} \tag{7.74}$$

Therefore, photocurrent density will be

$$J = \sigma E = n_0 e \mu E$$

or
$$J = \frac{L}{AN} e \mu E \tag{7.75}$$

which shows that photocurrent is directly proportional to illumination level (L).

2. If illumination level L is high, n_0 will be high. So $n_0 \gg N$. Hence, Eq. (7.73) becomes

$$\frac{L}{A} = n_0^2$$

or
$$n_0 = \left(\frac{L}{A}\right)^{1/2} \tag{7.76}$$

which is the same steady state current carrier concentration in the absence of traps.

When light incident on the crystal is switched off, $L = 0$. Therefore from Eq. (7.72), we get

$$\frac{dn}{dt} = -An(n + N)$$

or
$$\frac{dn}{n(N + n)} = -A dt$$

Integrating both sides, we get

$$-\int \frac{dn}{n(n + N)} = A \int dt$$

$$\frac{1}{N} \log \frac{n + N}{n} = At + C$$

At $t = 0$, $n = 0$

$$\therefore \qquad \frac{1}{N} \log \frac{n_0 + N}{n_0} = C$$

$$\therefore \qquad \frac{1}{N} \log \frac{n + N}{n} = At + \frac{1}{N} \log \frac{n_0 + N}{n_0}$$

or
$$\frac{1}{N}\left[\log \frac{n + N}{n} - \log \frac{n_0 + N}{n_0}\right] = At$$

or
$$\log \frac{n + N}{n} - \log \frac{n_0 + N}{n_0} = N At$$

If the number of traps per unit volume $N \gg n$, then

$$\log \frac{N}{n} - \log \frac{N}{n_0} = NAt$$

or

$$\log \frac{N}{n} \times \frac{n_0}{N} = NAt$$

or

$$\log \frac{n_0}{n} = NAt$$

or

$$\log \frac{n_0}{n} = -NAt$$

or

$$n = n_0 \, e^{-NAt} \tag{7.77}$$

At

$$t = t_0 = \frac{1}{NA} \tag{7.78}$$

$$\therefore \qquad n = \frac{n_0}{e}$$

Therefore, the response time in the absence of traps is $t_0 = \dfrac{1}{NA}$ in which carrier concentration reduces to $1/e$ of its steady state.

Equation (7.78) shows the response time is reduced by the presence of traps. Further

$$\sigma = n_0 e \mu \tag{7.79}$$

From Eqs. (7.74) and (7.79), we get $\sigma = \dfrac{L}{AN} e\mu$

which shows that the presence of traps reduced the photoconductivity.

7.9.3 Photoconduction In Semiconductors

The conductivity of a semiconductor increases when the radiation is made to incident on it. This phenomenon is called *photoconductive effect*. This may be explained as follows:

The conductivity (σ) of a semiconductor increases with the increase of concentration of charge carriers as is clear from the following relation

$$\sigma = (n_e \mu_e + n_h \mu_h)e$$

where n_e is the free electron concentration per m^3, n_h is the hole concentration per m^3, μ_e is the free electron mobility, μ_h is the hole mobility, and e is the electronic charge $= 1.6 \times 10^{-19}$ C.

When the radiant energy falls on a semiconductor, the covalent bonds get broken and electron-hole pairs in excess of those generated thermally are created. These increased charge-carriers decreases the resistance or increase the conductivity of the material and, hence, such a device is called a *photoresistor* or a *photoconductor*.

Figure 7.19 represents the energy level diagram of semiconductor having both acceptor and donor impurities (i.e., of *P–N* junction).

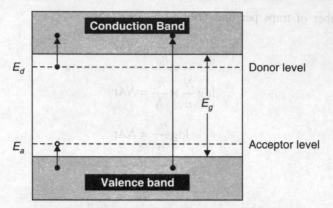

Figure 7.19   Energy level diagram of semiconductor.

If the photon of light is allowed to fall on the specimen, the following transitions are possible:

1. An electron may jump from valence band to conduction band, thus creating an electron-hole pair. This process is called *intrinsic excitation*.
2. A donor electron may jump into conduction band or a valence electron may jump into an acceptor state. These two transition are called the *impurity excitations*.

As the density of states in the conduction and valence bands is much greater than the density of impurity states, therefore the *photoconductivity is mainly due to intrinsic excitation*.

Spectral Response

The minimum energy of a photon required for intrinsic excitation is the forbidden gap energy (E_g) of the semiconductor material. The threshold long wavelength of the material is defined as λ_c corresponding to the energy gap, E_g and is given by

$$E_g = \frac{hc}{\lambda_c}$$

i.e.

$$\lambda_c = \frac{hc}{E_g}$$

If E_g is expressed in eV and λ_c in micron, we get

$$\lambda_c = \frac{6.62 \times 10^{-34} \times 3 \times 10^8}{E_g \times 1.6 \times 10^{-19}}$$

$$= \frac{1.264}{E_g} \times 10^{-6} \text{ m}$$

$$= \frac{1.24}{E_g} \text{ } \mu m$$

For silicon $E_g = 1.1$ eV, therefore, $\lambda_c = 1.13$ µm at room temperature. For germanium $E_g = 0.72$ eV, therefore, $\lambda_c = 1.73$ µm at room temperature.

Figure 7.20 represents the spectral sensitivity curves for Si and Ge. It is noted that the long wavelength limit is slightly greater than the calculated values of λ_c. This is due to the impurity excitations. As the wavelength decreases from value λ_c, the response increases, becomes maximum and again decreases.

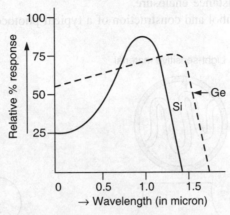

Figure 7.20 Spectral sensitivity curve.

Photoelectric Current

If the potential difference is applied across the semiconductor bar and light is allowed to fall on it, the charge carriers are generated and those move under the influence of applied potential difference. The charge carriers, which do not undergo recombination, reach the conducting contacts at the ends of the bar and hence, constitute electric current. This current is given by

$$i = \frac{eP_rT}{T_i}$$

where P_r is the rate of change carriers produced by light, T is the average life time of the nearly created carriers, T_i is the average transit time for the carriers to reach the conducting contacts, and e is the charge on each carrier $= 1.6 \times 10^{-19}$ C.

7.9.4 Photoconductive Cell

The photoconductive cell is a two-terminal device whose terminal resistance will vary (linearly) with the intensity of light. For various reasons, it is frequently called a *photoresistive device*.

The photoconductive materials most frequently used include cadmium sulphide (CdS) and cadmium selenide (CdSe). Both materials respond rather slowly to changes in light intensity. The peak spectral response time of CdS units is about 100 ms and 10 ms for CdSe cells. Another important difference between the two materials is their temperature sensitivity. There are large change in the resistance of a CdSe cell with change in ambient temperature, but the resistance of CdS remains relatively stable. The spectral response of CdS cell closely matches that of the human eye, and the cell is, therefore, often used in applications where human vision is a factor, such as street light control or automatic iris control for cameras.

The essential elements of photoconductive cell are the ceramic substrate, a layer of photoconductive material, metallic electrodes to connect the devices into a circuit and a moisture resistance enclosure.

The circuit symbol and construction of a typical photoconductive cell are shown in Figure 7.21.

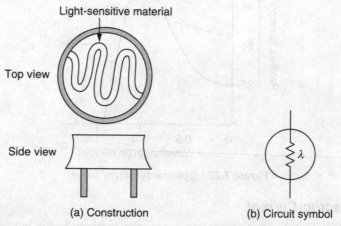

Figure 7.21 Photoconductive cell.

The light-sensitive material is arranged in the form of a long strip, zigzagged across a disc-shaped base with protective sides. For added protection, a glass or plastic cover may be included. The two ends of the strip are brought out to connecting pins below the base.

Characteristics

The illumination characteristics of a typical photoconductive cells are shown in Figure 7.22 from which it is obvious that when the cell is not illuminated, its resistance may fall to a few hundred ohms. Note that the illumination characteristics are logarithmic to cover a wide range of resistance and illumination that are possible. Cell sensitivity may be expressed in terms of the cell current for a given voltage and given level of illumination.

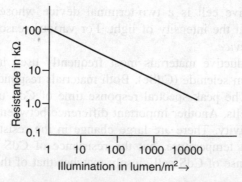

Figure 7.22 Illumination characteristics of a typical photoconductive cell.

Drawback

The major drawback of the photoconductivite cells is that temperature variations cause substantial variations in resistance for a particular light intensity. Therefore, such a cell is unsuitable for analog applications.

Applications

Such cells find wide use in industrial and laboratory control applications.

1. **For relay control:** The photoconductive cell is used for relay control as visualized in Figure 7.23. When the cell is illuminated, its resistance is low and the relay current is maximum. When the cell is dark, its high resistance reduces the current down to a level too low to energize the relay. Resistance R is included to limit the relay current to the desired level when the resistance of the cell is low.

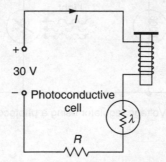

Figure 7.23 Photoconductive cell for relay control.

2. **As a switch transistor:** Photoconductive cells are used to switch transistor ON and OFF, as shown in Figure 7.24. When the cell shown in Figure 7.24(a) is dark, the transistor base is biased above its emitter level, and the device is turned ON. When the cell is illuminated, the lower resistance of the cell in series with R biases the transistor base voltage below its emitter level. Thus, the device is turned OFF. Figure 7.24(b) shows a circuit in which the transistor is ON, when the cell is illuminated and the transistor is turned OFF when the cell is dark.

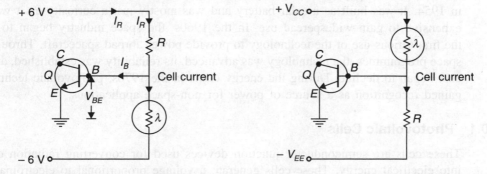

(a) Circuit to switch transistor OFF, when photoconductive cell is illuminated

(b) Circuit to switch transistor ON, when photoconductive cell is illuminated

Figure 7.24 Circuit diagram of photoconductive cell as switch transistor.

3. **As a voltage regulator:**   Voltage regulator employing a photoconductive cell is shown in Figure 7.25. This system is formed by a photoconductive cell, lamp and a resistor R_1. If the input voltage V_{in} falls in magnitude for any reason, the brightness of the lamp would also decrease. The decrease in illumination would cause increase in resistance R_λ of the photoconductive cell to maintain output voltage V_{out} at its rated level as determined by the voltage divider rule, i.e.

$$V_{out} = \frac{R_\lambda V_{in}}{R_1 + R_\lambda} \tag{7.80}$$

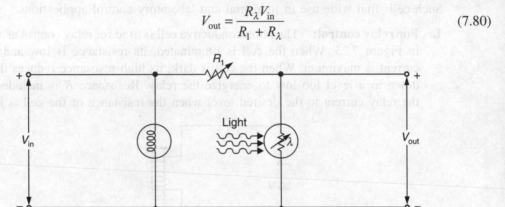

Figure 7.25   Voltage regulator using a photoconductive cell.

7.10   PHOTOVOLTAICS

Photovoltaics is the direct conversion of light into electricity at the atomic level. Some materials exhibit a property known as the photoelectric effect that causes them to absorb photons of light and release electrons. When these free electrons are captured, an electric current results that can be used as electricity.

The photoelectric effect was first noted by a French physicist Edmond Becquerel in 1839, who found that certain materials would produce small amounts of electric current when exposed to light. In 1905, Albert Einstein described the nature of light and the photoelectric effect on which photovoltaic technology is based, for which he later won a Nobel Prize in physics. The first photovoltaic module was built by Bell Laboratories in 1954. It was built as a solar battery and was mostly just a curiosity as it was too expensive to gain widespeared use. In the 1960s, the space industry began to make the first serious use of the technology to provide power abroad spacecraft. Through the space programmes, the technology was advanced, its reliability was established, and the cost began to decline. During the energy crisis in the 1970s, photovoltaic technology gained recognition as a source of power for non-space applications.

7.10.1   Photovoltaic Cells

These cells are semiconductor junction devices used for converting radiation energy into electrical energy. These cells generate a voltage proportional to electromagnetic radiation intensity and are called the *photovoltaic cells* because of their voltage generating capability.

Example: Silicon solar cell, which converts radiant energy of the sun into electrical energy.

Selenium and silicon are the most commonly used material for solar cells, though GaAs, InAs and CdS are also used.

Consider a *P–N* junction as depicted in Figure 7.26 with a resistive load. Even when there is no bias applied to the diode, an electric field exists in the space charge region as shown in Figure 7.26. Incident photon illumination can generate electron-hole pairs (EHPs) in the space charge region that will be swept out generating photocurrent I_λ in the reverse bias direction as illustrated.

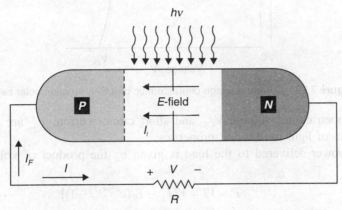

Figure 7.26 *P–N* junction solar cell with resistive load.

The photocurrent I_λ causes a voltage across the resistive load which biases forward the *P–N* junction and, therefore, a forward bias current I_F as shown in Figure 7.26. The resultant *P–N* junction current in the reverse bias direction is given by

$$I = I_\lambda - I_F = I_\lambda - I_S(e^{eV/k_BT} - 1) \tag{7.81}$$

where the ideal diode equation has been employed. As the diode becomes forward biased, the magnitude of the electric field in the space charge region reduces, but does not reduce to zero or charge direction. The photocurrent is always in the reverse bias direction and the resultant current is also always in the reverse bias direction.

The short circuit condition occurs when load resistance $R = 0$. The current in this case is called the short circuit current or

$$I = I_{SC} = I_\lambda \tag{7.82}$$

The open circuit condition occurs, when load resistance $R \to \infty$. The resultant current is zero and the voltage developed is *the open circuit voltage*. The photocurrent is just balanced by the forward biased junction current and, therefore, we have

$$I = 0 = I_\lambda - I_S[e^{eV_{OC}/k_BT} - 1] \tag{7.83}$$

The open circuit voltage V_{OC} can, therefore, be determined by equation

$$V_{OC} = V_T \ln\left(1 + \frac{I_\lambda}{I_S}\right) \tag{7.84}$$

The *I–V* characteristics is shown in Figure 7.27.

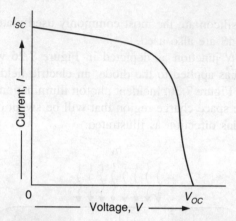

Figure 7.27 Current–voltage characteristic of a *P–N* junction solar cell.

The open circuit voltage V_{OC} and short circuit current I_{SC} are determined for a given level of light by the cell properties.

The power delivered to the load is given by the product of voltage and current, therefore,

$$P = VI = V[I_\lambda - I_S(e^{eV/k_BT} - 1)] \tag{7.85}$$

The power delivered to the load will be maximum of its derivative is zero, i.e.

$$\frac{dP}{dV} = 0$$

or

$$I_\lambda - I_S\left(e^{\frac{eV_m}{k_BT}} - 1\right) - I_SV_m\frac{e}{k_BT}e^{\frac{eV_m}{k_BT}} = 0 \tag{7.86}$$

where V_m is the voltage that makes the power delivered to the load maximum. Equation (7.86) may be rewritten as

$$\left(1 + \frac{V_m}{V_T}\right)e^{eV_m/k_BT} = 1 + \frac{I_\lambda}{I_S} \tag{7.87}$$

The value of V_m can be found by trial and error method. It is to be noted that the maximum power delivered to the load illustrated by the shaded rectangle in Figure 7.28 is less than the $V_{OC}I_{SC}$ product. The ratio $\dfrac{V_m I_m}{V_{OC} I_{SC}}$ is called the *fill factor* and is a figure of merit for solar cell design.

The worthnoting point is that J_S is a function of semiconductor doping concentrations. With the increase in doping concentration, J_S decreases, which results in increase of open circuit voltage V_{OC}. However, as the open circuit voltage V_{OC} is a function of log of I_λ and I_S, the V_{OC} is not a strong function of these parameters.

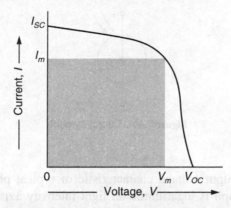

Figure 7.28 Maximum power rectangle of solar cell *I–V* characteristic.

The construction and cross-section of a typical power solar cell for use as an energy converter are given in Figure 7.29. The surface layer of *P*-type material is extremely thin so that light can penetrate to the junction. The nickel-plated ring around the *P*-type material is the positive output terminal, and the plating at the bottom of the *N*-type material is the negative output terminal. Power solar cells are also available in flat strip form for efficient coverage of available surface areas. The circuit symbol is often used for a photovoltaic cell (Figure 7.30.)

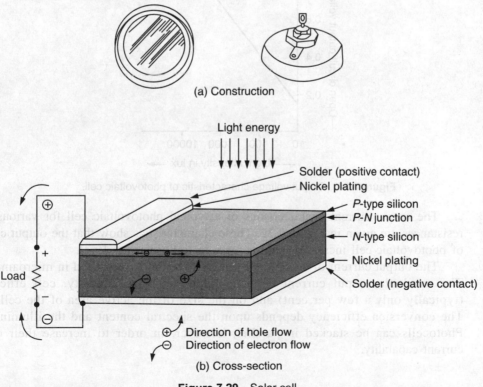

(a) Construction

Light energy

Solder (positive contact)
Nickel plating
P-type silicon
P-N junction
N-type silicon
Nickel plating
Solder (negative contact)

Load

⊕ Direction of hole flow
⊖ Direction of electron flow

(b) Cross-section

Figure 7.29 Solar cell.

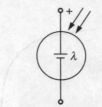

Figure 7.30   Circuit symbol

Characteristics

The open circuit output voltage characteristic of typical photovoltaic cell is given in Figure 7.31, the graph is logarithmic on light intensity axis. This characteristic shows that the cell is more sensitive for low light levels, because a small change in light intensity (say from 10 lux to 100 lux) can produce same increase in output voltage as a large change in light intensity (say from 100 lux to 1000 lux) at higher light intensity level.

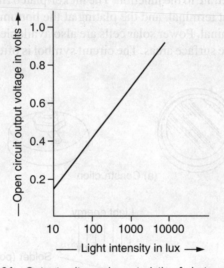

Figure 7.31   Output voltage characteristic of photovoltaic cell.

The output current characteristics of a typical photovoltaic cell for various load resistances are given in Figure 7.32. These characteristics show that the output current of photovoltaic cell increases with increase in light intensity level.

The output current of such a cell is very low and is measured in microamperes. The available output current depends upon the light intensity, cell efficiency (typically only a few per cent) and on the size of the active area of the cell face. The conversion efficiency depends upon the spectral content and the illumination. Photocells can be stacked in parallel, however, in order to increase their output current capability.

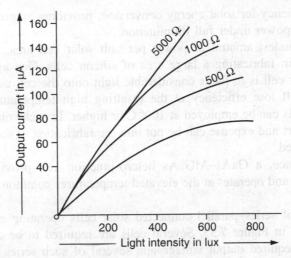

Figure 7.32 Output current characteristics of photovoltaic cell.

Applications

- The photovoltaic cell can be operated satisfactorily over a wide range of temperature (say from 100°C to 125°C). The temperature variations have little effect on short circuit current, but affect the open circuit output voltage considerably. There variations may be of the order of a few mV per °C in output voltage. An individual solar cell generates an open circuit voltage of about 500 mV (depending on light intensity when active). However, photocells can be connected in series to increase the available terminal voltage.

- The advantages of such devices are their ability to generate a voltage without any bias and their extremely fast response, i.e. these devices can be used as energy converters directly.

- Multiunit silicon photovoltaic devices may be used for sensing light in applications, such as reading punched cards in the data processing industry.

- Gold-doped germanium cells with controlled spectral response characteristic act as photovoltaic devices in the infrared region of the spectrum and may be used as infra-red detectors.

- Solar cells are extremely used as a source of power for many space satellites. The array of junctions can be distributed over the surface of the satellite or can be contained in solar cell "paddles" attached to the main body of the satellite. Applications of solar cells are not restricted to outer space.

- It is possible to have useful power from the sun in terrestrial applications using solar cells, even through the solar industry is reduced by atmosphere. About 1 kW/m^2 is available in a particularly sunny location, but not all of this solar power can be converted into electrical power.

- Much of the photon flux is at energies less than the E_g of the cell, and is not absorbed. High energy photons are strongly absorbed and the resulting electron-hole pairs may recombine at the surface. A well made silicon cell can have

10% efficiency for solar energy conversion, providing approximately 100 W/m² of electrical power under full illumination.

- This is modest amount of power per unit solar cell area, considering the effort involved in fabricating a large area of silicon cells. One approach to have more power per cell is to focus considerable light onto the cell using mirrors. Although silicon cell lose efficiency at the resulting high temperature, GaAs and related compounds can be employed at 100° C or higher. In such solar concentrator system more effort and expense can be put into the fabrication of solar cell, as fewer cells are required.

- For instance, a GaAs–AlGaAs hetero-junction cell provides good conversion efficiency and operates at the elevated temperatures common to solar concentration systems.

- A group of series–parallel connected solar cells operating as a battery charger is visualized in Figure 7.33. Several cells are required to be connected in series to give the required output voltage and several of such series connected groups are to be connected in parallel to provide the necessary output current.

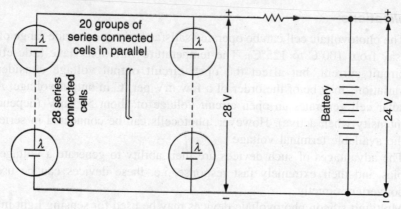

Figure 7.33 Group of series–parallel connected solar cells operating as a battery charger.

- Though photoconductive cell can be used for many of the same purposes that photovoltaic cells are used, but they can not be employed as energy sources. Photoconductive cells are preferred to photovoltaic cells when very sensitive response to changing light condition is required.

- When fast response is necessary, photovoltaic cells are preferred to photoconductive cells. Likewise, if a photocell is to be rapidly switched ON and OFF, photovoltaic cells are preferred because they can be switched at higher frequency than photoconductive cells. As a rough rule of thumb, photoconductive cell cannot be successfully switched at frequency higher than about 1 kHz, whereas photovoltaic cells can be switched successfully at frequencies upto about 100 kHz and sometimes higher even.

SOLVED NUMERICAL PROBLEMS

PROBLEM 7.1 Find the velocity of electrons in copper with Fermi energy of 7 eV. If the relaxation time of electrons is 3×10^{-14}s, find the mean free path.

Solution Given $E_F = 7.0$ eV $= 7.0 \times 1.6 \times 10^{-19}$ J, $\tau = 3 \times 10^{-14}$ s

The mean free path is $\lambda = v_F \times \tau$. We also know that $\frac{1}{2} mv_F^2 = E_F$

$$\therefore \qquad v_F = \sqrt{\frac{2E_F}{m}} = \sqrt{\left(\frac{2 \times 7.0 \times 1.9 \times 10^{-19}}{9.1 \times 10^{-31}} \right)} = 1.57 \times 10^6 \text{ m/s}$$

$$\therefore \qquad \lambda = v_F \times \tau = 1.57 \times 10^6 \times 3 \times 10^{-14} = 4.71 \times 10^{-8} \text{ m} = 471 \text{ Å}$$

PROBLEM 7.2 The thermal and electrical conductivities of copper at 20°C are 380 Wm^{-1}K^{-1} and 5.87×10^7 Ω^{-1}m^{-1}K^{-1} respectively. Calculate the Lorentz number.

Solution Given $\sigma = 5.87 \times 10^7$ Ω^{-1}m^{-1}, $T = 20$ °C $= 293$ K, $L = ?$
From Weidemann–Franz law,

$$\frac{K}{\sigma} = LT \quad \Rightarrow \quad L = \frac{K}{\sigma T} = \frac{380}{5.87 \times 10^7 \times 293} = 2.21 \times 10^{-8} \text{ W}\Omega\text{K}^{-2}$$

PROBLEM 7.3 The density of zinc is 7.13 g/cm^3 and its atomic weight is 65.4. Calculate its Fermi energy. The effective mass of a free electron in zinc crystal is 7.7×10^{-31} kg and the Avogadro's number is 6.02×10^{23} atoms/g atom.

Solution The electronic configuration of zinc ($Z = 30$) is $1s^2, 2s^2, 2p^6, 3s^2, 3p^6, 3d^{10}, 4s^2$.

This shows that each atom has 2 valence electrons which contributes to electron gas.

$$\text{Volume of zinc} = \frac{\text{Mass of 1 g atom}}{\text{Density}} = \frac{65.4}{7.13} = 9.17 \text{ cm}^3$$

The number of Zn atoms in this volume is 6.02×10^{23}. Thus number of atoms per unit volume is:

$$N = \frac{6.02 \times 10^{23}}{9.17} = 6.56 \times 10^{22} \text{ per cc}$$

Number of free electrons per unit volume is $2 \times 6.56 \times 10^{22} = 13.1 \times 10^{22}/\text{m}^3$

($\because$ Each atom contributes 2 valence electrons)

The Fermi energy of Zn is given by

$$E_{F(0)} = \frac{\hbar^2}{2m} \left(\frac{3N}{8\pi} \right)^{2/3}$$

$$= \frac{(6.63 \times 10^{-34})^2}{4 \times 3.14 \times 3.14 \times 2 \times 7.7 \times 10^{-31}} \left[\frac{3 \times 13.1 \times 10^{22}}{8 \times 3.14} \right]^{2/3}$$

$$= 17.78 \times 10^{-19} \text{ J} = 11 \text{ eV}$$

PROBLEM 7.4 Determine the temperature at which there is one percent probability that a state with energy 0.25 eV above the Fermi energy will be occupied by an electron.

Solution $$F(E) = \frac{1}{e^{(E-E_F)/k_BT} + 1} \implies \frac{1}{100} = \frac{1}{e^{0.25/k_BT} + 1}$$

Putting $$x = \frac{0.25}{k_BT}, \text{ then } 0.01 + 0.01\, e^x = 1 \implies e^x = 99$$

or $$x = \frac{0.25}{k_BT} = 2.303 \log_{10} 99$$

Hence $$T = \frac{0.25 \times 1.6 \times 10^{-19}}{2.303 \times \log_{10} 99 \times 1.38 \times 10^{-23}} = 630 \text{ K}$$

PROBLEM 7.5 Calculate the energy gap of a crystal which is transparent only for light of wavelength greater than 12345 Å, [Given $h = 6.6 \times 10^{-34}$ Js, $c = 3 \times 10^8$ m/s]

Solution   The energy of 12345 Å photon is:

$$E = \frac{hc}{\lambda} = \frac{(6.6 \times 10^{-34}) \times (3 \times 10^8)}{12345 \times 10^{-10}} \cong 1.6 \times 10^{-19} \text{ J} = 1 \text{ eV}$$

PROBLEM 7.6 A photocell cathode is coated with material having a work function of 3.5 eV. The photocell is irradiated of frequency 4×10^{15} Hz. Find the velocity of the emitted electrons.

Solution   We know that

$$E = h\nu$$

$$= 6.625 \times 10^{-34} \text{ J s} \times 4 \times 10^{15} \frac{1}{\text{s}}$$

$$= 25.5 \times 10^{-19} \text{ J}$$

We have $$E = h\nu = \phi + \frac{1}{2}mv^2$$

or $$25.5 \times 10^{-19} = 3.5 \times 1.6 \times 10^{-19} + \frac{1}{2} \times 9.1 \times 10^{-31} v^2$$

or $$v = 2.08 \times 10^6 \text{ m/s}$$

PROBLEM 7.7 A cadmium sulphide photodetector is irradiated over a receiving area of 4×10^{-6} m^2 by a light of wavelength 0.4×10^{-6} m and intensity 200 Wm^{-2}. Assuming that each quantum generates an electron-hole pair, calculate the number of pairs generated per second.

Solution   Given

$$\text{Wavelength of light } \lambda = 0.4 \times 10^{-6} \text{ m}$$
$$\text{Area of the crystal} = 4 \times 10^{-6} \text{ m}^2$$
and $$\text{Intensity of light} = 200 \text{ Wm}^2$$

Then, the number of photons $= \dfrac{\text{Intensity}}{h\nu} \times \text{Area}$

$$= \frac{200 \times 4 \times 10^{-6} \lambda}{hc}$$

$$= \frac{200 \times 4 \times 10^{-6} \times 0.4 \times 10^{-6}}{6.626 \times 10^{-34} \times 3 \times 10^{8}}$$

$$= 1.6098 \times 10^{15}$$

Since we know that each quantum will generate on electron-hole pair.

Therefore, the number of pairs generated per second $= 1.6 \times 10^{15}$.

PROBLEM 7.8   Determine the open circuit voltage of a silicon P–N junction solar cell at 300 K with the following parameters

$$I_\lambda = 12.5 \text{ mA}, \ I_S = 2.54 \times 10^{-11} \text{ A}$$

Solution   Open circuit voltage $V_{OC} = V_T \ln\left(1 + \dfrac{I_\lambda}{I_S}\right)$

where $V_T = k_B T = 26 \text{ mV}$

Then

$$V_{OC} = 26 \times 10^{-3} \ln\left[1 + \frac{12.5 \times 10^{-3}}{2.4 \times 10^{-11}} \right]$$

$$= 521.84 \times 10^{-3} \text{ A}$$

$$= 5.22 \times 10^{-5} \text{ A}$$

PROBLEM 7.9   The electrical conductivity of a semiconductor increases when electromagnetic radiation of wavelength shorter than 2480 nm is incident on it. Find the band gap of the semiconductor.

Solution   The bandgap is the minimum energy required to push an electron from the valence band to conduction band. Thus

$$\Delta E = h\nu = \frac{hc}{\lambda}$$

$$= \frac{6.64 \times 10^{-34} \times 3 \times 10^{8}}{2480 \times 10^{-9}} \text{ J} = 8.02 \times 10^{-20} \text{ J}$$

$$= \frac{8.02 \times 10^{-20}}{1.6 \times 10^{-19}} \text{ eV} = 0.5 \text{ eV}$$

/// **EXERCISES** ///

THEORETICAL QUESTIONS

7.1 What is free electron theory? Derive the expression for conductivity of metals on the basis of Drude–Lorentz theory.

7.2 Discuss quantum theory of free electrons and explain the following:
  (a) Fermi level
  (b) Density of states
  (c) Fermi–Dirac distribution function

7.3 Obtain an expression for energy levels in one-dimensional free electron gas.

7.4 How do you define Fermi energy and Fermi level? Derive an expression for Fermi energy.

7.5 Differentiate the quantum and classical theories of free electron.

7.6 Give Drude's theory of conduction.

7.7 Write notes on:
  (a) Energy levels and wave function of free electron in a box
  (b) Density of states in one dimension
  (c) Drude's theory of conduction
  (d) Fermi energy, Fermi velocity and Fermi temperature

7.8 Explain Kronig–Penney model for the motion of an electron in a periodic potential. Show from E–k graph that materials can be classified into conductors, insulators and semiconductors.

7.9 Give Kronig–Penney model of electron in a periodic potential. What are its consequences?

7.10 Discuss the origin of energy bands in solids. How can you distinguish between metals, semiconductors, and insulators on the basis of energy bands?

7.11 What is the relation of Fermi energy with temperature?

7.12 How does photoconductivity is explained in insulating crystals? Give suitable model and also discuss the factors which affect the photoconductivity.

7.13 How do photoconduction cell and photovoltaic cell differ? Explain.

7.14 What is photoconductivity? Discuss simple model of a photoconductor. Show that sensitive photoconductors should have long response time.

7.15 Using a simple model, explain the phenomenon of photoconductivity in insulating solids. How do traps contribute to the phenomenon? Explain.

7.16 Differentiate between photoconducting and photovoltaic cells.

7.17 Define the terms photoconductivity, sensitivity of a photoconductor and traps.

7.18 Discuss briefly the phenomenon and applications of photoconductivity.

7.19 What are solar cells? Describe in detail their construction, working, characteristics and applications.

7.20 Explain a photovoltaic cell and give its characteristics. What is 'fill factor'?

7.21 What are photovoltaic cells and how are they fabricated? Discuss their main characteristics?

7.22 Discuss energy conversion process is photovoltaic cell. Also give and explain the characteristics of a photovoltaic cell.

NUMERICAL PROBLEMS

P7.1 The atomic weight of sodium is 23 and its density 0.97 g/cc. Calculate the value of Fermi energy at 0 K. **(Ans. 3.153 eV)**

P7.2 The density of zinc is 7130 kg/m^3 and its atomic weight is 65.4. Calculate the Fermi energy at 0 K. **(Ans. 6.66 eV)**

P7.3 If $F(E) = 0.2$ at 300 K and Fermi energy of a metal is 3.2 eV, find energy at 300 K. **(Ans. 3.24 eV)**

P7.4 There are 2.54×10^{22} free electrons per cm^3 in sodium. Calculate its Fermi energy, Fermi velocity and Fermi temperature. **(Ans. 3.1 eV, 1.05×10^6 m/s and $T_F = 3.6 \times 10^4$ K)**

P7.5 Show that the average kinetic energy of an electron is 60% of Fermi energy at absolute zero.

P7.6 In germanium the energy gap is about 0.75 eV. Show that the crystal behaves as a transparent medium only for light of wavelength above 16533 Å.

P7.7 When yellow light from a sodium lamp falls over a photocell, a negative potential of 0.30 V is needed to stop all the electrons from reaching the collector. Find the potential needed to stop the electrons if light with $\lambda = 4500$ Å is used. **(Ans. $V_2 = 0.953$ V)**

P7.8 The electrical conductivity of a semiconductor increases when electromagnetic radiation of wavelength shorter than 2480 nm is incident on it. Find the bandgap of the semiconductor. **(Ans. $\Delta E_g = 0.5$ eV)**

P7.9 The energy of photoelectrons emitted from a sensitive plate is 1.40 eV. If the wavelength of the incident light is 192 nm, calculate the threshold wavelength. **(Ans. $\lambda_0 = 2.46 \times 10^{-7}$ m)**

P7.10 The band energy of a semiconductor is 1.9 eV. Calculate the wavelength of radiation emitted when a conduction band electron combines with a valence band hole in the material. **(Ans. $\lambda = 6250$ Å)**

P7.11 In a photocell, a copper surface was irradiated by light of wavelength 1849 Å, the stopping potential was found to be 2.72 eV. Calculate the maximum energy of photoelectrons. **(Ans. $I_{max} = 2.72$ eV)**

MULTIPLE CHOICE QUESTIONS

MCQ 7.1 At temperature above absolute zero, for $E \ll E_F$, the Fermi–Dirac function approaches:
    (a) $e^{-E/k_B T}$ (b) Zero
    (c) Unity   (d) Infinity

MCQ 7.2 The relation between average drift velocity v_d of electron in a metal is related to the applied electric field E and collision time τ as

    (a) $\dfrac{m}{ek\tau}$ (b) $\sqrt{\dfrac{m}{eE\tau}}$

    (c) $\sqrt{\dfrac{eE\tau}{m}}$ (d) $\dfrac{eE\tau}{m}$

MCQ 7.3 Which of the following relation gives Wiedemann–Franz law?

(a) $\dfrac{K}{\sigma} = LT$

(b) $K - \sigma = LT$

(c) $\dfrac{K}{\sigma} = \dfrac{L}{T}$

(d) $K\sigma = \dfrac{L}{T}$

MCQ 7.4 The general expression for Fermi energy of metal at 0 K is (n is concentration of electron)

(a) $E_{F_{(0)}} \propto n^{3/2}$

(b) $E_{F_{(0)}} \propto n^{2/3}$

(c) $E_{F_{(0)}} \propto n^{1/3}$

(d) $E_{F_{(0)}} \propto n^3$

MCQ 7.5 The quantum mechanical expression for Lorentz number is

(a) $\dfrac{\pi}{3}\left(\dfrac{k_B}{e}\right)^2$

(b) $\dfrac{\pi^2}{3}\left(\dfrac{k_B}{e}\right)^2$

(c) $\dfrac{\pi}{3}\left(\dfrac{e}{k_B}\right)^2$

(d) $\dfrac{\pi^2}{3}\left(\dfrac{e}{k_B}\right)^2$

MCQ 7.6 According to Bloch theorem, the solution of the Schrödinger wave equation for an electron moving in periodic potential is of the form

(a) $\psi_k(r) = e^{ikr}U_k(r)$

(b) $\psi_k(r) = e^{ikr}$

(c) $\psi_k(r) = \dfrac{e^{ikr}}{U_k(r)}$

(d) $\psi_k(r) = e^{ikr}U(kr)$

MCQ 7.7 In Kronig–Penney model if p << 1 [$p = \hbar k$], the energy to the lowest band is

(a) $E = \dfrac{p^2 m}{a^2}$

(b) $E = \dfrac{h^2 p}{4\hbar^2 ma^2}$

(c) $E = \dfrac{a^2 \hbar p}{m}$

(d) $E = \dfrac{ma^2 p}{\hbar}$

MCQ 7.8 In the Kronig–Penney model, if there is no potential barrier, then
(a) There are forbidden energy regions
(b) There are not forbidden energy regions
(c) All values of energy are not allowed
(d) There is a periodic dependence of E on k

MCQ 7.9 The phenomenon of increase of conductivity of an insulator, when light radiation falls on it, is called
(a) photoelectric effect
(b) Compton effect
(c) photoconductivity
(d) Frenkel's effect

MCQ 7.10 The direct conversion of light into electricity at atomic level is known as
(a) photoconductivity
(b) photovoltaics
(c) photoelectric effect
(d) none of these

MCQ 7.11 A photocell is illuminated by a intense source placed 1 m away. When the same source is placed 2 m away, the electrons emitted from the photocathode
(a) carry one quarter of their previous energy
(b) carry quarter of their previous momentum
(c) are one quarter in number
(d) are half in number

MCQ 7.12 Which of the following statements is wrong for photoconductive cells
(a) they have a forward biased P–N junction
(b) they are made of a single photosensitive material
(c) they are also called photoresistors
(d) they have a high dark to light resistance ratio

MCQ 7.13 For obtaining maximum power from a solar cell, it should be operated on
(a) horizontal part of the curve   (b) the knee voltage of V–I characteristics
(c) falling portion of V–I characteristics (d) any part of the V–I characteristics

MCQ 7.14 The expression for λ_g corresponding to energy gap ΔE_g is

(a) $\lambda_g = \dfrac{hc}{\Delta E_g}$

(b) $\lambda_g = \dfrac{hc}{\lambda}$

(c) $\lambda_g = \dfrac{hc}{E_F}$

(d) $\lambda_g = \dfrac{\hbar c}{\lambda}$

MCQ 7.15 The fill factor of a photovoltaic cell is expressed as

(a) $1 + \dfrac{I_\lambda}{I_S}$

(b) $\dfrac{V_m I_m}{V_{OC}\, I_{SC}}$

(c) $I_S[1 - e^{eV/k_B T}]$

(d) $I_S[e^{V/V_T} - 1]$

where symbols have their usual meanings.

Answers

7.1 (c)	**7.2** (d)	**7.3** (a)	**7.4** (b)	**7.5** (b)	**7.6** (a)	**7.7** (b)
7.8 (b)	**7.9** (c)	**7.10** (b)	**7.11** (c)	**7.12** (a)	**7.13** (b)	**7.14** (a)
7.15 (b)						

Superconductivity

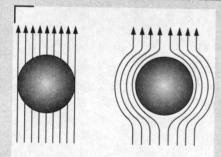

▌8.1   INTRODUCTION

Superconductivity has been intriguing subject since its discovery by Kammerlingh Onnes in 1911. The superconducting state of solid, which presents no resistance to the passage of electricity and repels a magnet, has not been a favourite ground for scientists interested in the study of solids, but is also of strong interest to technocrats because of the range of applications. We have been beneficiaries of superconducting technology. NMR imaging employed in medical diagnostics instead of the traditional CATSCAN employing X-rays uses large superconducting magnets. Many electronic and magnetic devices using superconducting substances have been fabricated. Although power transmission or energy storage using superconducting wire has not become a reality, there is every likelihood that there will be high speed levitating trains (Maglev trains) using superconducting technology. All these have been made with materials which become superconducting at liquid helium temperatures. The discovery of a new copper oxide superconductor ($T_C \sim 30$ K) by Bednorz and Müller in 1986 opened up new and exciting potentials. Today, we have many ceramic oxide materials, which become superconducting well above liquid nitrogen temperature.

There is every chance that some day one may discover materials, which become superconducting even at higher temperatures, possible close to room temperature. It would be truly remarkable, if we can have large superconducting magnet which at best would require chilled water for cooling. There are various potential applications of high temperature superconductors, although there are certain challenging problems that need to be solved before the ceramic oxide superconductors are exploited to full advantage.

It was discovered by Kammerlingh Onnes in 1911 that at a certain temperature and often within a very narrow temperature range, resistivity of certain metals becomes very small. He observed that the resistance of mercury falls by a factor 500 from its melting point (233 K) to 4.2 K, i.e., it becomes 10^{-5} Ω at 4.2 K. This absence of electrical resistance below a certain temperature was termed as *superconductivity* by Onnes.

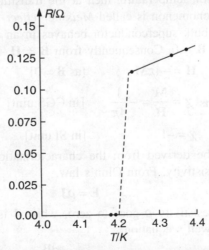

Figure 8.1   Resistance versus temperature of mercury as taken by Onnes in 1911.

During the next several years, Onnes and his collaborators showed that many metallic elements, e.g., lead, tin and iridium become superconductors at low temperatures. Superconductivity is not a unique for rare metallic elements and innumerable alloys. Only, it requires low temperatures.

Onnes also found, in 1913, that there is appearance of resistance in superconductors, which are brought into a material at a threshold value of the magnetic field.

This disappearance of superconductivity above a certain magnetic field was the first hint that superconductivity and magnetism were closely connected.

Superconductivity is absent in the ferromagnetic metals, monovalent metals and rare-earth elements except lanthanum (La), which has entirely empty 4*f* electronic shell. Note that magnetic impurities have a deterious effect on superconductivity in their host solid.

8.2 EXPERIMNENTAL SURVEY: MISCELLANEOUS TOPICS

The dc electrical resistivity in superconducting state is zero, one must expect persistent electrical currents to flow within such superconducting material. This fact has actually been observed by File and Mill who concluded that the decay time of supercurrent in a solenoid of $Nb_{0-75}Zr_{0.25}$ is not less than 10^5 years. Thus the existence of persistent current is a proof of occurrence of superconducting state.

It has been found that the superconducting properties of metals can be changed by varying temperature, magnetic field, magnetic stress, frequency of excitation of applied electric fields, impurities and with atomic structure, size and isotopic mass of the specimen. Some effects are discussed as follows.

8.2.1 Meissner's Effect

Meissner and Ochsenfeld found that if a superconductor is cooled in a magnetic field to below the transition temperature, then at the transition, the line of induction **B** is pushed out. This phenomenon is called *Meissner effect*.

It shows that a bulk superconductor behaves in an applied external field **H** as if inside the specimen **B** = 0. Consequently from **B** = **H** + 4π**M**, we get

$$\mathbf{H} = -4\pi\mathbf{M} \qquad [\text{as } \mathbf{B} = 0]$$

So the susceptibility is $\chi = \dfrac{\mathbf{M}}{\mathbf{H}} = -\dfrac{1}{4\pi}$ [in CGS uint]   (8.1)

or $\qquad\qquad\qquad \chi = -1 \qquad$ [in SI unit]   (8.2)

This result cannot be derived from the characterisation of a superconductor as a minimum of zero resistivity. From Ohm's law,

$$\mathbf{E} = \rho\mathbf{J} \tag{8.3}$$

We see that if the resistivity ρ goes to zero, while **J** is held finite, then **E** must be zero. But from Maxwell's equation,

$$\nabla \times \mathbf{E} = -\frac{\partial \mathbf{B}}{\partial t} \tag{8.4}$$

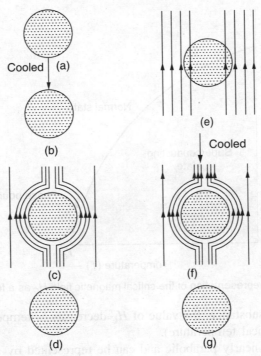

Figure 8.2   Magnetic behaviour of a superconductor, in (a) → (b); the specimen goes superconducting on cooling in the absence of an applied magnetic field; in (c), the magnetic field is applied to the superconductor; in (d), the magnetic field is removed; in (e) → (f), the specimen becomes superconducting on cooling in an external magnetic field *H* and in (g), the applied field is removed. The field inside a superconductor is seen to be zero, independent of its history.

which will imply that

$$\frac{\partial \mathbf{B}}{\partial t} = 0 \qquad (8.5)$$

This result predicts that the flux through the metal cannot change on cooling through the transition. The Meissner effect contradicts this result and suggests that perfect diamagnetism is an essential property of the superconducting states:

$\mathbf{E} = 0$ (from the absence of resistivity) and $\mathbf{B} = 0$ (from Meissner effect)

8.2.2   Effect of Magnetic Field: The Critical Magnetic Field

'The minimum applied magnetic field necessary to destroy superconductivity and restore the normal resistivity is called the critical magnetic field H_C.'

When the applied magnetic field exceeds, the critical field exceeds the critical value H_C, and the superconducting state is destroyed, and the material goes into the normal state. H_C depends on the temperature. The critical magnetic field H_C versus temperature plot is visualised in Figure 8.3. A specimen is superconducting below the curve and normal above the curve.

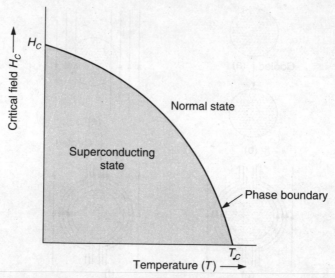

Figure 8.3   Schematic representation of the critical magnetic field H_C as a function of temperature.

For a given substance, the value of H_C decreases as temperature increases from $T = 0$ K to T_C (critical temperature).

The curve is nearly parabolic and can be represented by

$$H_C = H_0 \left(1 - \frac{T^2}{T_C^2} \right) \tag{8.6}$$

where H_0 is the critical field at 0 K.

Thus, the field has its maximum value H_0 at $T = 0$ K.

At $T = 0$ K,
$$H_C = H_0 \left(1 - \frac{0}{T_C^2} \right) = H_0$$

At $T = T_C$,
$$H_C = H_0 \left(1 - \frac{T_C^2}{T_C^2} \right) = 0$$

Equation (8.6) tells us the phase boundary between the normal and superconducting state.

EXAMPLE 8.1   The critical field for niobium is 1×10^5 A/m at 8 K and 2×10^5 A/m at 0 K. Calculate the critical temperature of the material.

Solution   Given $T = 8$ K, $H_C = 1 \times 10^5$ A/m, $H_0 = 2 \times 10^5$ A/m

We know that

$$H_C = H_0 \left(1 - \frac{T^2}{T_C^2} \right) \Rightarrow \frac{H_C}{H_0} = 1 - \frac{T^2}{T_C^2}$$

or $\quad \dfrac{T}{T_C} = \sqrt{\left(1 - \dfrac{H_C}{H_0}\right)}\quad$ or $\quad T_C = \dfrac{T}{\sqrt{\left(1 - \dfrac{H_C}{H_0}\right)}} = \dfrac{8\ \text{K}}{\sqrt{1 - \dfrac{1 \times 10^5}{2 \times 10^5}}} = 11.31\ \text{K}$

8.2.3 Isotope Effect

The critical temperature of superconductors varies with the isotopic mass. The transition temperature T_C is found to be proportional to the reciprocal of the square root of their respective isotopic masses:

$$T_C \propto 1/\sqrt{M} \qquad \text{or} \qquad T_C\sqrt{M} = \text{Constant}$$

$$T_C M^{\alpha} = \text{Constant} \tag{8.7}$$

Here $\alpha = \dfrac{1}{2}$.

A heavier isotopic mass lowers the lattice vibrations. The Debye temperature θ_D of the phonon spectrum is also proportional to $1/\sqrt{M}$ so that

$$\theta_D \sqrt{M} = \text{Constant} \tag{8.8}$$

Equations (8.7) and (8.8) indicate the lattice vibrations or the electron–phonon interaction are deeply involved in superconductivity. There is no other reason for the superconducting transition temperature to depend on the number of neutrons in the nucleus.

8.2.4 Persistent Current

When a superconducting material is placed in a magnetic field at transition temperature, then an induced current flows in the material such that it opposes the applied field and ejects external magnetic lines of forces outside the specimen as shown in Figure 8.4.

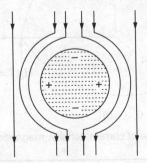

Figure 8.4 Illustration for persistent current.

Since the resistivity of superconductor in superconducting phase is zero, the current due to induction in specimen persists as long as the field is on. Hence, due to induced current, specimen acquires a negative magnetic moment, which proves the diamagnetic property of a superconductor or Meissner effect.

The induced current (which is permanent and continuous due to zero resistance) produced in superconductor at transition temperature in presence of field, is called *persistent current*.

8.2.5 Critical Current Density (Silsbee's Rule)

If a superconducting material carries a current such that the magnetic field which it produces is equal to H_C, the superconductivity disappears. The current density **J** at which the superconductivity disappears is called the *critical current density J_C*.

In 1916, Silsbee observed that for any value of $J < J_C$, the current can sustain itself, whereas for values of $J > J_C$, the current cannot sustain itself. This effect is known as *Silsbee rule* or *Silsbee effect*.

A superconducting ring of radius r ceases to be a superconductor, when the current is:

$$I_C = 2\pi r H_C \tag{8.9}$$

Critical current density $J_C = \dfrac{I_C}{\text{Area}}$

$$= \frac{2\pi r H_C}{\pi r^2} \Rightarrow J_C = \frac{2H_C}{r} \tag{8.10}$$

As H_C depends on temperature, hence J_C also depends on temperature. Also $J_C \propto 1/r$, therefore, J_C sets a definite limit to the size of the current that can pass through a superconducting ring, without destroying its superconducting.

The dependence of J, H and T is shown in Figure 8.5.

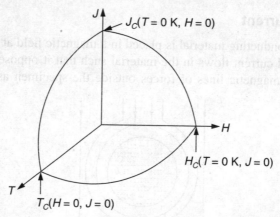

Figure 8.5   Dependence of current density, magnetic field and temperature.

EXAMPLE 8.2 Calculate the critical current which can flow through a long thin superconducting wire of aluminium of diameter 10^{-3} m. The critical magnetic field for Al is 7.9×10^3 A/m.

Solution   Given $H_C = 7.9 \times 10^{-3}$ A/m and $r = 5 \times 10^{-4}$ m.
Critical current $I_C = 2\pi r H_C = 2 \times 3.14 \times (5 \times 10^{-4} \text{ m}) \times (7.9 \times 10^3 \text{ A/m}) = 24.806$ A

8.2.6 Magnetic Levitation

The distinction between perfect conductors and superconductors, and also the Meissner effect, can be demonstrated by the magnetic levitation experiment. The experiment can be performed in two ways:

First method	Second method
We place a superconductor in a field-free region and then bring a magnet above it. The approaching magnet induces current in the superconductor according to Faraday's law, and the magnetic field of this current holds up the magnet. Since the material is superconducting, the current continues and the magnet remains levitated indefinitely.	We placed the magnet on the superconductor at a temperature $T > T_C$, and then cool the system. As the temperature falls below T_C, current starts flowing in the superconductor (to expel the applied magnetic field). This current lifts the magnet up and holds it levitated above the superconductor. This demonstrates the *Meissner effect.*

8.2.7 Entropy

In all superconductors, entropy decreases significantly on cooling below critical temperature. Entropy is a measure of randomness or disorderness of a system. The experiments show that the superconducting state is more ordered state than the normal state as shown in Figure 8.6.

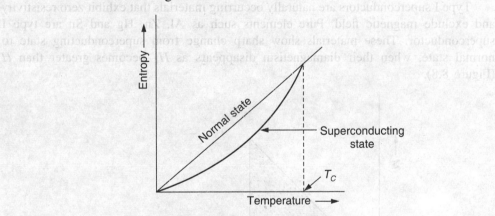

Figure 8.6 Entropy versus temperature plot.

8.2.8 Specific Heat

In normal metals, specific heat varies with absolute temperature as given by the following relation:

$$C_n(T) = \gamma(T) + \beta T^3 \tag{8.11}$$

Here the first term is specific heat of electrons in metals, and the second term is the contribution of lattice vibrations at low temperature.

But in superconducting state, specific heat is not linear with temperature, but varies exponentially as shown in Figure 8.7.

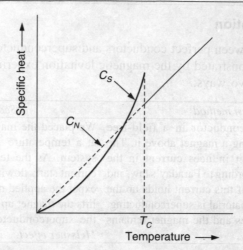

Figure 8.7   Temperature verses specific heat plot of a superconductor.

8.3   TYPES OF SUPERCONDUCTORS

Superconductors are classified as type I and type II superconductors based on their diamagnetic response.

Type I superconductors are naturally occurring materials that exhibit zero resistivity and exclude magnetic field. Pure elements such as Al, Zn, Hg and Sn are type I superconductor. These materials show sharp change from superconducting state to normal state, when their diamagnetism disappears as H_C becomes greater than H (Figure 8.8).

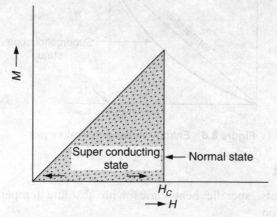

Figure 8.8   Type I superconductor.

The materials that show incomplete Meissner effect form type II superconductors. A number of alloys have been found to exhibit superconductivity. Unlike type I superconductors, these materials have much higher critical fields. As a result, they are capable of carrying much higher current densities, while remaining in superconducting state. Type II superconductors have high mechanical strength and hence are also called *hard superconductor.*

Since a type II superconductor does not totally exclude magnetic field, it is characterised by two critical fields H_{C_1}, and H_{C_2} as shown in Figure 8.9.

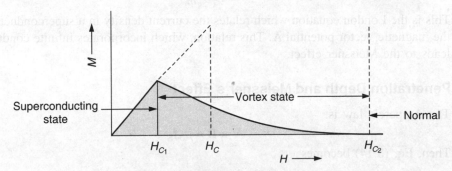

Figure 8.9 Type II superconductor.

When, the critical field reaches H_{C_1} or low critical field, filament within the material reaches to normal state. At this stage, normal state filaments within the materials are surrounded by filaments of superconducting state. This mixed state is called *vortex state.*

8.4 ELECTRODYNAMICS OF SUPERCONDUCTORS: LONDON'S EQUATIONS

The Meissner effect establishes the superconductors as a distinct equilibrium thermodynamic phase. The London's brothers, F. London and H. London in 1935 argued that in this phase, when an external electromagnetic field is applied, the system of electrons responds in a characteristic way, producing a certain electrical current density.

If the electrons in a superconductor move freely in an electric field, **E** then

$$\frac{d\mathbf{v}}{dt} = \frac{e\mathbf{E}}{m} \tag{8.12}$$

So

$$\frac{d}{dt}(n_s e\mathbf{v}) = \left[\frac{n_s e^2}{m}\right]\mathbf{E}$$

where n_s is the number density of such electrons, and e is the charge carried by each. But $n_s e\mathbf{V} = \mathbf{J_S}$, the electrical current density

Then

$$\frac{d}{dt}(\mathbf{J_s}) = -\frac{n_s e^2}{m}\left(\frac{\partial \mathbf{A}}{\partial t}\right) \tag{8.13}$$

Since $\mathbf{E} = -\left(\dfrac{\partial \mathbf{A}}{\partial t}\right)$. Integrating Eq. (8.13), we get

$$\mathbf{J_s} = -\frac{n_s e^2}{m}\mathbf{A} + C$$

The constant C depends on the initial conditions, i.e., the history of superconductor. London's brothers argued that $C = 0$, so that

$$\mathbf{J_s} = -\frac{n_s e^2}{m}\mathbf{A} \tag{8.14}$$

This is the London equation which relates the current density in a superconductor to be the magnetic vector potential $\mathbf{A}$. This relation, which incorporates infinite conductivity, leads to the Meissner effect.

8.4.1 Penetration Depth and Meissner's Effect

The Ampere's law is:

$$\nabla \times \mathbf{B} = \mu_0 \mathbf{J_S}$$

Then, Eq. (8.14) becomes

$$\nabla \times (\nabla \times \mathbf{B}) = -\left[\frac{n_s e^2}{m}\right]\mu_0 \mathbf{B}$$

or

$$\nabla(\nabla \cdot \mathbf{B}) - \nabla^2 \mathbf{B} = -\left[\frac{n_s e^2}{m}\right]\mu_0 \mathbf{B}$$

$$-\nabla^2 \mathbf{B} = -\left[\frac{n_s e^2}{m}\right]\mu_0 \mathbf{B} \qquad [\because \quad \nabla(\nabla \cdot \mathbf{B}) = 0] \tag{8.15}$$

Consider semi-infinite superconductor within the half space $x > 0$. A magnetic field B_e acts on it (i.e., in the region $x > 0$, there is a uniform magnetic field B_e). We take the direction of the field to be the z-axis.

So outside the superconductor [Figure 8.10] and inside the superconductor, we expect the magnetic field to be still along the z-direction and to depend on x, i.e.,

$$\mathbf{B}(x) = \mathbf{e_Z}B(x) \tag{8.16}$$

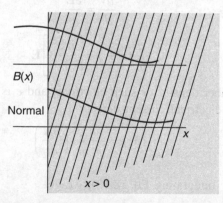

Figure 8.10   A superconductor is the half space $x > 0$ (shaded) placed in an external field H_C. The magnitude of the field $B(x)$ and the current density are shown.

Then Eq. (8.14) becomes

$$\frac{d^2 B(x)}{dx^2} = \frac{\mu_0 n_s e^2}{m} B(x) \tag{8.17}$$

This has a solution:

$$B(x) = (\mu_0 B_e) \exp\left(\frac{-x}{\lambda}\right) \tag{8.18}$$

where

$$\lambda = \sqrt{\frac{m}{n_s e^2 \mu_0}} \tag{8.19}$$

Here we choose the solution which matches with the outside field at $x = 0$ because any discontinuity at $x = 0$ means an infinite surface current. Also because of finiteness of energy, we choose the exponentially decaying rather than the exponentially growing solution. The decay length λ is a measure of the distance to which the magnetic field goes in and is called the *London's penetration depth*. Deep inside the superconductor, for $x \geq \lambda$, the field B is zero. We thus have a detailed theory of Meissner effect.

Let $\mu_0 B_e = B_0$, then $B(x) = B_0 e^{-x/\lambda}$

When $\qquad\qquad x = \lambda, \qquad \dfrac{B(x)}{B_0} = \dfrac{1}{e} \tag{8.20}$

Hence the penetration depth is defined as the distance into the superconductor at which the magnitude of magnetic field $B(x)$ falls to $1/e$ of its value at the surface B_0.

8.4.2 Drawback of the Theory

Though the London's theory explains the two conditions, i.e. **E** = 0 and **B** = 0 characterising the superconducting state, but it does not give any insight into the underlying electronic processes in superconductors. Bardeen, Copper and Schrieffer (BCS) theory gives the better understanding of the phenomenon of superconductivity.

The variation of penetration depth may also be expressed as:

$$\lambda(T) = \frac{\lambda(0)}{\sqrt{1 - \dfrac{T^4}{T_C^4}}} \tag{8.21}$$

where $\lambda(T)$ is the value at temperature T, and $\lambda(0)$ is the value at absolute zero.

EXAMPLE 8.3 Find London's penetration depth for lead having superconducting electron density of 3×10^{28} m^{-3} at 5 K. The transition for lead is 7.22 K.

Solution Given $m = 9.1 \times 10^{-31}$ kg, $n_s = 3 \times 10^{28}$ m^{-3}

$$\mu_0 = 4\pi \times 10^{-7} \text{ H/m}, e = 1.6 \times 10^{-19} \text{ C}$$

Then $\qquad\qquad \lambda^2 = \dfrac{9.1 \times 10^{-31}}{4\pi \times 10^{-2} \times 3 \times 10^{28} \times (1.6 \times 10^{-19})^2}$

$$\Rightarrow \qquad\qquad \lambda \cong 30 \text{ nm}, \ \lambda(0) \cong 30 \text{ mn}$$

For evaluating $\lambda(T) = \lambda_0 \left[1 - \left(\dfrac{T}{T_C} \right)^4 \right]^{-1/2} = 30 \text{ nm} \left[1 - \left(\dfrac{5}{7.22} \right)^4 \right]^{-1/2} = 34.18 \text{ nm}$

8.5 ELEMENTS OF BCS THEORY

In 1957, Bardeen, Cooper and Schrieffier gave a theory to explain the phenomenon of superconductivity, which is known as BCS theory.

The BCS theory is based upon the formation of Cooper pairs, which is purely a quantum mechanical concept. During the flow of current in a superconductor, when an electron comes near a positive ion core of the lattice, it experiences an attractive force because of opposite charge polarity between electron and the ion core. The ion core will be displaced from the position due to this interaction which is called *lattice distortion*. Now, an electron that comes near that place, will also interact with the distorted lattice, which tends to reduce the energy of the electron. This process is looked upon as equivalent to interaction between the two electrons via the lattice.

The lattice vibrations are quantised in terms of *phonons*. Thus the process is called *electron–lattice electron interaction via phonon field*. Because of the reduction of energy during the interaction, it is treated as equivalent to establish an attractive force between the two electrons, which is shown by Cooper to be maximum, if the two electrons have equal and opposite spins and opposite momenta. The attractive force thus established may exceed the Coulomb repulsive force between the two electrons at temperatures below the critical temperature, leading to the formation of Cooper pairs.

Cooper pair is a bound pair of electrons formed by the interaction between the electrons with opposite spin or momentum in a phonon field.

As per quantum mechanical rules, a wave function could be associated with a Cooper pair by treating it as a single entity. Such a wave function has a property that it extends over a fairly large volume with finite value for its amplitude all over the region. As a result, the wave functions associated with similar Cooper pairs start overlapping. Typically for a given Cooper pair, such an overlapping may extend over 10^6 other pairs. Thus it extends virtually over the entire volume of the superconductor. This leads to a union of vast number of Cooper pairs in which motion of all such pairs are under strong correlation—one adding the motion of the other. It results in an effect equivalent to the entire union moving as a single unit. The resistance encountered by any single Cooper pair is simply overcome by a cooperative action of other pairs in the union. Thus we can define superconductivity as follows:

When the electrons flow in the form of Cooper pairs in materials, they do not encounter any scattering and the resistance factor vanishes or in other words, conductivity becomes infinity, which is named as superconductivity.

In addition to the isotope effect, the idea that the electron–lattice interaction plays an important role in superconductivity is supported by the fact that the best of the conductors such as gold, silver and copper do not exhibit superconductivity. The reason attributed is that the electrons in those metals move so freely in the lattice that the electron–lattice interaction is virtually absent. These rules out the possibility of formation of Cooper pairs and also that of occurrence of superconductivity in the materials.

8.6 FLUX QUANTIZATION

In 1957, A. Abrikosov predicted that magnetic flux within a superconducting ring is quantised. The magnetic flux enclosed by superconducting ring is given by

$$\phi = n\left(\frac{h}{2e}\right) = n\phi_0 \tag{8.22}$$

where n is an integer and $\phi_0 = \dfrac{h}{2e} = 2.067 \times 10^{-15}$ weber (T m^2)

It is known as *flux quantum* or *fluxoid* or *fluxon*.

Flux quantization is special property of superconductors and the magnetic flux quantization does not occur in ordinary solenoid.

8.7 JOSEPHSON EFFECTS

In quantum mechanics, a particle is described by a wave function, and it has a finite probability of tunneling through barrier. I. Griever demonstrated that an electron can tunnel through a thin insulator barrier. A tunnel current exists, when the electron wavelength is comparable to or greater than the thickness of the barrier. Josephson, in 1962, showed that tunneling of Cooper pairs was possible similar to tunneling of normal unpaired electrons.

A schematic arrangement of Josephson junction is shown is Figure 8.11. It consists of two superconductors separated by a thin barrier insulating layer (10–20 Å). The effects observed by Josephson are as follows:

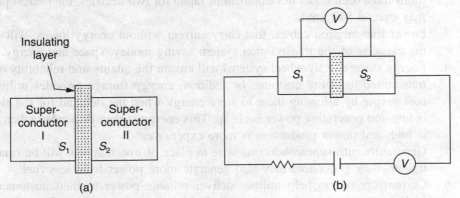

Figure 8.11 (a) Josephson junction, (b) Explanation of Josephson effect.

The dc Josephson Effect

According to this effect, a dc current flows across the junction, when no voltage is applied across it due to tunneling of Cooper electrons.

The ac Josephson Effect

When a dc voltage is applied across a Josephson junction, an ac current is produced. The frequency of the ac current is $2eV/h$ where V is the dc voltage. By measuring the frequency and voltage, the value of e/h can be determined. Hence the effect is utilised to measure e/h precisely. For example, when $V = 1$ μV, an ac current of 483.6 MHz is developed across the junction.

Macroscopic Quantum Interference

This effect describes influence of applied magnetic field on supercurrent flowing through the junction. According to this effect, if a dc magnetic field is applied through a superconducting circuit containing two junctions, the maximum supercurrent shows interference effects which depend on the intensity of the magnetic field.

8.8 APPLICATIONS OF SUPERCONDUCTIVITY

The capability of retaining superconductivity at high current densities make superconductor useful in a variety of applications in electric power engineering, transportation, medical diagnostics and microelectronics

Some futuristic applications of superconductors include transmission lines that carry power without resistance, medical diagnostic equipment that eliminate the need of surgery, levitating trains and so on. Some of the current uses of superconductors and some that hold the most promise for near future are as follows:

1. Magnetic resonance imaging (MRI) machines enhance medical diagnostics by imaging internal organs often eliminating the need for invasive surgeries. MRIs, which are currently made from low temperature superconductors, will be smaller and less expensive when made with high temperature superconductors (HTS).
2. Maglev trains that seem to float on air because of superconducting magnets. These trains have been under development in Japan for two decades, the nearest prototype may exceed 547 km/h.
3. Power transmission cables, that carry current without energy losses, will increase the capacity of the transmission system saving money, space and energy.
4. Energy storage in fly wheel systems will ensure the quality and reliability of power transmitted to utility customs. In addition, energy storage provides utilities with cost saving by allowing them to store energy when the demand for the electricity is low and generating power is cheap. This energy is then dispended when demand is high and power production is more expensive.
5. Generators with superconducting wire in place of iron magnets will be smaller and lighter. New generators may also generate more power from less fuel.
6. Current controllers help utilities deliver reliable power to their customers. HTS fault current limiters detect abnormally high current in utility. Then they reduce the fault current so that the system equipment can handle it.
7. Motors made with superconducting wire will be smaller and more efficient.

SOLVED NUMERICAL PROBLEMS

PROBLEM 8.1 Prove that Meissner's effect contradicts the Maxwell's equation.

Solution The resistivity of a superconductor is zero below T_C, i.e.,

$$\rho = 0 \tag{i}$$

From Ohm's law, we have

$$\mathbf{J} = \sigma \mathbf{E} \quad \text{and} \quad \mathbf{E} = \frac{1}{\sigma} \mathbf{J} = \rho \mathbf{J} \tag{ii}$$

From Eqs. (i) and (ii), we have

$$\mathbf{E} = 0 \quad \text{or} \quad |\mathbf{E}| = E = 0 \tag{iii}$$

From Maxwell's third equation,

$$\nabla \times \mathbf{E} = -\frac{\partial \mathbf{B}}{\partial t} \tag{iv}$$

From Eqs. (iii) and (iv), we have

$$\frac{\partial \mathbf{B}}{\partial t} = 0 \quad \Rightarrow \quad \mathbf{B} = \text{Constant} \tag{v}$$

But according to Meissner's effect for superconductor, $\mathbf{B} = 0$. Hence Meissner's effect contradicts the Maxwell's equation.

PROBLEM 8.2 A superconducting material has a critical temperature of 3.7 K and a magnetic field of 0.0306 T at 0 K. Find the critical field at 2 K.

Solution Given $H_0 = 0.0306$ T, $T_C = 3.7$ K, $T = 2$ K
   We know that the critical magnetic field is:

$$H_C = H_0 \left[1 - \left(\frac{T}{T_C} \right)^2 \right] = 0.0306 \left[1 - \left(\frac{2}{3.7} \right)^2 \right]$$

$$= 0.0306 \, [1 - 0.2921] = 0.0217 \text{ T}$$

PROBLEM 8.3 Calculate the critical current for a superconducting wire of lead having a diameter of 1 mm at 4.2 K. Critical temperature for lead is 7.18 K and $H_C (0) = 6.5 \times 10^4$ A/m.

Solution Given $H_0 = 6.5 \times 10^4$ A/m, $T_C = 7.18$ K, $r = 0.5 \times 10^{-3}$ m, $T = 4.2$ K, $I_C = ? \ H_C = ?$
   We have critical magnetic field:

$$H_C = H_0 \left[1 - \frac{T^2}{T_C^2} \right] = 6.5 \times 10^4 \left[1 - \left(\frac{4.2}{7.18} \right)^2 \right] = 4.276 \text{ A/m}$$

We also know that critical current is:

$$I_C = 2\pi r H_C = 2 \times 3.14 \times 0.5 \times 10^{-3} \times 4.276 \times 10^4 = 134.26 \text{ A}$$

PROBLEM 8.4 The critical temperature for metal with isotopic mass 199.5 is 4.185 K. Calculate the isotopic mass of metal when the critical temperature falls to 4.133 K.

Solution   Given $T_{C_1} = 4.185$ K, $M_1 = 199.5$, $T_{C_2} = 4.133$ K, $M_2 = ?$

We know from isotopic effect, the critical temperature varies with isotopic mass:

$$T_C \sqrt{M} = \text{Constant}$$

Then $T_{C_1} \sqrt{M_1} = T_{C_2} \sqrt{M_2}$ or $T_{C_1}^2 M_1 = T_{C_2}^2 M_2$

$$\Rightarrow \qquad M_2 = M_1 \left(\frac{T_{C_1}}{T_{C_2}} \right)^2 = 199.5 \times \left(\frac{4.185}{4.133} \right)^2 = 199.5 \times 1.013 = 204.55$$

PROBLEM 8.5 If a Josephson junction has a voltage of 8.50 μV across its terminals, calculate the frequency of the alternating current. [Planck's constant $h = 6.626 \times 10^{-34}$ Js].

Solution   Given $V_0 = 8.50$ μV $= 8.5 \times 10^{-6}$ V,

Then frequency of alternating current is:

$$\phi = \frac{2eV_0}{\hbar} = \frac{2 \times 1.6 \times 10^{-19} \text{ C} \times 8.5 \times 10^{-6} \text{ V} \times 2\pi}{6.626 \times 10^{-34} \text{ Js}} = 7.35 \times 10^9 \text{ Hz}$$

///// **EXERCISES** /////

THEORETICAL QUESTIONS

8.1 What are superconductors? Explain the occurrence of superconductivity. What are the different types of superconductors? Explain them in detail.

8.2 Explain the property of superconducting materials in detail.

8.3 What is Meissner's effect? Explain.

8.4 What is superconductivity? Describe the effect of
(a) magnetic field
(b) frequency
(c) isotopes on superconductors

8.5 Distinguish between type I and type II superconductors. Explain how the penetration depth varies with
(a) magnetic field strength
(b) temperature

8.6 Show that London's equation leads to Meissner's effect. Hence explain that thin film do not exhibit Meissner's effect.

8.7 How the various observable phenomena of superconductors such as Meissner's effect, magnetic field effect can be explained by BCS theory? How will you use the property of superconductors in calibration of very small voltages?

8.8 Explain why good conductors cannot be superconductors.

8.9 What is flux quantisation? Describe Josephson effects and their applications.

8.10 Write notes on the applications of superconducting materials.

NUMERICAL PROBLEMS

P8.1 A superconducting lead has a critical temperature of 7.26 K at zero magnetic field and a critical field of 8×10^5 A/m at 0 K. Find critical field at 5 K.

(**Ans.** 4.2×10^5 A/m)

P8.2 Estimate the London penetration depth from the following data:

Critical temperature = 3.7 K
Density = 7.3 g cm^{-3}
Atomic weight = 118.7
Effective mass m^* = 1.9 m [where m = mass of free electron]

(**Ans.** $\lambda = 381$ Å)

P8.3 The penetration depth λ of Hg at 3.5 K is about 750 Å. Estimate the value of (a) λ and (b) n_s (superconducting electron density) as T tends to 0 K. (**Ans.** (a) 530 Å (b) 10^{28}/m^3)

P8.4 A superconducting tin has a critical temperature of 3.7 K in zero magnetic field and a critical field of 0.0306 T at 0 K. Find the critical field at 2 K. (**Ans.** 0.0217 T)

P8.5 Using normal isotopic effect, calculate the transition temperature for samples of ^{201}Hg and ^{202}Hg and ^{204}Hg given that the transition temperature for ordinary mercury, of an average atomic mass 200.59 is 4.153 K. (**Ans.** 40149 K, 4.138 K, 4.118 K)

P8.6 The London's penetration depths for lead at 3 K and 7.1 K are respectively 396 Å and 1730 Å respectively. Calculate the penetration depth at 0 K. (**Ans.** 7.19 K, 390 Å)

P8.7 A specimen of V$_3$Ga has a transition temperature 14.5 K. At 14 K, the critical field is 1.4×10^5 A/m and at 13 K, 4.2×10^5 A/m. Determine critical field at 4.2 K.

(**Ans.** 14.5 K, 20×10^5 and 18.3 A)

P8.8 London's penetration depth for a sample at 5 K and 7 K are 41.2 nm and 180.3 nm respectively. Find its transition temperature and penetration depth at 0 K.

(**Ans.** $\lambda(0) = 29.5$ nm)

P8.9 Two lead isotopes of atomic masses 206 and 210 have T_C values 7.193 K and 7.125 K respectively. Calculate the value of β. (**Ans.** $\beta = 0.4962$)

P8.10 For a superconductor $T_C = 3$ K and $n_S = 10^{28}$/m^3. Find the penetration depth at 0 K and 1 K. (**Ans.** $\lambda(0) = 530$ Å and $\lambda(1) = 533$ Å)

MULTIPLE CHOICE QUESTIONS

MCQ 8.1 Two essential and independent properties of a superconducting material are
(a) Zero resistivity and ferromagnetic   (b) Zero resistivity and diamagnetic
(c) High resistivity and paramagnetic   (d) High resistivity and ferromagnetic

MCQ 8.2 The critical magnetic field H_C at temperature T is

(a) $H_0 \left[1 - \left(\dfrac{T}{T_C} \right)^2 \right]$

(b) $H_0 \left[1 - \left(\dfrac{T_C}{T} \right)^2 \right]$

(c) $H_0 \left[1 - \left(\dfrac{T}{T_C} \right) \right]$

(d) $H_0 \left[\left(\dfrac{T}{T_C} \right)^2 - 1 \right]$

MCQ 8.3 Meissner's effect is related to
   (a) Optical properties of superconductors
   (b) Magnetic properties of superconductors
   (c) Thermal properties of superconductors
   (d) Electrical properties of superconductors

MCQ 8.4 The London's penetration depth is given by

   (a) $\lambda = \sqrt{\dfrac{n_s}{m\mu_0 e^2}}$ (b) $\lambda = \sqrt{n_s m \mu_0 e^2}$

   (c) $\lambda = \sqrt{\dfrac{m}{n_s m \mu_0 e^2}}$ (d) $\lambda = \sqrt{\dfrac{n_s m \mu_0 e^2}{m}}$

MCQ 8.5 For a material to be considered as a superconductor, it has to exhibit
   (a) Only zero resistivity ($\rho = 0$)
   (b) Only Meissner's effect ($\mathbf{B} = 0$) inside the superconductor
   (c) Zero resistivity and Meissner's effect both
   (d) Only Josephson effect

MCQ 8.6 The tunneling of Cooper pairs between two superconductors separated by an insulator even in the absence of applied voltage between the superconductors is known as
   (a) Josephson effect   (b) ac Josephson effect
   (c) dc Josephson effect   (d) London's effect

MCQ 8.7 When a material makes transition from the normal to the superconducting state, it pushes out the magnetic flux from its interior. This phenomenon is known as
   (a) Meissner effect   (b) Magnetic levitation
   (c) Josephson effect   (d) London effect

MCQ 8.8 A vortex state is observed in
   (a) Type I superconductor
   (b) Type II superconductor
   (c) Both type I and type II superconductors
   (d) Neither type I nor type II superconductor

MCQ 8.9 The transition temperature of mercury is
   (a) 4.2 K   (b) 7.5 K
   (c) 20 K   (d) 12 K

MCQ 8.10 Maglev vehicles are constructed, based on
   (a) Gravitation effect   (b) Electrical effect
   (c) Meissner's effect   (d) None of these

Answers

8.1	(b)	**8.2**	(a)	**8.3**	(b)	**8.4**	(c)	**8.5**	(c)	**8.6.**	(a)	**8.7**	(a)
8.8	(b)	**8.9**	(a)	**8.10**	(c)								

CHAPTER 9

X-Rays

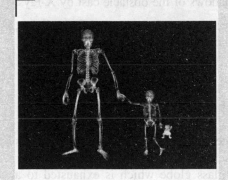

9.1 INTRODUCTION

The discovery of X-rays was accidently made by a German Professor, Röntgen, in 1895, while studying the discharge of electricity through gases at very low pressures. Röntgen found that a fluorescent screen, in the vicinity, coated with zinc sulphide (ZnS) or barium platinocyanide became luminous, even when covered with black paper. A photographic plate completely wrapped in dark paper was also affected by the discharge. Röntgen also found that when heavy objects like a plate of iron or a bunch of keys were interposed between the screen and the tube, they cast their shadows on the fluorescent screen, providing that fluorescence was due to some radiations coming out of the discharge tube at very low pressures. Because of the unknown (X) nature of these rays at the time of Röntgen, they were called *X-rays*. They are also known as *Röntgen rays*. Röntgen also found that X-rays were produced, when fast moving electrons (cathode rays) were stopped suddenly by an obstacle. The chief distinguishing properties of X-rays then known were as follows:

1. Luminescence of barium platinocyanide or zinc sulphide screen
2. Penetration through light objects like wood, paper, flesh, etc. which are opaque to light
3. Affecting photographic plates and recording shadows of the obstacle cast by X-rays.

As we know, now these properties of X-rays have been extensively used in medicine, surgery; and also have been put to wide use in industry.

9.2 PRODUCTION OF X-RAYS

9.2.1 By Röntgen Tube

The original Röntgen tube was rather a crude piece of apparatus. Many improvements had to be effected, since then, to get more intense beam of X-rays.

The modern X-rays tube consists of a large glass globe which is exhausted to a pressure of 10^{-4} mm of Hg as shown in Figure 9.1.

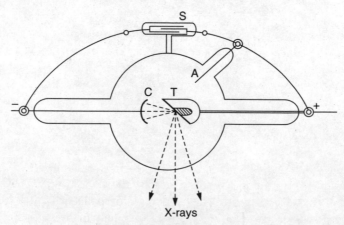

Figure 9.1 Röntgen X-rays tube.

A cathode C is a concave piece of aluminium that focuses the cathode ray on a target T which is placed at its centre of curvature. It is to be noted that the focusing spot should be sharp enough, otherwise the heating will be large and the anticathode will be melt. The anticathode or target is used with its face inclined at an angle 45° to the incident cathode ray beam, so that X-ray beam may be emitted in a perpendicular direction. A potential difference of about 100000 V is applied between the anticathode and the cathode. A subsidiary electrode called the anode is also used and is connected to the anticathode electrically. S is a softening device which emits gas on heating, when the X-ray tube becomes very hard for operation.

Aluminium is chosen for the cathode, since it produces the least amount of sputtering and thus preventing, the blackening of the glass bulb with a thin layer of the metal of the cathode. Platinum, tantalum and tungsten have high atomic number, high melting point, high electrical conductivity and low vapour pressure at high temperatures and are suitable materials for anticathode. Tungsten is the best. A heavy block of this metal is mounted on a thick copper rod, or alternatively, a small brick of the metal is inlaid in a heavy block of copper. The anticathode is cooled by radiation, by providing radiating fins of very thick copper which are connected to the anticathode rod just outside the X-ray tube. A more efficient device, however, is to make the anticathode tube hollow, and cool it by a current of cold water. Some cooling device is necessary for a long running of the tube, as large quantity of heat is generated by a cathode rays beam. Only a small fraction of the electrical energy, 0.1% to 0.2%, is converted into X-rays.

The function of the subsidiary anode is complex, but when used, it has following effects:

1. The life of the tube is prolonged.
2. The discharge becomes smoother and hence the intensity of the X-ray beam is constant.
3. The efficiency of the tube is increased, i.e., more of the energy of the cathode ray is converted into X-rays under like conditions.
4. The damage of the tube, when run with an induction coil, whose voltage is not quite unidirectional, is avoided.

The behaviour of the tube depends upon the pressure of the gas inside the tube and the applied potential. If the vacuum is comparatively low, say of the order of 10^{-4} cm, a comparatively low voltage is needed for the discharge, and the penetrating power of X-rays produced is less and they are *soft X-rays*. If the vacuum is high, of the order of 10^{-5} cm, a comparatively high voltage is needed to run the tube and the resulting X-rays have large penetrating power and are said to be *hard X-rays*.

During operation of X-ray tube, the positive ions of the gas are removed, and hence the gas filled tubes tend to become more evacuated or harder and greater potential is needed to run the discharge, and the emitted X-rays become harder. If the pressure falls considerably, the tube may even refuse to allow the discharge. For this purpose, softening devices are employed. At first, a rod of palladium was used, one-half of which was inside the tube. On heating the outer half of the rod in a Bunsen flame, the occluded hydrogen is liberated and enters the X-ray tube and thus increases its conductivity. A better softening device is a side tube containing platinised asbestos in which gases have been occluded when the tube becomes too hard; the discharge passes through the side tube, liberating some occluded gases and the tube soft again [Figure 9.1].

The main defects of the Röntgen tubes are:

1. The intensity of X-ray beam cannot be controlled.
2. The quality of X-rays cannot be controlled.
3. The softening device is not perfect, and there is a limit to the introduction of ions by this method.
4. The cathode is likely to be damaged, when the tube is run by an induction coil, by the reverse current which may produce great amount of heat on the cathode and cause excitation of cathode rays in the opposite direction.

The defects have been removed in the Coolidge X-ray tube which employs a heated filament as the cathode and is self rectifying.

9.2.2   By Coolidge Tube

The Coolidge tube was designed by Coolidge in 1912. The essential parts of this tube are shown in Figure 9.2.

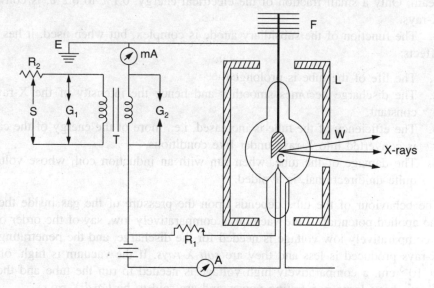

Figure 9.2   Coolidge tube.

It consists of a hot cathode which produces streams of electrons that can be used to control both the intensity and the quality of the X-rays. The cathode is heated by 4–12 V battery or by a separate tapping from the transformer T. The filament current is adjusted by a rheostat R_1 and is read by an ammeter A. By regulating this current, the intensity of X-ray beam can be altered. A larger filament current produces a more intense X-ray beam. The intensity of X-rays is measured in Röntgen, which is the quantity of X-ray radiation in 1 cc of air produces 1 esu of charge of either sign by ionization. The electron beam is focused on the target by surrounding the filament by a cylindrical tube S, called the *shield*, which is charged negative with respect to the filament. Such an arrangement is called the *electron gun*.

The anticathode or target is cooled by providing it with radiating fin F which consists of a thick copper spiral, radiating heat to the surrounding air. In some types of tubes, water cooling arrangement is provided. The high voltage is taken from a large step up transformer T. The X-ray tube works as its own rectifier. The current passes only when the anticathode is positive with respect to filament, since the emitted electrons can move only from the filament (negative) to the anticathode (positive) and not vice-versa. The anode current through the tube is read by miliammeter (mA) and is regulated by rheostat R_2. This regulates the applied voltage across the tube and controls the quality of X-rays produced. A large voltage or current in mA produces harder and more penetrating X-rays.

A spark gap G_1, about 1 mm in length, is placed across the primary of the transformer to prevent any damage due to break down in this circuit. A similar adjustable spark gap G_2 is provided across the secondary so that a discharge may pass across it. In this case, a few number of thermions are emitted to conduct current through Coolidge tube. This prevents arcing between the electrodes.

The tube is enclosed in a lead box L about 5 mm thick and is provided with a small window W through which the X-rays emerge.

In an improved design of tube, called the rotating anode type, the cooling arrangement is more effective. In this case, the anode consists of a tungsten disc, which is rotated at a high speed on the application of magnetic field. Thus each part of the target is bombarded and heated for a small fraction of a revolution and cools during the rest of its journey. By this arrangement, large tube currents are possible and the X-rays produced can be both intense and hard. These are generally used for short intense exposures as for radiographs of the heart. The highest voltage X-ray tube employs about a million volts, and produces X-rays of 10^{-10} cm wavelength.

EXAMPLE 9.1 Electron bombarding the anode of a Coolidge tube produces X-rays of wavelength 1 Å. Find the energy of each electron at the moment of impact. $h = 6.63 \times 10^{-34}$ Js, $e = 1.6 \times 10^{-19}$ J.

Solution $E = eV = h\nu = \dfrac{hc}{\lambda} \Rightarrow 1.6 \times 10^{-19} \, V = \dfrac{6.63 \times 10^{-34} \times 3 \times 10^8}{1 \times 10^{-10}}$

$$\Rightarrow V = 1.24 \times 10^4 \, V$$

9.3 PROPERTIES OF X-RAYS

The X-rays are the pulses of short wavelengths and are of the same nature as light. Both are electromagnetic waves and travel in space with the velocity (3×10^8 m/s).

X-rays resemble light in the following properties:

1. Both travel in straight lines with the speed of light.
2. Both affect the photographic plate.
3. Both are not deviated by electric and magnetic fields. This distinguishes X-rays from cathode rays.
4. Both liberate photoelectrons when they fall on certain metals.

Apparently, X-rays differ from light in the following properties:

5. The phenomenon of reflection, refraction, interference, diffraction and polarisation of X-rays cannot be easily demonstrated. Later experiments have, however, shown that X-rays do exhibit these properties under special circumstances.

   The difference, however, is due to the shortness of wavelength of X-rays, which is about 1000th part of the wavelength of visible light and not because of any essential difference in their nature. [Wavelengths of X-rays lie between 1000 Å and 0.06 Å, while wavelengths of light are between 7900 Å and 3900 Å].

   In 1923, Compton demonstrated an experiment of X-ray beam from well polished glass surface at nearly grazing incidence. Seigbahn obtained refraction of an X-ray beam and its dispersion into a spectrum with the help of an obtuse angled prism.

   The phenomenon of interference and diffraction of X-rays was demonstrated by Laue, who allowed a beam of X-rays to fall on a crystal, which works as an optical grating of very short grating element and gives rise to reflected and diffracted beams. The direct diffraction effect from a plane transmission grating was demonstrated by Compton and Wadlund in 1921.

X-rays resemble cathode rays in the following properties:

6. X-rays ionize the gas through which they pass.
7. X-rays cause luminescence in many substances, e.g., zinc sulfide, barium platinocyanide, cadmium, tungstate, etc.
8. X-rays, however, are stopped by heavier substances such as bones, sheets of iron, lead, etc., and cast shadows when obstacles fall in their way.

   The heavier atoms have greater stopping power of X-rays than lighter atoms. It is this property, which is made use of in radiography (or X-ray photography) and is employed to detect fractures, presence of foreign matter like bullets inside the human body. Lead is opaque to X-rays and a thickness of 1 mm of lead is quite sufficient to stop X-rays of medium hardness.

9. X-rays produce secondary X-rays when they strike metals.
10. X-rays have a baneful effect on the human body. Slight exposure produces reddening of the skin, like sun burn, while prolonged exposure produces surface sores and causes injuries in the tissues of the body. The white corpuscles in the blood are killed by a prolonged exposure to X-rays and may develop into gangrene. To shield the effect of X-rays, the operators used lead lined aprons and lead glasses.

9.4 SECONDARY RADIATION

When X-rays fall on a substance, a portion is transmitted and reduce intensity I obeying the relation $I = I_0 \, e^{\mu x}$, where I_0 is the original intensity, μ is the linear coefficient of absorption and x is the depth of penetration. Beside this phenomenon, the substance itself acts as a source of secondary radiations, which can be grouped into following four different varieties:

1. *Characteristic* or *homogeneous X-rays,* whose wavelength depends on the nature of the target material provided that the primary beam contains wavelength of greater penetrating power.
2. Scattered radiation forms *continuous X-ray spectra.* They are of the same nature as the primary X-rays and are practically independent of target material.
3. Scattered radiation due to Compton effect consists of *modified X-rays.* In this case, a change of wavelength occurs as a result of scattering.
4. *Corpuscular rays,* which consist of high velocity electrons. They are independent of the nature of the substance but depend on the quality of primary rays.

9.4.1 Characteristic X-rays

These radiations are uniformly spread in all directions and are emitted only when the primary beam possesses rays of shorter wavelengths. A spectrum obtained from molybdenum target has been represented in Figure 9.3.

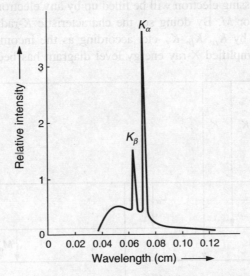

Figure 9.3 Spectra from molybdenum target.

It is evident from Figure 9.3 that the intensity is distributed continuously over a wide range of wavelength ending abruptly at a finite minimum wavelength λ_{min}. There is a sharp increase in the intensity of radiation at some specific wavelengths. These peaks constitute the line spectrum or characteristic X-radiation. These lines are sharp and their wavelengths and number depend on the target material. Using potential difference of 35 kV, Barkla (a scientist) obtained two lines K_α and K_β at wavelengths of 0.71 Å and 0.63 Å. These lines are regarded as K-lines. In 1911, Barkla also discovered a series of L-lines of smaller penetrating power, i.e., of greater wavelength. Soon after, M-lines and N-lines of correspondingly higher wavelengths were discovered.

With the advent of high resolving power X-ray spectrometer, it was found that the *K*-series usually consists of fine lines. Similarly, it was soon established that *L*, *M*, *N* lines also have a fine structure.

The characteristic X-ray spectra can easily be explained with the help of energy level diagrams. If the incident electron is sufficiently energetic, it may knock out electrons from the innermost shell (*K* shell). Let the minimum energy associated with the incident electron be E_K, when it ejects an electron from *K* shell with minimum kinetic energy, i.e., with zero velocity. Although *K* shell has two electrons, yet the energy required for the purpose is the same for both the electrons. Thus the *K*-level energy is single and can be represented by E_K.

The colliding electron may as well expel an electron from *L* shell and pass out leaving the atom as a positive ion. Out of eight electrons in the *L* shell, these are three possible values of E_L, denoted by E_{LI}, E_{LII}, E_{LIII}. The number of energy levels for *M* shell is five and so on.

The characteristic X-radiation is emitted as a result of electronic transitions between the different energy levels. A selection rule, however, governs such transitions. The permitted transitions are in conformity with conditions $\Delta l = \pm 1$ and $\Delta j = 0$ or ± 1. As stated previously, consider the release of a *K* shell electron from an atom. The vacancy created by this missing electron will be filled up by any electron from other higher energy levels such as *L* or *M*. By doing so, the characteristic *K*-radiations are emitted. These lines are denoted by K_α, K_β, K_γ, etc. according as the incoming electron hails from *L*, *M*, *N* shells. A simplified X-ray energy level diagram has been shown in Figure 9.4.

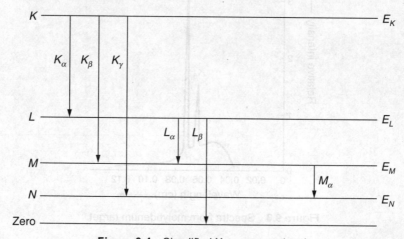

Figure 9.4 Simplified X-ray energy level.

The frequencies of the K_α, K_β lines are given by

$$\nu(K_\alpha) = \frac{E_K - E_L}{h} \tag{9.1}$$

$$\nu(K_\beta) = \frac{E_K - E_M}{h} \tag{9.2}$$

and so on.

The fine structure of these lines taking into account the splitting of energy levels and selection rule for allowed transition has been shown in Figure 9.5. The frequency of K_{α_1} line for instance, may be expressed as:

$$v(K_{\alpha_1}) = \frac{E_K - E_{L_{III}}}{h} \tag{9.3}$$

Since the electron is accommodated in K level from L_{III} level.

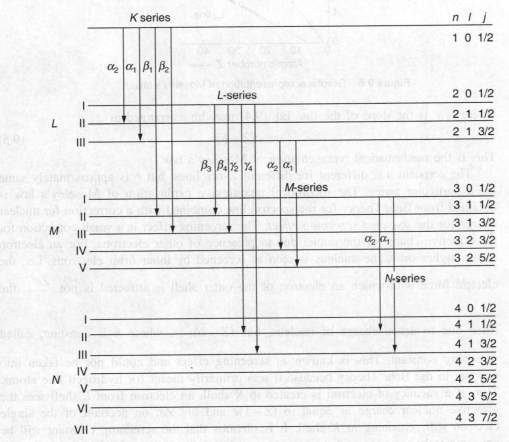

Figure 9.5 Energy level diagram.

Moseley's Law

A systematic and comprehensive study of the characteristic X-ray spectra of large number of element was done by Moseley during the years 1913–1915. He found that for a particular line, the root of frequency $\sqrt{v}$ varies linearly with the atomic number Z of element used as a target. This is called the *Moseley's law* and is conveniently represented by Moseley's diagram (Figure 9.6).

The equation of a particular straight line can be represented as:

$$\sqrt{v} = \sqrt{a}(Z - b) \tag{9.4}$$

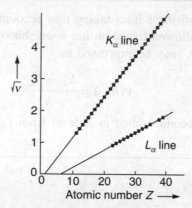

Figure 9.6 Graphical representation of Moseley's law.

where $\sqrt{a}$ is the slope of the line Eq. (9.4) may be rearranged as

$$v = a(Z - b)^2 \qquad (9.5)$$

This is the mathematical representation of Moseley's law.

The constant a is different for different X-ray lines, but b is approximately same for a particular series. The theoretical quantitative explanation of Moseley's law is obtained from Bohr Theory for the spectral line combined with a correction for nuclear charge for the so-called *screening effect*. The screening effect is a small correction for the effective charge of the nuclei due to presence of other electrons. For an electron in the higher orbit, the nucleus is seen as screened by inner orbit electrons, i.e., the electric force with which an electron of the outer shell is attracted is not $\dfrac{Ze^2}{r^2}$, the force due to actual charge of nucleus, but $(Z - b)\dfrac{e^2}{r^2}$, where b is constant, called screening constant. This is known as screening effect and could not be taken into account in the Bohr Theory because it was primarily meant for hydrogen like atoms. Thus if a vacancy of electron is created in K shell, an electron from L shell sees the effective nuclear charge as equal to $(Z - 1)e$ and not Ze; on account of the single electron still remaining in K shell. It is obvious that the screening constant will be approximately same for particular series, e.g. b is 1 for K-series lines, 7.4 for L-series and so on. Taking this into account [i.e., replacing Z by $(Z - b)$], one can write general formula for the frequency v of X-ray line for any element using Bohr Theory as follows:

$$v = R(Z - b)^2 \left[\frac{1}{n_1^2} - \frac{1}{n_2^2} \right] \qquad (9.6)$$

where R is Rydberg constant (109700 cm^{-1} or 3.291×10^{15} Hz), Z is the atomic number of element emitting the X-ray line, b is the screening constant and n_1 and n_2 are principal quantum numbers of electronic states involved in the transition. Since for a particular line, n_1 and n_2 are constant (e.g. for K-line $n_1 = 1$, $n_2 = 2$ and for L line $n_1 = 2$, $n_2 = 3$ and so on), one can write

$$v = a(Z - b)^2 \qquad (9.7)$$

where

$$a = R\left(\frac{1}{n_1^2} - \frac{1}{n_2^2}\right)$$

This is Moseley's law. The quantitative agreement between the values a and b given by above theory is reasonably good as obtained by Moseley. Following are some examples which justify the explanation:

Example 1 *For K line from Cu, Moseley found $\lambda = 1.54$ Å.*

Thus experimental value of frequency v is:

$$v = \frac{c}{\lambda} = \frac{3 \times 10^8}{1.54 \times 10^{-10}} = 1.95 \times 10^{18} \text{ Hz}$$

Now for the K_α line, $n_1 = 1$, $n_2 = 2$ and $b = 1$.

$$v_{K_\alpha} = R(Z - 1)^2\left(\frac{1}{1^2} - \frac{1}{2^2}\right) = \frac{3}{4}R(Z - 1)^2$$

But for Cu, $Z = 29$, therefore,

$$(v_{K_\alpha})_{Cu} = \frac{3}{4} \times 3.291 \times 10^{15} \times (28)^2 = 1.94 \times 10^{18} \text{ Hz}$$

It is in good agreement with the experimental limit.
Similar example is given below:

Example 2 *Wavelength for L_α for platinum is 1.32 Å.*

The experimental value of v is:

$$v = \frac{c}{\lambda} = \frac{3 \times 10^8}{1.32 \times 10^{-10}} = 2.27 \times 10^{18} \text{ Hz}$$

Now, for the L_α line, $n_1 = 2$, $n_2 = 3$ and $b = 7.4$

$$v_{K_\alpha} = R(Z - 7.4)^2\left(\frac{1}{2^2} - \frac{1}{3^2}\right) = \frac{5}{36}R(Z - 7.4)^2$$

But for platinum, $Z = 78$

$$\therefore \quad (v_{K_\alpha})_{Pt} = \frac{5}{36} \times 3.291 \times 10^{15} \times 70.6 \times 70.6 \text{ Hz} = 2.28 \times 10^{18} \text{ Hz}$$

In this case, agreement is good within the experimental limit.

Use of Moseley's law: The researches carried out by Moseley attracted the attention of physicists as well as chemists towards atomic number. It was first experiment to show the supremacy of the atomic number over the atomic weight in the construction of atoms. The development of the entire periodic table is based on the atomic number which in turn finds its origin in Moseley's law. This led to the discovery of unknown elements. Further the difference between the atomic weight and atomic number was explained in the number of neutrons, a fundamental constituent of nucleus.

Moseley's law can be obtained from Bohr theory. Thus in a way this law supports the Bohr theory.

Note: The only drawback of the Moseley's law is that by this method, we cannot distinguish isotopes of an element as they give rise to same frequencies in X-ray region.

9.4.2 Continuous X-ray Spectrum

When we critically examine the line spectra as represented in Figure 9.7, we find spectra, which are obtained in the background of continuous radiation of relatively weak intensity.

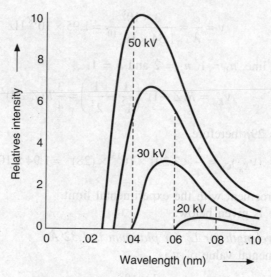

Figure 9.7 Continuous spectra (X-rays).

In other words, the continuous spectrum is superimposed on characteristic spectra obtained from the X-ray tube. The distribution of energy for such spectrum was thoroughly studied by Urey, who obtained different curves for different target potentials (as shown in Figure 9.7), all of which resembled curves of Wien's displacement law for black body radiation.

The interesting features of such a spectrum are the following:

1. The spectrum disappears abruptly at a certain minimum wavelength λ_{min} depending upon the applied voltage across the tube irrespective of the target material used.
2. As the applied voltage is increased, the short wavelength limit λ_{min} shifts toward a smaller value.

If V be the target potential and ν_{max} corresponds to the highest frequency, i.e., to the short wavelength limit λ_{min}, then we have

$$eV = h\nu_{max} = \frac{hc}{\lambda_{min}}$$

or $$\lambda_{min} = \frac{hc}{eV} \qquad (9.8)$$

Thus the short wavelength limit varies inversely as the potential difference across the tube.

3. The wavelength having the maximum intensity is also shifted towards smaller wavelengths as the applied voltage is increased.

4. The maximum intensity for each spectrum increases with the increase in the applied voltage.

It is possible to determine the value of the Planck's constant using Eq. (9.8). Duane and Hunt performed carefully a series of experiments on the determination of λ_{min} for different applied voltages and obtained value of Planck's constant to be $h = 6.256 \times 10^{-34}$ Js, accepting the standard value of electronic charge e.

The value of Planck's constant h thus determined, agrees quite well with that obtained from the photoelectric effect.

Origin of Continuous X-Ray Spectrum

According to electromagnetic theory, radiation occurs continuously when high speed electrons are decelerated as they move into the vicinity of nuclei of an atom of the target material. If the electron looses an amount of energy ΔE in passing nucleus, the frequency of the emitted radiation becomes $\Delta E/h$. Thus the source of this continuous background radiation is the deceleration of electrons caused by the atomic nuclei of the target. This is often called *bremsstrahlung* (from German words *breme*: brake, and *strahlung*: radiation.)

EXAMPLE 9.2 A potential difference of 10 kV is used to accelerate electrons in an X-ray tube. What is the minimum wavelength of X-ray produced?

Solution To produce X-rays with minimum wavelength (or maximum frequency and hence maximum energy) all the kinetic energy of an electron must go into producing single X-ray photon. We know

$$\lambda_{min} = \frac{hc}{eV} = \frac{6.626 \times 10^{-34} \text{ Js} \times 3.0 \times 10^8 \text{ m/s}}{1.602 \times 10^{-19} \text{ C} \times 10.0 \times 10^3 \text{ V}} = 1.24 \times 10^{-10} \text{ m} = 0.124 \text{ nm}$$

9.5 REFLECTION OF X-RAYS

The reflection of an X-ray beam on the surface of the crystal (Figure 9.8), which consists of symmetrical arranged atoms, has been explained by a simple theory advanced by W.H. Bragg and W.L. Bragg. According to them, when a beam of X-rays falls on a crystal, each atom in that crystal begins to act as a source of secondary waves and the resultant wavefront is obtained according to Huygen's construction in optics. As the atoms in the crystal are arranged in a regular pattern, a beam of X-rays will be reflected from plane of atoms, just as a beam of light is reflected from a plane mirror. The reflections will also take place at successive layers and the resultant intensity is due to their combination. In order that the intensity may be large, the reflected rays must combine in the same phase; and hence the path difference between the reflected waves from any one layer and that from the next layer must be an exact multiple of a wavelength.

Bragg's Law

Let BA be an incident ray and AB the incident wavefront as shown in Figure 9.8. AC is reflected ray from the second layer of atoms and in order that the intensity of the resultant ray AC be large, the path difference between them should be an exact multiple of a wavelength. The path difference is:

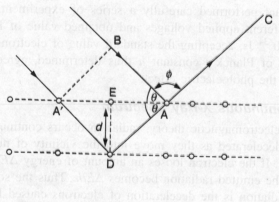

Figure 9.8 Reflection of X-rays at a crystal surface.

$$= A'D + DA - BA = 2DA - 2AE \cos \theta$$

$$\equiv \frac{2d}{\sin \theta} - \frac{2d}{\tan \theta} \times \cos \theta = \frac{2d}{\sin \theta}[1 - \cos^2 \theta]$$

$$= 2d \sin \theta = n\lambda$$

where θ is the angle between the incident ray and the surface of the crystal. This angle is called the *glancing angle* and d is the distance between the successive layers of atoms inside the crystal. This equation is called *Bragg's equation* and this is also known as Bragg's law.

For maximum intensity,

$$2d \sin \theta = n\lambda \tag{9.9}$$

Bragg's Spectrometer

In Bragg's spectrometer, the intensity of the X-rays, reflected at various angles, is measured by the degree of ionization, they produce in a gas enclosed in an ionization chamber I (Figure 9.9). The ionization chamber may be replaced by a photographic camera, and the X-ray lines will then get photographed. A narrow X-ray beam with the help of slits S_1 and S_2 with fine holes in them, is led on crystal C which is mounted on the spectrometer table that rotates about a vertical axis. Its position is determined

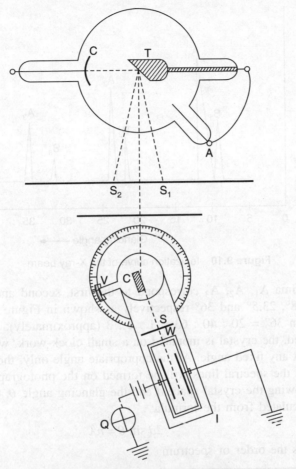

Figure 9.9 Bragg's spectrometer.

by a vernier V. Instead of the usual telescope, the moveable arm carries an ionization chamber or a photographic plate and its position is read by second vernier. A thin aluminium window W admits the reflected beam whose width is regulated by an adjustable slit. One of the plates of the ionization chamber is charged to high potential and is connected to quadrant electrometer or tilted gold leaf electroscope, while the other plate is earthed, and the intensity of the X-ray beam is determined. A graph is obtained as shown in Figure 9.10.

To render the chamber completely absorbing, it is filled with SO_2 when the absorption is two times as much as with air. When highly penetrating X-rays are used, methyl bromide is employed.

The angle is gradually increased and ionization curve is plotted. We find maxima at sine (angle), which have two times or three times the sine (angle) corresponding to first maxima.

$$2d \sin \theta_n = n\lambda \qquad (9.10)$$

where $n = 0, 1, 2, 3, 4, \ldots$.

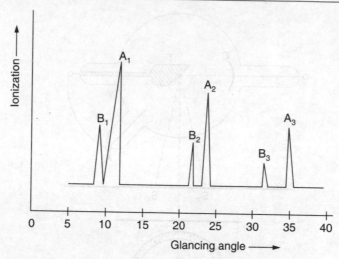

Figure 9.10 Ionization curve of the X-ray beam.

The maxima A_1, A_2, A_3 correspond to the first, second and third spectra, occur at angles 11.8°, 23.5° and 36° respectively as shown in Figure 9.10. Now sin 11.8°: sin 23.5° : sin 36° = 20 : 40 : 63 = 1 : 2 : 3 (approximately). When a photographic camera is used, the crystal is mounted on a small clock-work, which makes a slightly rotation about any fixed angle. At the appropriate angle only, the X-ray beam will get reflected, and the spectral lines will be formed on the photographic plates.

Thus knowing the crystal space d and the glancing angle θ, the wavelength of the X-rays is calculated from the formula:

$$2d \sin \theta = n\lambda$$

where n gives the order of spectrum.

EXAMPLE 9.3 X-rays from a tube operated at 50 kV are analysed with a Bragg's spectrometer using a calcite crystal cut along the cleavage plane. If the grating space of calcite be given as 3.029 Å, calculate the smallest angle between the crystal plane and the X-ray beam at which shortest wavelength produced by the tube can be detected.
Given 1 eV = 1.6×10^{-19} J, $h = 6.623 \times 10^{-34}$ Js.

Solution We have shown that

$$\lambda_{min} = \frac{hc}{eV}$$

Now, $2d \sin \theta = \lambda_{min} = \dfrac{hc}{eV}$

$\Rightarrow \quad \sin \theta = \dfrac{1}{2d}\left(\dfrac{hc}{eV}\right) = \dfrac{6.623 \times 10^{-34} \times 3 \times 10^{8}}{2 \times 3.029 \times 10^{-10} \times 50 \times 10^{3} \times 1.6 \times 10^{-19}} = 0.04053$

$\therefore \qquad \theta = \sin^{-1}(0.04053) = 2.3°$

9.6 ABSORPTION OF X-RAYS

The energy of an X-ray beam depends upon its penetrating power. Hard X-rays have larger penetrating power than soft X-rays. Again, the radiations of the K-series are 300 times harder than those of L-series, which are harder than those of M-series. Atoms of low atomic weight emit only the K-series, of medium atomic weight give rise to K- and L-series and of high atomic weight liberate K-, L- and M-series. Although it is very difficult to get their K-series excited. Now, these can be distinguished from each other by their penetrating power.

If a beam of X-rays of intensity I passes through a length dx of any material, its intensity on emergence is decreased by dI. For any given material, the amount of absorbed intensity is $\alpha I dx$, where α is called the *absorption coefficient*.

Then $\qquad -dI = \alpha I dx \qquad$ or $\qquad \dfrac{dI}{I} = \alpha dx \qquad$ or $\qquad \log I = -\alpha x + A$

But $x = 0$, $I = I_0$ and hence $A = \log I_0$

or
$$I = I_0 e^{-\alpha x} \qquad (9.11)$$

where I be the intensity of emergent rays after traverse the distance x.

A better way to measure the quantity of X-rays with respect to absorption coefficient (α) is the ratio $\left(\dfrac{\alpha}{\rho}\right)$, where ρ is the density of the material. The $\dfrac{\alpha}{\rho}$ is known as the mass absorption coefficient. The amount of absorption is also dependant of traversed materials mass. The amount of absorption also depends on the mass of the material traversed. The mass absorption coefficient varies directly as the atomic weight. The curve between the $\log_e$ (atomic weight) and $\log\left(\dfrac{\alpha}{\rho}\right)$ gives two straight lines as shown in Figure 9.11, one for K-radiation and another for L-radiation.

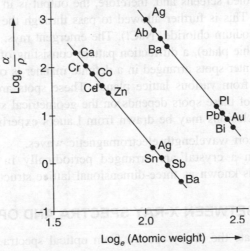

Figure 9.11 Variation of mass absorption coefficient with atomic weight.

9.7 X-RAY DIFFRACTION (XRD) AND LAUE'S EXPERIMENT

We know that a wave shows diffraction effect only if the grating element is of the order of wavelength of wave. For visible range a grating having grating element of the order of wavelength of light may be constructed, but for X-rays ($\lambda = 1$ Å), it is not possible to construct a grating having grating element of the order of 1 Å. Von Laue in 1913 suggested that crystals have regular and periodic arrangement of atoms with spacing of the order of 1 Å, therefore, can be used to study the diffraction effect of X-rays.

Just after the discovery of X-rays two things were put forward regarding the nature of X-rays. According to first, the X-rays were supposed to be made up of high speed particles having high penetrating power but in second, these X-rays were regarded as high frequency electromagnetic waves.

The first successful experiment showing the wave nature of these rays was performed by Von Laue. He allowed X-rays to pass through a crystal and obtained a diffraction pattern on a photographic plate. The experimental arrangement is shown in Figure 9.12.

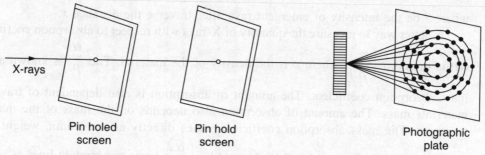

Figure 9.12 Experimental arrangement for photographing Laue spot.

In Laue's experiment, the X-rays emerging from the X-ray tube are allowed to pass through the pin holed screens and, therefore, the output is in the form of a narrow fine pencil of X-rays. This is further allowed to pass through the thin crystal of either zinc sulfide (ZnS) or sodium chloride (NaCl). The emergent rays, when allowed to fall on a screen (photographic plate), a diffraction pattern consisting of a central spot surrounded by many other fainter spots arranged in a definite manner is obtained. This is due to the X-ray diffraction from various lattice planes. These spots are known as *Laue's spots*. The arrangement of these spots depends on the geometrical structure of the crystal.

The two conclusions may be drawn from Laue's experiment:

1. X-rays are short wavelength electromagnetic waves.
2. The atoms in a crystal are arranged periodically in a regular fashion. This arrangement is known as three-dimensional lattice structures.

9.8 DIFFERENCE BETWEEN X-RAY SPECTRA AND OPTICAL SPECTRA

We can differentiate the X-ray spectra from optical spectra in the following manner

Table 9.1 Difference between X-ray Spectra and Optical Spectra

S.No.	X-ray spectra	Optical spectra
1.	X-ray spectra are produced due to transition of the tightly bound electrons among the inner completely filled shells.	The optical spectra originate due to transition of outermost electrons of the atom.
2.	X-ray spectra vary smoothly from element to element in periodic table.	The optical spectra show periodic changes.
3.	X-ray spectra originate due to removal of an electron from the inner shell of atom.	The electron from the outer shell is merely raised to a higher level and not completely removed.
4.	X-ray emission spectra consist of discrete lines and X-ray absorption spectra consist of continuous region bounded by sharp edge.	In the optical spectra, the emission and absorption spectra are identical.

9.9 PRACTICAL APPLICATIONS OF X-RAYS

X-rays have varied applications. Its applications can be classified as follows:

1. Radiography (or shadow photograph of X-rays, which are stopped by heavy substances)
2. Radio therapy (use of X-rays to cure certain disease by killing certain types of tissue)
3. Scientific research (use of X-rays in many scientific activities)

Radiography

Within a few weeks of the discovery of X-rays by Röntgen in 1895, X-rays began to be used in surgical operation in Vienna. In 1897, Morton in New York exhibited an X-ray picture of the entire skeleton of a living and fully clothed adult.

For radiograph of the elementary tract, the patient is fed with a meal containing $BaSO_4$. For gall bladder, tetra-iodo phenolphthalein is administered through the mouth; and for bronchi, injection of lipidol; and for kidneys, injection of NaI have been used. Radiographs have in this way been used for diagnosing diseases like tuberculosis of the lung, defects of the digestive system, etc.

Radiographs are very useful in surgery for the detection of:

(i) Fracture
(ii) Foreign matter like bullets
(iii) Formation of bones and stones in internal organs and observation of healing of broken bone

Radiographs have also been used in engineering for:

(i) Testing of metal castings, moulds and weldings
(ii) Detecting the faults, cracks, flaws, gas pockets in finished products (These must be avoided particularly in shaft of airplane engines)

Radiographs have been used in the detective departments for:

(i) Detection of explosives and other contraband goods concealed in leather or wooden cases
(ii) Detection of gold and silver in the body of smugglers and thieves
(iii) Testing of genuineness of old documents, pearls, diamonds and gems
(iv) Examining the content of parcels

Radiographs are used in industry for:

(i) Examination of rubber tyres, golf balls, wood, wireless and valves
(ii) Testing of homogeneity of timber and uniformity of insulating materials
(iii) Detecting the presence of pearls in oysters

X-ray use become a common practice among art authenticators. Not only does it unlock secrets underneath paintings, but helps to establish authenticity. Types of paper, materials, preparatory sketches, changes to composition, and other clues can be discovered through the use of an X-ray to prove the nature and origin of a painting.

Radio Therapy

Intractable skin diseases and malignant growths have been cured by controlled exposures of X-rays. An attempt to treat cancer in a similar way has also been made. Under X-ray treatment, chromosomes show important changes leading to new hereditary characteristics in the progeny thus actually causing variations of form in the off springs of insects and animals, called mutation. White corn plants without chlorophylls have been obtained by the process of mutation induced by X-rays. An over-exposure has a baneful effect on animal and plant life, and must be avoided. Too strong exposures on human body cause sterilisation.

Scientific Research

X-rays have been used for investigating the structure of crystals, structure and properties of atoms and to investigate the arrangement of atoms, molecules in complex molecular structures. By Fourier analysis of X-ray diffraction pattern, it is possible by present day techniques to build a single molecular structure of a compound and see it as if we were actually handling a molecule.

SOLVED NUMERICAL PROBLEMS

PROBLEM 9.1   Calculate the wavelength of X-rays produced, when the potential difference is 12.4 kV.

Solution   Given $V = 12400$ V

We know that $\lambda = \dfrac{hc}{eV}$

$$= \frac{12400}{V}\,\text{Å} = \frac{12400}{12400}\,\text{Å} = 1\,\text{Å}$$

PROBLEM 9.2 The voltage of a X-ray tube is 30 kV. Calculate the maximum frequency of the emitted X-ray.

Solution Given $V = 30$ kV = 30000 V

We know $E_{max} = h\nu_{max}$ $\Rightarrow$ $eV = h\nu_{max}$ $\Rightarrow$ $\nu_{max} = \dfrac{eV}{h}$

$$\nu_{max} = \frac{1.6 \times 10^{-19} \times 30000}{6.623 \times 10^{-34}} \text{ Hz} = 7.24 \times 10^{8} \text{ Hz}$$

PROBLEM 9.3 An X-ray tube operated at 40 kV emits continuous X-ray spectrum with a short wavelength limit $\lambda_{min} = 0.310$ Å. Calculate the value of Planck's constant.

Solution Given $V = 40$ kV = 40000 V, $\nu_{max} = 0.310$ Å; $e = 1.6 \times 10^{-19}$ C

We know that $\nu_{max} = \dfrac{hc}{eV}$ $\Rightarrow$ $h = \dfrac{eV\lambda_{min}}{c}$

$$h = \frac{1.6 \times 10^{-19} \times 4.0 \times 10^{4} \times 0.310 \times 10^{-10}}{3 \times 10^{8}} \text{ Js} = 6.16 \times 10^{-34} \text{ Js}$$

PROBLEM 9.4 A Coolidge tube operates at 50 kV. Find (i) the maximum velocity of electrons striking the anticathode; (ii) minimum wavelength of X-rays generated.

Solution Given $V = 50$ kV = 50000 V, $e = 1.6 \times 10^{-19}$ C and $m = 9.1 \times 10^{-31}$ kg

(i) We have $\dfrac{1}{2} mv_{max}^2 = eV$ $\Rightarrow$ $v_{max} = \sqrt{\dfrac{2eV}{m}}$

$$\Rightarrow \qquad v_{max} = \sqrt{\left(\frac{2 \times 1.6 \times 10^{-19} \times 50 \times 10^{3}}{9.1 \times 10^{-31}}\right)} = 1.326 \times 10^{-19} \text{ m/s}$$

(ii) $\lambda_{min} = \dfrac{hc}{eV} = \dfrac{6.623 \times 10^{-34} \times 3 \times 10^{8}}{1.6 \times 10^{-19} \times 50000} = 0.248 \times 10^{-10}$ m = 0.248 Å

PROBLEM 9.5 The K_α X-ray line from molybdenum has a wavelength of 0.7078 Å. Calculate the wavelength of the K_α-line of copper.

Solution Given $Z_1 = 42$, $\lambda = 0.7078$ Å, $Z_2 = 29$
From Moseley's law,

$$\nu \propto (Z-1)^2 \quad \text{or} \quad \frac{c}{\lambda} \propto (Z-1)^2$$

or $\qquad (Z_1 - 1)^2 \lambda = \text{Constant} \Rightarrow (Z_1 - 1)^2 \lambda = (Z_2 - 1)^2 \lambda'$

Thus $\qquad (42 - 1)^2 \times 0.7078 = (29 - 1)^2 \lambda' \Rightarrow \lambda' = 1.517$ Å

PROBLEM 9.6 The spacing between principal planes of NaCl crystal is 2.82 Å. It is found that first order Bragg's reflection occurs at an angle 10°. What is the wavelength of X-rays?

Solution Given $d = 2.82 \times 10^{-10}$ m, $n = 1$, $\theta = 10°$, $\lambda = ?$
From Bragg's law,

$$2d \sin \theta = n\lambda \implies \lambda = \frac{2d \sin \theta}{n} = \frac{2 \times (2.82 \times 10^{-10}) \sin 10°}{1} = 0.98 \times 10^{-10} \text{ m}$$

PROBLEM 9.7 The distance between adjacent atomic planes in calcite is 0.300 nm. Find the smallest angle of Bragg's scattering for 0.030 nm X-ray.

Solution Given $d = 0.300$ nm $= 3 \times 10^{-10}$ m, $\lambda = 3.0 \times 10^{-11}$ m, $n = 1$
From Bragg's law,

$$2d \sin \theta = n\lambda \implies \sin \theta = \frac{n\lambda}{2d}$$

$$= \frac{1 \times 3.0 \times 10^{-11}}{2 \times 3 \times 10^{-10}} = 0.05 \implies \theta = \sin^{-1}(0.05) = 2.87°$$

PROBLEM 9.8 While comparing the wavelength of two monochromatic X-ray lines, it is found that line A gives a first order Bragg's reflection maximum at a glancing angle 30° to the smooth face of a crystal. Line B of wavelength $\lambda_B = 0.97$ Å gives a third order reflection maximum at a glancing angle 60°. Find the wavelength of line A(λ_A).

Solution For line A,

$$2d \sin \theta = n\lambda_A$$
$$2d \sin 30° = \lambda_A \tag{i}$$

Here $n = 1$, then $2d \sin \theta = \lambda_A$
For line B,

$$2d \sin \theta = n\lambda_B$$

Here $n = 3$, then $2d \sin \theta = 3\lambda_B$

or $$2d \sin 60° = 3 \times 0.97 \times 10^{-10} \tag{ii}$$

From Eq. (ii), $$d = \sqrt{3} \times 0.97 \times 10^{-10} \text{ m} = 1.68 \text{ Å}$$

Then from Eq. (i) $$\lambda_A = 2 \times 1.68 \times 10^{-10} \sin 30° = 1.68 \text{ Å}$$

PROBLEM 9.9 Calculate the glancing angle of the <110> plane of simple cubic crystal ($a = 2.814$ Å) corresponding to second order diffraction maxima for the X-rays of wavelength 0.710 Å.

Solution For nth order diffraction maximum for X-rays of wavelength λ from lattice planes of spacing d, the glancing angle θ is given by

$$2d \sin \theta = n\lambda \tag{i}$$

The distance between consecutive lattice planes defined by miller indices (*hkl*) in a cubic lattice is:

$$d_{hkl} = \frac{a}{\sqrt{h^2 + k^2 + l^2}}$$

Given $\qquad\qquad a = 2.814 \text{ Å}, h = 1, k = 1, l = 0$

Then $\qquad\qquad d_{110} = \frac{2.814}{\sqrt{1^2 + 1^2 + 0}} = \frac{2.814}{\sqrt{2}} \text{ Å} = 1.989 \text{ Å}$

Also we have $n = 2$, and $\lambda = 710 \text{ Å}$
Substituting these values in Eq. (i), we get

$$\Rightarrow \qquad\qquad 2 \times 1.989 \times 10^{-10} \sin\theta$$
$$= 2 \times 0.710 \times 10^{-10}$$
$$\Rightarrow \qquad\qquad \theta = \sin^{-1}(0.3569)$$
$$= 20°55'$$

PROBLEM 9.10 Electrons are accelerated in TV through a potential difference of about 10 kV. Find the highest frequency of electromagnetic waves emitted when these electrons strikes the screen of the tube. Recognise the waves.

Solution Given $V = 10 \text{ kV} = 10000 \text{ V}$
We know that $E = h\nu = eV$

and $\qquad\qquad \nu_{max} = \frac{eV}{h} = \frac{1.6 \times 10^{-19} \times 10000}{6.626 \times 10^{-34}} = 2.4 \times 10^{18} \text{ Hz}$

On the basis of its frequency, it is X-rays.

///// **EXERCISES** /////

THEORETICAL QUESTIONS

9.1 Describe and explain how X-rays are produced? State the properties of these rays and describe briefly experiments to demonstrate them.

9.2 Describe and explain briefly the production of X-rays by Coolidge tube.

9.3 Describe the properties and important applications of X-rays.

9.4 Write an essay on the production, properties, nature and uses of X-rays.

9.5 Describe a method of production of X-rays. Indicate briefly how their wavelength has been determined.

9.6 Describe X-ray spectrometer devised by Bragg. Deduce the necessary formulae. Describe Moseley's work emphasing its importance.

9.7 Describe an experimental arrangement to determine the wavelength of X-rays and discuss clearly the theory underlying the experiment.

9.8 Derive Bragg's formula for X-ray diffraction. Explain how the distance the distance between successive reflecting planes in rock-salt crystal is estimated for measuring X-ray wavelength by Bragg's method.

9.9 What are the characteristic and continuous X-rays? How is their production account for?

9.10 Write an account of the production and properties of X-rays. What is evidence for their electromagnetic nature?

NUMERICAL PROBLEMS

P9.1 An X-ray tube works at 18 kV. Find maximum speed of electron striking the anticathode.
(**Ans.** $V = 8 \times 10^5$ m/s)

P9.2 An X-ray machine uses an accelerating potential of 50 kV. Calculate the shortest wavelength present in X-ray produced. (**Ans.** $\lambda_{min} = 0.248$ Å)

P9.3 X-ray spectrum of cobalt ($Z = 27$) contains strong K_α-line of wavelength 0.1785 nm and weak K_α lines having wavelengths of 0.2285 nm and 0.1537 nm due to impurities. Using Moseley's law, calculate the atomic numbers of the two impurities and identify their nature. The screening constant σ for K-series is unity. (**Ans.** $Z = 24$(chromium), $Z = 29$(copper))

P9.4 Calculate the energy of electrons that produce Bragg's diffraction of first order at an angle 22°, when incident on crystal with interplaner spacing. (**Ans.** 1.99×10^{16} eV)

P9.5 Calculate the wavelength of X-rays emitted from the following data:
  (i) Glancing angle for NaCl for first order spectrum = 11.8°
  (ii) Atomic weight of Na = 23
  (iii) Atomic weight of Cl = 35.5
  (iv) Density of NaCl = 2.17 g/cc
  (v) Mass of H_2 atom = 1.64×10^{-24} g
(**Ans.** 1.15 Å)

P9.6 Calculate the largest wavelength, that can be analysed by rock-salt crystal of spacing $d = 2.824$ Å in (i) first order and (ii) second order. (**Ans.** (i) 5.64 Å, (ii) 2.82 Å)

P9.7 In a Laue photograph of an fcc crystal, whose unit cell has an edge of length 4.50 Å, what is the minimum distance from the centre of the pattern at which reflections can occur from the planes of maximum spacing, if the potential difference across the X-ray tube is 50 kV and the distance of the film from the crystal is 5.0 cm? (**Ans.** 0.97 cm)

P9.8 Potassium chloride (KCl) has a set of crystal planes separated by a distance $d = 0.31$ nm. At what glancing angle to these planes could be first order Bragg maximum occur for X-rays of wavelength 0.05 nm? (**Ans.** $\theta = 4.6°$)

P9.9 The Bragg's angle corresponding to the first order reflection from (1, 1, 1) planes in a crystal is 30°, when X-rays of wavelength 1.75 Å are used. Calculate inter-atomic spacing.
(**Ans.** $a = 3.031$ Å)

P9.10 The glancing angles of reflection for K_α X-rays from palladium are 5.4° from (100) plane; 7.6° from (110) plane and 9.4° from (111) plane. From the above data, determine the cubic lattice structure of the crystal.

$$\left(\textbf{Ans.} \ \frac{1}{d_1} : \frac{1}{d_2} : \frac{1}{d_3} = 0.0914 : 0.1323 : 0.1633 = 1 : \sqrt{2} : \sqrt{3} :, \text{simple cubic} \right)$$

MULTIPLE CHOICE QUESTIONS

MCQ 9.1 X-ray are:
    (a) Negatively charged particles   (b) A stream of neutrons
    (c) Electromagnetic radiation   (d) Positively charged particles

MCQ 9.2 X-rays can be deflected
    (a) By a magnetic field
    (b) By an electric field
    (c) By an electric field as well as a magnetic field
    (d) Neither by an electric field nor by a magnetic field

MCQ 9.3 The constant a in Moseley's law is
    (a) Screening constant   (b) Rydberg constant
    (c) Atomic constant   (d) None of these

MCQ 9.4 The wave number of the characteristic radiation
    (a) Increases with increasing atomic number of the anode
    (b) Decreases with increasing atomic number of the anode
    (c) May be (a) and (b) both
    (d) None of these

MCQ 9.5 As the wavelength of X-rays is smaller than that of visible light, the speed of X-rays in vacuum is
    (a) Larger than that of visible light   (b) Same as that of visible light
    (c) Smaller than that of visible light   (d) None of these

MCQ 9.6 According to the Moseley's law, the frequency of the characteristic radiation is proportional to the square of
    (a) Atomic number of the element   (b) Atomic weight of the element
    (c) Smaller than that of visible light   (d) None of these

MCQ 9.7 In a Coolidge tube, the electrons are focused onto target by using
    (a) Transformer   (b) An electron gun
    (c) Magnetic lens   (d) A curved cathode

MCQ 9.8 X-rays and γ-rays are both electromagnetic waves. Which of the following statement is true?
    (a) In general, X-rays have larger wavelength than that of γ-rays
    (b) γ-rays have smaller frequency than that of X-rays
    (c) X-rays have smaller wavelength than that of γ-rays
    (d) Wavelength and frequency of X-rays are both larger than those of γ-rays

MCQ 9.9 For harder X-rays
    (a) The frequency is higher   (b) The photon energy is higher
    (c) The wavelength is higher   (d) The intensity is higher

MCQ 9.10 Laue's experiment proves
    (a) X-rays are charged particles   (b) X-rays are electromagnetic radiation
    (c) The wavelength of X-rays is very high   (d) X-rays will not penetrate through metal

Answers

9.1 (c)   **9.2** (d)   **9.3** (a)   **9.4** (a)   **9.5** (b)   **9.6** (a)   **9.7** (b)
9.8 (a)   **9.9** (a), (b) **9.10** (b)

CHAPTER **10**

Lasers

CHAPTER OUTLINES

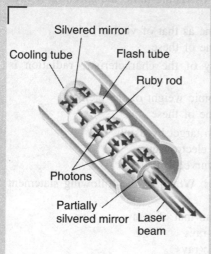

Silvered mirror
Cooling tube   Flash tube
Ruby rod
Photons
Partially silvered mirror   Laser beam

- ⊙ Introduction
- ⊙ Boltzmann's Distribution
- ⊙ Interaction of Light with Matter
- ⊙ Einstein's Relations
- ⊙ Requirements for Laser Action
- ⊙ Population Inversion and Pumping
- ⊙ Metastable States
- ⊙ Cavity Resonator
- ⊙ Threshold Condition for Laser Action
- ⊙ Pumping Schemes
- ⊙ Coherence
- ⊙ Properties of Laser Light
- ⊙ Various Types of Lasers
- ⊙ Applications of Lasers

10.1 INTRODUCTION

The name LASER is an acronym of 'Light Amplification by Stimulated Emission of Radiation'. Laser is a device that produces an intense, concentrated and highly parallel beam of light. Sometimes, lasers are also called 'Optical Masers'. Masers were realised before the lasers. MASER stands for 'Microwave Amplification by Stimulated Emission of Radiation'. Infact laser is an optical oscillator which generates light energy. The principle of laser was put forward by A.H. Schawlow and C.H. Townes in 1958. The first laser was invented by T.H. Maiman in 1960. The emission of extremely energetic and monochromatic radiation from laser makes it highly useful for science and technology. High capacity optical communication systems are realised only due to availability of lasers. This device has opened many new branches of research which ultimately serves mankind in form of technology.

10.2 BOLTZMANN'S DISTRIBUTION

Any material medium is made of atoms, and these atoms occupy suitable energy levels. To understand the distribution of atoms in thermal equilibrium situation, consider two energy levels with energies E_1 and E_2. Further assume that there are N_1 atoms in state E_1 and N_2 in state E_2. Here for simplicity, let us assume that E_1 represents ground level and E_2 is excited energy level. The population of atoms is defined as number of atoms per unit volume that occupy a given energy level. The Boltzmann's law gives the number of atoms in an energy state. Therefore,

$$N_1 = e^{-E_1/k_B T} \tag{10.1}$$

and

$$N_2 = e^{-E_2/k_B T} \tag{10.2}$$

where k_B is Boltzmann's constant and T represents temperature. If a group of atoms is in thermal equilibrium, the relative population N_2/N_1 may be given by

$$\frac{N_2}{N_1} = e^{-(E_2 - E_1)/k_B T} \tag{10.3}$$

Suppose a light of frequency v interacts with the group of atoms, then $\Delta E = E_2 - E_1 = hv$, and we have

$$\frac{N_2}{N_1} = e^{-\Delta E/k_B T} = e^{-hv/k_B T} \tag{10.4}$$

For given energy levels, the fraction N_2/N_1 depends only on the temperature of the systems.

The typical frequency of visible light is $v \approx 5 \times 10^{14}$ Hz, and this corresponds to energy difference $\Delta E \approx 2$ eV. The thermal energy at room temperature (300 K) may be estimated as $k_B T \approx 0.026$ eV. Since $\Delta E >> k_B T$, therefore, from Eq. (10.4), N_2/N_1 comes out of negligible amount. This is equivalent to say that most of atoms are in ground state at room temperature. As T increases, the number N_2 increases, and in limiting case when $T \to \infty$, the population N_2 becomes equal to N_1. Thus in thermal equilibrium position, we can never have energy state E_2 more populated than energy state E_1. To achieve state E_2 more populated, also known as population inversion, system must be pushed into a non-equilibrium state.

10.3 INTERACTION OF LIGHT WITH MATTER

Let us assume a container containing atoms of a gas. These atoms collide with walls of container and other atoms due to thermal energy. This may lead some atoms to higher energy levels. Now, if light is incident on this container, photons start interacting with atoms of gas also. Albert Einstein postulated that in this radiation field the interaction of photon with atom may gives rise to three distinct phenomena, which are shown in Figure 10.1.

10.3.1 Absorption of Photons

Consider a system with two energy levels E_2, excited state and E_1, ground state. In any system, normally atoms are in ground state. When a photon of energy $h\nu(= \Delta E)$ impinges on the system, an atom in state E_1 can absorb the photon energy and be excited to state E_2. The probability of absorption is proportional to the radiation intensity of light and also to the number of atoms N_1 currently in the ground state [Figure 10.1(a)].

10.3.2 Spontaneous Emission

If nothing happens to the atoms in excited state, it will return to level E_1 by releasing a photon of energy $h\nu$. This occurs without any external stimulation and, therefore, known as spontaneous emission. For a group of atoms, there is no fixed phase relationship between emitted photons. In other words, they are incoherent and isotropic. In the absence of other processes, the number of atoms in the excited state at time t is given by

$$N_2(t) = N_2(0) \exp\left(-\frac{t}{\tau}\right) \tag{10.5}$$

where $N_2(0)$ is the number of excited atoms at time $t = 0$ and τ is the lifetime of the transition between two states [Figure 10.1(b)].

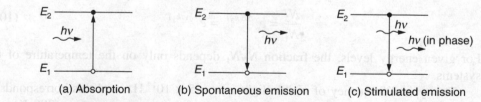

| (a) Absorption | (b) Spontaneous emission | (c) Stimulated emission |

Figure 10.1   The three phenomena arising due to interaction of light with matter.

10.3.3 Stimulated Emission

The atom can also be induced to make a downward transition from the excited level to the ground level by an external stimulation. If a photon of energy $h\nu(=\Delta E)$ impinges on the system, while the atom is in excited state, the atom is stimulated to make a transition to ground state by emitting a photon of energy $h\nu$. The original photon is not absorbed by the atom, and so we have two photons. In a system of atoms, these two photons may stimulate another two atoms to go to ground level by giving off two photons. Now, we have four photons. Clearly this process may ultimately result in a large number of photons. The emitted photons are in phase with the incident photons, and the resultant emission is known as stimulated emission [Figure 10.1(c)].

The rate at which stimulated emission occurs is proportional to the number of atoms N_2 in the excited state and the radiation density of the light. In stimulated emission, the emitted photons have same direction, phase, frequency and plane of polarisation as that of stimulating photons so we have a very intense beam of light. This stimulated emission process is responsible for laser action.

10.4 EINSTEIN'S RELATIONS

The stimulated emission process can be explained only from quantum ideas. The idea of stimulated emission was introduced by A. Einstein. In 1917, he demonstrated that the rates of the three transition processes (absorption, spontaneous emission and stimulated emission) were related mathematically. Einstein developed this by considering the atomic system in thermal equilibrium such that the rate of upward transition (absorption) must be equal to the rate of downward transitions (spontaneous and stimulated).

Consider a two-level system with energies E_1 and E_2 such that $E_2 > E_1$, and density of atoms in states are N_1 and N_2 respectively. The rate r_{12} of upward transition or absorption is proportional to both N_1 and the spectral density $\rho(\nu)$ of the radiation energy at frequency ν. Thus

$$r_{12} = B_{12} N_1 \rho(\nu) \tag{10.6}$$

where B_{12} is constant of proportionality and is known as Einstein's coefficient of absorption. The rate of spontaneous emission depends on number of atoms N_2 in excited state and spontaneous life time τ_{21}. The spontaneous emission rate is equal to the product of N_2 and $1/\tau_{21}$. It is given as:

$$(r_{21})_{sp} = N_2 \frac{1}{\tau_{21}}$$

or

$$(r_{21})_{sp} = N_2 A_{21} \tag{10.7}$$

where A_{21} ($\approx 1/\tau_{21}$) is Einstein's coefficient of spontaneous emission.

The rate of stimulated emission is proportional to density of atoms N_2 in excited state and spectral density $\rho(\nu)$ of impinging radiation energy. This rate is given as:

$$(r_{21})_{st} = B_{21} N_2 \rho(\nu) \tag{10.8}$$

where B_{21} is proportionality constant, known as Einstein's coefficient of stimulated emission. The total transition rate from level E_2 to level E_1 is:

$$r_{21} = (r_{21})_{sp} + (r_{21})_{st}$$

or

$$r_{21} = N_2 A_{21} + B_{21} N_2 \rho(\nu) \tag{10.9}$$

In equilibrium situation, upward transition rate must be equal to downward transition rate, i.e.

$$r_{12} = r_{21}$$

$$B_{12}N\rho(v) = N_2 A_{21} + N_2\rho(v)B_{21} \tag{10.10}$$

or

$$(B_{12}N_1 - N_2 B_{21})\,\rho(v) = N_2 A_{21}$$

or

$$\rho(v) = \frac{N_2 A_{21}}{B_{12}\,N_1 - N_2 B_{21}}$$

Thus

$$\rho(v) = \frac{A_{21}/B_{21}}{\left[\left(\dfrac{N_1}{N_2}\right)\left(\dfrac{B_{12}}{B_{21}}\right) - 1\right]}$$

Substituting the value of $\dfrac{N_2}{N_1} = e^{-hv/k_B T}$, we get

$$\rho(v) = \frac{A_{21}/B_{21}}{\left[\left(\dfrac{B_{12}}{B_{21}}\right)e^{hv/k_B T} - 1\right]} \tag{10.11}$$

The system taken into consideration is in thermal equilibrium and, therefore, its radiations may be considered as black body radiation. The Planck's radiation formula for such a case is:

$$\rho(v) = \frac{8\pi h v^3}{c^3}\frac{1}{(e^{hv/k_B T} - 1)} \tag{10.12}$$

Comparing Eqs. (10.11) and (10.12), we obtain

$$\frac{A_{21}}{B_{21}} = \frac{8\pi h v^3}{c^3} \tag{10.13}$$

and

$$\frac{B_{12}}{B_{21}} = 1 \tag{10.14}$$

Equations (10.13) and (10.14) are known as Einstein's relations for A and B coefficients.

Example: Let us take the example of hydrogen gas to demonstrate the effect of temperature on population in energy states. Consider the gas to be monoatomic. For hydrogen,

$$E_1 = -13.6\ \text{eV}$$

and

$$E_2 = -3.39\ \text{eV}$$

So $E_2 - E_1 = 10.21$ eV at room temperature $T = 300$ K

Therefore

$$k_B T = 1.38 \times 10^{-23} \times 300\ \text{J}$$

$$= \frac{1.38 \times 10^{-23} \times 300}{1.6 \times 10^{-19}}\ \text{eV}$$

$$= 0.025\ \text{eV}$$

Now,
$$\frac{N_2}{N_1} = e^{-(E_2-E_1)/k_BT}$$
$$= e^{-\frac{10.21}{0.025}} \approx e^{-408} \approx 0$$

This shows that at room temperature, almost all atoms are in ground state E_1 and no atom exists in state E_2.

If temperature is raised to 6000 K, then
$$k_BT = 0.516 \text{ eV}$$
and
$$\frac{N_2}{N_1} \approx e^{-19.8}$$
$$\frac{N_2}{N_1} \approx \frac{4}{10^{10}}$$

This data shows that four atoms out of 10^{10} of ground state are excited to energy state E_2. It occurs due to thermal energy given to the system.

10.5 REQUIREMENTS FOR LASER ACTION

For laser action, the stimulated emission has to dominate over other processes. Now, we shall compare rate of stimulated emission with absorption rate and spontaneous emission rates.

The ratio of Eqs. (10.8) and (10.7) gives

$$\frac{\text{Stimulated emission rate}}{\text{Spontaneous emission rate}} = \frac{N_2\rho(v)B_{21}}{N_2A_{21}} = \frac{B_{21}}{A_{21}}\rho(v) \tag{10.15}$$

Again, the ratio of Eqs. (10.8) and (10.6) results in

$$\frac{\text{Stimulated emission rate}}{\text{Absorption rate}} = \frac{N_2\rho(v)B_{21}}{N_1\rho(v)B_{12}} = \frac{N_2}{N_1} \quad \text{(as } B_{12} = B_{21}) \tag{10.16}$$

Thus, to make stimulated emission dominant requires:

Large $\rho(v)$: The presence of large number of photons is required. This is achieved by enclosing the emitted radiations in a cavity usually between two reflectors. The repeated motion of radiation between reflectors creates large radiation density $\rho(v)$ and hence stimulated emission process grows.

Large ratio of B_{21} to A_{21}: The large value of B_{21}/A_{21} also helps stimulated emission to dominate over other processes. In other words, it is necessary to have A_{21} small, that is, life time of upper state must be large. This is achieved by choosing the upper level as a metastable level.

Making $N_2 > N_1$: In thermal equilibrium situation, $N_1 > N_2$. To make $N_2 > N_1$ is equivalent to invert the equilibrium state. The condition $N_2 > N_1$ is termed as population inversion. This is achieved by pumping the atoms of ground state to upper state. The pumping can be done by various methods.

EXAMPLE 10.1 Calculate the ratio of the stimulated emission rate to the spontaneous emission rate for an incandescent lamp operating at a temperature of 1000 K. It may be assumed that the average operating wavelength is 0.5 μm.

Solution: The operating frequency is:

$$v = \frac{c}{\lambda} = \frac{3 \times 10^8}{0.5 \times 10^{-6}} = 6.0 \times 10^{14} \text{ Hz}$$

$$\frac{\text{Stimulated emission rate}}{\text{Spontaneous emission rate}} = \frac{B_{21}}{A_{21}} \rho(v)$$

$$= \frac{c^3}{8\pi h v^3} \cdot \frac{8\pi h v^3}{c^3} \frac{1}{(e^{hv/k_B T} - 1)}$$

$$= \frac{1}{(e^{hv/k_B T} - 1)}$$

$$= \frac{1}{\exp\left(\dfrac{6.6 \times 10^{-34} \times 6.0 \times 10^{14}}{1.38 \times 10^{-23} \times 1000}\right) - 1}$$

$$= \frac{1}{\exp(28.8) - 1}$$

Since $e^{28.8}$ is very large in comparison to one, we may neglect one and, therefore,

$$\frac{\text{Stimulated emission rate}}{\text{Spontaneous emission rate}} = \frac{1}{\exp(28.8)}$$

$$= e^{-28.8} \approx 3 \times 10^{-13}$$

The ratio of stimulated emission rate and spontaneous emission rate is 3×10^{-13}.

10.6 POPULATION INVERSION AND PUMPING

The essential requirement to get more stimulated emission is to have a larger population of excited atoms in the system so that photons may have more interactions with the excited atoms. Under thermal equilibrium condition, lower energy level E_1 contains more atoms than level E_2 as shown in Figure 10.2(a). For optical amplification, we have to get larger population of atoms in the excited state. This condition is termed as population inversion, and it is illustrated in Figure 10.2(b). In population inversion condition, any randomly emitted photon may trigger stimulated emission of photons. These stimulated photons induce more stimulated emissions and this process continues. These emitted photons are coherent and, therefore, intense beam of light is obtained.

The process of getting population inversion is referred as pumping, and device utilised for this purpose is known as pump. There are various types of pumping such as optical pumping, chemical pumping and electrical pumping.

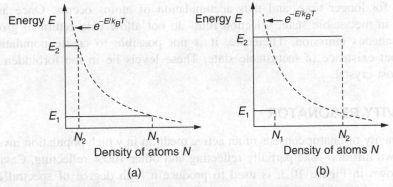

Figure 10.2   Population in energy levels (a) in thermal equilibrium condition (b) in population inversion condition.

Optical Pumping

To produce a larger population in the excited state, the atoms are exposed to optical energy. The photon energy $h\nu$ must be equal to energy difference between ground state and excited state. The ground state atoms jump to excited state by absorbing photons. This method is referred to as optical pumping. The optical pumping is normally utilised in solid state lasers to achieve population inversion, for example, xenon flash lamp is used in ruby laser.

Chemical Pumping

The energy released in making or breaking of chemical bonds may be utilised for pumping the ground state atoms. If we get pumping energy from chemical reactions, then pumping is termed as chemical pumping. The exothermic reactions are useful for chemical lasers.

Electrical Pumping

The ground state atoms can be excited to upper level by using electrical pumping. The atoms are excited by electron impact by passing an electric discharge through the system. This pumping is usually used in gas and semiconductor lasers.

10.7   METASTABLE STATES

The population inversion can not be achieved in a two-level system because probable rate of induced absorption and emission is almost equal. Thus laser action is impossible in a two-level system.

An atom can be excited to a higher level by supplying energy to it. The atom remains in excited level for approximately 10^{-8}s before jumping down to lower level with emission of a photon. In order to achieve the population inversion condition, the atom must reside in higher level for longer time. In other words, the state must be of longer life time. A metastable is such a state for which life time is approximately in range 10^{-6}s to 10^{-3}s. Obviously the metastable state allows excited atoms to remain there for longer time, and thus accumulation of atoms occurs. Once an atom finds itself in metastable state, selection rules do not allow it to return to ground state by spontaneous emission. Therefore, it is not possible to create population inversion without existence of metastable state. These levels lie in the forbidden band gap of the host crystal.

10.8   CAVITY RESONATOR

The cavity resonator consists of an active medium in which population inversion occurs and two mirrors—one partially reflecting and other 100% reflecting. Cavity resonator, as shown in Figure 10.3, is used to produce a high degree of spectral density $\rho(v)$. This is achieved by placing active medium between two reflecting mirrors.

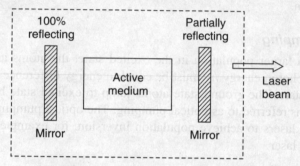

Figure 10.3   Schematic of cavity resonator.

Like electronic oscillator, laser is an optical oscillator. The electronic oscillator is basically an amplifier in which a part of output is fed back to input in right phase. Some circuit conditions results in stable output in such systems. When a light beam passes through a suitable medium with population inversion, it is amplified by the stimulated emission process. Thus we have optical amplification. To realise an oscillator (optical), a part of output light must be coupled with input (here active medium). This coupling or feedback is achieved by reflecting mirrors.

A spontaneously emitted photon may start laser action. When such a photon travels along medium, the stimulated emission process starts. Such photons travelling from one mirror are amplified which pass through active medium. This amplified wave is reflected from other mirror and again passes through active medium which results in further amplification. This to and fro motion of light produces a large number of stimulated photons. Ultimately when the amplification balances the losses in the cavity, the laser beam emerges out. The waves of light travelling along both directions (to and fro motion) between mirrors interfere, and a standing wave pattern is formed. If we consider the waves as plane waves, then total phase charge in completing one round trip must be integral multiple of 2π, that is,

$$\frac{2\pi}{\lambda}(L+L) = n2\pi \tag{10.17}$$

where L is distance between mirrors, λ is wavelength of light in the active medium and n is integral number such that $n = 1, 2, 3, \dots$.

Equation (10.17) may also be written as:

$$L = n\frac{\lambda}{2} \qquad (10.18)$$

By using well known relation $c = v\lambda$, we may also have

$$v = n\frac{c}{2L} \qquad (10.19)$$

Equation (10.19) gives the discrete frequencies in which system may resonate (oscillate). Thus we have many modes of oscillations and n gives mode number. Normally laser oscillates at continuum of frequencies about the frequency given by Eq. (10.19). Therefore, we have a pulse width. When losses are compensated by active medium, then resultant laser emission is extremely narrow and is limited due to the presence of spontaneous emission. In general, mirrors of cavity are spherical which helps in focusing the radiation. For simplicity, here plane mirrors are considered.

▌10.9   THRESHOLD CONDITION FOR LASER ACTION

The steady state condition for laser oscillation may be achieved by keeping balance between gain and losses. In addition to population inversion between energy level necessary for amplification, a minimum gain within amplifying medium is a must for initiating and sustaining the laser action. Consider a medium, in which population inversion has been established, is placed between two mirrors having reflectivity R_1 and R_2 (Figure 10.4). Let us assume that all losses in the system, except due to mirrors, are included in a single loss coefficient per unit length α.

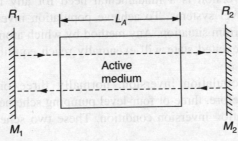

Figure 10.4   Round trip motion of waves for laser action.

Suppose light wave of intensity I_0 starts from mirror M_1. During its propagation towards M_2, some loss and gain occur. The intensity of beam at mirror M_2 is given as

$$I = I_0 e^{-\alpha L_A} e^{g L_A} \qquad (10.20)$$

where g is gain per unit length due to stimulated emission and L_A is length of active medium. The reflectivity of mirror M_2 is R_2 so after reflection from mirror M_2, the intensity of beam that moves towards M_1 is:

$$I_{M_2} = I_0 e^{-\alpha L_A} e^{g L_A} R_2 \qquad (10.21)$$

Keeping in mind the discussion of loss and gain, the intensity of beam at M_1 that becomes ready to move towards M_2 is:

$$I_{M_1} = I_0 e^{-\alpha L_A} e^{gL_A} R_2 e^{-\alpha L_A} e^{gL_A} R_1 \qquad (10.22)$$

Equation (10.22) gives intensity of beam after one round trip. For steady sate,

$$I_{M_1} = I_0 \qquad (10.23)$$

In other words, gain $(= I_{M_1}/I_0)$ must be unity for steady state. Any increment in gain will start the laser action. Equation (10.23) may be written as:

$$R_1 R_2 e^{(g-\alpha)2L_A} = 1$$

or

$$e^{(g-\alpha)2L_A} = \frac{1}{R_1 R_2}$$

or

$$(g - \alpha)2L_A = \ln \frac{1}{R_1 R_2}$$

The threshold gain per unit length may be given as:

$$g_{th} = \alpha + \frac{1}{2L_A} \ln \frac{1}{R_1 R_2} \qquad (10.24)$$

Equation (10.24) is known as the threshold condition for laser action. The second term in this equation represents losses due to mirrors. This condition may be used to determine the value of pumping energy for laser action.

10.10   PUMPING SCHEMES

The population inversion is a fundamental need for any laser system. It cannot be achieved in two-level systems. To achieve population inversion, the system must be kept in non-equilibrium situation. Any method by which atoms are excited from ground state to the upper excited state will eventually reach equilibrium by the de-excitation process.

To achieve population inversion, normally three or four energy levels are necessary and, therefore, three-or four-level pumping schemes are generally utilised for establishing population inversion condition. These two schemes are widely employed in laser systems.

10.10.1   Three-level Pumping Scheme

Consider a three-level system as shown in Figure 10.5. Level E_1 is ground state, E_2 is metastable state and level E_3 is excited state. The lifetime of metastable state is much larger and this characteristic helps in achieving population inversion.

Initially, the system of atoms is in thermal equilibrium and majority of the atoms will be in the ground state. When atoms in ground state are pumped by means of optical energy, electrical energy or chemical energy the atoms are excited to level E_3. This transition is also known as pump transition $(E_1 \rightarrow E_3)$. Now, the atoms in E_3 decay

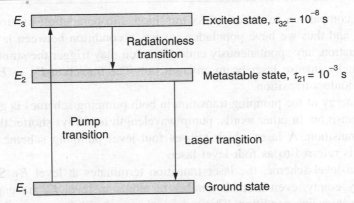

E_3 — Excited state, $\tau_{32} = 10^{-8}$ s

Radiationless transition

E_2 — Metastable state, $\tau_{21} = 10^{-3}$ s

Pump transition

Laser transition

E_1 — Ground state

Figure 10.5 A three-level pumping scheme.

to level E_2 quickly. The transition $E_3 \rightarrow E_2$ is radiationless transition, and released energy is transferred to lattice of the host material.

The laser emission is obtained when atoms go to level E_1 from level E_2 under stimulated emission condition. The lifetime of transition $E_3 \rightarrow E_2$ is quite small in comparison to life time of transition $E_2 \rightarrow E_1$. Therefore, the population of excited atoms in the level E_2 becomes larger, and thus we have population inversion condition between E_2 and E_1. In this situation, any spontaneously emitted photon of frequency $v(=(E_2 - E_1)/h)$ may start stimulated emission process.

The laser, which utilised three levels for its operation, is termed as three-level laser. In such lasers, the atoms in ground state may absorb photons emitted from laser transition which reduces beam intensity. This makes the three-level laser less efficient.

10.10.2 Four-level Pumping Scheme

The four-level pumping scheme consists of four energy levels as shown in Figure 10.6. The pump transition excites the atoms in the ground state level E_1 into the pump band E_4. This excitation can be carried out by using optical, electrical or chemical energy of pump sources. The transition of atoms from E_4 to E_3 is fast and radiationless. The emitted energy goes to crystal lattice. Since the life time of laser transition is much

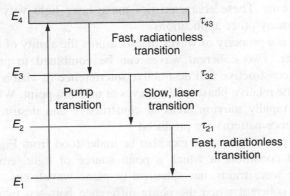

E_4 — τ_{43}

Fast, radiationless transition

E_3 — τ_{32}

Pump transition

Slow, laser transition

E_2 — τ_{21}

Fast, radiationless transition

E_1 —

Figure 10.6 A four-level pumping scheme.

longer in comparison to radiationless transition, the population of atoms in level E_3 increases, and thus we have population inversion condition between levels E_3 and E_2. In this situation, any spontaneously emitted photon may trigger the stimulated emission process, and a intense laser beam emerges out. The transition $E_2 \rightarrow E_1$ is also a fast and radiationless transition.

The energy of the pumping transition in both pumping schemes is greater than that of laser transition. In other words, pump wavelength is always shorter than wavelength of laser transition. A laser which utilises four-level pumping scheme for population inversion is referred to as four-level laser.

In four-level scheme, the laser transition terminates at level E_2. Since this level is virtually empty, even a small number of atoms in level E_3 are capable to create population inversion condition. This reduces the need of large pump power, and hence enhances the efficiency of laser. The terminal point of laser transition in three-level system is ground state where we have larger number of atoms. Therefore, to achieve population inversion condition between level E_2 and E_1, more pump power is needed. This reduces the laser efficiency in three-level lasers.

Normally three-level lasers are pulse lasers, while four-level lasers are continuous wave lasers. In three-level lasers, after the commencement of stimulated emission, the population inversion condition reverts to normal population condition. For next emission, population inversion condition has to be established again. Therefore, we have output in form of pulses. In four-level lasers, system does not revert to normal condition and, therefore, population inversion condition is maintained which results in continuous wave at output.

10.11   COHERENCE

The coherence between two light sources is defined as the existence of a constant phase relation between them. It is of two kinds, namely spatial coherence and temporal coherence. In lasers, the stimulated photon has exactly the same frequency, direction and state of polarisation as that of stimulating photon. These characteristics come under spatial coherence. The stimulated photons also have exactly the same phase and speed as that of stimulating photon. The temporal coherence covers these characteristics of emitted photons. These characteristics make laser light very useful for scientific, industrial and many other applications.

Coherence is a property of waves that measures the ability of the waves to interfere with each other. Two coherent waves can be combined to produce an unmoving distribution of constructive and destructive interference (a visible interference pattern) depending on the relative phase of the waves at meeting point. When incoherent waves are combined, rapidly moving areas of constructive and destructive interference (no visible interference pattern) are produced.

The two types of coherence can also be understood from Figure 10.7. This figure shows an ideal condition in which a point source of light, emitting infinitely long monochromatic wave trains, has spherical or plane wavefronts. The emergent waves are temporally coherent when the phase difference between two fixed points x_1 and x_2 spaced any distance along any ray is independent of time or phase difference at

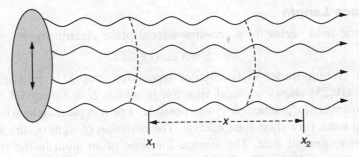

Figure 10.7 Coherent waves from a laser.

a single point measured for a fixed time interval does not vary with time. For spatial coherence, the phase difference for any two points in a plane normal to the ray direction must be independent of time.

The temporal coherence time is the time over which the beam wave fronts remain equally spaced or field remains sinusoidal and spatial coherence length is distance over which the beam wavefronts remain flat. Infact whole effort in design of a laser is to make longer coherence time and length. Figure 10.8 makes the above statements more clear.

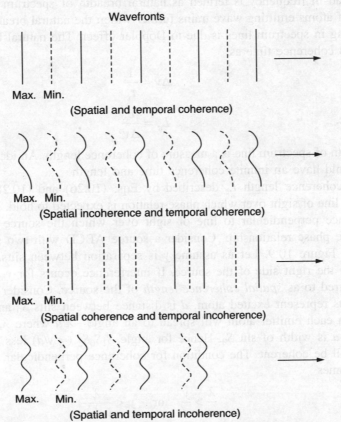

Wavefronts

Max. Min.
(Spatial and temporal coherence)

Max. Min.
(Spatial incoherence and temporal coherence)

Max. Min.
(Spatial coherence and temporal incoherence)

Max. Min.
(Spatial and temporal incoherence)

Figure 10.8 Illustrations of coherence and incoherence.

Coherence Length

The electric field vector for a one-dimensional plane electromagnetic wave is:

$$E = A \cos(\omega t - kx) \tag{10.25}$$

where A is amplitude, ω is angular frequency and k is propagation constant. Equation (10.25) shows an ideal situation is which E is sinusoidal for any x at all times. However, in practice, it is not possible. For a typical light source, the field is sinusoidal for a very short time interval. The emission of light occurs when an excited atom goes to ground state. The average life time of an atom in the state of radiation is of the order of 10^{-8} s. The field is sinusoidal only for this duration. In case of laser beam, the field remains sinusoidal for extremely long time. The duration for which field remains sinusoidal is referred to as coherence time τ_c of the optical beam and, therefore, for light waves travelling with speed c, the coherence length may be defined as:

$$L_c = c\tau_c \tag{10.26}$$

The Fourier analysis of wave train shows that it contains waves of frequencies spread on both sides of central frequency. Therefore, on place of getting only central frequency from a source, we get a frequency spread $\Delta\nu$ about the central frequency ν_0. This spread is natural. It is not due to the observing instrument or any other inefficiency. The spread in frequency is termed as natural breadth of spectrum line. The thermal motion of atoms emitting wave trains further widen the natural breadth of lines. Infact broadening in spectrum lines is due to Doppler effect. The natural breadth $\Delta\nu$ can be related to coherence time as:

$$\Delta\nu = \frac{1}{\tau_c} \tag{10.27}$$

Therefore,

$$L_c = \frac{c}{\Delta\nu} \tag{10.28}$$

Thus width of spectrum line is a measure of coherence length. An ideal monochromatic wave would have an infinite coherence time and length.

The coherence length L_c described by Eqs. (10.26) and (10.28) is the distance along the line of sight over which phase-relation is expected to hold. Now, we consider the distance perpendicular to line of sight over which the source shows coherence or definite phase relationship. Consider a source ABCD with two slits S_1 and S_2 as shown in Figure 10.9. Let us assume y is separation between slits, and interference occurs on the right side of the source. If interference occurs for $y_{max} = L_s$, then this L_s is referred to as *spatial coherence length* of the source. Consider Figure 10.9(a) in which dots represent excited atom, d is distance between slits S_0 and S_1 and S_2. The light from each emitter atom will spread to an angle $\sim\lambda/a$ where λ is wavelength of light and a is width of slit S_0. Hence for angle $S_1 S_0 S_2$ ($\approx y/d$) less than λ/a, both S_1 and S_2 will be coherent. The condition for coherence perpendicular to propagation of light becomes

$$\frac{\lambda}{a} > \frac{y}{d} \quad \text{or} \quad y < \frac{\lambda}{(a/d)} \tag{10.29}$$

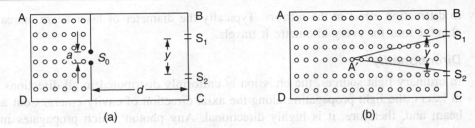

Figure 10.9 Two sources for discussing spatial coherence.

Here we can control (or maximise) the value of y with help of a and d.

In case of Figure 10.9(b), slits are directly on the source. The phase relation at S_1 and S_2 will be governed by the path difference $AS_1 \sim AS_2$. Here A is radiation emitting atom. Due to large number of atoms present, the position of A will change which results in different phase relationships at S_1 and S_2. Since no constant phase relationship exists at S_1 and S_2, so there is little spatial coherence.

In brief, laser can be taken as a device which increases coherence length L_c along line of sight and spatial coherence length L_s perpendicular to line of sight. Increasing L_c results in successive wave trains that fit in phase and enhancing L_s means controlling the atoms in a way to emit radiations in phase. A point source of negligible diameter (negligible a) emits spatially coherent light. The spatial coherence of light will increase as it travels away from the source and becomes like a plane wave. Light from distant stars, though far from monochromatic character, has extremely high spatial coherence.

10.12 PROPERTIES OF LASER LIGHT

Laser radiation is electromagnetic in radiation. Some important characteristics of laser emission are as follows:

Monochromaticity

Laser light is monochromatic that is laser emits light at a single wavelength (or frequency). Ideally no light is monochromatic. In practice, any source of light has some spread in wavelength. In other words, many other wavelength components are present along with the wavelength of interest. The monochromaticity of a light is defined as ratio of central wavelength λ to wavelength spread $\Delta\lambda$ about central wavelength, i.e.,

$$\text{Monochromaticity} \equiv \frac{\lambda}{\Delta\lambda}$$

It is also known as quality factor Q for laser emission. Generally wavelength spread of traditional monochromatic sources is about 1000 Å, while the wavelength spread of laser emission is about 10 Å.

Divergence

Laser light is highly directional in nature. The light from ordinary sources diverges as it travels in space, while the spread of laser beam is much smaller. This spread is due to diffraction effect at the mirrors. Typically the diameter of laser beam increases by about 1 mm for every kilometre it travels.

Directionality

In ordinary light sources, the emission is uniformly distributed in all directions, while in lasers, the light propagating along the axial direction of cavity emerges out as laser beam and, therefore, it is highly directional. Any photon which propagates in other direction is eliminated inside the cavity.

Intensity

The intensity of laser beam is very high as compared to normal light sources. Since the laser power is concentrated in a beam of small diameter (few mm) so beam intensity becomes tremendously large.

Coherence

The laser light is highly coherent both spatially and temporally. Each and every ray of laser beam bears the same phase relationship with each other at all times. The laser emission is due to stimulated emission of photons. These photons are identical in every respect and hence resultant beam is coherent.

EXAMPLE 10.2 If light of wavelength (λ) 660 nm has wave trains 20 λ long, what will be the coherence time?

Solution The coherence time is:

$$\tau_c = \frac{L_c}{c}$$

$$= \frac{20\lambda}{c}$$

$$= \frac{20 \times 660 \times 10^{-9}}{3.0 \times 10^8}$$

$$\tau_c = 4.4 \times 10^{-14} \text{ s}$$

The coherence time is 4.4×10^{-14} s.

EXAMPLE 10.3 Find the Q-factor of white light if wavelength of white light is 400 nm to 700 nm.

Solution $\Delta\lambda = (700 - 400)$ nm

or $\Delta\lambda = 300$ nm

and $\lambda = \dfrac{700 + 400}{2}$ nm

                                $= 550$ nm

So
$$Q = \frac{\lambda}{\Delta\lambda} = \frac{550}{300} = 1.8$$

The Q-factor of white light is 1.8.

▌10.13 VARIOUS TYPES OF LASERS

Various types of lasers may be classified on the basis of active medium used for laser action. In general, lasers are divided into four broad categories, namely solid state lasers, gas lasers, semiconductor lasers and chemical lasers. Here some laser systems are discussed which fall in these categories.

10.13.1 Ruby Laser

The ruby laser comes under solid state lasers category. The outcoming beam from ruby laser cavity consists of pulses separated by small time interval, therefore, it is pulse laser. It is a first successful laser developed by T.H. Maiman in 1960. A three-level pumping scheme is utilised for laser operation in ruby lasers.

The ruby rod used in lasers is made by fusing aluminium oxide (Al_2O_3) and chromium oxide (Cr_2O_3). The amount of chromium ions (Cr^{3+}) in the ruby rod is about 0.05 per cent. These chromium ions are responsible for laser action and provide pink colour to the ruby rod. The ruby crystal is about 0.60 cm in diameter and 4 cm in length. Both ends of ruby rod is polished and silvered; one end is heavily coated, while the other is lightly coated with silver. This ruby crystal is placed within the helical-shaped xenon flash tube which is essentially used for optical pumping. The whole arrangement is shown in Figure 10.10. The xenon flash tube produces white light on application of power supply.

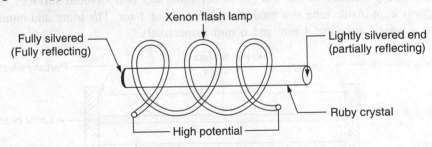

Figure 10.10 Schematic diagram of ruby laser.

The energy level diagram of ruby laser is illustrated in Figure 10.11. When flash of light falls on the ruby rod, photons are absorbed by chromium ions in ground level. These ions are excited to level E_3. The excited ions pass down to metastable state giving part of their energy to crystal lattice. The life time of this state is higher so population of chromium ions increases in the metastable state. In this way, population inversion condition is established. Now, a spontaneously emitted photon between level E_2 and E_1 may trigger stimulated emission process which results in the production of

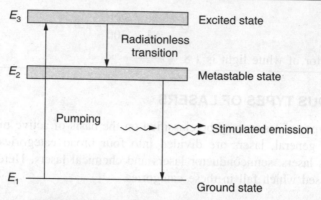

Figure 10.11 Energy level diagram of ruby laser.

coherent photons. These photons induce further emission of photons which are in phase with earlier photons. This process is repeated, and a intense beam of laser light emits from partially reflecting end of ruby rod.

10.13.2 Helium–Neon (He–Ne) Laser

The helium–neon (He–Ne) laser was the first successful gas laser. It was put into operation by Ali Javan, William Benret and Donald Herriott in 1961. He–Ne laser is a continuous wave laser consisting of helium and neon gases. This laser is widely used in optics laboratories. A four-level pumping scheme is utilised for establishing population inversion in helium–neon lasers.

The schematic arrangement for He–Ne laser is depicted in Figure 10.12. The laser assembly is consist of a high voltage power supply, nearly 1-m long glass tube containing mixture of He and Ne in 5:1 ratio and two silvered surfaces. The He–Ne gas is kept inside tube at a total pressure of about 1 tor. The inner and outer diameters of a typical tube are 4 mm and 6 mm respectively.

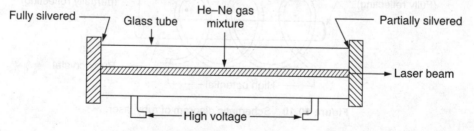

Figure 10.12 Schematic diagram of He–Ne laser.

The energy level diagram of helium–neon laser is shown in Figure 10.13. In Figure 10.13(a), energy levels of helium are depicted, while Figure 10.13(b) energy levels depicts of neon gas. When power is switched on, helium atoms are excited to level F_2 by an electron impact. This process is unable to excite the neon atoms. The F_2 state of helium is of longer life time, and energy allotted to this state is almost equal to E_4 level of neon.

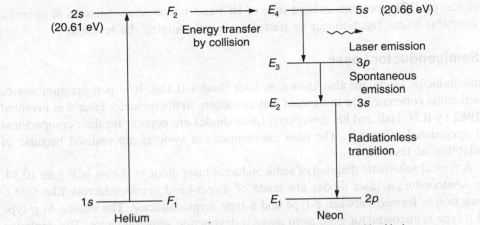

Figure 10.13   The relevant energy levels for a He–Ne laser.

The energy of helium atoms in this state can easily be transferred to neon atoms when they collide. This collision excited the neon atoms to E_4 level, and population inversion is established between E_4 and E_3 levels of neon gas. Now, any spontaneously emitted photon between E_4 and E_3 can trigger the stimulated emission process at 6329 Å wavelength (red colour). The photons travelling along the axis of tube are reflected back and forth between end mirrors and rapidly build into an intense beam of laser light. This beam emerges out from lower reflecting end.

State E_4 of neon atoms is a metastable state and, therefore, it is populated. The transition $E_3 \rightarrow E_2$ is spontaneous, and radiationless transition occurs between E_2 and E_1. State E_2 is also a metastable state so neon atoms tend to accumulate here. These atoms must be brought quickly to level E_1 otherwise number of ground state atoms reduces which ultimately reduces the level of population inversion. The rapid collision with walls of tube may help in depopulating quickly the level E_2. For this purpose, the diameter of tube is kept smaller.

10.13.3 Chemical Lasers

These are lasers which utilise energy from chemical reactions for their operation. Such lasers are referred to as chemical lasers. When laser action is achieved in a chemical reaction, the chemist gains new information about the distribution of energy at the time of reaction.

The reaction between hydrogen and chlorine in an explosion provides necessary energy for pumping in a chemical laser. Such a laser produces laser beam in infrared region of the electromagnetic spectrum. The mechanism responsible for this laser is briefly given as follows:

$$H_2 + Cl \rightarrow HCl + H$$
$$H + Cl_2 \rightarrow HCl^* + Cl + Energy$$
$$HCl^* \rightarrow HCl + \nu$$

Here star on HCl represents excited state of HCl and ν represents emission of photons. In chemical lasers, reaction can be started by merely mixing the reactants.

10.13.4 Semiconductor Laser

Semiconductor lasers are also known as laser diodes (LDs). It is p–n junction device which emits coherent light in forward bias condition. Semiconductor laser was invented in 1962 by R.N. Hall and his coworkers. Laser diodes are popular for their compactness and operational efficiency. The fiber communication systems are realised because of availability of laser diodes.

A typical schematic diagram of semiconductor laser diode is shown in Figure 10.14. The semiconductor laser diodes are made of direct-band semiconductors like GaAs. A junction is formed between p-type and n-type semiconductor. The doping in p-type and n-type semiconductor is so as to make it degenerate semiconductor. The depletion region between two types of semiconductor works as active region.

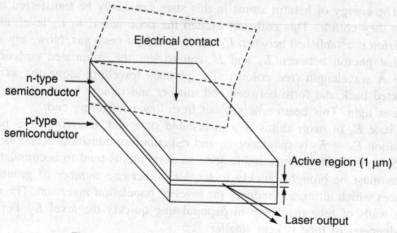

Figure 10.14 Schematic of semiconductor laser.

In semiconductor laser diodes, population inversion occurs in conduction band. When laser diode is forward biased, electrons from n-side and holes from p-side diffuse into the depletion region where both carriers recombine. The result of recombination in direct band semiconductors is production of light photons. Under forward biased condition, when drive current through electrical connections reaches threshold level, population inversion results, and a coherent radiation emerges out. A graph between optical output and drive current is given in Figure 10.15.

If P_e is electrical power used to generate optical power P_o, then efficiency of laser or quantum efficiency η is defined as:

$$\eta = \frac{P_o}{P_e}$$

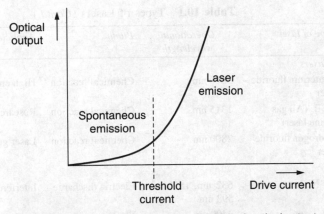

Figure 10.15 A plot between drive current and optical output.

To improve the operation of semiconductor laser diodes in terms of line width, output power, threshold current, quantum well structures are quite advantageous. When thickness of active layer becomes 10 nm or less, quantum effects come into play. The quantum well structure means location of carriers is restricted to one dimension. This structure reduces number of carriers needed for population inversion and, therefore, threshold current is reduced.

There are thousand kinds of lasers, but most of them are used for specialised works. A list of various types of lasers, their operational wavelength, gain medium and applications are tabulated in Table 10.1.

Table 10.1 Types of Lasers

S.No.	Type of laser	Operational wavelength	Pump	Applications
Solid state lasers				
1.	Ruby laser	695 nm	Flash lamp	Visible light laser, holography
2.	Nd: YAG laser	1320 nm	Flash lamp	Surgery, research, material processing
3.	Neodymium glass laser	1062 nm	Laser diode or flash lamp	Production of high power
4.	Yb: YAG laser	1030 nm	Laser diode	LIDAR, optical refrigeration
5.	Er: YAG laser	2940 nm	Flash lamp or laser diode	Dentistry
6.	Ho: YAG laser	2100 nm	Laser diode	Kidney stone removal
Semiconductor lasers				
1.	Laser diode	400 nm–20000 nm	Electrical current	Holography, welding, optical communication
2.	In Ga As P	1000 nm–2100 nm	Electrical current	Medical, telecommunication
3.	Vertical cavity surface emitting laser (VCSEL)	850 nm–1500 nm	Electrical current	Telecommunication

(Contd.)

Table 10.1 Types of Lasers (*Contd.*)

S.No.	Type of laser	Operational wavelength	Pump	Applications
	Chemical lasers			
1.	Deuterium fluoride laser	3800 nm	Chemical reaction	High energy production
2.	AGIL (All gas iodine laser)	1315 nm	Chemical reaction	Research, weapons
3.	Hydrogen fluoride laser	2800 nm	Chemical reaction	Laser guided weapons
	Gas lasers			
1.	He–Ne laser	632 nm, 1520 nm, 593 nm	Electric discharge	Interferometry, holography
2.	CO_2 laser	9400 nm	Electric discharge	Cutting, surgery
3.	Argon laser	488 nm, 514 nm, 528 nm	Electric discharge	Retinal surgery, lithography
4.	Nitrogen laser	337 nm	Electric discharge	Measuring air pollution, scientific research

10.14 APPLICATIONS OF LASERS

The large intensity, narrow bandwidth and narrow angular spread make laser an extremely useful tool for a variety of applications. Lasers are used in industrial electronics, communication, medical and environmental researches.

Medical Application

Laser is used in eye surgery where retinal tissue is cauterized to weld detached retinas. For this purpose, laser is aligned to be focused by the pupil of the eye on the area to be treated. The laser energy is adjusted to melt the tissue. Lasers are also used to treat a patient suffering from myopia.

Communication

The communication field is revolutionised by laser application. The fiber optic communication system has enormous capacity in comparison to other available means of communication. The use of lasers in fiber communication increases the data rate. The higher data rate and capacity is able to accommodate many more channels of message simultaneously.

Environment

Environment related researches can be carried out in better way by using lasers. It is very useful in the measurement of atmospheric pollutants such as dust, smoke and fly ash. The laser remote sensing method permits the measurement of pollutants by collecting the samples and chemical analysis. Normally pulsed lasers are used for this kind of work.

Holography

Holography is a method to develop holograms. A hologram is photograph of a three-dimensional object that records the details of both amplitudes and phases of light received from different parts of the three-dimensional object. Holography requires temporally and spatially coherent light. Therefore, laser finds application in this area. Holography is utilised to store large data, e.g. a one cm^3 crystal may accommodate the information contained in a library of million books.

Welding and Cutting

The energetic and highly collimated beam of lasers can be used for welding and cutting. The automobile industry makes extensive use of lasers for computer controlled welding. The stainless steel handles can be welded on copper cooking pots. The conventional technique cannot do this job because of large difference in conductivities of two metals.

Laser Printers

Now-a-days laser printers are extensively used for printing. In laser printers, normally semiconductor laser (AlGaAs laser) at 760 nm wavelength is used. Laser printers employ a semiconductor laser and xeroxography technique.

SOLVED NUMERICAL PROBLEMS

PROBLEM 10.1 What is ratio of number of atoms in excited state and lower state in a laser that produces light of wavelength 6328 Å at 27 °C?

Solution According to Boltzmann's relation,

$$\frac{N_2}{N_1} = e^{-(E_2 - E_1)/k_B T}$$

$$E_2 - E_1 = \frac{hc}{\lambda} = \frac{6.6 \times 10^{-34} \times 3 \times 10^8}{6328 \times 10^{-10}} = 3.1 \times 10^{-9} \, \text{J}$$

So

$$\frac{N_2}{N_1} = \exp\left[\frac{-3.1 \times 10^{-9}}{1.32 \times 10^{-23} \times 300}\right] \approx 10^{-33}$$

The ratio of atoms in two states is 10^{-33}.

PROBLEM 10.2 The coherence length of sodium line is 2.5 cm. Find out the coherence time.

Solution Coherence time $\tau_c = \dfrac{L_c}{c}$

$$= \frac{2.5 \times 10^{-2}}{3 \times 10^8}$$

∴ Coherence time $\tau_c = 0.8 \times 10^{-10}$ s

PROBLEM 10.3 A ruby laser emits 0.1 J of light of wavelength 720 nm. How many minimum number of chromium ions are there in ruby?

Solution The emitted energy is equal to number of photons multiplied by energy of one photon, i.e.,

$$E = N \times h\nu$$

So

$$N = \frac{E}{h\nu} = \frac{E\lambda}{hc}$$

$$N = \frac{0.1 \times 720 \times 10^{-9}}{6.6 \times 10^{-34} \times 3 \times 10^8} = 3.6 \times 10^{17}$$

Minimum numbers of chromium ions are 3.6×10^{17}.

PROBLEM 10.4 In a Michelson interferometer using He–Ne laser, fringes are remained visible when the path difference was increased upto 8 m. The wavelength of laser light is 11.5×10^{-7} m. What is the lower limit for the coherent times? Also find the spectral width and Q of the line.

Solution Coherence length L_c will be greater than, 8 m.

and

$$\text{Coherence time } \tau_c = \frac{L_c}{c}$$

$$> \frac{8}{3 \times 10^8} = 2.7 \times 10^{-8} \text{ s}$$

Coherence time must be greater than 2.7×10^{-8} s

$$\text{Spectral width } \Delta\lambda = \frac{\lambda^2}{L_c} = \frac{(11.5 \times 10^{-7})^2}{8}$$

$$= 1.6 \times 10^{-3} \text{ Å}$$

The spectral width is 1.6×10^{-3} Å.

The Q of line $= \dfrac{L_c}{\lambda} = 7 \times 10^6$

Clearly Q of line is 7×10^6.

EXERCISES

THEORETICAL QUESTIONS

10.1 What do you understand by term coherence? How it is achieved in laser systems?

10.2 Develop relation between Einstein's coefficients.

10.3 Discuss construction and working of Ruby laser.

10.4 Describe the need of resonant cavity in a laser. Find the threshold condition for laser action.

10.5 Give complete description of three-level and four-level pumping schemes.

10.6 Give details of spatial and temporal coherence.

10.7 Explain the principle and working of He–Ne laser.

10.8 With the help of diagram, discuss the working of semiconductor laser diodes.

10.9 Give brief account of chemical lasers. Write necessary reaction for getting energy to excite atoms.

10.10 What are the characteristics of lasers? Give some applications of lasers.

10.11 What do you mean by population inversion? How it is necessary for laser action?

Numerical Problems

P10.1 The radio of population of two states is 1×10^{-30}. Find the wavelength of light emitted due to transition at 330 K temperature. **(Ans. 6320 Å)**

P10.2 The energy difference between two states of a powerful laser is about 0.12 eV. What will be the wavelength of radiation? **(Ans. 10 μm)**

P10.3 What will be the ratio of population of two states of He-Ne laser that produces light of wavelength 700 nm at 27 °C? **(Ans. 5.9×10^{-29})**

P10.4 The length of ruby crystal in a ruby laser is 150 mm and gain per centimetre of the crystal is 0.0005. If one mirror is fully reflecting, what will be reflectance of partially reflecting mirror? **(Ans. 98.5%)**

P10.5 The coherence length of sodium D_2 line is 2.5 cm. Find out spectral width of the line of wavelength 6×10^{-7} m. **(Ans. 14×10^{-12} metre)**

Multiple Choice Questions

MCQ 10.1 The process responsible for laser emission is
    (a) Spontaneous emission   (b) Stimulated emission
    (c) Absorption   (d) None of these

MCQ 10.2 The high spectral density in lasers is obtained due to
    (a) Resonant cavity   (b) Flash lamp
    (c) Pumping arrangement   (d) Applied potential difference

MCQ 10.3 The most important characteristic of laser light is
    (a) Polarisation   (b) Stimulated emission
    (c) Directionality   (d) Coherence

MCQ 10.4 Semiconductor laser diodes are realised by using
    (a) Indirect band semiconductors   (b) Nondegenerate semiconductors
    (c) Direct band semiconductors   (d) Metals

MCQ 10.5 In He–Ne laser, the He atoms are used
    (a) To enhance efficiency of laser   (b) To provide energy to Ne atoms
    (c) For high directionality   (d) For getting coherent light

MCQ 10.6 The population inversion in lasers is obtained at
    (a) Excited state   (b) Ground state
    (c) Metastable state   (d) None of these

MCQ 10.7 In chemical lasers, the population inversion is achieved by utilising energy obtained from
    (a) Flash lamp   (b) Chemical reaction
    (c) Driving current   (d) Collision of atoms

MCQ 10.8 The concept of stimulated emission is given by
    (a) A. Einstein   (b) T.H. Maiman
    (c) Ali Javan   (d) Newton

MCQ 10.9 In radiationless transition, the released energy goes to
    (a) Atmosphere   (b) Lattice crystal
    (c) None   (d) Pump

MCQ 10.10 A coherent light means it is
    (a) Temporally coherent   (b) Spatially coherent
    (c) Both (a) and (b)   (d) Highly directional

Answers

  10.1 (b)   **10.2** (a)   **10.3** (d)   **10.4** (c)   **10.5** (b)   **10.6** (c)   **10.7** (b)

  10.8 (a)   **10.9** (b)   **10.10** (c)

Optical Fibers

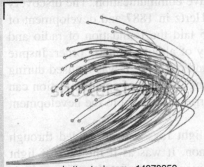

11.1 INTRODUCTION

Mankind had the necessity for communication throughout the history. Initially communication was done through signals, voice or primitive forms of writing. Different ways to exchange information over long distances were in use. Out of these, use of pigeons and smoke signals for communication is well known.

The invention of telegraph by Samuel F.B. Morse in 1838 ushered in a new epoch in communication—the era of electrical communication. This kind of communication relies on modulation and demodulation techniques. In this technique, the increase in carrier frequency increases the available transmission bandwidth and hence provides larger information capacity. Thus the trend in electrical communication system development was to employ progressively higher frequencies which resulted in the birth of radio, television, radar and microwave links.

In 1876, the invention of telephone by Alexander Grahm Bell was a major revolution in the field of electrical communication, and analog electrical techniques were to dominate the communication systems for a century or so. The use of coaxial cables in place of wire pairs increased system capacity considerably. In coaxial cable systems, frequency dependent cable losses were major bandwidth limiting factors. This kind of limitation led to the development of microwave communication. The discovery of existence of electromagnetic waves by Heinrich Hertz in 1887 and development of first radio by Guglielmo Marchese Marconi in 1885 laid the foundation of radio and microwave systems operating at the carrier frequency of about gigahertz order. Inspite of having communication system with high bit rate and capacity, it was realised during the second half of the twentieth century that increasing demand of communication can only be fulfilled if optical waves were used as the carriers. This led to the development of light wave communication systems.

In year 1870, John Tyndall demonstrated that light rays can be guided through a water jet due to total internal reflection phenomenon. It was as if the rays of light were trapped within the stream of water. This phenomenon is depicted in Figure 11.1. The preservation of light inside the stream of water (or inside any bend medium) can be understood from Snell's law.

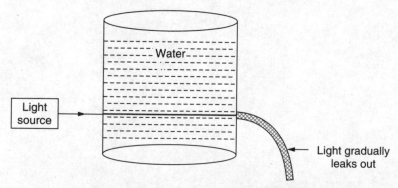

Figure 11.1 Light guided down a water jet.

The light wave communication system needs a guiding medium because light transmission in the atmosphere is severely affected by the disturbances such as rain, snow, fog, dust and atmospheric turbulences. Optical fibers are used for this purpose.

In 1953, Narinder Kapany of Indian origin, working at the Imperial College of Science and Technology, London developed a fiber with cladding, thus giving birth to the structure used in today's optical fiber. Infact Kapany coined the term 'fiber optics'. The proposal for optical communication through optical fibers fabricated from glass to avoid degradation of the optical energy by atmosphere was made almost simultaneously in 1966 by K.C. Kao and E.A. Hockham and A. Wertz. The attenuation of optical signal in fibers was major problem. Improvement in fiber fabrication technology yielded fibers with very small attenuation (about 0.2 dB/km). This led to world-wide effort for development of fiber optic communication systems. The developments in field of fiber optic communication technology open the door of information super highway.

11.2   ADVANTAGES OF OPTICAL FIBER COMMUNICATION

Optical fiber is a guiding medium through which light wave could be transmitted. It is hair-thin glass/plastic structure consisting of central core surrounded by a cladding of slightly lower refractive index. The propagation of light takes place because of phenomenon of total internal reflection occurring at the core–cladding interface.

Fibers exhibit extremely low loss of about 0.2 dB/km and have capacity of carrying tremendous amount of information. There are several advantages and attractive features of fibers for using them in communication systems:

Enormous Bandwidth

According to communication theory, larger the carrier frequency, greater the information carrying capacity. The optical carrier frequency, typically 10^{15} Hz (near IR region), yields far greater transmission bandwidth than coaxial cable system. The bandwidth available with fiber optic systems is more than 100 GHz. It is possible to transmit thousands of voice signals over a fiber simultaneously. The use of wavelength division multiplexing technique can increase further the information carrying capacity of fiber optic systems.

Small Size and Weight

Fibers have small diameter of order of a few micrometre which is not greater than human hair. Hence they are smaller and much lighter than copper cables. The reduction in size and weight makes the fibers very useful transmission medium for ships, aircrafts, missiles and satellites.

Greater Safety

Optical fibers carry light not electricity, and they are made of electrically insulator material like glass or plastic. This property makes optical fibers suitable for transmission in electrically hazardous environments because they create no spark at short circuits. Optical fibers can be repaired even when transmission is on.

Low Cost

Fibers are made of common natural resources like sand and plastic. Therefore, in comparison with copper or iron conductors, they offer lower cost. Further the use of fibers help in conserving the resources.

No Crosstalk

In conventional communication system, sometimes a call is heard in the background of other one. This crosstalk is due to stray of signals from one circuit to another. Optical fibers are made of dielectric materials and so free from electromagnetic or radiofrequency interference. Hence optical fibers can be routed in an electrically noisy environment. Moreover, it is easy to ensure that there is no optical interference between fibers.

11.3   FIBER FABRICATION

There are many requirements in selection of material for fabrication of fibers. It must be possible to make long, thin and flexible fibers from material. The variation of refractive index inside the optical fiber is a fundamental necessity in the fabrication of fiber for light transmission. For this purpose, two different materials, which are transparent to light, over operating wavelength range (800 nm to 1700 nm) are required. Further the material must exhibit low optical loss. Materials that satisfy these requirements are glasses and plastics. Normally glass fibers are used for long distance communication because it exhibits lower signal loss. Plastic fibers are used in short distance application and in abusive environment.

In general, two methods are used in the fabrication of optical fibers. These are vapour-phase deposition and direct-melt methods.

11.3.1   Vapour-phase Deposition Method

Silica-rich glasses of optimal optical properties are produced by vapour-phase technique. In this technique, highly pure vapours of metal halides such as $SiCl_4$, $GeCl_4$, BCl_3 and SiF_4 react with oxygen and form powder of SiO_2. The refractive index modification is achieved through vapour-phase dopants like TiO_2, GeO_2 and Al_2O_3. The dopant concentration may be varied slowly to produce a graded index fiber or maintained to give a step-index fiber. The powder of SiO_2 is transformed to a homogeneous glass rod by heating it without melting. This glass rod is known as preform. Its diameter is typically around 10–20 mm. Optical fibers are made from this preform by using the equipment shown in Figure 11.2. The rotational speed of the drum at the bottom determines the thickness of the fiber. Elastic coating is done on fiber immediately after it is drawn to save the fiber from contamination.

$$SiCl_4 \;+\; O_2 \;\xrightarrow{\text{Heat}}\; SiO_2 \;+\; 2Cl_2$$
$$\text{(Vapour)}\quad\text{(Gas)}\qquad\qquad\text{(Solid)}\quad\text{(Gas)}$$

$$GeCl_4 \;+\; O_2 \;\xrightarrow{\text{Heat}}\; GeO_2 \;+\; 2Cl_2$$
$$\text{(Vapour)}\quad\text{(Gas)}\qquad\qquad\text{(Solid)}\quad\text{(Gas)}$$

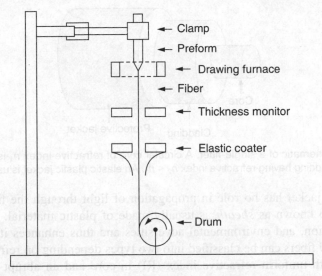

Figure 11.2 Schematic diagram of fiber-drawing machine.

11.3.2 Direct-melt Method

The direct-melt method adopts traditional glass-making procedures. The preparation of ultra pure material of required constitution is the first stage of this process. The next stage is to melt the pure material to form homogeneous bubble-free multi-component glass. Variation in refractive index is achieved by changing the composition of material in molten phase. The melt is cooled and shaped into long rod of glass. From this rod, fine optical fibers are drawn through fiber-drawing techniques.

11.4 TYPES OF FIBERS

An optical fiber is a dielectric waveguide usually of circular cross-section. It guides the information in form of light energy in direction of its axis. The transmission characteristics of fiber are generally determined by its structure. The propagation of light along fiber can be described in terms of modes which are set of guided electromagnetic waves. Each guided mode is a pattern of electric and magnetic field distributions that is repeated along the fiber at equal intervals. Modes which satisfy the homogeneous wave equation and boundary condition at the surface of fiber can propagate along the fiber.

The structure and the refractive index profile is normally uniform along the length of fiber. The refractive index distribution along direction normal to length varies in such a way that the index is relatively higher in the central region, known as the *core*, with respect to the surrounding region known as the *cladding*. The cladding is also covered by a protective jacket as shown in Figure 11.3. The cladding is made sufficiently thick so that propagating light cannot interact with the protective jacket.

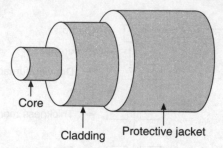

Core

Cladding Protective jacket

Figure 11.3 Schematic of a single-fiber. A circular core of refractive index n_1 is surrounded by a cladding having refractive index $n_2 < n_1$. An elastic plastic jacket is used for protection.

Hence the jacket has no role in propagation of light through the fiber. The protection jacket, also known as *sheath*, is usually made of plastic material. It protects the fiber from abrasion, and environmental adversities and thus enhances its tensile strength.

Optical fibers can be classified into two types depending on refractive index profile. A fiber with uniform refractive index (RI) in core and an abrupt change (or step) at the core–cladding boundary is termed as a step-index (SI) fiber. In the second type of fiber, the refractive index of core varies as a function of the radial distance from centre of the fiber. This type of fiber is called graded-index (GI) fiber. These are shown in Figure 11.4.

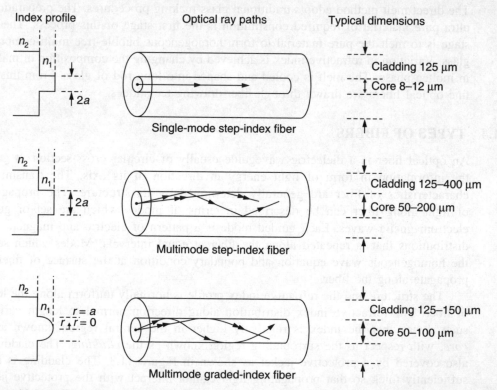

Figure 11.4 Index profile, ray paths and typical dimensions of single-mode and multimode step-index and graded-index fibers.

On the basis of modes of light propagation, both the step-index and graded index fibers can further be classified into single-mode fiber (SMF) and multimode fiber (MMF). The single-mode fiber or mono-mode fiber supports only one mode of propagation, whereas a multimode fiber supports many modes. Optical fibers can also be classified into three categories on the basis of material used for making core and cladding. Glass fibers are made of glass core and glass cladding, whereas plastic fiber consists of plastic core and cladding. The third one is plastic clad silica (PCS) fibers. Such fibers are composed of silica core and polymer (plastic) cladding.

The classification of optical fibers is given in Figure 11.5.

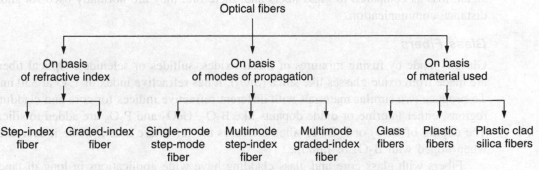

Figure 11.5 Types of optical fibers.

The single-mode step-index fiber has the distinct advantage of low inter modal dispersion or inter modal distortion as only one mode is transmitted. Inter modal dispersion is broadening of light pulses due to different group velocities of modes as they propagate along the fiber. Obviously multimode step-index fiber offers considerable inter modal dispersion because of presence of many modes. This phenomenon restricts maximum attainable bandwidth with such fibers. However, for lower bandwidth applications, multimode fibers have several advantages over single-mode fibers.

1. It is easier to couple optical power from spatially incoherent optical sources like LEDs.
2. Larger core radii facilitate easier coupling to sources.
3. Lower tolerance requirements on fiber connectors.

Single-mode step-index and multimode graded-index fibers are compared in Table 11.1.

Table 11.1 Comparison of Single-mode Step-index Fiber and Multimode Graded-index Fiber

S. No.	Single-mode step-index fiber	Multimode graded-index fiber
1.	The core radius of single-mode fibers is small, and they can support only one mode.	Multimode fibers have comparatively larger core radii, and they support many modes.
2.	Optical sources used are normally lasers.	LEDs may be used as optical sources.
3.	Single-mode fibers are expensive but more efficient.	Multimode fibers are comparatively cheap.
4.	Refractive index of core is constant.	Refractive index of core decreases as we go from centre of core to core–cladding interface.

11.5   FIBER MATERIALS

The materials used for fiber production must satisfy the following requirements:

1. The material must exhibit low loss.
2. It must be possible to make long, thin and flexible fibers.
3. It must be possible to change refractive indices of core and cladding regions.

The materials, which satisfy these requirements, are normally glass and plastic. In general, fibers are made of glass consisting of silica (SiO_2). Plastic fibers exhibit greater signal loss as compared to glass fibers and, therefore, they are normally used for short distance communication.

Glass Fibers

Glass is made by fusing mixtures of metal oxides, sulfides or selenides. Optical fibers are made from oxide glasses like silica (SiO_2). It has refractive index of 1.45 at 850 nm. To produce two similar materials with different refractive indices for core and cladding regions, either fluorine or oxide dopants like B_2O_3, GeO_2 and P_2O_5 are added to silica. The doping of GeO_2 or P_2O_5 to silica increases the refractive index, while it decreases when doped with B_2O_3 or fluorine.

Fibers with glass core and glass cladding have wide applications in long distance optical fiber links. Some typical parameters of glass fibers are given as under.

Attenuation coefficient : 0.20 dB/km – 2 dB/km

Core diameter   : 5 μm – 10 μm for SM
                        40 μm – 100 μm for MM

Cladding diameter   : Typically 125 μm for SM
                        150 μm – 400 μm for MM

Plastic Clad Silica Fibers

The plastic clad silica fibers are consist of silica core and plastic cladding. They may be multimode step-index or multimode graded-index. These fibers are less expensive in comparison to silica core and silica cladding fibers, but exhibit greater loss. Due to larger core diameter, low cost and large area light sources can be used for launching optical power in these fibers. Therefore, these fibers are suitable for short distance low data rate fiber optic systems. Typical parameters for plastic clad silica fibers are given as follows:

Attenuation coefficient : 1 dB/km – 100 dB/km
Core diameter   : 50 μm – 100 μm
Cladding diameter   : 120 μm – 150 μm

Plastic Fibers

The demand for delivering high-speed services directly to work station led to the development of plastic fibers for use in customer premises. Plastic fibers are cheaper to produce, but they exhibit higher signal loss. The large core diameter of such fibers allows the use of multimode sources for launching the optical power. Typical characteristic parameters are given as under.

Attenuation coefficient : 50 dB/km – 300 dB/km

Core diameter   : 200 μm – 600 μm

Cladding diameter : 450 μm – 1000 μm

11.6   OPTICAL PROPAGATION THEORY IN FIBERS

To study the propagation characteristics of light inside the optical fiber, the geometrical ray theory may be used. This model of light propagation provides a good approximation to the light acceptance and guiding properties of optical fiber when core radius a is large as compared to free space wavelength λ of light wave, and relative core–cladding index difference Δ is not very small. The advantage of ray approach is that, compared with the exact electromagnetic wave (modal) analysis, it gives a more direct physical interpretation of the light propagation in a fiber. The wavelength λ can be combined with core radius a and Δ to get a composite parameter known as normalised frequency or V-number of the optical fiber.

$$V = \frac{2\pi}{\lambda} a \sqrt{n_1^2 - n_2^2} \tag{11.1}$$

Here n_1 and n_2 are refractive indices of core and cladding regions respectively. For step-index fibers, the values of n_1 and n_2 are constant, but in case of graded-index fibers, n_2 is constant, while n_1 varies. In this case, n_1 represents highest value of refractive index. A new parameter Δ can be defined as:

$$\Delta \approx \frac{n_1^2 - n_2^2}{2n_1^2} \tag{11.2}$$

Since n_1 is normally very close to n_2, therefore, we have

$$\Delta \approx \frac{n_1 - n_2}{n_1} \tag{11.3}$$

The V-number can be written as:

$$V = \frac{2\pi}{\lambda} an_1 \sqrt{2\Delta} \approx \frac{2\pi}{\lambda} an_2 \sqrt{2\Delta} \tag{11.4}$$

The geometrical ray theory can be used to discuss light propagation inside fibers for $V > 10$. This theory does not work well for $V \leq 10$. In this case, electromagnetic wave theory has to be used for describing propagation characteristics in optical fibers.

EXAMPLE 11.1   Find out the V-number of a silica fiber with core and cladding refractive indices 1.48 and 1.46 respectively. The core diameter of fiber is 62.5 μm and operating wavelength is 1300 nm.

Solution V-number of fiber is:

$$V = \frac{2\pi}{\lambda} a \sqrt{n_1^2 - n_2^2}$$

$$= \frac{2 \times 3.34}{1300 \times 10^{-9}} \times 31.25 \times 10^{-6}[(1.48)^2 - (1.46)^2]^{1/2}$$

$$V = 36$$

V-number of fiber is 36.

11.6.1 Ray Theory Model

The refractive index of a medium is defined as ratio of the velocity of light in free space to velocity of light in the medium. When a ray of light is incident on interface separating two mediums of different refractive indices, the reflection and refraction of ray occur. If a ray is travelling from a medium with higher index n_1 to a medium of lower index n_2, it bends towards interface after refraction. The Snell's law relates the angles of incidence ϕ_1 and refraction ϕ_2 as follows:

$$n_1 \sin \phi_1 = n_2 \sin \phi_2$$

or
$$\frac{\sin \phi_1}{\sin \phi_2} = \frac{n_2}{n_1} \tag{11.5}$$

As n_1 is greater than n_2, the angle ϕ_2 must be greater than ϕ_1. Thus when angle of refraction is 90° and the refracted ray emerges parallel to the interface, the incidence angle must be less than 90°. This angle of incidence is known as *critical angle of incidence* ϕ_c and given as:

$$\sin \phi_c = \frac{n_2}{n_1} \qquad \text{for } \phi_2 = 90° \tag{11.6}$$

In this situation, at angles of incidence greater than the critical angle, the light is reflected back into originating medium. This event is known as *total internal reflection* which is responsible for propagation of light inside the fiber. These phenomenon are shown schematically in Figure 11.5. The propagation of light inside fiber via a series of total internal reflections at the interface of core and cladding is shown in Figure 11.6.

The light ray, shown in Figure 11.6, is called *meridional ray* as it passes through the axis of the fiber core. Meridional ray is simplest to discuss the transmission properties of optical fiber. This ray is of two types: bound rays that are trapped in the core and propagate along the fiber axis according to laws of geometrical optics and unbound rays that are refracted out of the fiber core.

There is another category of ray, known as *skew ray*, exists which is transmitted without passing through the fiber axis. Skew rays are not confined to a single plane, but instead follow a helical-type path along the fiber as shown in Figure 11.7. Skew rays tend to propagate only in the annular region near the outer surface of the core and do not fully utilise the core as a transmission medium. They are complementary to meridional rays and increase the light-gathering capacity of the fiber. For most

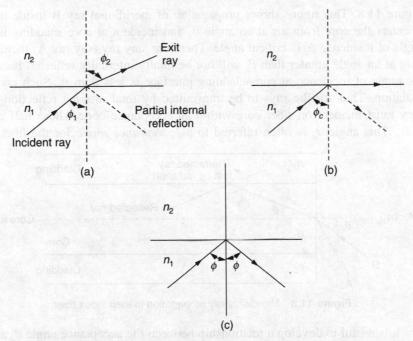

Figure 11.6 Light ray incident at interface of two medium (a) refraction, (b) limiting case when ϕ_1 becomes ϕ_c and (c) total internal reflection when $\phi > \phi_c$.

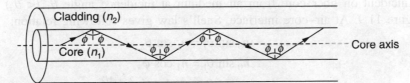

Figure 11.6 Propagation of ray of light in a perfect optical fiber.

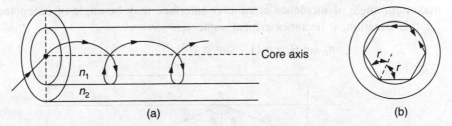

Figure 11.7 Ray representation of skew rays in fiber (a) skew rays propagating along a fiber, and (b) cross-sectional view of the fiber.

communication design purposes, their analysis is not necessary. The examination of meridional rays will suffice for this purpose.

Acceptance Angle and Numerical Aperture

Let us now consider the propagation of meridional rays in a step-index fiber. The geometry concerned with launching a light ray into an optical fiber is shown in

Figure 11.8. This figure shows propagation of meridional ray B inside the fiber. Ray B enters the core from air at an angle θ_a and incident at core–cladding interface. The angle of incidence ϕ_c is critical angle. Therefore, any ray (say ray A) incident into fiber core at an angle greater than θ_a will not be totally internally reflected because for that, the angle of incidence at core–cladding interface is less than ϕ_c. Such rays are lost by radiation. Thus for the rays to be transmitted by total internal reflection within fiber, they must incident on fiber core within an acceptance cone whose half conical angle is θ_a. This angle θ_a is often referred to as *acceptance angle* for the fiber.

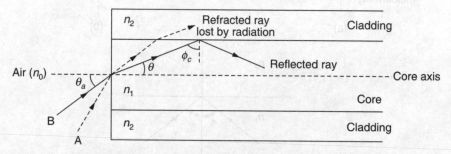

Figure 11.8 Meridional ray propagation in step index fiber.

It is useful to develop a relationship between the acceptance angle θ_a and refractive indices of the all three media (core, cladding and air) involved. This relationship provides definition of numerical aperture (NA) of the fiber. Consider a ray of light is incident on fiber core from air medium at incidence angle θ_1 ($< \theta_a$) as shown in Figure 11.9. At air–core interface, Snell's law gives following relation:

$$n_0 \sin \theta_1 = n_1 \sin \theta_2 \tag{11.7}$$

or

$$n_0 \sin \theta_1 = n_1 \cos \phi$$

or

$$n_0 \sin \theta_1 = n_1 (1 - \sin^2 \phi)^{1/2} \tag{11.8}$$

where angle $\phi(= 90° - \theta_2)$ is greater than critical angle at core–cladding interface. The maximum angle of incidence at air–core interface may be angle of acceptance θ_a, and for this situation, ϕ becomes critical angle ϕ_c.

$$n_0 \sin \theta_a = n_1 (1 - \sin^2 \phi_c)^{1/2} \tag{11.9}$$

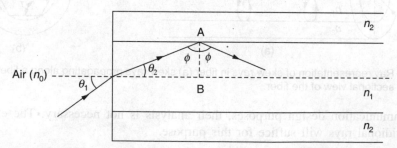

Figure 11.9 Ray transmission inside the fiber when launched at an angle less than acceptance angle.

$$n_0 \sin \theta_a = n_1 \left(1 - \frac{n_2^2}{n_1^2} \right)^{1/2} \qquad \left\{ \text{since } \sin \phi_c = \frac{n_2}{n_1} \right\}$$

$$n_0 \sin \theta_a = (n_1^2 - n_2^2)^{1/2} \tag{11.10}$$

This equation defines numerical aperture of fiber for meridional rays. The incident meridional rays will propagate inside the fiber over the range $0 \le \theta_1 \le \theta_a$.

$$\mathrm{NA} = n_0 \sin \theta_a = (n_1^2 - n_2^2)^{1/2} \tag{11.11}$$

The numerical aperture describes light-gathering capability of the optical fiber. It is a dimensionless quantity having numerical values less than unity and ranges from 0.14 to 0.50. Typically, for communication fibers, its value is between 0.1 and 0.2, and non-communication application requires NA more than 0.4. The numerical aperture may also be given in terms of the relative refractive index difference Δ between core and cladding as:

$$\Delta = \frac{n_1^2 - n_2^2}{2n_1^2}$$

$$\approx \frac{n_1 - n_2}{n_1} \qquad \text{(for } \Delta \ll 1\text{)}$$

Hence numerical aperture can also be given as:

$$\mathrm{NA} = n_1 \, (2\Delta)^{1/2} \tag{11.12}$$

EXAMPLE 11.2 The refractive indices of the core and cladding of a silica fiber are 1.48 and 1.46 respectively. What is the critical angle of propagation? Find the NA of fiber also.

Solution The numerical aperture of fiber is given by

$$\mathrm{NA} = \sqrt{n_1^2 - n_2^2}$$

$$= [(1.48)^2 - (1.46)^2]^{1/2}$$

Numerical aperture = 0.24
The critical angle of propagation inside the fiber is:

$$\theta_2 = 90° - \phi_c$$

So

$$\sin \theta_2 = \sin(90° - \phi_c)$$

$$= \cos \phi_c$$

$$= \sqrt{1 - \sin^2 \phi_c}$$

$$\sin \theta_2 = \sqrt{1 - \frac{n_2^2}{n_1^2}}$$

$$\sin \theta_2 = \sqrt{1 - \left(\frac{1.46}{1.48}\right)^2}$$

$$\theta_2 = \sin^{-1} \left[1 - \left(\frac{1.46}{1.48}\right)^2\right]^{1/2}$$

Critical angle of propagation $\theta_2 = 9.43°$.

11.6.2 Electromagnetic Mode Theory

Fibers are normally cylindrical dielectric waveguides which operate at optical frequencies. The transmission properties of an optical waveguide are determined by its structural characteristics. The propagation of light along a waveguide can be described in terms of a set of guided electromagnetic waves known as the modes of the waveguide. Each guided mode is a pattern of electric and magnetic field distributions that is repeated along the fiber at equal intervals. Only those modes can propagate inside fiber for which the electromagnetic waves satisfy the homogeneous electromagnetic wave equation and boundary condition at the waveguide surfaces.

An improved model for the propagation of light energy in fibers can be obtained by using electromagnetic wave theory. The geometrical optics provides a good understanding in many situations, but it has number of limitations. Electromagnetic wave approach is necessary for knowledge of coherence, interference and field distribution inside fiber. Electromagnetic wave analysis can be carried out by solving Maxwell's equations subject to the cylindrical boundary condition at the interface between core and cladding of optical fiber.

Let us qualitatively examine the appearance of modal field patterns in a planar dielectric slab waveguide. Figure 11.10 shows field distribution for some lower-order guided modes which are solutions of Maxwell's equations for the slab waveguide. The order of a mode is equal to the number of field zero across the waveguide. The electric fields of guided modes are not totally confined to core region, but they extend partially into cladding region also. The variation of fields inside central region is harmonic, and they decay outside this region exponentially.

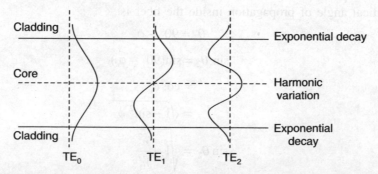

Figure 11.10 Electric field distribution of some lower-order guided modes.

▌11.7   BASIC ELEMENTS OF FIBER OPTIC COMMUNICATION SYSTEMS

The key elements of fiber optic communication system are: a transmitter consisting of light source and its drive circuitry, optical fiber cable and a receiver consisting of a photo detector plus amplification and signal-restoring circuitry. Additional components include optical amplifiers, connectors, splicers, couplers and regenerators. The schematic arrangement of optical fiber communication system is shown in Figure 11.11. Information source provides an electrical signal to the transmitter comprising an electrical stage which modulates the light wave carrier. The optical source may be either a laser diode (LD) or light emitting diode (LED). The electrical input signals to the transmitter circuitry can be either analog or digital. For high data rate system (> 1 Gb/s), direct modulation of the source can lead to unacceptable signal distortion. In such cases, external modulator is used. After an optical signal is launched into a fiber, it will become progressively attenuated and distorted with increasing distance. To restore the signal-shape characteristics, regenerators are employed. The design of an optical receiver is inherently more complex than that of the transmitter, since it has to interpret the content of the weakened and degraded signal received by the photodetector. Photodiodes (PIN or avalanche) and, in some instances, phototransistors and photoconductors are utilised for detection of the optical signal. Here some characteristics of light sources and detectors are given in brief.

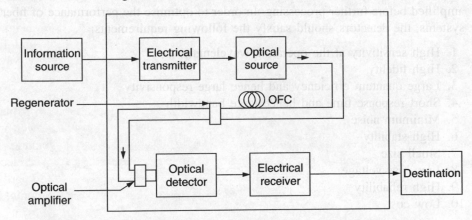

Figure 11.11   Schematic diagram of an optical fiber communication system.

11.7.1   Light Sources

The fundamental function of an optical source is to convert electrical signal into an optical signal in an efficient manner which allows the light output to be effectively launched or coupled into the optical fiber. In optical fiber communication systems, the optical sources must fulfill the following requirements:

1. The size and configuration must be compatible to the optical fiber so that the maximum light is coupled into the fiber.
2. Ideally the light output should be highly directional.

3. Source output should be linear with the electrical input signal to minimise the distortion and noise.
4. They should emit light at wavelengths where the fiber has low losses and low dispersion and where the detectors are efficient.
5. They should have wide bandwidth and large power output.
6. They should have a very narrow line-width in order to minimise the dispersion.
7. They must be capable of maintaining a stable output which is largely unaffected by changing in ambient conditions.

No source satisfies all the above-mentioned requirements. Semiconductor injection lasers and light emitting diodes (LEDs) partially meet above requirements and are used in optical fiber communication systems. Semiconductor lasers are based on the stimulated emissions. They provide a highly coherent output and have higher power output, higher modulation bandwidths and narrower spectral width than LEDs. In contrast to lasers, LEDs make use of the spontaneous emission of carriers and are relatively cheaper. They are best suited for local area networks. Laser diodes, on the other hand, are used in applications involving high data rates and long distances.

11.7.2   Photodetectors

A photodetector converts an optical signal into an electrical signal which is then amplified before further processing. In order to optimise the performance of fiber optic systems, the detectors should satisfy the following requirements:

1. High sensitivity at the operating wavelengths
2. High fidelity
3. Large quantum efficiency and hence large responsivity
4. Short response time and hence large bandwidth
5. Minimum noise
6. High stability
7. Small size
8. Low bias voltage
9. High reliability
10. Low cost

The photodetectors used in fiber-optic systems are: avalanche photodiode (API) and PIN photodiode. For short wavelength systems (800–900 nm), silicon material is used, whereas germanium or InGaAs material is used for long wavelength systems.

The PIN photodiodes have responsivity of the order of 0.5 A/W, a rise time of about 1 ns and a good frequency response up to about 1 GHz. The required bias voltage lies between 5–10 volts. They are relatively inexpensive and easy to use. The disadvantages with these diodes are low sensitivity and poor signal-to-noise ratio.

The responsivity and rise time of APDs are of the order of 15 A/W and 2 ns respectively. They have better gain and sensitivity. Though somewhat expensive, they provide inherent front end gain of 50–500 without sacrificing speed. On the minus side, they require high bias voltage and are sensitive to temperature variations, and thus require temperature compensating networks.

11.8 ATTENUATION IN OPTICAL FIBERS

The signal attenuation (or fiber loss) is a major factor which limits the performance of fiber-optic systems. It plays a major role in determining the maximum repeaterless distance between the transmitter and receiver. The main mechanisms responsible for signal attenuation inside the fiber are material absorption, material scattering, macro and micro-bending losses as shown in Figure 11.12.

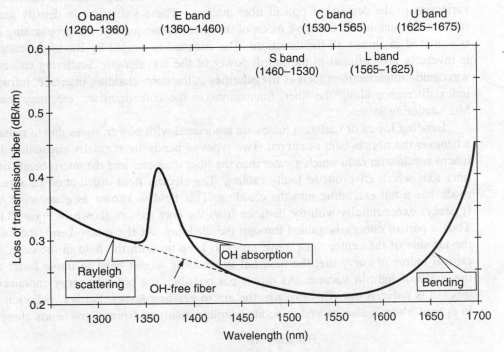

Figure 11.12 Typical loss profile of optical fibers.

Material absorption losses can be divided into intrinsic and extrinsic absorption losses. Optical fibers are normally made of silica-based glasses. As light passes through the fiber, it may be absorbed by one or more major components of glass. This is known as *intrinsic absorption*. Impurities within the fiber material also absorb light resulting into extrinsic absorption. Intrinsic absorption results from the electronic and vibrational resonances associated with specific molecules of glass. For pure silica, the electronic resonances in the form of absorption bands have been observed in ultraviolet range, while vibrational resonances have been observed in infrared range. Extrinsic absorption results from the impurities present in the fiber. Transition metal ions such as Cr^{3+}, Cu^{2+}, Fe^{2+}, Fe^{3+}, Ni^{2+}, Mn^{2+}, etc. absorb energy strongly in the wavelength region 600–1600 nm. Thus the impurity content of these ions should be below 1 ppb (part per billion) in order to have losses below 1 dB/km. Another source of extrinsic absorption is OH^-ions (hydroxyl ion) which are incorporated in the fiber during the manufacturing process. Though the vibrational resonance of OH^- ions peaks at

2730 nm, its overtones produces strong absorption at 1380 nm, 950 nm and 720 nm. Typically, the OH⁻ ions concentration should be less than 10 ppb to obtain loss which is less than 10 dB/km at 1380 nm.

The losses due to scattering are categorised as Rayleigh and Mie-scattering losses. Both these scattering mechanisms cause linear transfer of optical power contained in the guided mode into a different mode. This transfer results in the attenuation of the power. Rayleigh scattering is a loss mechanism arising from the microscopic variations in the density of optical fiber material. These variations in density lead to random fluctuation of refractive index of the core. Consequently, the propagating light is scattered in almost all the directions. The attenuation caused by Rayleigh scattering is inversely proportional to the fourth power of the wavelength. Scattering caused by waveguide imperfections such as irregularities at the core–cladding interface, refractive index difference along the fiber, fluctuations in the core diameter, etc. may lead to Mie-scattering losses.

Bending losses or radiation losses are associated with power losses due to radiation whenever the fiber is bent or curved. Two types of bends are normally encountered: (a) macro-bends with radii much greater than the fiber diameter and (b) micro-bends of the fiber axis which arise due to faulty cabling. The electric field distribution of a guided mode has a tail extending into the cladding. This field is known as *evanescent field*. It decays exponentially with the distance from the core axis as shown in Figure 11.13. Thus a part of energy is guided through the cladding. At the macro-bend, the tail on the far side of the centre must move faster to keep up with the field in the core. At a critical radius of curvature, the field tail in cladding would have to move faster than the speed of light in vacuum. As this is not possible, the optical energy contained in this tail is radiated out. This explains the macro-bending losses. Further, radiation loss in optical fibers is also caused by mode coupling resulting from micro-bends along the

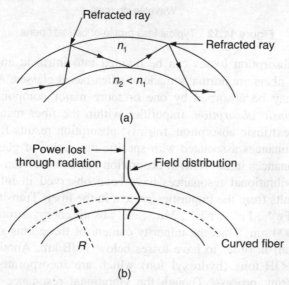

Figure 11.13 Mechanism of microbending loss in (a) multimode and, (b) single-mode fibers.

length of the fiber. Such micro-bends are caused by manufacturing defects which are in the form of non-uniformities in the core radius or in the lateral pressure created by cabling the fiber. The effect of mode coupling can be significant for multimode fibers. Micro-bend losses in single mode fibers can also be excessive if proper care is not taken to minimise them.

11.8.1   Attenuation Unit

The power of light decreases exponentially as it travels along the length of fiber. Consider light is travelling along z-direction and P_o is optical power at $z = 0$, then power P_z at distance z is:

$$P_z = P_o e^{\alpha z}$$

where $\alpha = \dfrac{1}{z} \ln\left(\dfrac{P_o}{P_z}\right)$

is the fiber attenuation coefficient given in units of per kilometre (km^{-1}). The attenuation coefficient in decibel per kilometre is given as:

$$\alpha(\text{dB/km}) = \frac{10}{z} \log_{10}\left(\frac{P_o}{P_z}\right)$$

The parameter α in dB/km is generally referred to as the fiber loss or the fiber attenuation.

EXAMPLE 11.3   An optical fiber of length 10 km is launched with power 2 µW. Calculate attenuation per unit length if 2 µW power is obtained at the end of fiber.

Solution   The attenuation per unit length in dB/km is:

$$\alpha = \frac{10}{z} \log_{10} \frac{P_o}{P_z}$$

$$= \frac{10}{z} \log_{10} \frac{2 \times 10^{-3}}{2 \times 10^{-6}}$$

$$\alpha = 3$$

The attenuation per unit length is 3 dB/km.

11.9   DISPERSION IN OPTICAL FIBERS

Dispersion is another important characteristic of optical fibers which affects the performance of fiber-optic communication systems. It causes broadening of transmitted light pulses as they travel along the fiber. For high data rates, when the pulses are closely spaced, the pulse broadening results in overlapping of adjacent pulses. The pulses become indistinguishable when overlapping is large. Thus, dispersion limits the number of pulses those can be transmitted in a given time and, thus the information carrying capacity of the fiber. The pulse broadening due to dispersion is maximum in the case of step-indexed multimode fibers, moderate in graded-indexed multimode fibers and least in single-mode fibers. The information carrying capacity, which is known as bandwidth-length product, of these fibers is about 20 MHz.km, 1 GHz.km and 100 GHz.km respectively.

There are two types of dispersion: intermodal dispersion and intramodal dispersion. Intramodal dispersion occurs in both multimode and single-mode fibers, while intermodal dispersion takes place only in multimode fibers. Pulse broadening due to intermodal dispersion results from the propagation delay differences between different modes as different modes travel with different group velocities. The bandwidth of such fibers is determined with the delay difference between slowest and fastest mode. This delay difference is comparatively large in case of step-index fibers. It is less in graded-index fibers.

Intramodal dispersion, also known as chromatic dispersion, results from the finite spectral line width of the optical sources. Since optical sources do not emit just a single frequency but a band of frequencies, there may be propagation delay differences between the different spectral components of the transmitted signal. Intramodal dispersion can further be divided into material dispersion and waveguide dispersion. Material dispersion is due to the dispersive properties of the waveguide material, and waveguide dispersion results due to guidance effects within the fiber structure. Pulse broadening due to material dispersion results from the different group velocities of the various spectral components launched into the fiber. It occurs when the phase velocity of a plane wave propagating in the fiber varies nonlinearly with wavelength. The material is said to exhibit material dispersion when the second differential of the refractive index with respect to wavelength is non-zero. The waveguide dispersion results from the variation in group velocity with wavelength for a particular mode. Material dispersion per unit length is directly proportional to the source line width.

SOLVED NUMERICAL PROBLEMS

PROBLEM 11.1 An optical fiber has core refractive index of 1.50 and a cladding refractive index of 1.47. Find out the critical angle, numerical aperture and acceptance angle in air for the fiber.

Solution At critical angle,

$$\sin \phi_c = \frac{n_2}{n_1}$$

$$\phi_c = \sin^{-1} \left(\frac{n_2}{n_1} \right)$$

$$= \sin^{-1} \left(\frac{1.47}{1.50} \right)$$

$$\phi_c = 78.5°$$

The critical angle is 78.5°.

Then $\qquad$ Numerical aperture NA $= (n_1^2 - n_2^2)^{1/2}$

$$= [(1.50)^2 - (1.47)^2]^{1/2}$$

Therefore, $\qquad\qquad\qquad\qquad$ NA $= 0.30$

The acceptance angle θ_a is related to numerical aperture as:

$$NA = \sin \theta_a$$

or $\qquad\qquad\qquad\qquad \theta_a = \sin^{-1}(NA)$

$$= \sin^{-1}(0.30)$$

$$\theta_a = 17.4°$$

The acceptance angle θ_a is $17.4°$.

PROBLEM 11.2 The numerical aperture of a step-index fiber is 0.22. If core radius of fiber is 25 µm, find the V-number of fiber for wavelength 1300 nm.

Solution The normalised frequency or V-number is given by

$$V = \frac{2\pi a}{\lambda}(NA)$$

$$= \frac{2 \times 3.14 \times 25 \times 10^{-6} \times 0.22}{1300 \times 10^{-10}}$$

$$V = 26.6$$

The V-number is 26.6.

PROBLEM 11.3 The refractive index of a core is 1.48 and that of cladding is 1.46. Under what condition will light be trapped inside the core?

Solution The condition is total internal reflection, and it can be achieved for critical angle of incidence.

$$\sin \phi_c = \frac{n_2}{n_1}$$

$$\phi_c = \sin^{-1}\left(\frac{1.46}{1.48}\right)$$

$$\phi_c = 80.6°$$

Critical angle of incidence is $80.6°$.

PROBLEM 11.4 Calculate the maximum transmission distance for a fiber link with an attenuation of 0.5 dB/km if the power launched is 1 µW and power received is 50 µW.

Solution The attenuation coefficient is:

$$\alpha = \frac{10}{z} \log_{10} \frac{P_o}{P_z}$$

So
$$z = \frac{10}{\alpha} \log_{10} \frac{P_o}{P_z}$$
$$= \frac{10}{0.5} \log_{10} \frac{0.5}{10}$$
$$= 20 \log_{10} \frac{5}{100}$$
$$\approx 26 \text{ km}$$

Maximum transmission distance is almost 26 km.

PROBLEM 11.5   Find out fractional index change for a fiber having core and cladding refraction indices 1.56 and 1.49 respectively.

Solution   Fractional index change is given as:

$$\Delta \approx \frac{n_1 - n_2}{n_1}$$
$$= \frac{1.56 - 1.49}{1.52}$$
$$\Delta = 0.041$$

Fractional index change is 0.041.

EXERCISES

THEORETICAL QUESTIONS

11.1 Define single-mode fiber and multimode fibers.

11.2 Discuss step-index and graded-index fibers. Give their advantages and disadvantages.

11.3 Define acceptance angle and numerical aperture of fiber. How they are related?

11.4 Describe different effects responsible for signal loss inside fiber.

11.5 Give classification of fibers and compare different types of fibers.

11.6 Discuss the dispersion mechanism inside the fiber. How this affects the performance of fiber?

11.7 Give the advantage of fiber communication over conventional way of communication.

11.8 What are the basic elements of a fiber link? Discuss in brief each of them.

11.9 Describe fiber bend losses with suitable diagram.

11.10 Give the ray theory for optical propagation inside the fiber.

NUMERICAL PROBLEMS

P11.1 An optical fiber has an acceptance angle 26.80°. Find out its numerical aperture.

(**Ans.** 0.45)

P11.2 Calculate the numerical aperture and the acceptance angle of an optical fiber with core and cladding indices 1.62 and 1.52 respectively.   (**Ans.** 0.56, 34°)

P11.3 For an optical fiber, the relative refractive index difference is 1%. Find the NA of the fiber if refractive index of core is 1.46.   (**Ans.** 0.21)

P11.4 A single mode fiber operating at 820 nm have core and cladding indices 1.48 and 1.47 respectively. Find the core radius necessary for this operation.   (**Ans.** 4 μm)

P11.5 Optical power input to an optical fiber of length 10 km is 100 μW. The output power obtained at end is 1 μW. Calculate the attenuation coefficient per unit length of fiber.
(**Ans.** 2 dB/km)

P11.6 A step-index fiber is made with core of index 1.52. The diameter of core is 29 μm and fractional index difference is 0.007. Find out the V-number if fiber is operated at wavelength of 1.3 μm.   (**Ans.** 4.05)

P11.7 A fiber of length 8 km is launched with optical power 120 μW. Calculate signal attenuation if 3 μW power is obtained at the end of fiber.   (**Ans.** 16 dB)

P11.8 A multimode step-index fiber has core refractive index 1.460, diameter 60 μm and cladding refractive index 1.447. Calculate number of guided modes when the operating wavelength is 0.75 μm.
(**Hint:** Number of guided modes = $V^2/2$)   (**Ans.** 45)

P11.9 A multimode step-index fiber is capable to support 1200 modes at operating wavelength 900 nm. If the numerical aperture of fiber is 0.20, then calculate the diameter of core.
(**Ans.** 7 μm)

P11.10 A multimode graded-index fiber has refractive index 1.52 at core axis, and acceptance angle in air is 8°. Find out the relative refractive index.   (**Ans.** 0.42%)

MULTIPLE CHOICE QUESTIONS

MCQ 11.1 The propagation of light inside fiber is based on the principle of
(a) Polarisation of light
(b) Diffraction of light
(c) Total internal reflection
(d) Interference of light

MCQ 11.2 For propagation of many modes inside fiber, fiber must be
(a) Single-mode fiber
(b) Multimode fiber
(b) Plastic fiber
(d) None of these

MCQ 11.3 The propagation of light in fiber occurs inside
(a) Cladding
(b) Core
(c) Jacket
(d) Between core and cladding

MCQ 11.4 The fractional index is defined as

(a) $\Delta = \dfrac{n_1 - n_2}{n_1}$

(b) $\Delta = \dfrac{n_1}{n_2}$

(c) $\Delta = \dfrac{n_2}{n_1}$

(d) $\Delta = \dfrac{n_1}{n_1 - n_2}$

MCQ 11.5 The pulse broadening inside fiber is due to
- (a) Dispersion
- (b) Attenuation
- (c) Bending loss
- (d) Interference of pulses

MCQ 11.6 Light-gathering capacity of a fiber is determined by
- (a) Dispersion effects
- (b) *V*-number
- (c) Numerical aperture
- (d) Attenuation effects

MCQ 11.7 The dispersion effect reduces the
- (a) Capacity of fiber
- (b) Enhance losses
- (c) Attenuation
- (d) None of these

MCQ 11.8 Rayleigh scattering coefficient R depends on the wavelength λ of light as
- (a) $R \propto \dfrac{1}{\lambda^2}$
- (b) $R \propto \dfrac{1}{\lambda^3}$
- (c) $R \propto \dfrac{1}{\lambda}$
- (d) $R \propto \dfrac{1}{\lambda^4}$

MCQ 11.9 The attenuation coefficient α is
- (a) Inversely proportional to length of fiber
- (b) Directaly proportional to length of fiber
- (c) Inversely proportional to square of length of fiber
- (d) Directly proportional to square of length of fiber

MCQ 11.10 The meridional ray
- (a) Crosses the fiber axis
- (b) Does not cross
- (c) Goes in straight way
- (d) Moves on a helical path

Answers

11.1 (c)	**11.2** (b)	**11.3** (b)	**11.4** (a)	**11.5** (a)	**11.6** (c)	**11.7** (a)
11.8 (d)	**11.9** (a)	**11.10** (a)				

Motion of Charged Particles in EM Fields

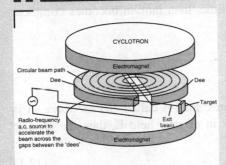

12.1 INTRODUCTION

There are several devices and experiments that use the motion of charge particle in electric or magnetic or both fields, i.e., cathode ray tube (CRT), cyclotron, synchrotron, mass spectrometer, scanning electron microscope (SEM), Transmission electron microscope, Hall effect, Thomson experiment, Millikan's oil drop experiment, Van Allen radiation belts, Aurora borealis, southern lights and various other phenomenon in plasma physics. The working of these devices or explanation of involved phenomenon can be well understood by the study of motion of charge particle in different fields under different conditions. In this chapter, we will study the details for a motion of charge under the effect of electromagnetic fields.

The motion of charge particles like electron, proton, etc. is influenced by the presence of electric, magnetic, electromagnetic, gravitational and radiation fields because they exert force on the charge particle. If the forces are known, then path of particle can be determined by applying Newton's law of motion. When a charge particle moves perpendicular to electric or magnetic or crossed electric and magnetic fields, then its path becomes parabolic or circular or cycloid respectively. But when the charge particle moves at an angle with magnetic field, its path becomes helical. If electric and magnetic fields vary with space and time, then motion becomes complex.

12.2 FORCE ON CHARGE PARTICLE

Electric Force

The force experienced by a charge particle in electric field is called *electric force*. It is denoted by $\mathbf{F_e}$. If a charge of amount q is present in electric field $\mathbf{E}$, then the electric force can be obtained as:

$$\mathbf{F_e} = q\mathbf{E} \tag{12.1}$$

Equation (12.1) suggests that the positive charge experiences a force along the direction of electric field, while a negative charge experiences force opposite to field as shown in Figure 2.1. The electric force working on charge particle may be stationary or dynamic, as Eq. (12.1) does not contain velocity term.

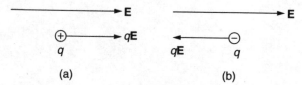

Figure 12.1 Force on charge particle in electric field.

Magnetic Force

It is a force that acts on the moving charge particle in magnetic field. It is denoted by $\mathbf{F_m}$. If a charge particle of q coulomb moves with velocity v in uniform magnetic field $\mathbf{B}$, then magnetic force can be determined as:

$$\mathbf{F_m} = q v \times \mathbf{B} \tag{12.2}$$

From Eq. (12.2), we can say that magnetic force is always perpendicular to both **v** and **B**. The direction of magnetic force can be determined with screw rule or Fleming's left hand rule (Figure 12.2). Let θ is an angle between v and **B**, then we can write

$$|\mathbf{F_m}| = qvB \sin \theta \qquad (12.3)$$

From Eq. (12.3), it is clear that no magnetic force acts on a charge particle, if it is either stationary ($v = 0$) or moving along the direction of magnetic field ($\theta = 0$). But the magnetic force have maximum value (qvB) when charge particle moves perpendicular to **B** (i.e., $\theta = 90°$).

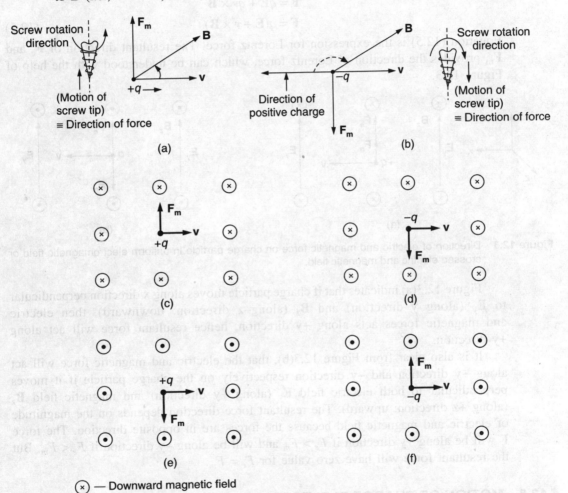

Figure 12.2 Direction of magnetic force in different condition of charge motion in magnetic field.

For the determination of direction of magnetic force, the Fleming's left hand rule is only applicable for perpendicular motion of charge, while screw rule can be applied for any case of charge motion in magnetic field.

Lorentz Force

The force experienced by a moving charge particle in presence of both electric and magnetic field is called *Lorentz force*. The Lorentz force **F** is the sum of electric and magnetic force:

$$F = F_e + F_m \qquad (12.4)$$

Let a particle of charge q is moving with velocity **v** in both electric field **E** and magnetic field **B**. Then Eq. (12.4) becomes

$$F = qE + qv \times B$$
$$F = q(E + v \times B) \qquad (12.5)$$

Equation (12.5) is the expression for Lorentz force. The resultant direction of F_e and F_m provides the direction of Lorentz force, which can be understood with the help of Figure 12.3.

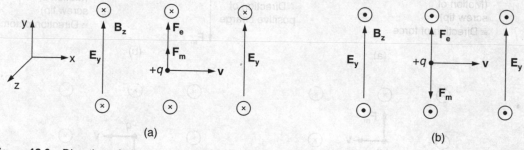

(a)   (b)

Figure 12.3   Direction of electric and magnetic force on charge particle in uniform electromagnetic field or crossed electric and magnetic field.

Figure 12.3(a) indicates that if charge particle moves along x direction perpendicular to E_y (along y direction) and B_z (along–z direction: downward), then electric and magnetic forces acts along +y direction, hence resultant force will act along +y direction.

It is also clear from Figure 12.3(b), that the electric and magnetic force will act along +y direction and –y direction respectively on the charge particle if it moves perpendicular to both electric field E_y (along +y direction) and magnetic field B_z (along +z direction: upward). The resultant force direction depends on the magnitude of electric and magnetic field because the forces are in opposite direction. The force **F** will be along +y direction if $F_e > F_m$ and will be along –y direction if $F_e < F_m$. But the resultant force will have zero value for $F_e = F_m$.

12.3   MOTION OF CHARGE IN ELECTRIC FIELD

12.3.1   Motion of Stationary Charge in Electric Field

Suppose a charge particle of mass m and charge q is released from rest ($u = 0$) in uniform electric field E, which is directed along x-axis and is produced by a potential difference V (Figure 12.4). Due to electric field, charge experiences a force and hence accelerated along x-direction. If acceleration is a_x, then

$$a_x = \frac{F}{m} = \frac{qE}{m} \qquad (12.6)$$

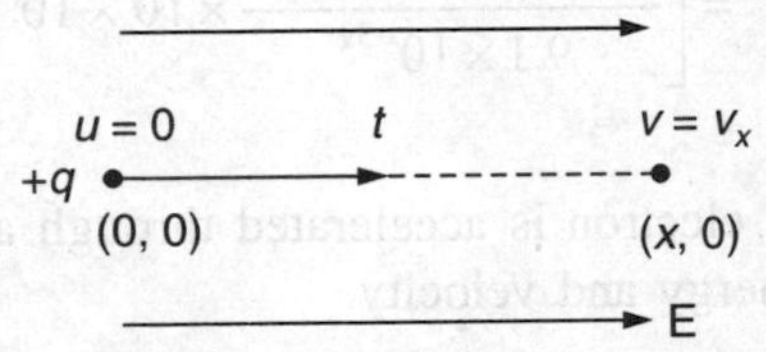

Figure 12.4 Stationary charge in uniform electric field.

Due to this acceleration, particle moves along the direction of electric field on a linear path. Let in time t, it travels a distance x and gains a velocity v_x. Then, from equation of kinematics, we have

$$v = u + at \implies v_x = a_x t = \frac{qE}{m} t \qquad (12.7)$$

and

$$s = ut + \frac{1}{2} at^2 \implies x = \frac{1}{2} a_x t^2 = \frac{qE}{m} t^2 \qquad (12.8)$$

and

$$v^2 = u^2 + 2as \implies v_x^2 = 2a_x x = \frac{2qE}{m} x \qquad (12.9)$$

Equations (12.7) and (12.8) provide the variation of velocity of charge particle with respect to position and time, while Eq. (12.7) shows the variation in displacement with time. If K is the kinetic energy of particle, when it leaves the electric field, then we have

$$K = \frac{1}{2} mv_x^2 \qquad (12.10)$$

From Eqs (12.9) and (12.10), we can write

$$K = \frac{1}{2} m \cdot \frac{2qE}{m} x = qEx \qquad (12.11)$$

Since the electric field is produced by a potential difference (p.d.) V, thus $V = Ex$. Now, Eq. (12.11) becomes

$$K = qV \qquad (12.12)$$

Equation (12.12) implies that if a charge particle q is accelerated by a p.d. V, then its kinetic energy will be qV.

Note: If a charge q is accelerated by a potential difference V, then we can write

$$K = qV \implies \frac{1}{2} mv^2 = qV \implies v = \sqrt{\frac{2qV}{m}}$$

If $q = e$, then $v = \sqrt{\frac{2eV}{m}}$.

EXAMPLE 12.1 An electron starting from rest travels a distance of 10 mm in uniform electric field of 3×10^3 N/C. Calculate the speed of the electron.
Solution

$$v = \left(\frac{2qE}{m}x\right)^{1/2} = \left[\frac{2 \times 1.6 \times 10^{-19} \times 3}{9.1 \times 10^{-31}} \times 10 \times 10^{-3}\right]^{1/2} = 3.25 \times 10^6 \text{ m/s}$$

EXAMPLE 12.2 An electron is accelerated through a potential difference of 2 kV. Calculate its kinetic energy and velocity.
Solution

$$K = qV = (1.6 \times 10^{-19}) \times (2 \times 10^3) = 3.2 \times 10^{-16} \text{ J}$$

$$v = \left(\frac{2qV}{m}\right)^{1/2} = \left(\frac{2 \times (1.6 \times 10^{-19}) \times (2 \times 10^3)}{9.1 \times 10^{-31}}\right)^{1/2} = 2.65 \times 10^7 \text{ m/s}$$

12.3.2 Motion of Charge in Transverse Electric Field

Suppose an uniform electric field E_y along y direction is produced by two horizontal plates having separation d, length L and potential difference V (Figure 12.5). Let a charge of mass m and charge q enters in this field along x direction (transverse direction) with a initial velocity v_x. Thus initial velocity (i.e., at $t = 0$) components along x, y and z directions become v_x, 0, and 0 respectively. Due to electric field, charge experiences a force along y direction, while no force acts along x and z directions. Thus an acceleration a_y acts along y direction, that changes its velocity along y direction, but x direction velocity remains unaltered. As a result, particle moves in xy plane.

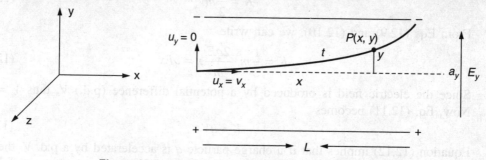

Figure 12.5 Motion of charge in transverse electric field.

If after time t, the co-ordinates of particle is P(x, y), then from the kinematics equation, we have

$$x = v_x t \quad \Rightarrow \quad t = \frac{x}{v_x} \tag{12.13}$$

and

$$y = u_y t + \frac{1}{2}a_y t^2 = 0 + \frac{1}{2}a_y t^2 = \frac{1}{2}a_y t^2 \tag{12.14}$$

Since we know that

$$F = qE$$

Thus

$$F_y = qE_y$$

or

$$ma_y = qE_y$$

$$a_y = \frac{qE_y}{m} \tag{12.15}$$

Putting values in Eq. (12.14) from Eqs. (12.13) and (12.15), we have

$$y = \frac{1}{2} \times \left(\frac{qE_y}{m}\right) \cdot \left(\frac{x}{v_x}\right)^2$$

$$\boxed{y = \frac{qE_y}{2mv_x^2} x^2} \tag{12.16a}$$

or

$$\boxed{y = \frac{qV}{2mdv_x^2} x^2} \quad \left(\because E_y = \frac{V}{d}\right) \tag{12.16b}$$

Equation (12.16) shows that the particle moves in a parabolic path in transverse electric field or within the region of charged plates, i.e., trajectory of particle will be parabola in transverse electric field (Figure 12.6). If y_1 is maximum vertical displacement at end of the field (at $x = L$), then from Eq. (12.16a), we have

$$y_1 = \frac{qE_y L^2}{2mv_x^2} \tag{12.17}$$

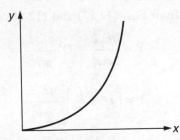

Figure 12.6   Trajectory of particle in transverse electric field.

After the end of applied field or region of electric field, charge particle emerges out tangentially to the parabolic path and moves on linear path. Let θ is the angle of emergence from the direction of initial motion of charge particle, then from Figure 12.7, we can write

$$\tan \theta = \frac{v_y}{v_x} = \frac{u_y + a_y}{v_x} = \frac{0 + a_y t}{v_x} = \frac{a_y t}{v_x}$$

$$\tan \theta = \left(\frac{qE_y}{m}\right) \times \frac{1}{v_x} \times \left(\frac{L}{v_x}\right) = \frac{qE_y L}{mv_x^2}$$

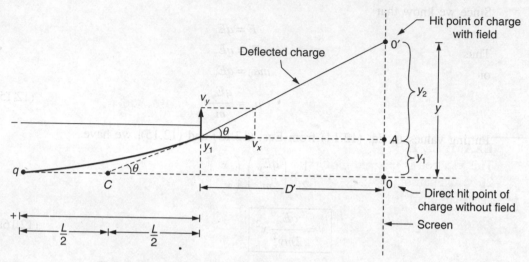

Figure 12.7 Motion of charge inside and outside the field region.

or

$$\tan \theta = \frac{qE_yL}{mv_x^2} \tag{12.18}$$

If there is screen at distance D' from the end of field, then the deflected charge particle with electric field will hit the screen at point O′ instead of direct hit point O (in absence of electric field). Let OO′ is equal to Y. From Figure 12.7, we can write

$$Y = y_1 + y_2$$

$$\therefore \qquad Y = y_1 + D' \tan \theta \tag{12.19}$$

Substituting the values from Eqs. (12.17) and (12.18) in Eq. (12.19), we have

$$Y = \frac{qE_yL}{2mv_x^2} + D' \cdot \frac{qE_yL}{mv_x^2}$$

$$Y = \left(D' + \frac{L}{2} \right) \cdot \frac{qE_yL}{mv_x^2} \tag{12.20}$$

Let D is distance from centre of plates C to screen. Hence $D = D' + \dfrac{L}{2}$. Now, Eq. (12.20) becomes

$$\boxed{Y = D\frac{qE_yL}{mv_x^2} = D \tan \theta} \tag{12.21}$$

Equation (12.21) provides an expression for the deflection of a charge particle on screen with effect of transverse electric field or charged deflecting plates. The deflection per unit applied voltage on deflecting plates is called **Electrostatic deflection sensitivity** (EDS).

From Eq. (12.21), we can write

$$Y = D \frac{qVL}{mdv_x^2}$$

$$\boxed{\text{EDS} = \frac{Y}{V} = \frac{D}{d} \frac{qL}{mv_x^2}} \qquad (12.22)$$

EXAMPLE 12.3 An electron is accelerated through a potential difference of 150 V. This electron is injected into a transverse electric field produced by the application of 20 V to a pair of parallel plates of length 10 cm and separation 1 cm. A screen is placed at 50 cm apart from the centre of applied electric field. Find the deflection on screen.

Solution Given that $V_A = 150$ V, $V = 20$ V, $L = 10$ cm, $d = 1$ cm and $D = 50$ cm

If the electric field is along y direction and electron is accelerated along x axis, then its velocity along x axis, v_x will be

$$v_x = \sqrt{\frac{2qV_A}{m}} = \sqrt{\frac{2 \times 1.6 \times 10^{-19} \times 150}{9.1 \times 10^{-31}}} = 7.26 \times 10^6 \text{ m/s}$$

Since this e^- is moving in electric field produced by p.d. of 20 V thus acceleration a_y in e^- along y direction will be

$$a_y = \frac{q}{m} \frac{V}{d} = \frac{1.6 \times 10^{-19}}{9.1 \times 10^{-31}} \times \frac{20}{1 \times 10^{-2}} = 3.516 \times 10^{14} \text{ m/s}^2$$

Let v_y: velocity in y direction after time t

$$v_y = a_y t = a_y \cdot \frac{l}{v_x} = 3.516 \times 10^{14} \times \frac{10 \times 10^{-2}}{7.26 \times 10^6} = 4.8 \times 10^6$$

If Y is deflection on screen, which is placed at D distance from the centre of electric field, then

$$Y = D \tan \theta = D \times \frac{v_y}{v_x} = 50 \times 10^{-2} \times \frac{4.8 \times 10^6}{7.26 \times 10^6} = 0.33 \text{ m}$$

EXAMPLE 12.4 An α-particle with charge 3.2×10^{-19} C and mass 6.68×10^{-27} kg is injected in transverse electric field of 40×10^3 V/m with velocity 7.26×10^6 m/s. If length of field is 10 cm then calculate the maximum deflection at end of field.

Solution At the end of electric field, the maximum deflection is given by following formula:

$$Y = \frac{qE_y L^2}{2mv_x^2}$$

$$Y = \frac{3.2 \times 10^{-19} \times 40 \times 10^3 \times (10 \times 10^{-2})^2}{2 \times 6.68 \times 10^{-27} \times (7.26 \times 10^6)^2} = \frac{0.93 \times 10^{12} \times 10^{-2}}{52.7 \times 10^{12}} = 0.0176 \times 10^{-2} \text{ m}$$

$$= 0.176 \text{ mm}$$

▌12.4 MOTION OF CHARGE IN MAGNETIC FIELD

12.4.1 Rest or Parallel Motion

When a charge particle is released in rest in a magnetic field or it is moving with uniform velocity along the field, then in both cases, the magnetic force acting on charge becomes zero, i.e.,

$$\mathbf{F_m} = q(v \times \mathbf{B})$$

$$\Rightarrow \qquad |\mathbf{F_m}| = F_m = qvB \sin \theta$$

$\Rightarrow$ If $v = 0$ or $\theta = 0$, then, $F_m = 0$

So, the state of rest or motion of charge remains unchanged in above two cases due to zero magnetic force.

12.4.2 Transverse Motion

When a charge particle enters in the magnetic field at the right angle to direction of field with certain initial velocity, then the magnetic force acts on it, which is perpendicular to both the direction of motion of charge and the direction of the field as shown in Figure 12.8. Due to transverse magnetic force, no work is done by the magnetic field on the charge particle. Thus the particle gains no kinetic energy, and hence its amplitude of velocity remains unchanged. The constant amplitude of velocity results constant magnetic force in magnitude. This magnetic force causes a continuous deflection of charge governed by Fleming's left hand rule (i.e., the direction of velocity and force continuously changes such that they are perpendicular to each point). This provides a centripetal force or acceleration to the charge. In the influence of this acceleration, charge particle moves in circular path. Hence we can say that motion of charge in transverse magnetic field is circular in nature.

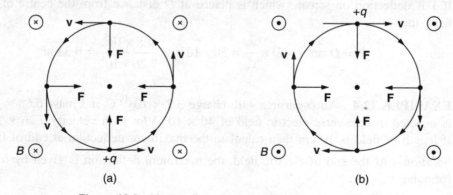

(a)   (b)

Figure 12.8 Motion of charge in transverse magnetic field.

Let a charge q enters in transverse magnetic field B with velocity v, then magnetic force F_m will be

$$F_m = qvB \sin \theta$$

$$F_m = qvB \sin 90° \qquad (\text{as } \theta = 90°)$$

$$F_m = qvB \qquad (12.23)$$

This force provides the centripetal force to the charge for moving in circular path of radius γ. Then

$$\text{Magnetic force} = \text{Centripetal force}$$

$$F_m = \frac{mv^2}{r} \qquad (12.24)$$

From Eqs. (12.23) and (12.24), we have

$$qvB = \frac{mv^2}{r}$$

$$\boxed{r = \frac{mv}{qB}} \qquad (12.25)$$

Equation (12.25) gives the radius of circular path traced by charge particle. It is clear from radius expression that radius of path is inversely proportional to magnetic field induction B. Thus for large B, radius of path will be small, while it will be large for low magnetic field induction B.

For circular motion, $v = r\omega$. Thus

$$\omega = \frac{v}{r} \qquad (12.26)$$

From Eqs. (12.25) and (12.26), we can write

$$\omega = \frac{qB}{m} \qquad (12.27)$$

The angular frequency ω with which a charge particle circulates in transverse magnetic field is called *Larmour frequency*.

Trajectory Equation

Let the charge q is moving along x direction with velocity $v_x = v\hat{\mathbf{i}}$ (at $t = 0$) and the magnetic field is along z direction $\mathbf{B} = B\hat{\mathbf{k}}$. If at any instant, velocity is $\mathbf{v} = v_x\hat{\mathbf{i}} + v_y\hat{\mathbf{j}} + v_z\hat{\mathbf{k}}$, then from equation of magnetic force, we have

$$\mathbf{F_m} = q(\mathbf{v} \times \mathbf{B})$$

$$m\mathbf{a} = q\begin{vmatrix} \hat{\mathbf{i}} & \hat{\mathbf{j}} & \hat{\mathbf{k}} \\ v_x & v_y & v_z \\ 0 & 0 & B \end{vmatrix}$$

$$m(a_x\hat{\mathbf{i}} + a_y\hat{\mathbf{j}} + a_z\hat{\mathbf{k}}) = q\hat{\mathbf{i}}\begin{vmatrix} v_y & v_z \\ 0 & B \end{vmatrix} + q\hat{\mathbf{j}}\begin{vmatrix} v_z & v_x \\ B & 0 \end{vmatrix} + q\hat{\mathbf{k}}\begin{vmatrix} v_x & v_y \\ 0 & 0 \end{vmatrix}$$

$$m\dot{v}_x\hat{\mathbf{i}} + m\dot{v}_y\hat{\mathbf{j}} + m\dot{v}_z\hat{\mathbf{k}} = qv_yB\hat{\mathbf{i}} - qv_xB\hat{\mathbf{j}} + 0\hat{\mathbf{k}} \qquad (12.28)$$

Equation (12.28) implies that

$$m\dot{v}_x = qv_yB \tag{12.29}$$

$$m\dot{v}_y = -qv_xB \tag{12.30}$$

$$m\dot{v}_z = 0 \tag{12.31}$$

Equation (12.31) shows that the velocity remains constant in z direction in the magnetic field. Since in this direction, particle has no initial velocity so it remains zero. Differentiating Eq. (12.29) and putting the value from Eq. (12.30), we get

$$m\ddot{v}_x = +qB\dot{v}_y = -qB\frac{qB}{m}v_x$$

$$\ddot{v}_x = -\left(\frac{qB}{m}\right)^2 v_x$$

$$\ddot{v}_x = -\omega^2 v_x \qquad \left(\text{as } \omega = \frac{qB}{m} \right) \tag{12.32}$$

This equation represents an equation of simple harmonic oscillator whose angular frequency is $\omega = \dfrac{qB}{m}$, called *Larmour* or *cyclotron frequency*.

Now, integrating Eqs. (12.29) and (12.30), we have

$$mv_x = qyB + C_1$$

or

$$m\dot{x} = qyB + C_1 \tag{12.33}$$

and

$$mv_y = -qxB + C_2$$

$$m\dot{y} = qxB + C_2 \tag{12.34}$$

Here C_1 and C_2 are the integration constants. That can be determined with initial conditions, i.e. at $t = 0$, $\dot{x} = v_x = v$, $\dot{y} = v_y = 0$, $x = 0$ and $y = 0$.

Under this condition, Eqs. (12.33) and (12.34) imply that $C_1 = mv$ and $C_2 = 0$.
Now, Eqs. (12.33) and (12.34) becomes

$$m\dot{x} = qyB + mv \tag{12.35}$$

and

$$m\dot{y} = -qyB \tag{12.36}$$

Divide Eq. (12.35) with Eq. (12.36),

$$\frac{\dot{x}}{\dot{y}} = \frac{qyB + mv}{-qxB}$$

$$\frac{dx}{dy} = \frac{qyB + mv}{-qxB}$$

$$-qBxdx = qBydy + mvdy \tag{12.37}$$

Integrating Eq. (12.37), we have

$$-\frac{qBx^2}{2} = \frac{qBy^2}{2} + mvy + C_3 \tag{12.38}$$

The integration constant C_3 becomes zero under initial condition, thus Eq. (12.38) becomes

$$-x^2 = y^2 + 2\frac{mv}{qB}y$$

or

$$x^2 + y^2 + 2\frac{mv}{qB}y = 0$$

or

$$x^2 + y^2 + \left(\frac{mv}{qB}\right)^2 + 2\frac{mv}{qB}y = \left(\frac{mv}{qB}\right)^2$$

or

$$x^2 + \left(y + \frac{mv}{qB}\right)^2 = \left(\frac{mv}{qB}\right)^2$$

Let

$$\boxed{r = \frac{mv}{qB}}$$

Then

$$\boxed{x^2 + (y+r)^2 = r^2} \tag{12.39}$$

It is the equation of circle. This proves that path of charge particle in transverse magnetic field is circular.

Deflection by Magnetic Field

From Eq. (12.25), we have found that the radius ($r = mv/qB$) of circular path in transverse magnetic field is inversely proportional to magnetic field induction. Thus radius of path will be large for small magnetic field induction. If the length of field is also small, then the charge is not able to complete its complete circular path, but it moves in circular arc within the magnetic field region. After the range of field, it moves in linear path in tangential direction to arc [Figure 12.9(a)]. If there is a screen in front of charge motion, one can get deflection of charge on screen from its original direction of motion [Figure 12.9(b)].

Let the charge q enters in transverse magnetic field B (small) with initial velocity v. The length of field is L. Due to magnetic force qvB, it moves in circular arc RST within the field and moves in linear path outside the field. Consider this deflected charge gives a deflection Y on the screen, placed at distance D from the centre of magnetic field [Figure 12.9(b)]. If angle of deflection is θ, then

From $\Delta MOO'$

$$\tan\theta = \frac{OO'}{MO} \quad \Rightarrow \quad OO' = MO\tan\theta$$

$$Y = D\tan\theta \tag{12.40}$$

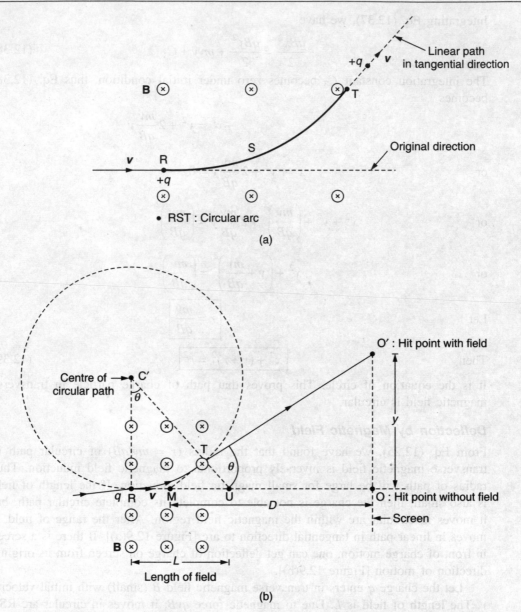

Figure 12.9 Motion of charge inside and outside of transverse magnetic field and deflection.

From the geometry, $\angle RC'U = \angle OMO'$. Then from $\Delta RC'U$, we can write

$$\tan \theta = \frac{RU}{RC'}$$

$$\tan \theta = \frac{L}{r} \tag{12.41}$$

From Eqs. (12.40) and (12.41), one can have

$$Y = D\frac{L}{r} = DL\frac{qB}{mv}$$

$$Y = D\frac{qBL}{mv} \tag{12.42}$$

Knowing the quantities in R.H.S. of Eq. (12.42), the deflection at screen due to magnetic field can be calculated. The deflection per unit magnetic induction is called **Magnetostatic deflection sensitivity** (MDS).

From Eq. (12.42), we can write

$$\text{MDS} = \frac{Y}{B} = D\frac{qL}{mv} \tag{12.43}$$

For $q = e$, $\boxed{\text{MDS} = D\frac{eL}{mv}}$

12.4.3 Motion at an Angle to B

Consider a charge particle has charge q and mass m. This charge enters in a magnetic field B at angle θ with a velocity v (Figure 12.10). The velocity can be resolved into two components: $V_{\parallel} = v \cos \theta$ and $V_{\perp} = v \sin \theta$. The component $V_{\parallel}$ is parallel to B, while $V_{\perp}$ is perpendicular to B. The component $V_{\parallel}$ remains unaffected due to no magnetic force along parallel direction. Thus the particle tries to continue its motion along B due to $V_{\parallel}$ component. The velocity component $V_{\perp}$ is perpendicular to direction B, thus the force $qV_{\perp}B$ exerts on the particle. This force tries to move the particle in a circle with a constant speed $V_{\perp}$. If r is the radius of circular path, then

$$r = \frac{mV_{\perp}}{qB} = \frac{mv \sin \theta}{qB} \tag{12.44}$$

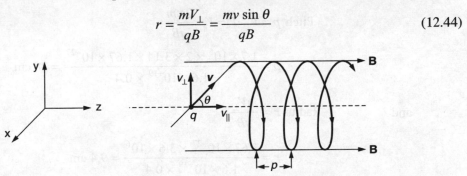

Figure 12.10 Motion of charge at oblique incidence to the magnetic field.

As a result, the charge particle has two simultaneous motion: circular motion in x-y plane and linear motion along z direction. The combination of these two motions results that the charge moves in a helical path.

If T is the time taken by charge in one revolution, then

$$T = \frac{2\pi r}{V_\perp} = \frac{2\pi m V_\perp}{qBV_\perp} = \frac{2\pi m}{qB} \qquad (12.45)$$

The distance travelled by the charge particle in one revolution in helical path is called *pitch* of helix. If p is pitch, then

$$p = V_\parallel T = v \cos \theta T$$

$$\boxed{p = \frac{2\pi m v \cos \theta}{qB}} \qquad (12.46)$$

Equation (12.46) is the expression for pitch of the helical path of charge.

EXAMPLE 12.5   A charge particle of charge 1.6×10^{-19} C and mass 3.34×10^{-27} kg is moving in a circular orbit of radius 20 cm in a magnetic field of 0.4 T, perpendicular to the velocity of particle. Calculate the orbital speed of the particle.

Solution $v = \dfrac{qBr}{m} = \dfrac{(1.6 \times 10^{-19}) \times 0.4 \times (20 \times 10^{-2})}{3.34 \times 10^{-27}} = 3.8 \times 10^6$ m/s

EXAMPLE 12.6   A magnetic field $B = 0.4$ T, is along the x axis. The proton enters in this field at angle 65° from x-axis with speed 4×10^6 m/s. The motion of proton is helical. Calculate the pitch of helix and the radius of trajectory.

Solution

$$V_\parallel = v \cos \theta = 4 \times 10^6 \times \cos 65° = 1.7 \times 10^6 \text{ m/s}$$

$$V_\perp = v \sin \theta = 4 \times 10^6 \times \sin 65° = 3.6 \times 10^6 \text{ m/s}$$

$$\text{Pitch } p = V_\parallel \cdot T = V_\parallel \cdot \frac{2\pi m}{qB}$$

$$p = \frac{1.7 \times 10^6 \times 2 \times 3.14 \times 1.67 \times 10^{-27}}{1.6 \times 10^{-19} \times 0.4} = 28 \text{ cm}$$

and $\text{Radius } r = \dfrac{mV_\perp}{qB}$

$$r = \frac{1.67 \times 10^{-27} \times 3.6 \times 10^6}{1.6 \times 10^{-19} \times 0.4} = 9.4 \text{ cm}$$

EXAMPLE 12.7   An electron enters in magnetic field of 8×10^{-4} T at right angle with speed 8×10^6 m/s. The length of field is 2 cm.

(a) Find the path of electron inside field region.

(b) If there is a screen at distance of 10 cm from centre of field, then calculate deflection of electron on screen.

Solution

(a) When a charge enters in magnetic field at right angle, its path becomes circular. Let r is radius of its path.

$$r = \frac{mv}{qB} = \frac{9.1 \times 10^{-31} \times 8 \times 10^{+6}}{1.6 \times 10^{-19} \times 8 \times 10^{-4}} = 5.68 \times 10^{-2} \text{ m} = 5.68 \text{ m}$$

Since radius is larger than the length of field, thus particle e moves in circular path inside magnetic field.

(b) $\because$ Deflection $y = D\dfrac{qBL}{mv}$

$$y = \frac{10 \times 10^{-2} \times 1.6 \times 10^{-19} \times 8 \times 10^{-4} \times 2 \times 10^{-2}}{9.1 \times 10^{-31} \times 8 \times 10^{6}}$$

$$y = 3.52 \times 10^{-2} \text{ m} = 3.52 \text{ cm}$$

12.5 MOTION OF CHARGE IN PRESENCE OF BOTH ELECTRIC AND MAGNETIC FIELD

The motion of charge in presence of both electric and magnetic field can be classified as (a) when fields are parallel or Antiparallel and (b) when fields are perpendicular to each other.

12.5.1 Fields are Parallel or Antiparallel

Consider that both the electric E and magnetic fields B are along z direction, and a charge particle of mass m and charge q enters in the field with velocity v at angle θ from z direction as shown in Figure 12.11.

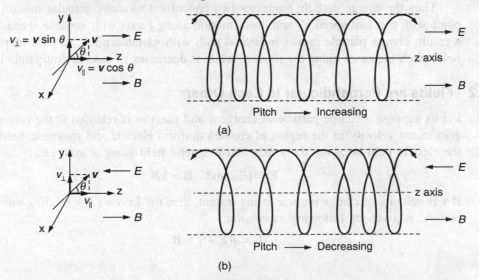

Figure 12.11 Motion of charge in (a) parallel (b) Antiparallel electric and magnetic fields.

The velocity can be resolved into two components: $V_{\parallel} = v \cos \theta$ and $V_{\perp} = v \sin \theta$. Since velocity component $V_{\perp}$ is perpendicular to B, a magnetic force acts on it, which provides a centripetal force to charge. Under this force, particle tries to move in circular path in xy plane. If r is the radius of circular path, then

$$r = \frac{mV_{\perp}}{qB} = \frac{mv \sin \theta}{qB} \qquad (12.47)$$

Since the velocity component $V_{\parallel}$ of charge is along z direction, thus no magnetic force acts on it. But an electrostatic force works on it due to electric field, i.e., an acceleration (+ for parallel field and – for Antiparallel field) works along z direction on the charge. If a is the magnitude of acceleration, then

$$a = \frac{qE}{m} \qquad (12.48)$$

This acceleration changes the $V_{\parallel}$ component of velocity with time. Let $V_{\parallel}'$ is the velocity of charge along z direction at time t, then from equation of kinematics, we can write

$$v = u \pm at$$

$$\boxed{V_{\parallel}' = V_{\parallel} \pm \frac{qE}{m} t} \qquad (12.49)$$

and
$$s = ut \pm \frac{1}{2} at^2$$

$$\boxed{Z = V_{\parallel} t \pm \frac{1}{2} \frac{qE}{m} t^2} \qquad (12.50)$$

Here Z is the linear distance travelled by charge along z direction. The positive and negative signs are for acceleration and retardation depending upon parallel or Antiparallel fields respectively.

Thus the charge particle possesses two concurrent motions: circular motion in xy plane with constant speed $V_{\perp}$ and linear motion along z axis with variable speed $V_{\parallel}$. As a result, charge particle moves in helical path with variable pitch. In case of parallel field, pitch increases along the motion, while it decreases in case of Antiparallel field.

12.5.2 Fields are Perpendicular to Each Other

Let us suppose a charge particle of charge q and mass m is released at the origin with zero initial velocity in the region of crossed uniform electric and magnetic fields. Let the electric field be along the y axis and magnetic field along z axis, i.e.,

$$\mathbf{E} = \hat{\mathbf{j}}E \quad \text{and} \quad \mathbf{B} = \hat{\mathbf{k}}B$$

If $\mathbf{v}$ is velocity of charge particle at any instant, then the Lorentz force acting on charge particle is given by following expression;

$$\mathbf{F} = q(\mathbf{E} + \mathbf{v} \times \mathbf{B}) \qquad (12.51)$$

Let $\mathbf{v} = v_x\hat{\mathbf{i}} + v_y\hat{\mathbf{j}} + v_z\hat{\mathbf{k}}$. Putting values of **E**, **v** and **B** in Eq. (12.51), we have

$$\mathbf{F} = qE\hat{\mathbf{j}} + q\begin{vmatrix} \hat{\mathbf{i}} & \hat{\mathbf{j}} & \hat{\mathbf{k}} \\ v_x & v_y & v_z \\ 0 & 0 & B \end{vmatrix}$$

$$\mathbf{F} = qE\,\hat{\mathbf{j}} + q(v_y B\hat{\mathbf{i}} - Bv_x\hat{\mathbf{j}} + 0\hat{\mathbf{k}})$$

$$\mathbf{F} = qv_y B\hat{\mathbf{i}} + (qE - qBv_x)\,\hat{\mathbf{j}} + 0\hat{\mathbf{k}} \qquad (12.52)$$

Thus the forces acting along x, y and z directions become

$$F_x = qv_y B$$

$$m\frac{dv_x}{dt} = qv_y B$$

$$m\frac{dv_x}{dt} = \frac{qB}{m}v_y \qquad (12.53)$$

and

$$F_y = qE - qBv_x$$

$$\frac{dv_y}{dt} = \frac{qE}{m} - \frac{qB}{m}v_x \qquad (12.54)$$

and

$$F_z = 0$$

$$\frac{dv_z}{dt} = 0 \qquad (12.55)$$

Equation (12.55) implies that, v_z is constant. If this component of velocity is zero initially, then motion of charge will be completely in xy plane. For the knowledge of trajectory of charge, the expressions along x axis or y axis [Eqs. (12.53) and (12.54)] must be solved.

Differentiating Eq. (12.54) w.r.t. t, we have

$$\frac{d^2v_y}{dt^2} = -\frac{qB}{m}\frac{dv_x}{dt}$$

Substituting value from Eq. (12.53) in it, we get

$$\frac{d^2v_y}{dt^2} = -\frac{q^2B^2}{m^2}v_y^2$$

$$\frac{d^2v_y}{dt^2} = -\omega^2 v_y^2 \qquad (12.56)$$

where $\omega = qB/m$ is the angular frequency of particle.

Equation (12.56) is similar to the general equation of S.H.M. whose solution can be written as:

$$v_y = A\sin(\omega t + \phi)$$

∵   At $t = 0$, $v_y = 0$, thus $\phi = 0$.

∴
$$v_y = A \sin \omega t \qquad (12.57)$$

Substituting this value of v_y in Eq. (12.54), we get

$$\frac{d}{dt}(A \sin \omega t) = \frac{qE}{m} - \frac{qB}{m} v_x$$

$$A\omega \cos \omega t = \frac{qE}{m} - \frac{qB}{m} v_x ; \; \omega = \frac{qB}{m}$$

or
$$v_x = \frac{E}{B} - A \cos \omega t \qquad (12.58)$$

Since $v_x = 0$ at $t = 0$, thus Eq. (12.58) provides that $A = E/B$.

So
$$v_x = \frac{E}{B}(1 - \cos \omega t) \qquad (12.59)$$

and
$$v_y = \frac{E}{B} \sin \omega t \qquad (12.60)$$

and
$$v_z = 0 \qquad (12.61)$$

Integrating Eq. (12.59), we get

$$\int \frac{dx}{dt} dt = \frac{E}{B} \int (1 - \cos \omega t) \, dt + C_1$$

$$x = \frac{E}{B}\left(t - \frac{\sin \omega t}{\omega} \right) + C_1 \qquad (C_1 = \text{Integration constant})$$

Since $x = 0$ at $t = 0$, thus $C_1 = 0$

Hence
$$x = \frac{E}{B}\left(t - \frac{\sin \omega t}{\omega} \right)$$

or
$$\boxed{x = \frac{E}{B\omega}(\omega t - \sin \omega t)} \qquad (12.62)$$

Integrating Eq. (12.60), we get

$$\int \frac{dy}{dt} dt = \frac{E}{B} \int \sin \omega t \, dt + C_2$$

$$y = -\frac{E}{B\omega} \cos \omega t + C_2 \qquad (C_2 = \text{Integration constant})$$

Since at $t = 0$, $y = 0$, thus $C_2 = E/B\omega$

∴
$$y = -\frac{E}{B\omega} \cos \omega t + \frac{E}{B\omega}$$

or
$$\boxed{y = \frac{E}{B\omega}(1 - \cos \omega t)} \qquad (12.63)$$

Equations (12.62) and (12.63) are the expressions of a cycloid. Thus charge particle moves in cycloidal path as shown in Figure 12.12. Cycloid is the curve traced out by a point on a circumference of circle rolling along a straight line. In present case, the radius of rolling circle is $E/B\omega$, while velocity at its centre is E/B ($v = r\omega$) along x-axis. The maximum displacement along y axis, will be twice of radius of rolling circle, that is justified from Eq. (12.63).

i.e., At $\omega t = \pi$, $y_{max} = 2E/B\omega$

$$y_{max} = 2 \times \text{Radius of rolling circle}$$

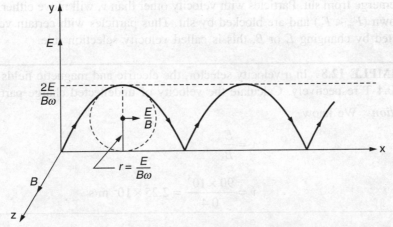

Figure 12.12 Motion of charge in crossed electric and magnetic field.

12.6 VELOCITY SELECTOR

There are many equipment and experiments where a source of charge particle with definite velocity is needed. This can be obtained by applying crossed electric and magnetic field. When a charge particle enters perpendicular to crossed electric and magnetic field (as shown in Figure 12.13), then the electric and magnetic forces act on it in opposite directions. When these forces are balanced, the particle motion is unaffected and continues its motion in the original direction.

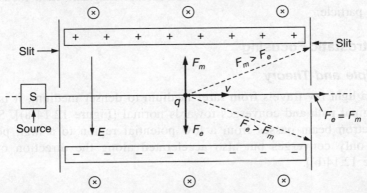

Figure 12.13 Arrangement of fields for selection of charge with constant velocity.

If charge particle of charge q moves with velocity v without any deviation from its path in crossed electric field E and magnetic field B, then

$$F_e = F_m$$
$$qvB = qE$$

$\Rightarrow$

$$\boxed{v = \frac{E}{B}}$$

Thus particles having a velocity v decided by both fields will move in a straight line and emerge from slit. Particles with velocity other than v, will move either up ($F_m > F_e$) or down ($F_m < F_e$) and are blocked by slit. Thus particles with certain velocity can be selected by changing E or B, this is called velocity selection rule.

EXAMPLE 12.8 In a velocity selector, the electric and magnetic fields are 90 kV/m and 0.4 T respectively. Calculate the velocity of undeflected charge particle.

Solution We know

$$v = \frac{E}{B}$$

$\therefore$

$$v = \frac{90 \times 10^3}{0.4} = 2.25 \times 10^5 \text{ m/s}$$

12.7   FOCUSING OF ELECTRON BEAM

It has been seen in earlier sections of this chapter that the motion of charge particles can be controlled by the application of electric or magnetic field. Thus these fields can also focus the diverging beam of charged particles originated from the sources. *The phenomenon of converging and focusing of a diverging beam of charge particles by means of electric or magnetic field is called electrostatic or magnetostatic focusing.* There are several equipment like cathode ray oscilloscope (CRO), mass spectroscope and electron microscope that require the focusing of the diverging beam of charge particle. Now, we will study how an electric or magnetic field focuses the beam of charge particle.

12.7.1   Electrostatic Focusing

Principle and Theory

When a light ray travels from rarer medium to denser medium, it is deviated from its original path and converges towards normal [Figure 12.14(a)]. Similarly, when an electron beam passes from a low potential region to a high potential region, it not only converges but also accelerated along the direction of electric field [Figure 12.14(b)].

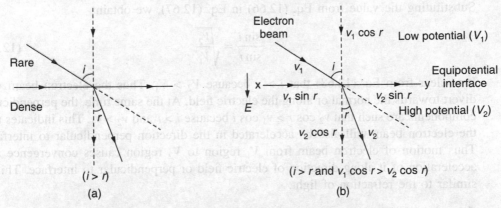

Figure 12.14 Refraction of (a) light and (b) electron beam.

Let an electron beam travels from a low potential V_1 at velocity v_1 to a high potential V_2 with velocity v_2. Then

Kinetic energy (K) of low potential region $= eV_1t$

$$\frac{1}{2}mv_1^2 = eV_1$$

$$v_1 = \sqrt{\frac{2eV_1}{m}} \tag{12.64}$$

and in high potential region, $K = eV_2$

$$\frac{1}{2}mv_2^2 = eV_2$$

$$v_2 = \sqrt{\frac{2eV_2}{m}} \tag{12.65}$$

From Eqs. (12.64) and (12.65), we get

$$\frac{v_1}{v_2} = \sqrt{\frac{V_1}{V_2}} \tag{12.66}$$

Here $V_2 > V_1$, thus $v_2 > v_1$.

Let XY be the equipotential interface between uniform electrostatic positive potentials V_1 and V_2 ($V_2 > V_1$). Suppose i and r are the angles of incidence and refraction of the electron beam at the interface XY. Velocities v_1 and v_2 can be resolved into parallel and perpendicular components to the interface [Figure 12.14(b)]. Since XY is an equipotential surface, so components of velocities parallel to XY ($v_1 \sin i$ and $v_2 \sin r$) will be equal. Thus

$$v_1 \sin i = v_2 \sin r$$

$$\frac{\sin i}{\sin r} = \frac{v_2}{v_1} \tag{12.67}$$

Substituting the value from Eq. (12.66) in Eq. (12.67), we obtain

$$\frac{\sin i}{\sin r} = \sqrt{\frac{V_2}{V_1}} \tag{12.68}$$

It is clear from Eq. (12.68) that $i > r$ because $V_2 > V_1$. Thus the electron beam will divert towards the normal or along the electric field. At the same time, the perpendicular components are such that $v_2 \cos r > v_1 \cos i$ because $i > r$ and $v_2 > v_1$. This indicates that the electron beam will also be accelerated in the direction perpendicular to interface. Thus motion of electron beam from V_1 region to V_2 region causes convergence and acceleration of it along direction of electric field or perpendicular to interface. This is similar to the refraction of light.

Arrangement

In electrostatic focusing, a series of cylindrical anodes A_1 and A_2 at increasing potentials V_1 and V_2 ($V_2 > V_1$) are used. These anodes are kept at highly positive with respect to cathode C and are separated by a gap. A high equipotential ring R is placed in the gap between two anodes. The electric line of forces will be curved between the anodes and a tangent which provides direction of resultant electric field (dotted lines in Figure 12.15). A perpendicular surface to direction of electric field forms an equipotential surface. Since electric line of forces are curved so equipotential surface will be curved. Such curved surfaces form an electrostatic lens, whose interface exists at ring R. The electrostatic lens formed by curved equipotential planes, diverts the diverging beam of electrons from its original path and focuses at point (Figure 12.15). It also accelerates the beam along the axes of cylindrical anodes. The focal length of electrostatic lens can be changed by altering the relative size and potentials of cylindrical anodes.

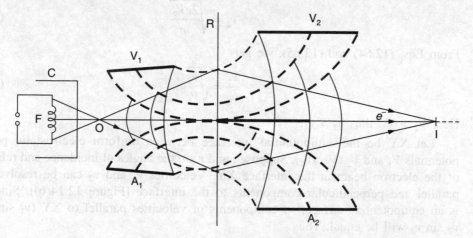

Figure 12.15   Electrostatic focusing.

12.7.2   Magneto-static Focusing

Principle

When a charge particle enters in uniform magnetic field at an angle, its path becomes helical. The pitch of helical path is independent of angle and is approximately constant, if particle enters in the field at very small angle. As a result, a uniform magnetic field converges a small angle diverging beam of electrons at the pitch length. This is known as *magneto-static focusing*.

Theory

Consider a particle of charge q and mass m enters in uniform magnetic field B at angle θ with velocity v. If p is pitch of its helical path, then

$$p = V_{\parallel}T$$

$$p = v \cos \theta \cdot \frac{2\pi m}{qB} = \frac{2\pi mv}{qB} \cos \theta \qquad (12.69)$$

Since

$$\cos \theta = 1 - \frac{\theta^2}{2} + \cdots$$

So, for small angles, $\cos \theta \approx 1$

Hence Eq. (12.69) becomes

$$\boxed{p \approx \frac{2\pi mv}{qB}} \qquad (12.70)$$

When the angle of divergence is small, all the charge particles are focused at distance equal to pitch length (Figure 12.16). This is called magneto-static focusing. The magnetic field behaves as lens because it focuses electron beam, hence it is called magnetic lens.

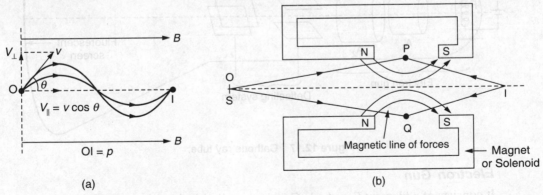

Figure 12.16   Magnetic focusing (a) uniform field (b) non-uniform field.

The focal length of magnetic lens ($p = O$?) can be varied by varying the intensity of magnetic field. A uniform magnetic field c$\cdot$.n also be generated by an electromagnet and controlled by a current flowing throug'.< coils.

Not only a uniform magnetic field but also non-uniform magnetic field can also focus the small diverging beam of charge particles. Busch first proposed that a non-uniform magnetic field along the axis of a short solenoid also has the properties of a magnetic lens. In Figure 12.16(b), there are two magnets or electromagnets having strong magnetic field at the narrow gaps P and Q, which are opposite to each other. The magnetic field is symmetrical about the axis SI. The magnetic field produced by magnets behaves as lens, which converges the diverging beam of electron (generated at O) at point I. The focal length of magnetic lens is of the order of few cm. This is the essential condition for getting large magnification in electron microscope (SEM or TEM), where an intense and fast beam of electrons is needed.

12.8 CATHODE RAY TUBE (CRT)

It is a vacuum tube in which electrons or beam of electron is generated, accelerated, focused and deflected by an electric or magnetic field. It is used to convert the electronic video signal into visual display. It is used in different equipments like in cathode ray oscilloscope (CRO), TV receivers, computer monitors, etc. The CRT has three major parts: (a) electron gun, (b) deflecting system and (c) a fluorescent screen (Figure 12.17).

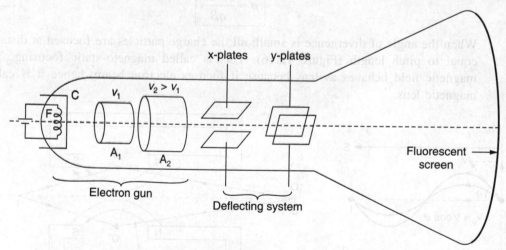

Figure 12.17 Cathode ray tube.

Electron Gun

It consists of a filament F, cathode C and two cylindrical anodes A_1 and A_2. The cathode is kept at negative potential, while anodes are kept at very high positive potentials such that $V_2 > V_1$. When the current flows in the filament, the electrons are emitted by the thermionic emission. These emitted electrons are linearised by the negative potential of cathode. Further, this slightly diverging electron beam is focused and accelerated by electrostatic focusing arrangement, i.e., by cylindrical anodes A_1 and A_2. Thus, we see that the arrangement from F to A_2 behaves as a gun, which produces an accelerated axial electron beam, hence called electron gun as shown in Figure 12.18.

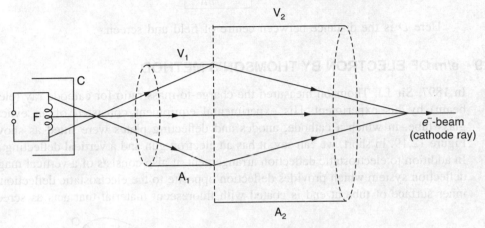

Figure 12.18 Electron gun.

Now, the electron beam generated by electron gun proceeds towards the deflecting system.

Deflecting System

The electrostatic deflection system consists of two set of parallel plates: X-plates and Y-plates. X-plates are the horizontal plates, while Y-plates are the vertical plates. When a potential difference is applied across the X-plates, it deflects the electron-beam along Y direction. Similarly, potential difference across Y-plates causes deflection of beam along X direction. Thus X-plates and Y-plates are termed as vertical and horizontal deflecting plates respectively. In case of a CRO, an internal horizontal sweep is generated; this takes the beam left to right (along X direction) at the desired frequency. The signal to be studied is given to the vertical deflecting plates. The deflected beam now proceeds towards the fluorescent screen.

Fluorescent Screen

The inner surface of the vacuum tube is coated with a fluorescent material, which acts as screen. On this screen, when an electron beam strikes, it emits a characteristic light. As a result, visual information of signal is obtained, which is implemented on the deflecting system.

Let in electrostatic deflecting system, the vertical deflecting plates have a potential difference V, separation d and length l. Suppose this potential difference produces a deflection Y along Y direction on screen, which is at distance D from the plate. Then

$$Y = \frac{qVlD}{mdv^2} = \frac{eVlD}{mdv^2}$$

Here m is the mass of electron and v is its velocity along axis of tube. If the deflection is produced by transverse magnetic field B, deflection Y becomes

$$Y = \frac{qBlD}{mv}$$

Here D is the distance between centre of field and screen.

12.9 *e/m* OF ELECTRON BY THOMSON'S METHOD

In 1897, Sir J.J. Thomson measured the charge-to-mass ratio for cathode ray (electron beam) by his experiment. His experimental equipment consisted of an evacuated glass tube, in which a cathode, anodes and deflecting plates were fitted as shown in Figure 12.19. In short, we can say, it has an electron gun and a vertical deflecting plate. In addition to electrostatic deflection arrangement, it also consists of a vertical magnetic deflection system which provides deflection opposite to the electrostatic deflection. The inner surface of tube at end is coated with fluorescent material that acts as screen.

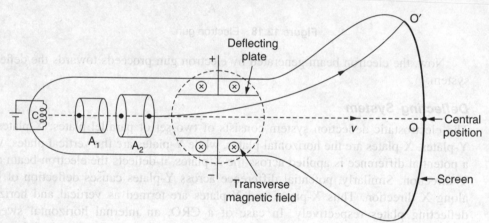

Figure 12.19 CRT with combined electrostatic and magneto-static deflecting arrangements.

Theory and Working

When current flows in cathode, then the electrons are produced or emitted by direct heating of cathode. These electrons are focused and accelerated by anodes A_1 and A_2 in form of an electron beam (cathode ray) that travels towards the deflecting system and then screen. In absence of any potential across the plate, the beam hits central position of screen and a spot is seen on the screen.

When a potential difference is applied across the deflecting plate, the produced transverse electric field causes a deflection in e^--beam. As a result, e^--beam strikes at new position O′ on the screen, i.e., position of spot is shifted from O to O′. If OO′ be Y, then

$$Y = \frac{eElD}{mv^2} \tag{12.71}$$

where E is the electric field between the plates that is produced by potential difference between them. The parameters l and D are the length of plate and distance from plates to screen. The terms e, m and v are the charge, mass and velocity of electron. Suppose a magnetic field is now applied perpendicular to both electric field and cathode ray such that spot O' moves towards O due to magnetic force acting opposite to electric force. Let at magnetic field B, spot goes from O' to O. Then this refers that the beam becomes undeflected due to balancing of forces, i.e.,

$$F_e = F_m$$
$$eE = evB$$
$$v = \frac{E}{B} \tag{12.72}$$

From Eqs. (12.71) and (12.72), we get

$$Y = \frac{eElD}{m} \times \frac{B^2}{E^2}$$

$$Y = \frac{e}{m} \cdot \frac{B^2 lD}{E}$$

$$\frac{e}{m} = \frac{YE}{B^2 lD} \tag{12.73}$$

If V is the potential difference across plates and d is the separation between them, then E will be equal to V/d. Hence Eq. (12.73) becomes

$$\frac{e}{m} = \frac{YV}{B^2 ldD} \tag{12.74}$$

From the right hand side parameters of Eq. (12.74), one can calculate the value of e/m for the cathode ray. Thomson discovered first time this value, that is, equal to 1.7×10^{11} C/kg.

12.10 MASS SPECTROMETER/SPECTROGRAPH

This is an instrument used to measure the relative abundance and masses of isotopes present in the sample. Most of the measurements are carried out with singly charged positive ions. When the ion beam is detected with the help of a photographic plate, it is called *mass spectrograph*. It is generally used for the precise measurement of relative masses of ions and for the determination of the masses of isotopes. When the ion beam is detected with help of an electrometer or electrical detector, it is called *mass spectrometer*. It is generally used for precise measurement of relative abundance of ions. It is extremely important for this instrument that it should be capable of focusing ions of different energies (velocity focusing) and different direction of motion (space focusing) at the same point or the line of detector. This combination of double focusing is the important feature of this instrument, but it is not possible to achieve good focusing for a wide range of ionic energies and directions. Now, we will discuss the

Bainbridge mass spectrograph, which is based on two principles: (a) velocity selector and (b) motion of charge in transverse magnetic field.

Bainbridge Mass Spectrograph

The diagram of this spectrograph is shown in Figure 12.20. It consists of sources of positive ions S situated above the slits. The ions under the study then pass through the slits S_1 and S_2 and move down into the electric field E_y produced by the potential difference across plates P_1 and P_2. In this region, there is also magnetic field B_z perpendicular to both electric field and direction of motion of charge. Thus ions move in crossed electric and magnetic fields. The electric and magnetic forces acting on ions will be in opposite directions. The ions for which, both forces are equal, will move without any deviation from their path, i.e.,

$$E_y q = q v B_z$$

$$v = \frac{E_y}{B_z} \tag{12.75}$$

Thus the ions having velocity v moves without deviation, while ions with other velocities are deviated from their path. This results that, the ions having velocity v emerge out from slit S_3, while other ions are stopped by this slit. This region of crossed field is called velocity selector. Below S_3, the ions proceed their motion in another region of magnetic field B_z', whose direction is similar to B_z, but there is no electric field. Due to transverse magnetic field, ions move in a circular path. If R be the radius of circular path, then

$$R = \frac{mv}{q B_z'} \tag{12.76}$$

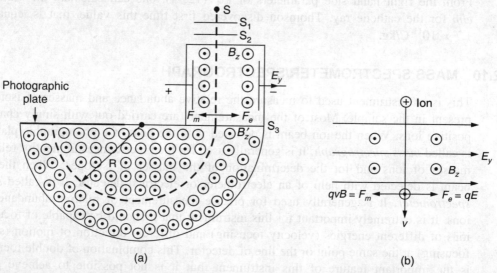

Figure 12.20   (a) Bainbridge spectrograph (b) Velocity selector.

From Eqs. (12.75) and (12.76), we get

$$R = \frac{mE_y}{qB_z B_z'} \tag{12.77}$$

Equation (12.77) suggests that the radius of ions is proportional to their masses if their charges and velocities are equal, i.e., $R \propto m$; if $v = E_y/B_z$, q and B_z are constant. So the ions of different isotopes are converged at different points on photographic plate. The selective abundance of the isotopes can be measured from the densities of photographic images they produced.

12.11   THE CYCLOTRON

The cyclotron is a device which is used to accelerate heavy particles like protons, deuterons, alpha particles, etc. It consists of two hollow semicircular metal boxes D_1 and D_2, called *dees*. A source of ions is located near the mid-point of the gap between the dees. The dees are insulated from each other and are enclosed in another vacuum chamber. These dees are connected to powerful radio frequency oscillator (RFO). The complete apparatus is placed between the pole-pieces of a strong electromagnet as shown in Figure 12.21(a) and (c). The magnetic field is perpendicular to plane of dees.

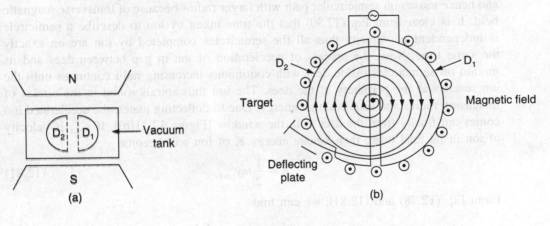

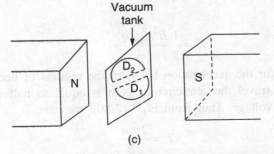

Figure 12.21   Cyclotron (a) side view (b) top view (c) front view.

Suppose a positive ion is released by the source at the centre of the chamber. At this instant, D_1 and D_2 are at negative and positive a.c. potentials respectively. The positive ion will be accelerated towards D_1 before entering it. If ion enters inside the dee with a velocity v, then its kinetic energy $(mv^2/2)$ will be equal to qV (i.e., $mv^2/2 = qV$; m and q are the mass and charge of ion). When the ion is inside the dee, it is not accelerated because this space is field free. Inside dee D_1, it moves in circular path due to transverse magnetic field B. If r and w are the radius and angular frequency of circular path, then

$$r = \frac{mv}{qB} \tag{12.78}$$

and

$$\omega = \frac{qB}{m} \tag{12.79}$$

Thus   Time taken in semicircular path $= \dfrac{T}{2} = \dfrac{2\pi}{2\omega} = \dfrac{\pi m}{qB}$ \hfill (12.80)

If the potentials of dees are changed at the same time as ion completes its one semicircular path in one dee, then ion is further accelerated in the gap between both dees due to negative potential of dee D_2. Now, ion enters in dee D_2 with greater speed and hence moves on semicircular path with larger radius because of transverse magnetic field. It is clear from Eq. (12.80) that the time taken by ion to describe a semicircle is independent of velocity thus all the semicircles completed by ion are on exactly the same time. Now, the process of acceleration of ion in gap between dees and its motion on semicircles inside dees with continuous increasing radii continues until the ion reaches at the periphery of the dees. The ion thus spirals round in the circles of increasing radius and acquires high energy. Due to deflecting plates, the accelerated ion comes out finally from the dees with the window [Figure 12.21(b)]. If v_{max} is velocity of ion in its final orbit, then kinetic energy K of ion will become

$$K = \frac{1}{2}mv_{max}^2 \tag{12.81}$$

From Eq. (12.78) and (12.81), we can find

$$K = \frac{1}{2}m \times \left(\frac{r_{max}\, qB}{m}\right)^2$$

$$K = \frac{1}{2}\frac{B^2 r_{max}^2\, q^2}{m} \tag{12.82}$$

Since the condition for the acceleration between the regions of dees is that the time taken by the ion to travel the semicircular path is equal to half of time period of oscillation of the ac voltage. Thus from Eq. (12.80), we have

$$T = \frac{2\pi m}{qB}$$

or
$$f = \frac{1}{T} = \frac{qB}{2\pi m} \qquad (12.83)$$

From Eqs. (12.82) and (12.83), we can write

$$K = 2\pi^2 r_{max}^2 f^2 m \qquad (12.84)$$

This expression provides the final energy of accelerated ion. The particles are ejected out of the cyclotron not in continuous pattern, but in pulsed stream.

If V is the alternating voltage applied at dees and n is the number of revolutions completed by ion before it reaches on periphery, then ion will be accelerated $2n$ times. Hence its energy will be

$$K = 2nqV$$
$$n = \frac{K}{2qV} \qquad (12.85)$$

With the help of Eq. (12.85), we can evaluate the number of revolution completed by the ion.

EXAMPLE 12.9 A cyclotron in which the flux density is 1.4 weber/m^2 is employed to accelerate protons. How rapidly should the electric field between the dees be reversed? Given that mass of proton = 1.67×10^{-27} kg and charge = 1.6×10^{-19} C.

Solution Let t is the time in which electric field or potential will be reversed between the dees.

$$t = \frac{\pi m}{qB} = \frac{3.14 \times 1.67 \times 10^{-27}}{1.6 \times 10^{-19} \times 1.4} = 2.342 \times 10^{-8} \text{ s}$$

EXAMPLE 12.10 Deuterons in a cyclotron describe a circle of radius 0.32 m just before emerging from dees. The frequency of applied e.m.f. is 10 MHz. Find out velocity of deuteron from cyclotron emerging out from cyclotron.

Solution Let v_{max} is the emerging velocity of deuteron from cyclotron.

$$\because \qquad r_{max} = \frac{m v_{max}}{qB}$$

$$\therefore \qquad V_{max} = \frac{qB r_{max}}{m}$$

Since, $\qquad f = \frac{qB}{2\pi m} \quad \Rightarrow \quad B = \frac{2\pi m f}{q}$

So $\qquad v_{max} = \frac{q}{m} \times \frac{2\pi m f}{q} \times r_{max} = 2\pi f\, r_{max}$

$$v_{max} = 2 \times 3.14 \times 10 \times 10^6 \times 0.32$$

$$v_{max} = 2.009 \times 10^7 \text{ m/s}$$

12.12 HALL EFFECT

The phenomenon of Hall Effect is also a consequence of motion of charge in both magnetic and electric field. If a piece of conductor (metal or semiconductor) carrying current is placed in a transverse magnetic field, then an electric field or a potential difference is produced inside the conductor in a direction normal to both current and magnetic field. This phenomenon is known as *Hall effect*. The developed electric field and potential are called *Hall field* and *Hall voltage*.

Explanation

Consider a specimen in the form of a rectangular cross-section carrying current I_x in the x direction. A uniform magnetic field B_z is applied along z axis as shown in Figure 12.22(a). If specimen is of conducting material or n-type semiconductor, then current flows due to motion of electron, but in p-type semiconductor and intrinsic semiconductor, current is carried by motion of holes and electrons–holes respectively. A moving charge in transverse magnetic field experiences a force normal to both current and magnetic field. Thus moving charges slightly shift towards y direction due to magnetic force ev_xB_z (Figures 12.22(b) and (c). This shift of charges produces a potential difference or an electric field E_y along y direction. The produced electric field also applies a force eE_y on charge. When both the forces become equal, the equilibrium occurs, i.e., upto this condition charges shift along y direction.

In equilibrium,

$$F_e = F_m$$
$$eE_y = ev_xB_z$$
$$E_y = v_xB_z \tag{12.86}$$

If J_x is the current density in x direction, then

$$J_x = \frac{I_x}{A} = nev_x \tag{12.87}$$

Here n is density of charge carrier (electron or hole or both) and $A(=$ breadth $\times$ thickness) is the cross-sectional area of specimen perpendicular to direction of J_x.

From Eqs. (12.86) and (12.87), we can write

$$\boxed{E_y = \frac{B_zJ_x}{ne}} \tag{12.88}$$

This is the expression for Hall field. The Hall effect is described by means of Hall coefficient R_H, defined in terms of J_x by the following relation:

$$R_H = \frac{E_y}{J_xB_z} \tag{12.89}$$

From Eqs. (12.88) and (12.89), we have

$$\boxed{R_H = \frac{1}{ne}} \tag{12.90}$$

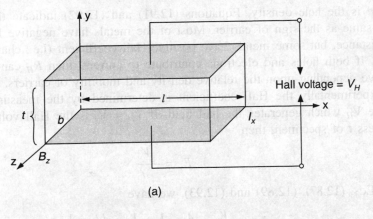

(a)

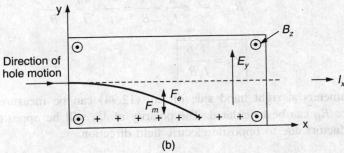

(b)

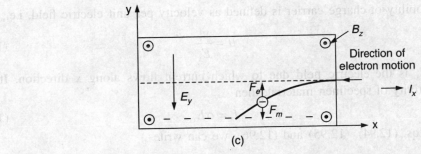

(c)

Figure 12.22 (a) Illustration of Hall effect (b) motion of hole in p-type semiconductor specimen (c) motion of electron in n-type semiconductor or metal specimen of Hall experiment.

For the conductor or n-type semiconductor, the electric field is developed in negative y direction. Hence from Eqs. (12.89) and (12.90), we can write

$$R_H = -\frac{E_y}{J_x B_z} = -\frac{1}{ne} \qquad (12.91)$$

In p-type semiconductor, the produced Hall field is along y direction, thus we have

$$R_H = \frac{E_y}{J_x B_z} = \frac{1}{pe} \qquad (12.92)$$

Here p is the hole density. Equations (12.91) and (12.92) indicate that the sign of R_H is same as the sign of carrier. Most of the metals have negative Hall coefficient or resistance, but some metals have positive Hall coefficient (i.e., charge carriers are holes). If both holes and electrons contribute to current, then R_H can be positive or negative depending upon the relative density and mobility of carriers.

Experimentally the Hall coefficient is determined by the measurement of Hall voltage V_H which generates the hall field. If $V_H = V_y$ is the Hall voltage across the thickness t of specimen, then

$$E_y = -\frac{V_y}{t} \qquad (12.93)$$

From Eqs. (12.87), (12.89) and (12.93), we have

$$R_H = \frac{V_y}{t} \times \frac{A}{I_x} \times \frac{1}{B_z} = \frac{V_y}{t} \times \frac{bt}{I_x} \times \frac{1}{B_z}$$

$$\boxed{R_H = \frac{V_y b}{I_x B_z}} \qquad (12.94)$$

The parameters at right hand side of Eq. (12.94) can be measured experimentally, and hence R_H can be calculated. The polarity of V_H will be opposite in n-and p-type semiconductors due to opposite electric field direction.

Mobility and Hall Angle

The mobility of charge carrier is defined as velocity per unit electric field, i.e.,

$$\mu = \frac{v_x}{E_x} \qquad (12.95)$$

Here E_x is the electric field due to which current flows along x direction. If σ is conductivity of specimen material, then

$$J_x = \sigma E_x \qquad (12.96)$$

From Eqs. (12.94), (12.95) and (12.96), we can write

$$\mu = \frac{v_x \sigma}{J_x} = \frac{v_x \sigma}{ne V_x} = \frac{\sigma}{ne}$$

$$\boxed{\mu = R_H \sigma} \qquad (12.97)$$

Thus mobility of charge carrier can be determined if the values of R_H and σ are known. There is another important parameter, with the help of which, μ can be determined. That is called *Hall angle* θ_H. It is defined by the following relation:

$$\tan \theta_H = \frac{E_y}{E_x} \qquad (12.98)$$

Putting values from Eqs. (12.89), (12.96) and (12.97) in Eq. (12.98), we have

$$\tan \theta_H = \frac{E_y \sigma}{J_x} = \frac{E_y}{J_x} \frac{\mu}{R_H}$$

$$\tan \theta_H = \frac{E_y}{J_x} \times \frac{J_x B_z \mu}{E_y}$$

$$\tan \theta_H = \mu B_z$$

$$\boxed{\mu = \frac{1}{B_z} \tan \theta_H} \tag{12.99}$$

Importance or Application of Hall Effect

1. The sign of Hall coefficient can be used to determine whether a given semiconductor is n- or p-type.
2. The carrier concentration can be evaluated when R_H is known (as $n = 1/R_H e$).
3. Since $\mu = R_H \sigma$, thus mobility of charge carrier can also be well determined with the help of R_H.
4. Since Hall voltage is proportional to magnetic flux density B_z and current I_x through a sample or specimen. The Hall effect can be used as the basis for design of a magnetic flux density meter.
5. Hall effect can be used to determine the power flow in an electromagnetic wave because V_H is proportional to product of E_y and B_z, i.e., magnitude of Poyning vector in e.m. wave.

EXAMPLE 12.11 A p-type semiconductor have Hall coefficient equal to 0.0125 m³/C. Find out the density of charge carrier in it.

Solution
$$R_H = \frac{1}{Ne}$$

$$N = \frac{1}{R_H e} = \frac{1}{0.0125 \times 1.6 \times 10^{-19}} = 50 \times 10^{-19}$$

$$N = 50 \times 10^{20} \text{ per m}^3$$

EXAMPLE 12.12 A semi conducting specimen of size 10 mm × 5 mm × 2 mm is used in Hall effect experiment. A current of 10 mA is flowing in it along its length. If 20 mV Hall voltage develops across its thickness, when it is placed in 0.5 T transverse magnetic field, then find Hall coefficient. [see Figure 12.20(a)]

Solution Given that $V_H = 20$ mV $= 20 \times 10^{-3}$ V, $b = 5$ mm $= 5 \times 10^{-3}$ m

$$I_x = 10 \text{ mA} = 10 \times 10^{-3} \text{ A and } B_z = 0.5 \text{ T}$$

Since
$$R_H = \frac{V_H}{I_x} \cdot \frac{b}{B_z}$$

$$R_H = \frac{20 \times 10^{-3} \times 5 \times 10^{-3}}{10 \times 10^{-3} \times 0.5} = 0.02 \text{ m}^3/\text{C}$$

SOLVED NUMERICAL PROBLEMS

PROBLEM 12.1 An electron has a velocity 10^8 m/s normal to a magnetic field of 0.1 Wb/m^2 flux density. Calculate radius of electron and its frequency.

Solution Given that $e = 1.6 \times 10^{-19}$ C, $m = 9.1 \times 10^{-31}$ kg, $v = 10^8$ m/s and $B = 0.1$ Wb/m^2, $r = ?$ and $f = ?$

$$r = \frac{mv}{qB} = \frac{9.1 \times 10^{-31} \times 10^8}{1.6 \times 10^{-19} \times 0.1} = \frac{9.1}{1.6} \times 10^{-31+28} = 5.69 \times 10^{-3} \text{ m} = 5.69 \text{ mm}$$

$$f = \frac{qB}{2\pi m} = \frac{1.6 \times 10^{-19} \times 0.1}{2 \times 3.14 \times 10^{-31} \times 9.1} = 0.028 \times 10^{11} = 2.8 \times 10^9 \text{ Hz} = 2.8 \text{ G Hz}$$

PROBLEM 12.2 An α-particle enters in a homogeneous magnetic field perpendicular to its velocity. The angular momentum of α-particle is 1.33×10^{-22} kg m^2/s. The induction of magnetic field is 0.025 Wb/m^2. Find K.E. of the particle. (mass of α-particle = 6.68×10^{-27} kg and charge on α-particle = 3.2×10^{-19} C).

Solution In transverse magnetic field, a charge particle moves on circular path. If I and ω are the moment of inertia and angular frequency of the charge particle, then its K.E. (k) will be

$$k = \frac{1}{2} I \omega^2 = \frac{1}{2} (I \omega) \cdot \omega$$

$$k = \frac{1}{2} L \cdot \frac{qB}{m}; \text{ as } L \text{ (angular momentum)} = I\omega \text{ and } \omega = qB/m$$

$$k = \frac{1}{2} \times \frac{1.33 \times 10^{-22} \times 3.2 \times 10^{-19} \times 0.025}{6.68 \times 10^{-27}} = 0.796 \times 10^{-16} \text{ J}$$

PROBLEM 12.3 An electron moving horizontally with a velocity of 1.7×10^7 m/s enters in a vertical uniform electric field of 3.4×10^4 V/m acting downwards. Find the vertical displacement of the electron in the field if its horizontal displacement is 3 cm.

Solution Given that $e = 1.6 \times 10^{-19}$ C, $m = 9.1 \times 10^{-31}$ kg, $V_x = 1.7 \times 10^7$ m/s,
$$E_y = 3.4 \times 10^4 \text{ V/m and } x = 3 \text{ cm} = 3 \times 10^{-2} \text{ m}$$

$$y = \frac{1}{2} a_y t^2 = \frac{1}{2} \frac{qE_y}{m} \cdot \left(\frac{x}{V_x}\right)^2$$

$$y = \frac{1}{2} \times \frac{(1.6 \times 10^{-19}) \times (3.4 \times 10^4)}{(9.1 \times 10^{-31})} \times \left(\frac{3 \times 10^{-2}}{1.7 \times 10^7}\right)^2$$

$$y = 0.926 \times 10^{-2} \text{ m} = 0.926 \text{ cm}$$

PROBLEM 12.4 In a CRT, deflection is produced by magnetic field. The length of field is 0.05 m. The distance of screen from end of field is 0.3 m. Calculate the magnetic field if deflection on screen is 1 cm. Given that anode voltage of CRT is 10^3 V.

Solution Given that $L = 0.05$ m, $q = e = 1.6 \times 10^{-19}$ C, $m = 9.1 \times 10^{-31}$ kg,

$$D = 0.3 + \frac{L}{2} = 0.3 + 0.025 = 0.325 \text{ m}$$

$$y = 1 \text{ cm} = 0.01 \text{ m}, V_A = 10^3 \text{ V}, B = ?$$

$\because \qquad\qquad y = Dl \dfrac{q}{m} \dfrac{B}{v}$

$\therefore \qquad\qquad B = \dfrac{ymv}{Dlq} = \dfrac{ym}{DLq}\sqrt{\dfrac{2qV_A}{m}} \qquad \left(\text{as } \dfrac{1}{2}mv^2 = qV_A\right)$

$$B = \frac{y}{DL}\sqrt{\frac{2mV_A}{q}}$$

$$B = \frac{0.01}{0.325 \times 0.05} \times \sqrt{\frac{2 \times 9.1 \times 10^{-31} \times 10^3}{1.6 \times 10^{-19}}} = 6.56 \times 10^{-5} \text{ Wb/m}^2$$

PROBLEM 12.5 A single ionized Ge atom enters in a Bainbridge mass spectrograph with a velocity 5×10^3 m/s. Calculate the radii of the paths followed by two abundant isotopes of masses 72 and 74 when the magnetic flux density is 0.3 Wb/m^2. Also calculate the linear separation of the lines on the photographic plate for those two isotopes.

Solution $\because \qquad r = \dfrac{mv}{qB}$

$$r_{72} = \frac{m_{72}v}{qB} = \frac{(72 \times 1.66 \times 10^{-27}) \times (5 \times 10^3)}{1.6 \times 10^{-19} \times 0.3} = 1.245 \text{ m}$$

$\therefore \qquad\qquad r \propto m$

$$\frac{r_{72}}{r_{74}} = \frac{m_{72}}{m_{74}}$$

$$r_{74} = \frac{m_{74}}{m_{72}} \times r_{72}$$

$$r_{74} = \frac{(74 \times 1.66 \times 10^{-27})}{(72 \times 1.66 \times 10^{-27})} \times 1.245 = 1.278 \text{ m}$$

Linear separation on photographic plate $= 2(r_{74} - r_{72}) = 2(1.278 - 1.245) = 0.066$ m.

PROBLEM 12.6   The electric field 3×10^5 V/m and magnetic field 0.2 Wb/m^2 are crossed to each other. If a charge moves without any deviation in this combined field, then find its velocity.

Solution $V = \dfrac{E}{B} = \dfrac{3 \times 10^5}{0.2} = 1.5 \times 10^6$ m/s

PROBLEM 12.7   An n-type germanium sample has donor density of 10^{21}/m^3. It has arranged in a Hall experiment having magnetic field of 0.5 T and the current density is 500 A/m^2. Find the Hall voltage if the sample is 3 mm wide.

Solution $\because \quad R_H = \dfrac{1}{Ne} = \dfrac{E_y}{J_x B_z} = \dfrac{V_H}{t J_x B_z}$

$\Rightarrow \qquad\qquad V_H = \dfrac{t J_x B_z}{Ne}$

$$V_H = \dfrac{(3 \times 10^{-3}) \times 500 \times 0.5}{10^{21} \times 1.6 \times 10^{-19}} = 468.75 \times 10^{-5} = 4.69 \times 10^{-3} \text{ V} = 4.69 \text{ mV}$$

PROBLEM 12.8   The resistivity of a doped silicon sample is 8.9×10^{-3} ΩM. The Hall coefficient for this sample is 3.6×10^{-4} m^3/C. Find the carrier density and mobility for this sample.

Solution   Since $R_H = \dfrac{1}{Ne}$

So $N = \dfrac{1}{R_H e} = \dfrac{1}{3.6 \times 10^{-4} \times 1.6 \times 10^{-19}} = 1.736 \times 10^{22}$ per m^3

$\because \qquad\qquad \mu = R_H \sigma = \dfrac{R_H}{\rho}$

$\therefore \qquad\qquad \mu = \dfrac{3.6 \times 10^{-4}}{8.9 \times 10^{-3}} = 0.041 \text{ m}^2\text{V}^{-1}\text{s}^{-1}$

PROBLEM 12.9   Calculate radius of the dees of a cyclotron capable of accelerating protons up to 3 MeV in presence of a magnetic field of 1.0 T.

Solution $\because \qquad E = \dfrac{1}{2} \dfrac{B^2 r_{max}^2 q^2}{m}$

$\therefore \qquad\qquad r_{max}^2 = \dfrac{2mE}{B^2 \times q^2} = \dfrac{2 \times 1.67 \times 10^{-27} \times 3 \times 10^6}{1 \times (1.6 \times 10^{-19})^2} = 6.26 \times 10^{-2}$ m

$\qquad\qquad r_{max} = 0.25$ m

The maximum radius of proton path will be equal to radius of dees.

PROBLEM 12.10   The radius of a cyclotron dee is 1 m and the applied magnetic field is 0.5 Wb/m². It is used to accelerate protons. Calculate the frequency of oscillating voltage needed to maintain resonance. What is the maximum energy gained by the proton?

Solution   Given that $R_{dee} = r_{max} = 1$ m, $B = 0.5$ Wb/m², $q = 1.6 \times 10^{-19}$ C, $m = 1.67 \times 10^{-27}$ kg, $f = ?$ and $E = ?$

$$f = \frac{qB}{2\pi m} = \frac{1.6 \times 10^{-19} \times 0.5}{2 \times 3.14 \times 1.67 \times 10^{-27}} = 0.076 \times 10^8 = 7.6 \times 10^6 \text{ Hz} = 7.6 \text{ MHz}$$

$$E = 2\pi^2 r_{max}^2 f^2 m = 2 \times (3.14)^2 \times 1 \times (7.6 \times 10^6)^2 \times 1.67 \times 10^{-27}$$

$$E = 1902.2 \times 10^{-15} \text{ J} = 1.902 \times 10^{-12} \text{ J} = 11.87 \times 10^6 \text{ eV} = 11.87 \text{ MeV}$$

PROBLEM 12.11   The magnetic field in a certain region of space is equal to $0.08\,\hat{\mathbf{i}}$ T. A proton is short into the field with velocity $(2 \times 10^5\,\hat{\mathbf{i}} + 3 \times 10^5\,\hat{\mathbf{j}})$ m/s. Calculate the radius and pitch of helical path of proton.

Solution $\mathbf{B} = 0.08\,\hat{\mathbf{i}}$ T $\Rightarrow$ $B_x = 0.08$ T

$$\mathbf{v} = 2 \times 10^5\,\hat{\mathbf{i}} + 3 \times 10^5\,\hat{\mathbf{j}} \quad \Rightarrow \quad v_x = 2 \times 10^5 \text{ m/s and } v_y = 3 \times 10^5 \text{ m/s}$$

$$r = \frac{mv_y}{qB} = \frac{1.67 \times 10^{-27} \times 3 \times 10^5}{1.6 \times 10^{-19} \times 0.08} = 0.039 \text{ m}$$

$$p = v_x T = v_x \cdot \frac{2\pi m}{qB} = \frac{2 \times 10^5 \times 2 \times 3.14 \times 1.67 \times 10^{-27}}{1.6 \times 10^{-19} \times 0.08} = 0.164 \text{ m}$$

EXERCISES

THEORETICAL QUESTIONS

12.1 What is Lorentz force?

12.2 Show that the motion of a charge in transverse electric field is parabolic.

12.3 Prove that the motion of a charge in transverse magnetic field is circular.

12.4 Describe in detail, the deflection produced in moving charge when it is placed in transverse (a) electric and (b) magnetic field.

12.5 What is electrostatic and magneto static deflection sensitivity.

12.6 When a charge moves at an angle to uniform magnetic field, its path is helical. Explain it.

12.7 Show that the kinetic energy of a charge, accelerated by potential difference V, is qV.

12.8 Find out the path of a charge, when it moves in crossed electric and magnetic fields.

12.9 Discuss the motion of a charge, when (a) its motion is parallel to both electric and magnetic field and (b) its motion is Antiparallel to electric field and parallel to magnetic field.

12.10 Show that the time period of circulating charge in transverse magnetic field is velocity independent.

12.11 Describe the principle and working of a velocity selector.

12.12 What is electrostatic and magneto-static focusing. Discuss in detail.

12.13 Explain the construction and working of a cathode ray tube.

12.14 Describe Thomson's experiment for the measurement of charge-to-mass ratio of cathode rays.

12.15 Write down a short note on Bainbridge mass spectrograph.

12.16 Describe the construction and working of a cyclotron and also show that kinetic energy of the particle in it is independent of the voltage applied.

12.17 What is Hall effect? Find the expression for (a) Hall voltage (b) Hall coefficient (c) Mobility and, (d) Hall angle.

12.18 What type of motion do you expect, when a charge is released in uniform electric field and why?

12.19 What is the explanation of positive and negative value of Hall coefficient?

NUMERICAL PROBLEMS

P12.1 An electron with speed of 3×10^6 m/s, travelling horizontally enters in upward electric field of 480 N/C. Find the vertical displacement and velocity of electron when it covers 50 mm horizontal distance. **(Ans. 1.17 cm, 1.4×10^6 m/s)**

P12.2 Find out frequency for a deuteron in a cyclotron with a flux density of 1.5 Wb/m^2.

(Ans. 1.14×10^7 Hz)

P12.3 An electron of energy 1 keV is to be restricted to a circular path of radius 0.7 m. What should be the magnetic field required for it. **(Ans. 1.07×10^{-4} T)**

P12.4 Calculate the cyclotron frequency of a charge particle ($q = 1.6 \times 10^{-19}$ C, $m = 3.3 \times 10^{-27}$ kg) in a magnetic field of 4 T. **(Ans. 2×10^8 radian/s)**

P12.5 An electron accelerated with a potential difference of 20 kV, moves in transverse magnetic field of 0.3 T. Find magnetic force on it. **(Ans. 4.02×10^{-12} N)**

P12.6 The Hall coefficient of a certain semi conducting specimen was found -7.35×10^{-5} m^3/c. Is the semiconductor n-type or p-type? The electrical conductivity of specimen is 200 Ω/m. Calculate density and mobility of charge carriers.

(Ans. 8.5×10^{22} per m^3, 1.47×10^{-2} m/Vs)

P12.7 The magnetic field $B = 0.4$ T is along the x axis. The velocity of a proton entering in this field is 4×10^6 m/s. The angle of entrance is 65° with magnetic field direction. What is the path of proton? If it is helical, then find radius and pitch of helical path.

(Ans. 9.4 cm, 28 cm)

P12.8 In a velocity selector, the electric field is 90 kV/m and magnetic field is 0.4 T. Calculate the velocity of the undeflected particles. **(Ans. 2.25×10^5 m/s)**

P12.9 A mixture of singly charged N^{20} and N^{22} is analysed in a Bainbridge mass spectrograph. The electric and magnetic fields in the velocity selector are 10^5 V/m and 0.5 T respectively.

The magnetic field in the chamber of mass spectrograph is 0.6 T. Calculate the linear separation of isotopes on the photographic plate. **(Ans. 0.416 m)**

P12.10 An α-particle of mass 16.65×10^{-27} kg and charge 3.2×10^{-19} C travels at right angle to magnetic field 0.2 Wb/m^2 with speed 6×10^5 m/s. Find radius, force and acceleration in it. **(Ans. 6.23 cm, 3.48×10^{-14} N, 5.77×10^{-22} m/s^2)**

P12.11 An electron moving horizontally with a velocity of 1.7×10^7 m/s enters a vertical uniform electric field of 3.4×10^4 V/m. When its horizontal displacement is 3 cm, find the magnified of magnetic field required to neutralise the vertical deflection. **(Ans. 2×10^{-3} Wb/m^2)**

P12.12 In a CRT, the horizontal plates have 3 cm length, separation of 0.75 cm and potential difference of 25 V. If anode potential is 800 V, then find the vertical deflection on screen, which is at 20 cm from the centre of plates. **(Ans. 1.34 cm)**

P12.13 The electrostatic deflection sensitivity of a CRT is 0.5 cm/V. Find out the potential difference across the plates if deflection on screen is 5 cm. **(Ans. 10 V)**

P12.14 In a CRT, deflection is produced by magnetic field. If 0.4 T of magnetic field causes a deflection of 0.4 cm on screen, find the magneto static deflection sensitivity.

(Ans. 0.01 m/T)

P12.15 A proton is to circulate the earth with speed of 1.0×10^6 m/s. Find the minimum magnetic field, which should be created at the equator for the purpose. The mass and charge of proton are 1.67×10^{-27} kg and 1.6×10^{-19} C respectively. The radius of earth is 6.37×10^6 m. **(Ans. 1.67×10^{-4} Wb/m^2)**

P12.16 The cydotron has radius 50 cm and magnetic flux density 1.5 Wb/m^2. Find the energy which may be imported in deuteron. **(Ans. 1.33×10^7 eV)**

P12.17 What is the velocity of an electron accelerated with a potential difference of 8 kV.

(Ans. 5.3×10^7 m/s)

P12.18 An electron starting from rest travels a distance 10 mm in a uniform electric field of 3×10^3 N/C. Calculate speed of the electron. **(Ans. 3.25×10^6 m/s)**

P12.19 An electric field of 3×10^3 V/m, produced by horizontal deflecting plates of CRT, produces a deflection of 6 cm on screen. Find electrostatic deflection sensitivity if separation of plate is 1 cm. **(Ans. 2×10^{-3} m/V)**

MULTIPLE CHOICE QUESTIONS

MCQ 12.1 The trajectory of a charge particle moving in transverse electric field is
(a) Circular (b) Cycloidal
(c) Straight line (d) Parabola

MCQ 12.2 An electron is accelerated with a potential V. If $V^{-1} = 0.352$ volt^{-1} and $e/m = 1.76 \times 10^{11}$ C/kg, then velocity of electron will be equal to
(a) 10^6 m/s (b) $\times 10^5$ m/s
(c) $\sqrt{2} \times 10^6$ m/s (d) $\sqrt{2} \times 10^5$ m/s

MCQ 12.3 The charge-to-mass ratio for an electron is 1.76×10^{11} C/kg. If an electron is released in electric field of magnitude 50 V/m, then acceleration in it will be
(a) 8.8×10^{14} m/s^2 (b) 6.2×10^{13} m/s^2
(c) 5.4×10^{12} m/s^2 (d) Zero

MCQ 12.4 The two electron beam having velocity ratio 1 : 2 are subjected separately in identical transverse electric field. What is the ratio of the deflections produced?
(a) 4 : 1
(b) 1 : 4
(c) 2 : 1
(d) 1 : 2

MCQ 12.5 When a charge moves in transverse magnetic field its path becomes circular. What will be the work done by the magnetic field on the charge?
(a) $qvBr$
(b) $2qvBr$
(c) Zero
(d) None of these

MCQ 12.6 The necessary condition for a moving charge, to be undeflected under velocity selector is
(a) $F_e > F_m$ or $F_e < F_m$
(b) $v = BE$
(c) $B = vE$
(d) $E = vB$

MCQ 12.7 Two electron beams having velocity ratio of 1 : 4 are subjected separately identical transverse magnetic fields. What is the ratio of deflections produced?
(a) 1 : 4
(b) 4 : 1
(c) 1 : 2
(d) 2 : 1

MCQ 12.8 The trajectory of a moving charge particle in crossed electric and magnetic fields is
(a) Circular
(b) Cycloidal
(c) Helical
(d) Straight line

MCQ 12.9 When a charge moves at an angle to parallel electric and magnetic fields, then it moves on helical path, whose pitch
(a) Increases with distance
(b) Decreases with distance
(c) Either decreases or increases with distance
(d) Remains unchanged with distance

MCQ 12.10 Bainbridge mass spectrograph is used
(a) For production of X-ray
(b) For production of cathode ray
(c) For determination of q/m
(d) For deflection of isotopes

MCQ 12.11 In CRT, the anodes are given at relatively increasing potentials with respect to cathode for
(a) Accelerate the e-beam
(b) Focus the e-beam
(c) Accelerate and focus the e-beam
(d) Generation of electrons

MCQ 12.12 If E_y, J_x and B are the Hall field, current density and magnetic flux density respectively, then Hall coefficient is given by
(a) $R_H = \dfrac{E_y}{J_x B}$
(b) $R_H = \dfrac{J_x}{E_y B}$
(c) $R_H = \dfrac{B J_x}{E_y}$
(d) $R_H = \dfrac{B E_y}{J_x}$

MCQ 12.13 If Hall coefficient is infinity, then material will be
(a) Conductor
(b) Insulator
(c) Intrinsic semiconductor
(d) Generation of electrons

MCQ 12.14   A cyclotron is used to accelerate
     (a) Charge particle   (b) A neutral particle
     (c) Photon   (d) Magnon

MCQ 12.15   For a cyclotron, the flux density of magnetic and its radius are 1 Wb/m^2 and 0.5 m respectively. The maximum energy that a proton ($m_p = 1.6 \times 10^{-27}$ kg) can acquire from this is nearly
     (a) 0.12 MeV   (b) 1.2 MeV
     (c) 12 MeV   (d) 120 MeV

MCQ 12.16   A cyclotron is operating at a frequency of 12×10^6 Hz. If charge-to-mass ratio for deuteron is 4.848×10^7 C/kg, the necessarily magnetic field to accelerate it, is
     (a) 16 T   (b) 1.6 T
     (c) 0.16 T   (d) 0.016 T

MCQ 12.17   The unit of Hall coefficient is
     (a) Vm^{2}A^{-2} Wb   (b) Vm^{3}A Wb^{-1}
     (c) Vm^{3}A^{-2} Wb^{-1} (d) Vm^{2}A Wb^{-1}

MCQ 12.18   If R_H, Q_H and σ are the Hall coefficient, Hall angle and conductivity of material, the correct relations for mobility are
     (a) $\mu = R_H\,\sigma,\ \mu = B_z \tan\theta_H$ (b) $\mu = R_H/\sigma,\ \mu = B_z \tan\theta_H$
     (c) $\mu = R_H\,\sigma,\ \mu = \tan\theta_H/B_z$ (d) $\mu = R_H/\,\sigma,\ \mu = \tan\theta_H/B_z$

MCQ 12.19   A charge particle of mass m and charge q enters in magnetic field B at right angle. Now, its path becomes circular. It L is its angular momentum, then energy of charge particle will be
     (a) Zero   (b) $\dfrac{LqB}{2m}$

     (c) $\dfrac{2m}{LqB}$ (d) $2mqLB$

MCQ 12.20   If a charge enters in a very low magnetic field with very high speed at right angle, its path within the field becomes
     (a) Circle   (b) Parabola
     (c) Circular arc   (d) Hyperbola

Answers

12.1 (d)	**12.2** (a)	**12.3** (b)	**12.4** (a)	**12.5** (c)	**12.6** (d)	**12.7** (b)
12.8 (b)	**12.9** (a)	**12.10** (d)	**12.11** (c)	**12.12** (a)	**12.13** (b)	**12.14** (a)
12.15 (c)	**12.16** (b)	**12.17** (c)	**12.18** (c)	**12.19** (b)	**12.20** (c)	

MCQ 12.14 A cyclotron is used to accelerate:

(a) Charge particles   (b) A neutral particle

(c) Photon   (d) Maynon

MCQ 12.15 For a cyclotron, the flux density of magnet and its radius are 1 Wb/m² and 0.5 m respectively. The maximum energy that a proton ($m_p = 1.5 \times 10^{-27}$ kg) can acquire from this is nearly:

(a) 0.12 MeV   (b) 1.2 MeV

(c) 12 MeV   (d) 120 MeV

MCQ 12.16 A cyclotron is operating at a frequency of 1.2×10^7 Hz. If charge-to-mass ratio for deuteron is 4.834×10^7 C/kg, the necessary magnetic field to accelerate it, is

(a) 16 T   (b) 1.6 T

(c) 0.16 T   (d) 0.010 T

MCQ 12.17 The unit of Hall coefficient is

(a) Vm A⁻¹ Wb⁻¹   (b) V m A Wb⁻¹

(c) m³ A⁻¹ Wb⁻¹   (d) Vm A Wb⁻¹

MCQ 12.18 If R_H, σ and μ are the Hall coefficient, Hall angle and conductivity of material, the correct relations for mobility are

(a) $\mu = R_H \sigma$; $\mu = B_H \tan \theta_H$ (b) $\mu = R_H \sigma$; $\mu = B_H \tan \theta_H$

(c) $\mu = R_H \theta_H$; $\mu = \tan \theta_H / B_H$ (d) $\mu = R_H \sigma$; $\mu = \tan \theta_H / B$

MCQ 12.19 A charge particle of mass m and charge q enters in magnetic field B at right angle. Now, its path becomes circular. If r is its angular momentum, then energy of charge particle will be:

(a) Zero   (b) $\dfrac{LqB}{2m}$

(c) $\dfrac{2m}{LqB}$ (d) $\dfrac{LqB}{m}$

MCQ 12.20 If a charge enters in a very low magnetic field with very high speed at right angle, its path within the field becomes:

(a) Circle   (b) Parabola

(c) Circular arc   (d) Hyperbola

Answers

12.1 (d)	12.2 (a)	12.3 (a)	12.4 (a)	12.5 (c)	12.6 (d)	12.7 (b)
12.8 (b)	12.9 (a)	12.10 (d)	12.11 (c)	12.12 (a)	12.13 (b)	12.14 (a)
12.15 (c)	12.16 (b)	12.17 (a)	12.18 (c)	12.19 (b)	12.20 (c)

Suggested Reading

Agrawal, B.K. and Prakash, Hari, *Quantum Mechanics*, Prentice-Hall of India, New Delhi, 1997.

Banerji, S. and Banerjee, Asit, *The Special Theory of Relativity*, PHI Learning, Delhi, 2012.

Bates, David Robert and Bederson, Benjamin, *Advanced in Atomic, Molecular and Optical Physics,* Vol. 26, Academic Press, San Deigo, 1990.

Beiser, A., *Perspectives of Modern Physics*, McGraw-Hill, New York, 1969.

Bertulani, C.A., *Nuclear Physics in a Nutshell*, Princeton, New Jersey, 2007.

Blatt, J.M. and Weasskopf, V.F., *Theoretical Nuclear Physics*, Wiley, New York, 1952.

Born, Max, Blin-Stoyle, Roger John and Radcliffe, J.M., *Atomic Physics*, Dover, New York, 1989.

Budker, Dmitry, Kimball, Derek F and DeMille, David P., *Atomic Physics*: An Exploration Through Problems and Solutions, Oxford University Press, Oxford, 2004.

Chow Tai, L., *Introduction to Electromagnetic Theory*: A Modern Perspective, Jones and Bartlett, Sudbury, 2006.

Datta, Somnath, *Introduction to Special Theory of Relativity*, Allied Publisher, New Delhi, 1998.

Dekker, A.J., *Solid State Physics*, Macmillan, New Delhi, 1981.

Demtröder, Wolfgang, *Molecular Physics*, Wiley, Weinheim, 2005.

Devanathan, V., *Nuclear Physics*, Narosa Publishing House, New Delhi, 2008.

Dunlap, R.A., *The Physics of Nuclei and Particle*, Thomson Learning, Cole, 2004.

Franz, J.H. and Jain, V.K., *Optical Communications*: Components and Systems, Narosa Publishing House, New Delhi, 2000.

Garg, J.B., *Nuclear Physics*: Basic Concepts, Macmillan, New Delhi, 2011.

Ghatak, A.K. and Thyagarajan, K., *Introduction to Fiber Optics*, Cambridge University Press, New Delhi, 1998.

Ghatak, Ajoy, *Einstein and the Special Theory of Relativity*, Viva Books, New Delhi, 2011.

Ghoshal, S.N., *Nuclear Physics*, S. Chand and Co., New Delhi, 2009.

Griffiths, D.J., *Introduction to Quantum Mechanics*, Prentice-Hall, New Jersey, 1995.

Hawking, Stephen, A Brief History of Relativity, Time Magazine, Amsterdam, December 31, 1999.

Hebbar, K.R., *Basics of X-Rays Diffraction and its Applications*, IK International, New Delhi, 2007.

Jackson, John David, *Classical Electrodynamics,* Wiley India, 1999.

Kakani, S.L., Kakani, Subhra, *Nuclear and Particle Physics*, Viva Books, New Delhi, 2008.

Kakani, S.L., *Solid State Physics*, New Age International, New Delhi, 2010.

Kaplan, I., *Nuclear Physics*, Narosa Publishing House, New Delhi, 1990.

Keiser, G., *Optical Fiber Communications*, 3rd ed., McGraw-Hill, New Delhi, 2000.

Kittel, C., *Introduction to Solid State Physics*, 8th ed., Wiley, New Delhi, 2005.

Kumar, Arun, *Introduction to Solid State Physics*, PHI Learning, Delhi, 2010.

Lal, H.B., *Introductory Nuclear Physics*, United Book Depot, Allahabad, 2005.

Laud, B.B., *Electromagnetics*, New Age International, New Delhi, 2011.

Lynton, E., *Superconductivity*, Methuen and Company, London, 1969.

Matveev, Aleksei Nikolaevich, *Molecular Physics*, Mir, Georgetown, 1985.

Maurer, A., *Lasers*: *Lightwave of the Future*, Arco Publishing Inc., Georgetown, 1982.

Mittal, V.K., Verma, R.C. and Gupta, S.C., *Introduction to Nuclear and Particle Physics*, PHI Learning, Delhi, 2009.

Morrison, M., *Understanding Quantum Mechanics*, Prentice-Hall, New Jersey, 1990.

Ramakrishnan, T.V. and Rao, C.N.R., *Superconductivity Today*: *An Elementary Introduction*, Universities Press, Hyderabad, 1999.

Resnik, R., *Introduction to Special Theory of Relativity*, Wiley Eastern, New Delhi, 1989.

Richtmyer, F.K., Kennard, E.H. and Cooper, J.N., *Introduction to Modern Physics*, 6th ed., McGraw-Hill, New York, 1969.

Saxena, B.S., Gupta, R.C. and Saxena, P.C., *Fundamentals of Solid State Physics*, Pragati Prakashan, Meerut, 2005.

Schoneberg, D., *Superconductivity*, Cambridge, London, 1954.

Schrieffer, J.R., *Theory of Superconductivity*, W.A. Benjamin, New York, 1964.

Senior, J.M., *Optical Fiber Communications*: *Principles and Practice*, Prentice-Hall of India, New Delhi, 2003.

Singh, Devraj, *Engineering Physics*, Vol. 1, 3rd ed., Dhanpat Rai and Co., New Delhi, 2013.

Singh, Devraj, *Engineering Physics*, Vol. 2, 3rd ed., Dhanpat Rai and Co., New Delhi, 2011.

Singh, Devraj, *Fundamentals of Engineering Physics*, Vol. 1 and Vol. 2, 2nd ed., Dhanpat Rai and Co., New Delhi, 2012.

Singh, Devraj, *Fundamentals of Optics*, PHI Learning, Delhi, 2010.

Singh, Devraj, Gautam, R.B. and Shukla, A.K., *Applied Physics*, 2nd ed., University Science Press, New Delhi, 2013.

Singh, Devraj, *Introductory Engineering Physics*, Dhanpat Rai and Co., New Delhi, 2011.

Singh, Devraj, Kumar, Jitendra and Tripathi, Sudhanshu, *Circuit Fundamentals and Basic Electronics*, IK International, New Delhi, 2013.

Singh, Devraj, *Principles of Engineering Physics*, Vol. 2, 2nd ed., Dhanpat Rai and Co., New Delhi, 2013.

Singh, V.K., Singh, Devraj and Singh, D.P., *Mechanics and Wave Motion*, IK International, New Delhi, 2013.

Smith, Frank H. and Smith, Deane K., *Industrial Applications of X-Ray Diffraction*, CRC Press, New York, 2000.

Srivastava, J.P., *Elements of Solid State Physics*, PHI Learning, Delhi, 2011.

Suryanarayana, C. and Norton, M.G., *X-Ray Diffraction, A Practical Approach*, Plenum, New York, 1998

Svelto, O., *Principles of Lasers*, Plenum Press, New York, 1976.

Tayal, D.C., *Electricity and Magnetism*, Himalayan Publication, Bombay, 2001.

Thyagarajan, K. and Ghatak, A.K., *Laser, Theory and Applications*, Macmillan, New Delhi, 1997.

Tinkham, M., *Introduction to Superconductivity*, McGraw-Hill, New York, 1964.

Tiwari, K.K., *Electricity, Magnetism with Electronics*, S. Chand and Co., New Delhi, 2005.

Upadhyay, J.C., *Mechanics*, Ram Prasad and Sons, Agra, 2003.

Warren, B.E., *X-Ray Diffraction*, Dover, New York, 1990.

Waseda, Yoshio, Matsubara, Eiichiro and Shinoda, Kozo, *X-Ray Diffraction Crystallography*, Springer, London, 2011.

Whelan, Colm T., *New Directions in Atomic Physics*, Kluwer Academic/Plenum Publishers, New York, 1999.

Wong Sameul, S.M., *Introductory Nuclear Physics*, PHI Learning, Delhi, 2010.

Zachariasen, W.H., *Theory of X-Ray Diffraction in Crystals*, Dover, New York, 2004.

Index